Advances in
ATOMIC, MOLECULAR, AND OPTICAL PHYSICS

VOLUME 35

Editors

BENJAMIN BEDERSON
New York University
New York, New York

HERBERT WALTHER
Max-Planck-Institut für Quantenoptik
Garching bei München
Germany

Editorial Board

P. R. BERMAN
University of Michigan
Ann Arbor, Michigan

K. DOLDER
The University of Newcastle-upon-Tyne
Newcastle-upon-Tyne
United Kingdom

M. GAVRILA
F.O.M. Instituut voor Atoom-en Molecuulfysica
Amsterdam
The Netherlands

M. INOKUTI
Argonne National Laboratory
Argonne, Illinois

S. J. SMITH
Joint Institute for Laboratory Astrophysics
Boulder, Colorado

Founding Editor

SIR DAVID BATES

ADVANCES IN

ATOMIC, MOLECULAR, AND OPTICAL PHYSICS

Edited by

Benjamin Bederson

DEPARTMENT OF PHYSICS
NEW YORK UNIVERSITY
NEW YORK, NEW YORK

Herbert Walther

UNIVERSITY OF MUNICH AND
MAX-PLANK-INSTITUT FÜR QUANTENOPTIK
MUNICH, GERMANY

Volume 35

ACADEMIC PRESS

San Diego   New York   Boston
London   Sydney   Tokyo   Toronto

This book is printed on acid-free paper. ∞

Copyright © 1995 by ACADEMIC PRESS, INC.

All Rights Reserved.
No part of this publication may be reproduced or transmitted in any form or by any means, electronic or mechanical, including photocopy, recording, or any information storage and retrieval system, without permission in writing from the publisher.

Academic Press, Inc.
A Division of Harcourt Brace & Company
525 B Street, Suite 1900, San Diego, California 92101-4495

United Kingdom Edition published by
Academic Press Limited
24-28 Oval Road, London NW1 7DX

International Standard Serial Number: 1049-250X

International Standard Book Number: 0-12-003835-8

PRINTED IN THE UNITED STATES OF AMERICA
95 96 97 98 99 00 BB 9 8 7 6 5 4 3 2 1

Contents

CONTRIBUTORS ... ix

Laser Manipulation of Atoms
K. Sengstock and W. Ertmer

I. Introduction ... 1
II. General Principles ... 3
III. Basic Manipulation Schemes ... 13
IV. Trapping and Cooling of Atoms ... 20
V. Manipulation Schemes Based on the Dipole Force ... 31
VI. Experiments with Trapped Atoms ... 35
VII. Final Remarks ... 39
References ... 40

Advances in Ultracold Collisions: Experiment and Theory
J. Weiner

I. Introduction ... 45
II. Scattering Length ... 46
III. Optical Control of Inelastic Collisions ... 48
IV. Trap-Loss Collisional Processes ... 65
V. Developments in Theory ... 70
VI. Future Directions ... 76
References ... 76

Ionization Dynamics in Strong Laser Fields
L. F. DiMauro and P. Agostini

I. Introduction ... 79
II. The Bound–Free Step ... 82
III. The Free–Free Step ... 92
IV. Strong-Field Double Ionization ... 108
V. Conclusion ... 116
References ... 118

Infrared Spectroscopy of Size Selected Molecular Clusters
U. Buck

    I. Introduction 121
    II. Experimental Methods 124
    III. Theoretical Methods 132
    III. Results 136
    IV. Conclusions 155
        References 159

Femtosecond Spectroscopy of Molecules and Clusters
T. Baumert and G. Gerber

    I. Introduction 163
    II. Experimental Setup 165
    III. Results and Discussion of Experiments in Molecular Physics 172
    IV. Results and Discussion of Experiments in Cluster Physics 188
    V. Conclusions 205
        References 206

Calculation of Electron Scattering on Hydrogenic Targets
I. Bray and A. T. Stelbovics

    I. Introduction 210
    II. Electron Scattering Theories for Hydrogenic Targets 211
    III. Convergent Close-Coupling Method 219
    IV. Electron–Hydrogen Scattering 234
    V. Electron Scattering on the He^+ Ion 241
    VI. Electron–Sodium Scattering 242
    VII. Concluding Remarks 250
        References 251

Relativistic Calculations of Transition Amplitudes in the Helium Isoelectronic Sequence
W. R. Johnson, D. R. Plante, and J. Sapirstein

    I. Introduction 255
    II. No-Pair Transition Amplitudes 258
    III. *S*-Matrix Theory for Decay Rates 270
    IV. Application of Perturbation Theory to Helium-like Ions 276
    V. Results and Comparisons 286
        Appendix: Useful Identities 326
        References 327

Rotational Energy Transfer in Small Polyatomic Molecules

H. O. Everitt and F. C. De Lucia

I.	Introduction to Rotational Energy Transfer	332
II.	State-Specific Rotational Energy Transfer—Principal Pathways	356
III.	Transfer to Nonprincipal Pathways: The Grouping of States	365
IV.	Near-Resonant Ro-Vibrational Energy Transfer	372
V.	The Physical Basis of Rotational Energy Transfer	376
VI.	The Future?	394
	References	397

SUBJECT INDEX 401

CONTENTS OF VOLUMES IN THIS SERIAL 409

Contributors

Numbers in parentheses indicate the pages on which the authors' contributions begin.

P. AGOSTINI (79), Département de Recherches sur les Atomes et les Molècules, Centre d'Etudes de Saclay, Gif Sur Yvette, France

T. BAUMERT (163), Physikalisches Institut der Universität, Am Hubland, D-97074 Würzburg, Germany

I. BRAY (209), Electronic Structure of Materials Centre, School of Physical Sciences, The Flinders University of South Australia, Adelaide 5001, Australia

U. BUCK (121), Max-Planck-Institut für Strömungsforschung, 37073 Göttingen, Germany

F. C. DE LUCIA (331), Department of Physics, Ohio State University, Columbus, Ohio 43210

L. F. DIMAURO (79), Chemistry Department, Brookhaven National Laboratory, Upton, New York 11973

W. ERTMER (1), Institut für Quantenoptik, Universität Hannover, D-30167 Hannover, Germany

H. O. EVERITT (331), U. S. Army Research Office, Research Triangle Park, North Carolina 27708

G. GERBER (163), Physikalisches Institut der Universität, Am Hubland, D-97074 Würzburg, Germany

W. R. JOHNSON (255), Department of Physics, University of Notre Dame, Notre Dame, Indiana 46556

D. R. PLANTE (255), Department of Physics, University of Notre Dame, Notre Dame, Indiana 46556

J. SAPIRSTEIN (255), Department of Physics, University of Notre Dame, Notre Dame, Indiana 46556

K. SENGSTOCK (1), Institut für Quantenoptik, Universität Hannover, D-30167 Hannover, Germany

A. T. STELBOVICS (209), Centre for Atomic, Molecular, and Surface Physics, School of Mathematical and Physical Sciences, Murdoch University, Perth 6150, Australia

J. WEINER (45), Department of Chemistry and Biochemistry, University of Maryland, College Park, Maryland 20742

LASER MANIPULATION OF ATOMS

K. SENGSTOCK and W. ERTMER

*Institut für Quantenoptik, Universität Hannover, Welfengarten 1,
D-30167 Hannover, Germany*

I. Introduction 1
II. General Principles 3
III. Basic Manipulation Schemes 13
    A. Atomic Beam Deceleration 13
    B. Atomic Beam Deflection 18
IV. Trapping and Cooling of Atoms 20
V. Manipulation Schemes Based on the Dipole Force 31
VI. Experiments with Trapped Atoms 35
    A. Optical Ramsey Spectroscopy on Mg Atoms 35
    B. Trapping of Single Atoms 37
VII. Final Remarks 39
    References 40

I. Introduction

From the beginning of this century there have been extensive investigations of the internal degrees of freedom of atoms. However, for a long time it was impossible to control the external degrees of freedom. In contrast, the past decade was characterized by the development of very versatile tools to manipulate and control the motion and temperature of free atoms by light pressure forces of nearly resonant laser light. During the last few years, spectacular results have been obtained, pushing the field of laser cooling to unforeseen new developments and opening totally new fields of applications.

The theoretical discussions and experimental results of these laser cooling techniques have intensively stimulated interest in a variety of meshed fields in physics like quantum optics, atom optics, atom interferometry, atom lithography, physics of correlated quantum effects, and high-precision measurements of fundamental constants, to name a few. Even the Gedanken experiments of the early days of quantum mechanics became feasible experimentally by the new preparation techniques for atomic ensembles.

The whole field is based on the "momenta of light quanta," discovered by Einstein (1917). Absorption and emission of these light quanta by atoms in gases may lead to a manipulation of the momentum and the momentum distribution of the atoms. In the early thirties Frisch succeeded in demonstrating experimentally for the first time the mechanical action of light pressure on atoms (Frisch, 1933). But those experiments suffered from the lack of intense, tunable and monochromatic light sources, which became available after the invention of the laser. Thus it was not surprising that in the mid-1970s the basic ideas of manipulation and cooling by laser light were derived for atoms (Hänsch and Schawlow, 1975) and ions (Wineland and Dehmelt, 1975). These first proposals introduced the so-called "Doppler cooling" mechanisms. The early theoretical discussions of laser cooling were based on the interaction of single radiative field modes with two level atoms. Under this assumption the expected lowest possible temperature is given by the "Doppler limit" (Gordon and Ashkin, 1980; Cook, 1980; Minogin, 1980) determined only by the natural linewidth of the relevant atomic transition.

Here and in the following the temperature of atoms is given by the kinetic energy distribution of the atomic ensemble. One half of the mean square of the velocity distribution $\langle v^2 \rangle$ times the atom's mass m corresponds to $\frac{1}{2} k_B T$ for each degree of freedom, where k_B is Boltzmann's constant. Thus in three dimensions the temperature is defined by

$$\tfrac{3}{2} k_B T = \tfrac{1}{2} m \langle v^2 \rangle. \tag{1}$$

For sodium atoms the Doppler limit gives a temperature of 240 μK, corresponding to some 10 recoil velocities of single photon emissions.

It was the group of Phillips who surprisingly measured temperatures below the Doppler limit (Lett et al., 1988) on laser-cooled sodium atoms. Since this discovery, considerable theoretical and experimental effort has focused on understanding the "new" mechanisms of laser cooling. These investigations revealed the more complex interactions between the rich inner structure of degenerated atomic states and the polarization gradients of the light fields. The achievable lowest temperatures with these so-called "sub-Doppler" mechanisms correspond to only a few single photon recoil velocities, e.g., 20 μK for sodium. But even the limit of one photon recoil was underscored with techniques based on "dark resonances" (Aspect et al., 1988) or "Raman cooling" mechanisms (Kasevich and Chu, 1992).

At the current state of laser cooling it is possible to cool an ensemble of atoms in three dimensions down to temperatures of a few tens of nanokelvins. Densities of laser trapped atoms up to 10^{11} cm^{-3} and clouds with 10^{10} atoms are currently the standard. Today more than 200 groups all over the world operate neutral atom traps or cold beams to study further cooling techniques and an enormous variety of applications.

With the manipulation techniques realized so far, a wide range of optimized phase space distributions can be created. In most of the experiments basic manipulation techniques reduce in a first step thermal velocities of the atoms. For example, the velocity distribution of an atomic beam is nearly monochromized and reduced down to a mean velocity of a few m/s by a counterpropagating laser beam within some 10 cm (corresponding to an acceleration of $10^5\, g$). This was first demonstrated by Prodan *et al.* (1985) and Ertmer *et al.* (1985). This process is based on the so-called "spontaneous light pressure force," by which photons from a laser beam are scattered by atoms, thus reducing the antiparallel velocity component of the atom by the momentum transfer of the individual photons. Afterward the slow atoms can be deflected (Nellessen *et al.*, 1989b), transversely compressed (Nellessen *et al.*, 1990), captured in an atom trap (Migdall *et al.*, 1985; Raab *et al.*, 1987), or a combination thereof. Also these techniques are based in principle on the spontaneous light pressure force. Once precooled and confined in a space region, the more sophisticated cooling schemes, like sub-Doppler cooling, can take over, damping atomic velocities down to the final temperature.

The main purpose of this article is to outline the development and achievements of the *manipulation* of atoms and to give examples for the design of optimized phase space distributions of cold atomic ensembles. Since the main emphasis of this article is an experimental point of view, only a short theoretical introduction is given in Section II with references to tutorials and reports on laser cooling theory. Section III describes, as examples for basic manipulation schemes, the deceleration and deflection of atomic beams with laser light. In Section IV principles and experimental realizations of the two- and three-dimensional confinement of neutral atoms, in particular the magnetooptical trapping configurations, are discussed. Some manipulation methods based on light gradient forces, the so-called "dipole forces," are then presented in Section V. In Section VI we give a few examples for applications with laser-trapped atoms.

In the fast developing field of laser manipulation only a snapshot of experiments can be described. For further information about theoretical and experimental aspects the reader is referred to textbooks of summer schools about laser cooling (Dalibard *et al.*, 1992; Arimondo *et al.*, 1992), special issues of journals dealing with laser cooling and trapping (Metcalf and van der Straten, 1994; Foot, 1991; Special Issue, 1985, 1989), and the references therein. For an overview of *related* fields which will not be discussed here the reader is referred to overview and tutorial articles, for example, in atom optics (Adams *et al.*, 1994), ion traps, and ion cooling (Walther, 1992, 1994) as well as in atom interferometry (Special Issue, 1992).

II. General Principles

Laser cooling and manipulation of free atoms are based on the principles of light matter interaction, well understood in QED. Nevertheless, the numerous degrees of freedom, present in complex cooling schemes, force theory to make a clever choice of assumptions and simplifications in order to take into account the details of the actual cooling scheme. This led and will lead to the development of new calculation methods (e.g., quantum Monte Carlo calculations (Zoller et al., 1987; Dalibard et al., 1992). In the following we give a brief review of basic laser cooling theories and the corresponding regimes in which they apply.

The discussion of light pressure and atom cooling starts with the definition of the Hamiltonian of the interacting systems:

$$H = H_A + H_V + V_{AL} + V_{AV}, \quad (2)$$

where

$$H_A = \frac{\mathbf{P}^2}{2m} + H_{int} \quad (3)$$

is the undisturbed atomic Hamiltonian including external degrees of freedom and H_{int} describes the internal structure of, for example, a two-level system:

$$H_{int} = \hbar\omega_A |e\rangle\langle e|. \quad (4)$$

H_V is the field Hamiltonian as the sum of the various field modes i:

$$H_V = \sum \hbar\omega_i (a_i^+ a_i + \tfrac{1}{2}). \quad (5)$$

The atom is coupled to the laser field modes $\mathbf{E}_L$, via V_{AL} and to all other modes of the radiation field, initially empty, via V_{AV}.

In this ansatz atom–atom interactions are neglected, although they play a central role in density and temperature limits of, for example, trapped atom clouds. Interactions like van der Waals forces and radiation-assisted atom–atom forces are typically introduced in a second step by an effective potential. A new approach, which incorporates atom–atom coupling, was developed by Lewenstein et al. (1994).

In laser cooling one is interested in the description of the dynamics of the external degrees of freedom of the atom. The important observables are the position $\mathbf{R}$, the momentum $\mathbf{P}$, and the electric dipole moment $\mathbf{d}$. The details of the radiation field dynamics, which are of special interest, for example, in cavity QED, are mostly neglected. As a consequence the field Hamiltonian is usually separated.

The coupling of the atom to the vacuum modes is essential as it is responsible for spontaneous emission. This process introduces friction and damping to the system—necessary for Doppler cooling—but also part of the fluctuations which, for example, limit final temperatures. In addition optical pumping, important for sub-Doppler cooling, is based on spontaneous emission.

This damping process is relevant only as long as the characteristic interaction times t_{int} is long compared with spontaneous emission times t_{sp}. Therefore a central ramification of laser cooling separates systems with short interaction times ($t_{\text{int}} \ll t_{\text{sp}}$) from systems coupled to the vacuum field reservoir. The first case usually guarantees a fully coherent interaction (e.g., coherent atomic wave beam splitters) and is widely used in atom optics (e.g., the Ramsey–Bordé atom interferometer, described in Section VI). However, coherence is possible for longer interaction times ($t_{\text{int}} \gg t_{\text{sp}}$) too, if spontaneous emission is suppressed, for example, by far-of-resonance interaction or dark state evolution.

If we first concentrate on interactions with damping, a fully quantum mechanical description of both the atoms and the radiation fields becomes necessary if the thermal energy $k_B T$ of the system is comparable to the characteristic lowest excitation energy of the system, for example, when the temperature reaches the order of a single photon recoil energy,

$$E_R = \frac{\hbar^2 k^2}{2m}. \tag{6}$$

Hence, phenomena like sub-Doppler cooling to lowest temperatures, optical lattices, or evaporative cooling demand a fully quantized description (see, e.g., Kazantsev *et al.*, 1990; Berman, 1991; Wallis, 1995). The method of quantum Monte Carlo calculations was introduced for an efficient control and description of, for example, three-dimensional sub-Doppler or sub-recoil mechanisms (Zoller *et al.*, 1987; Dalibard *et al.*, 1992). The results are in agreement with observed atom dynamics.

For the majority of the other cases in which the averaged action of individual scattering processes describes the exchange of energy and momentum correctly within an error of only single recoil energies, a semiclassical approach is adequate. The assumption that the atomic wave packet is sufficiently well localized in position space and that momentum space allows for descriptions as close as possible to classical concepts for the center of mass motion of the atom is justified when the spatial spread ΔR is small compared with λ_L and when the momentum spread ΔP does not lead to a Doppler shift $\Delta P / m \cdot k_L$ larger than the natural width Γ.

In this limit it is sufficient to consider first the change of the mean values of **P** and **R**, thus to derive mean forces. The fluctuations on those forces are then treated in a second step.

Essential derivations from this model arise in the case of the existence of degenerated ground state levels. Optical pumping becomes important and therefore an *additional* characteristic *internal* time scale appears. Again, the comparison of time scales subdivides the mechanisms and simplifies the analysis of the different regimes. For systems with nondegenerated ground state *and* $t_{\text{spont}} \ll t_{\text{int}}$ the fast internal variables can be adiabatically eliminated and the derivation of reduced equations for the external variables is straightforward as discussed below.

For multilevel systems the additional internal time scales may reach, for example, external damping times or become even longer. In this case the elimination of the internal variables is no longer possible, the calculation becomes more difficult. This is the regime in which additional mechanisms like sub-Doppler cooling forces take over or even the oscillation of atoms in wavelength-scaled potential wells appears significantly. Before we consider this regime we will shortly discuss the forces and limits of cooling relevant for two-level systems.

For two-level atoms the generalized optical Bloch equations are derived from the Heisenberg equations of motion (Kazantsev et al., 1985; Stenholm, 1986). Mean values for the appearing forces lead to a resulting force term, F_m, acting on the atom, containing two parts:

$$F_m = F_{diss} + F_{react}. \tag{7}$$

The reactive response F_{react} can be interpreted in classical terms as the in-phase component of the mean atomic dipole moment with the driving laser field, the dissipative part F_{diss} being the quadrature component.

The mean energy absorbed during the interaction is related to the dissipative part, whereas the reactive part does not change the mean energy of the atom (but the momentum!). The dissipative force is connected to the absorption and spontaneous emission of photons and is therefore often called spontaneous force F_{sp}.

F_{react} is connected to the redistribution of photons among the various plane wave modes forming a laser wave (and thus disappears in a single *plane* wave). A force, F_{react}, arises only in the presence of an intensity gradient and may be derived—neglecting the fluctuations—from a potential. F_{react} is usually called the dipole force.

The above mean force model is extensively discussed in the early work about light pressure forces (Cook, 1980; Gordon and Ashkin, 1980; Minogin et al., 1981; Andreev et al., 1981), and detailed calculations for different laser field geometries can be found, for example, in Letokhov and Minogin (1981) and in Minogin and Letokhov (1987). Closed representations are also given in Stenholm (1986) and Cohen-Tannoudji (1992).

For a given detuning, $\delta = \omega_L - \omega_A$, of the laser frequency ω_L from the unperturbed atomic resonance frequency ω_A, a two-level atom with velocity v relative to the laboratory frame of the laser field experiences an effective detuning due to the Doppler shift (in the first order)

$$\delta_{eff} = \delta - kv. \tag{8}$$

In the case of a plane wave the atom experiences an average force in the direction of the **k** vector of the plane wave:

$$F_{spon} = \hbar k \frac{\Gamma}{2} \frac{\frac{I}{I_0}}{1 + \frac{I}{I_0} + \left(\frac{2\delta_{eff}}{\Gamma}\right)^2}. \tag{9}$$

Here Γ denotes the spontaneous decay rate of the upper state (the lowest state is supposed to be stable), I is the laser intensity, and I_0 is here called the saturation intensity for polarized light, reading

$$I_0 = \frac{\pi \hbar c}{3 \lambda^3 \tau}. \tag{10}$$

We define the saturation parameter as

$$S = \frac{I/I_0}{1 + \left(\frac{2\delta_{\text{eff}}}{\Gamma}\right)^2}. \tag{11}$$

The Rabi frequency is then given by

$$\frac{\Omega^2}{\Gamma^2} = \frac{I}{2I_0}, \tag{12}$$

and the connection to the saturation parameter reads

$$S = \frac{2\Omega^2}{4\delta^2 + \Gamma^2}. \tag{13}$$

(It should explicitly be mentioned that in the literature some authors use different definitions which refer to $2I_0$ as the saturation intensity or to I/I_0 as the saturation parameter.)

In an intuitive picture, the spontaneous force is simply the sum of successive absorptions of photon momenta $\hbar \mathbf{k}_L$ followed by spontaneous emission with isotropic emission probability in space and therefore vanishing mean momentum distribution per time interval. As a function of the detuning, F_{sp} has a Lorentzian shape centered around $\delta = 0$. For low intensities F_{sp} is proportional to the intensity I but with increasing intensity saturates at a maximum value, $F_{\text{sp max}}(I \to \infty)$,

$$F_{\text{sp max}} = \hbar k \frac{\Gamma}{2}. \tag{14}$$

According to a typical upper-state lifetime of 10^{-8} s this force may result in very large acceleration values up to 10^5 times the gravitational acceleration. The large accelerations together with the dissipative character favor the spontaneous force for thermal atomic beam manipulation and cooling like beam stopping, focusing, or deflection or three-dimensional confinements.

Concerning now the reactive part of the light force, the dipole force acts from an intuitive point of view on the induced atomic dipole moment in an inhomogeneous field (Ashkin, 1970a). For a field with local intensity gradient $\nabla I_{(v)}$ the dipole force reads

$$F_{\text{dip}} = -\frac{\hbar \delta_{\text{eff}}}{8} \frac{\frac{\nabla I}{I_0}}{1 + \frac{I}{I_0} + \left(\frac{2\delta_{\text{eff}}}{\Gamma}\right)^2}. \tag{15}$$

The dipole force depends on the laser detuning δ_{eff} and has the shape of a Lorentzian dispersion curve. For detuning $\delta < 0$ ($\omega_L < \omega_A$, "red" detuning) the reactive force pushes the atom toward the regions of higher intensities; for $\delta > 0$ ("blue" detuning) it repels the atom from regions of higher intensities.

In contrast to the spontaneous force, the dipole force does not saturate and increases with $|\nabla I|$. In strong gradient fields like evanescent light fields (Section VI) the dipole force can overcome the spontaneous force by a factor of up to several 10^3.

Finally the dipole force can be expressed, in the simplified picture of nonfluctuating forces, as the gradient of a potential, U, given as

$$F_{\text{dip}} = -\nabla U, \tag{16}$$

with

$$U(r) = \frac{\hbar \delta_{\text{eff}}}{2} \ln\left[1 + \frac{I/I_0}{1 + \left(\frac{2\delta_{\text{eff}}}{\Gamma}\right)^2}\right]. \tag{17}$$

In the high-intensity limit the dipole force is readily interpreted in the "dressed atom" picture (Cohen-Tannoudji and Reynaud, 1977). We refer to the excellent article of Dalibard and Cohen-Tannoudji (1985). A discussion of manipulation schemes based on dipole forces follows in Section V.

We now focus on the fluctuations of the forces derived earlier. These fluctuations introduce noise in the atomic motion and are responsible for a diffusion of atomic momentum limiting, for example, the lowest possible temperatures in laser cooling. In the simplest picture atomic recoils introduced by spontaneous emission events randomize the atomic motion, similar to Brownian motion. This was already pointed out by Einstein (1917). Thus theoretical attempts recall in a first step results of classical Brownian motion as a basis for specific radiation force problems (see, e.g., Cohen-Tannoudji, 1992, and references therein).

To discuss this diffusion process in the frame of a typical configuration of laser cooling, we switch at this point to an explanation of the so-called "optical molasses." It consists of two counterpropagating laser beams slightly detuned to the red (Hänsch and Schawlow, 1975) and was the basic idea for a fully controlled damping of atomic motion to zero velocities. For a free atom moving in this standing wave the counterpropagating wave is Doppler shifted closer to resonance and exerts a stronger radiation pressure

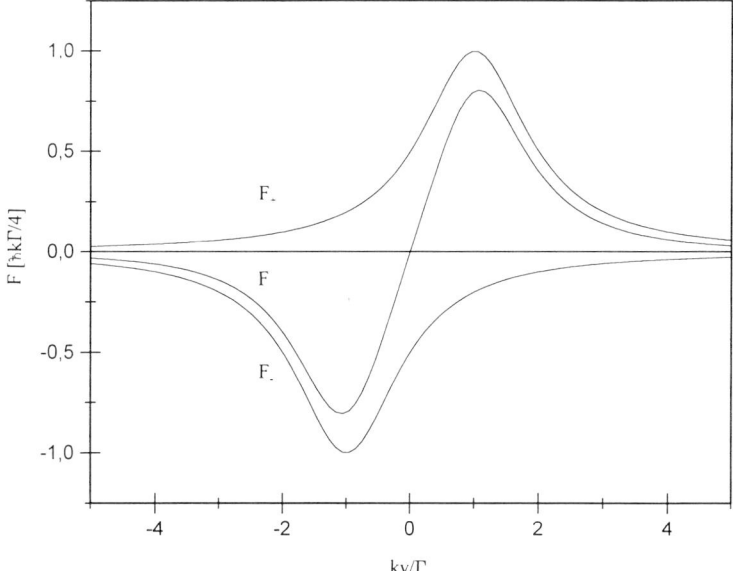

FIG. 1. Radiation pressure force versus velocity, for the case of a weak standing wave. For low saturation the resulting force F is the sum of the individual forces $F_+ + F_-$ of the two running waves.

on the atom than the copropagating wave, thus damping the atomic velocity.

Figure 1 shows the mean force experienced by an atom in this one-dimensional molasses configuration in the limit of low saturation as the sum of the two spontaneous forces from the two laser beams. Around zero velocities the slope is linear in the first order. Another important factor is the velocity capture range Δv of such a process. For a two-level system it is obviously determined by the natural width of the atomic excited state

$$k\Delta v \sim \Gamma. \tag{18}$$

Three orthogonal pairs of such laser beams were called optical molasses by the Bell Labs group in 1985, because the atomic motion is damped like in a highly viscous medium (Chu et al., 1985).

Atoms captured in the molasses region undergo many spontaneous emission cycles and therefore perform a random walk in the spatial region in which the laser beams overlap. The temperature of the atoms in the molasses is determined by the equilibrium between cooling and heating due to diffusion.

For two-level atoms from the Langevin equation one derives the spatial diffusion coefficient D_x, an estimate of the time required for atoms to leave the molasses region, and a final lowest temperature, T, as the equilibrium between cooling and heating (Stenholm, 1983). The time evolution of the mean square momentum of a Brownian particle is then given by

$$\frac{d\langle p^2 \rangle}{dt} = -\frac{2\alpha}{m}\langle p^2 \rangle + 2D_p, \tag{19}$$

where α is the constant friction cefficient of the spontaneous damping force $F_{sp} = -\alpha v$ and D_p is the momentum diffusion coefficient. After a time, $\tau = m/\alpha$, the particle reaches a steady state corresponding to the equilibrium

$$mv^2 = D_p/\alpha = k_B T, \tag{20}$$

giving the final temperature of the ensemble.

However, even in thermal equilibrium the motion of the atoms is still diffusive; the mean square of the spatial distribution increases linearly in time,

$$\frac{d\langle x^2 \rangle}{dt} = 2D_x, \tag{21}$$

where the spatial diffusion coefficient D_x is connected to the momentum diffusion coefficient D_p and the friction coefficient α by

$$D_x = D_p/\alpha^2. \tag{22}$$

The friction and diffusion coefficients for a molasses configuration can for a two-level system be expressed in terms of the laser parameters (see, e.g., Drewsen et al., 1994) and read

$$\alpha = \hbar k^2 \frac{8|\delta|\Gamma}{4\delta^2 + \Gamma^2} S \tag{23}$$

and

$$D_p = \hbar^2 k^2 \Gamma S, \tag{24}$$

and therefore the spatial diffusion coefficient in the limit $|\delta| > \Gamma$ becomes

$$D_x = \frac{\Gamma}{k^2}\left(\frac{\delta}{\Gamma}\right)^4 \frac{I}{I_0}. \tag{25}$$

The damping as well as heating increases with increasing laser power. The fascinating long time that a two-level atom with a typical lifetime of 10 ns of the upper state needs to diffuse 1 cm in a molasses configuration with $I/I_0 = \frac{1}{6}$ is 7.5 s!

In the low saturation limit the final temperature depends on the detuning according to

$$K_B T = \frac{\hbar \Gamma}{4} \frac{1 + \frac{I}{I_0} + \left(\frac{2\delta}{\Gamma}\right)^2}{\frac{2\delta}{\Gamma}}. \tag{26}$$

The minimum is reached for $\delta = -\Gamma/2$ and is normally called the Doppler limit:

$$k_B T_{\text{Doppler limit}} = \hbar\Gamma/2 = \hbar/2 \, 1/\tau_{\text{spont}}. \qquad (27)$$

Here τ_{spont} denotes the natural lifetime of the upper state.

For atomic systems with a degenerated ground state in a red detuned standing wave with orthogonal linear or circular polarizations, the two-level model fails; the sub-Doppler mechanisms reveal a much steeper force profile, and diffusion is coupled to different time scales. The final possible temperatures are much lower than the Doppler limit, and diffusion times of the atom out of the molasses region are much longer than those for two-level atoms.

As discussed above, sub-Doppler cooling mechanisms are based on the interaction of multilevel atomic systems with polarization gradients, introducing additional internal time scales, especially optical pumping times. Usualy two major mechanisms are separated (Dalibard and Cohen-Tannoudji, 1989), according to two essentially different polarization gradient configurations in a laser standing wave. In one case the polarization of the light beams is linear but *orthogonal* (lin ⊥ lin); in the other case the counterpropagating beams are circular polarized but with different orientation ($\sigma^+-\sigma^-$).

The lin ⊥ lin laser configuration exhibits a strong gradient of ellipticity along the laser propagation axis on a length scale of a fraction of the wavelength. For a *degenerated* ground-state atom, the optical pumping between different ground state levels and thus the internal atomic state depends on the polarization of the pumping light. The internal motion of a moving atom has to respond to the variations of the laser polarization in the moving frame. Because this internal optical pumping time is for low saturation longer than the external time of motion over one wavelength, the configuration leads to strong *nonadiabatic* effect. Due to the spatial dependence of the light shifts along the standing wave, the laser field additionally generates strong potential wells, also on the wavelength scale.

The deciding effect, producing the damping, is that the spontaneous emission probability is — due to optical pumping — higher at the potential hills than in the valleys. Thus a moving atom climbs a potential hill, looses kinetic energy, emits a photon, and jumps to another ground-state level, for which the light shift potential is *at its minimum*.

On average, an atom more often climbs up a hill than runs down, therefore transferring kinetic energy to potential energy and to the reemitted photon. The lin ⊥ lin sub-Doppler cooling is therefore called "Sisyphus cooling" in analogy to Greek mythology.

The combination of intensity and polarization gradient is essential. For counter-propagating laser waves with the same polarization, for example, linear parallel linear, only the gradient of intensity appears. Thus no spatial dependence for individual pumping processes can produce damping; no sub-Doppler cooling appears.

The second sub-Doppler mechanism is also based on the existence of several ground-state levels and polarization gradients. But in the $\sigma^+-\sigma^-$ configuration the spatial variation of the polarization corresponds to a corkscrew and gives rise to quite different physical mechanisms. Those are based on "motion-induced orientation" and again on the nonadiabatic following of the internal degrees of freedom of the atom. The orientation leads to different absorption probabilities for σ^+ and σ^- photons from the two counter-propagating laser beams, resulting in a net force. For a detailed explanation of sub-Doppler cooling the reader is referred to the literature which stimulated the field of laser cooling [see, e.g., Special Issue (1989), esp. articles by Dalibard and Cohen-Tannoudji (1989) and Ungar et al. (1989); see also Weiss et al. (1989); Dalibard et al. (1992); Berman (1991); Javanainen (1991, 1992); Molmer et al. (1991); Molmer (1991); Nienhuis et al. (1991); Werner et al. (1992, 1993); Wallis et al. (1993); Wallis (1995)].

One of the major predictions of the theory is that — for low saturation — the temperature is proportional to the light shift induced by the molasses beams; therefore

$$k_B T \sim \Omega^2/|\delta|, \tag{28}$$

where the Rabi frequency is proportional to the laser intensity. This dependence was explored in early experiments at NIST (Lett et al., 1988; Phillips et al., 1989) and later confirmed in numerous experiments (see e.g., Salomon et al., 1990).

Sub-Doppler cooling in optical molasses to temperatures corresponding to only a few recoil velocities is now state of the art in a huge number of labs for different kinds of atoms offering a split ground-level structure suitable for these cooling mechanisms (e.g., alkaline and earth alkaline elements like Li, Na, Rb, Cs, Mg, Ca, and Ba as well as metastable states of rare gases He, Ne, Ar, Kr, and Xe). One important question of laser cooling is, of course, what are the lowest possible temperature and accordingly the largest possible de Broglie wavelength? In the usual schemes fluorescence cycles never cease; there is always at least a single photon-induced random walk. It is then impossible to reach a temperature lower than the recoil cooling limit. There is, however, no fundamental limit, preventing laser cooling from the recoil limit.

Up to now two cooling schemes have allowed for sub-recoil temperatures. The one method traps atoms in velocity-selective dark resonances (Aspect et al., 1988); the other method uses sequences of stimulated Raman and optical pumping pulses (Kasevich and Chu, 1992). Detailed presentations are given in Aspect et al., 1989, Cohen-Tannoudji et al., 1992, and Ol'shanii, 1991. Both methods were first demonstrated in one dimension and then extended to two (Lawall et al., 1994; Davidson et al., 1994) and three dimensions (Davidson *et al.*, 1994). Further proposals for sub-recoil cooling methods exist (Wallis and Ertmer, 1989). Currently sub-recoil cooling is one of the most inspiring developments in the field of laser cooling of atoms and ions.

III. Basic Manipulation Schemes

A. ATOMIC BEAM DECELERATION

Optical molasses and most of the other cooling schemes have a limited capture range in momentum space, thus demanding already slow atoms (velocity equal to some tens up to a few hundred m/s) in order to work. In addition, direct atomic beam applications such as laser spectroscopy where the interaction time can be increased with decreasing atomic velocity profit from slow atoms. Atomic beam deceleration was thus one of the first experimental investigations of laser cooling in the early 1980s and stimulated the following steps of beam manipulation, for example, focusing, trapping, and channeling.

After the pioneering works (Prodan *et al.*, 1985; Ertmer *et al.*, 1985), which experimentally invented atomic beam slowing, laser beam deceleration became the standard technique to reduce thermal velocity distributions down to the level at which further cooling techniques take over. The principle scheme (see Fig. 2) utilizes the spontaneous force to reduce the longitudinal velocities of the atoms by a counter-propagating laser beam. It should be mentioned that the transverse velocity distribution of an atomic beam typically corresponds to a few millikelvins already.

The spontaneous force of the counterpropagating beam decelerates the atoms. But a fixed-frequency laser is — as a consequence of the Doppler effect — resonant with only a narrow velocity group; the width is determined by the homogeneous linewidth of the transition (natural width and saturation broadening). For sodium atoms this corresponds to an interval of 6 m/s (FWHM of the Lorenzian resonance curve for low saturation), in contrast to the thermal velocity distribution typically ranging from zero to a few thousand m/s. A counter-propagating laser beam thus only burns a hole in

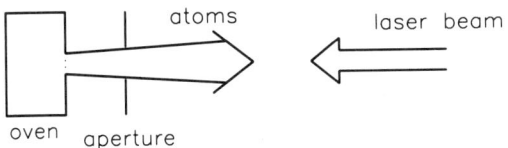

FIG. 2. Principle scheme of atomic beam deceleration with a counter-propagating laser beam.

the velocity distribution (Balykin et al., 1980, 1984; Andreev et al., 1982) (see Fig. 3, second curve).

Among the different approaches to solve this problem two became standard methods: "chirp slowing" (Ertmer et al., 1985) and "Zeeman slowing" (Prodan et al., 1985). In the former the laser frequency is swept *in time* to compensate synchronously for the changing Doppler shift during the deceleration; in the latter the atomic transition frequency is modified *in space* by an inhomogeneous magnetic field which shifts the energy levels by the Zeeman effect.

The choice of the desired method depends on the details of the atomic line structure and the requirements of the subsequent experiment. The chirp method is rather easy to implement, particularly with diode lasers which allow for an extremely fast frequency chirp by ramping the diode current. At the end of each sweep the laser frequency is set back to the start frequency $v_L(0)$ and the next sweep begins according to $v_L(t) = v_L(0) + \beta t$. However, such an experiment necessarily has a pulsed nature, and the spatial distribution of the atomic velocities at the end of each sweep tends to be rather complicated. This is a direct consequence of the different stopping distances of different initial velocity classes and the different drift periods during individual sweeps.

In a Zeeman slower all atoms reach their final velocities at the same position at the end of the slowing magnet, producing a continuous beam of slow atoms. This scheme, however, requires a well-designed magnet; especially the field slope at the end is critical for the final velocity width. If atoms are not suddenly forced out of resonance by a strong field gradient, the velocity distribution becomes considerably broader. A sharp magnetic field strength increase at the end of the magnet reduces the "postcooling" problem, but typically the final longitudinal temperature is higher for Zeeman cooling than for sweep cooling.

In general a counter-propagating laser beam, either swept in time or magnetic field assisted, may cool the longitudinal beam temperature to the Doppler limit. A transformation to the rest frame of the atom (Salomon and Dalibard, 1988) leads to the same forces as those in the molasses configuration (Section II; Fig. 1), decelerating the atoms into a "moving" velocity interval. Thus the results discussed in the context of Doppler cooling — final temperature and diffusion — still hold. The saturation behavior of the spontaneous force restricts the maximum sweep rate $\beta = \partial \delta/\partial t$ or the maximum magnetic field gradients, depending on the natural linewidth Γ of the atomic cooling transition.

For a given laser intensity the chirp rate β must be consistent with the achievable deceleration given by the actual saturation S (Section II) according to

$$\beta_{max} = h/(m\lambda^2 \tau(2 + I_0/I)). \tag{29}$$

If the sweep rate (field gradient) is higher than β_{max}, the locking of atomic velocity (kv) to the actual frequency breaks down, and the atoms walk out of resonance at the actual velocity.

This criterion leads to the following typical parameters for the deceleration of sodium.

Most probable velocity in a beam ($T = 700$ K)	$v_{pr} = 900$ m/s
Resonance wavelength	$\lambda = 589$ nm
Lifetime of the excited state	$\tau = 16$ ns
Recoil velocity due to the absorption of one photon: $v_R = \hbar k/m$	3 cm/s
Number of photons to stop an atom with velocity v_{pr}	30,000
Stopping distance (at one-half of the maximum deceleration, corresponding to saturation $s = 1$)	48 cm
Maximum acceleration for $s \to \infty$	10^6 m/s$^2 = 10^5 g$

The curves in Fig. 3 illustrate typical velocity distributions at the end of one sweep for different stopping laser frequencies. Corresponding measurements for sodium (Nellessen et al., 1989a) are given in Fig. 4. For a proper choice of parameters even beam reversal is possible (Ertmer et al., 1985). Alternative methods, to compensate for the changing Doppler shift apply an optical "white-light spectrum" to the atoms, covering the entire Doppler width (Moi, 1984; Liang and Fabre, 1986; Hoffnagle, 1988), or compensate the Doppler detuning by the Stark effect, analogous to Zeeman slowing.

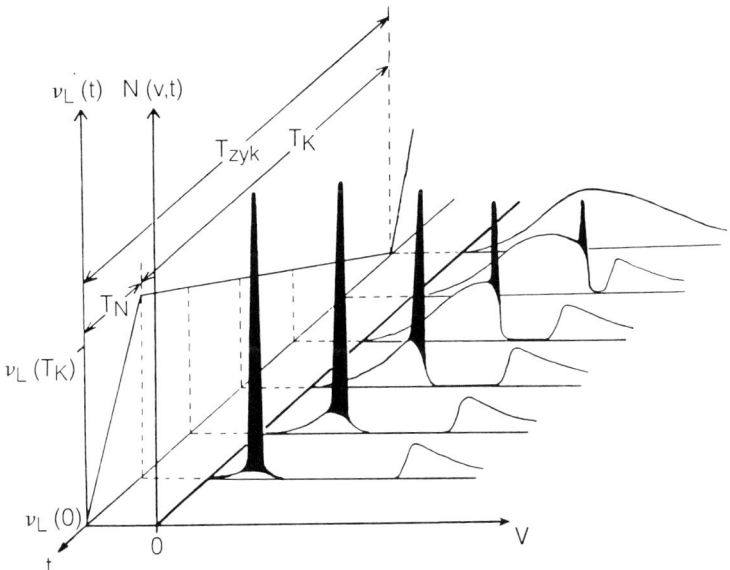

FIG. 3. Longitudinal velocity distributions of atoms during the time T_K of one sweep of the decelerating laser frequency $v_L(t)$. The first curve (background) shows the Maxwell–Boltzmann distribution of the thermal atomic beam. During the time T_N the laser frequency is set back, and the next sweep cycle starts.

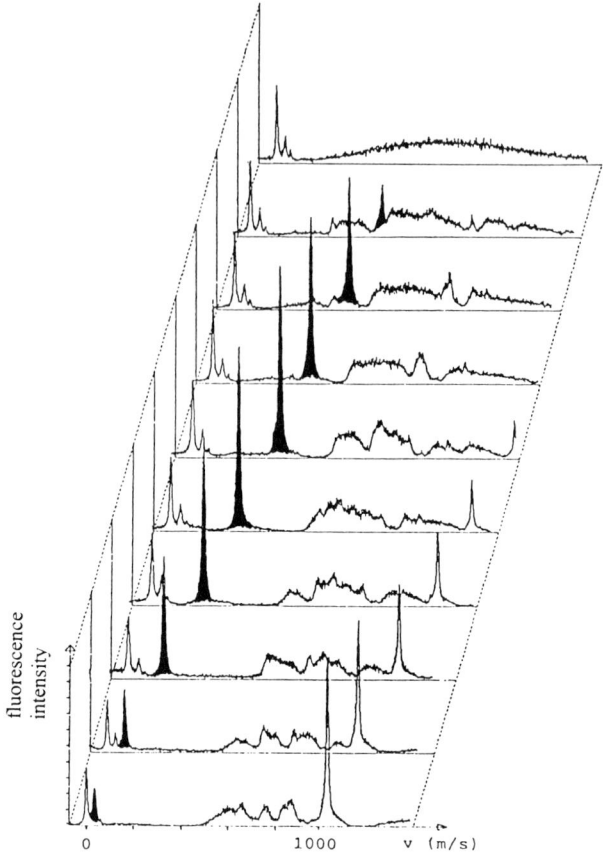

FIG. 4. Measured velocity distributions of a decelerated sodium atomic beam according to the principle time evolution of Fig. 3. The three peaks at $v = 0$ arise from the fluorescence induced by a zero marker laser beam. The complex peak structure at higher velocities is a consequence of the sideband chirp method in combination with pump processes.

In a very interesting experiment Ketterle et al. (1993) applied a "photon gas" to the atomic beam by diffuse laser radiation, coming from all directions. In that case during deceleration the atom automatically picks up the photons with the longitudinal **k**-vector component, just compensating the actual Doppler shift. But of course in this method (as well as in white-light cooling) actually more laser power is necessary to generate the same saturation on the whole spectrum.

Up to now we have not mentioned the necessity to overcome optical pumping during the manipulation process for atoms with a split ground state. This is normally taken care of by an additional laser frequency in the cooling laser beam, thus repumping atoms from lower states not accessible to the main resonance transition frequency.

In the case of sweep slowing, this frequency obviously has to be swept synchronously with the main frequency. This can be achieved, for example, through electrooptic modulation. In this scheme the sidebands are used for repumping and are automatically swept synchronously with the chirp of the carrier (Ertmer et al., 1985). As a consequence the various laser frequencies influence higher velocity classes during one sweep and thus produce a multiple peaked final velocity distribution, as shown in Fig. 4.

Complications with optical pumping disappear for atoms with a non-degenerated ground state. An interesting example for this is magnesium^{24} which offers a closed two-level system well suited for laser manipulation. It connects the $^1S_0$ ground state and the $^1P_1$ excited state with a transition frequency of 285 nm. Special interest in this particular element has arisen from its level scheme which is well suited to study optical frequency standards having the advantages of laser-cooled atoms. The Mg^{24} intercombination line at 457 nm has been discussed as a possible clock transition (Ertmer and Penselin, 1986).

Due to the short lifetime of 2 ns of the upper state of the cooling transition and the large recoil velocity of 5.6 cm/s, Mg atoms with an initial velocity of 1200 m/s can be stopped within 11 cm (Sterr et al., 1992). Therefore, a permanent-magnet Zeeman slower is best suited for this task. Since the $^1S_0$ ground state is not degenerate, stray fields at the exit of the magnet do not influence the decelerated atoms. In this way Mg atoms can be decelerated and precooled in a compact setup and subsequently can be captured in a magnetooptical trap for ultra-high-precision spectroscopy (Sengstock et al., 1993) (see Section VII).

In the above paragraph only the influence of the mean force in longitudinal direction was investigated. Transversely to the beam direction diffusion leads to heating, expanding the beam. Obviously the photon emission momenta also add up to zero mean value, but the mean square value will increase with the increasing number, N_{sp}, of spontaneously emitted photons according to

$$\langle p^2 \rangle = \hbar k \sqrt{N_{sp}}. \tag{30}$$

For typical parameters in the sodium experiments the transverse velocity spread reaches approximately 5 m/s which must be compared with the final longitudinal velocity of a few up to some 10 m/s to calculate the exit divergence of the cold part of the beam.

The transverse heating reduces the density and thus the brightness of the laser-decelerated beam and therefore limits the loading efficiency of atom traps which capture the slow atoms with a restricted spatial capture range (see Section V). Efficient reduction of transverse heating is possible with the help of additional transverse two-dimensional molasses or magnetooptical compression zones, as described in Section V.

B. Atomic Beam Deflection

The spontaneous force light pressure is also demonstrated quite well by the deflection of an atomic beam irradiated transversely with resonant laser light. During the time of flight through the laser beam (Fig. 5) atoms experience the spontaneous force; they gain additional transverse momentum and are deflected.

In fact the first demonstration of light pressure was the small deflection of a sodium atomic beam, illuminated from one side with light from a sodium vapor lamp, performed by Frisch (1933). After the development of intense laser light sources higher deflection angles (Schieder *et al.*, 1972) became feasible.

The special interest of laser deflection is given by applications demanding a slow, cold atomic beam, free of additional laser radiation (e.g., surface interactions). For that purpose a thermal beam is first laser decelerated. The cold part is then selectively deflected out of the main beam and the cooling laser light, thus generating a nearly monochromatic, pure atom source (Nellessen *et al.*, 1989a, b).

In the simplest configuration with perpendicular irradiated plane wave (see Fig. 5), the additional transverse momentum is obviously limited to the velocity v_R at which the additional Doppler shift kv_R shifts the atom out of resonance, similar to the "hole burning" effect in atomic beam deceleration. If the length of the interaction region is long enough, $l \geq \hbar v_{\text{long}}/E_R$ (E_R denotes for the recoil energy), for example, for previously laser decelerated atoms, atoms reach v_R, and this arrangement cools the deflected atoms also transversely to the Doppler limit.

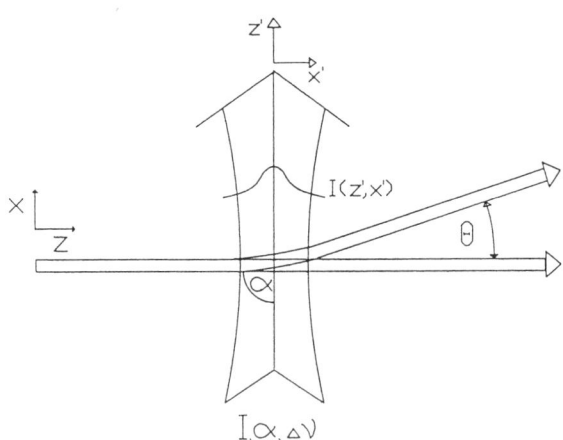

FIG. 5. Scheme for atomic beam deflection with a laser beam (intensity, I; detuning, Δv) irradiated under the angle α relative to the atomic beam axis (α, nearly 90°).

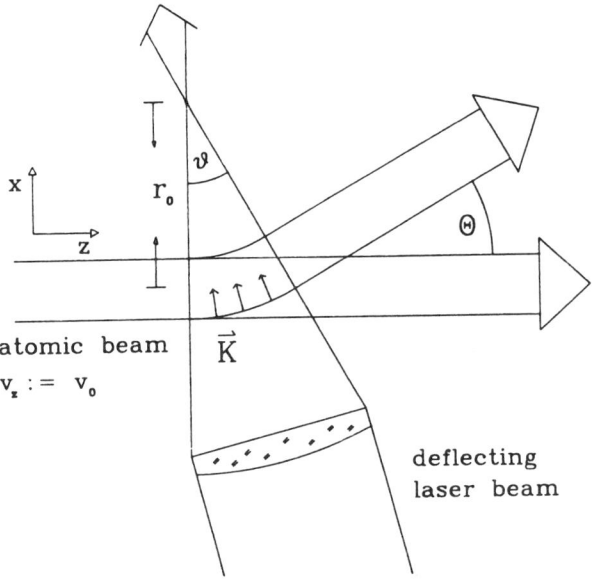

FIG. 6. Laser deflection with a focused laser beam. For a proper choice of laser parameters, atoms with velocity $v_z = v_0$ follow a circle with radius r_0.

For sodium, v_R corresponds to 6 m/s; thus, even for the slow part of a decelerated beam, the deflection angles are rather small. To overcome this limitation, it is — as in atomic beam deceleration — necessary to compensate for the change in the Doppler shift as the laser acts on the atom.

A purely geometrical solution of this problem (without changing the laser frequency or shift atomic levels in time) is given in Fig. 6. A converging laser beam (focused by a cylindrical lens) is irradiated toward the atoms (Ashkin, 1970b). For a proper choice of parameters the beam will be deflected in such a way that the actual **k** vector of the light field stays perpendicular (Doppler free) to the local direction of propagation (Nellessen et al., 1989b).

The atoms follow an orbit in which the local spontaneous force just compensates the centrifugal force.

$$F_{\text{spont}}(r_0) = \frac{mv^2}{r_0}. \tag{31}$$

If this condition is not exactly fulfilled the deflection process itself stabilizes this trajectory. Atoms which tend to higher values of r get closer to resonance and are thus redirected to r_0 and vice versa. In the case in which the light field is formed by a cylindrical lens, the intensity distribution $I(v)$ is proportional to $1/r$; thus the angle of deflection is independent of the radius (for low saturation), and atom are transversely cooled to the Doppler

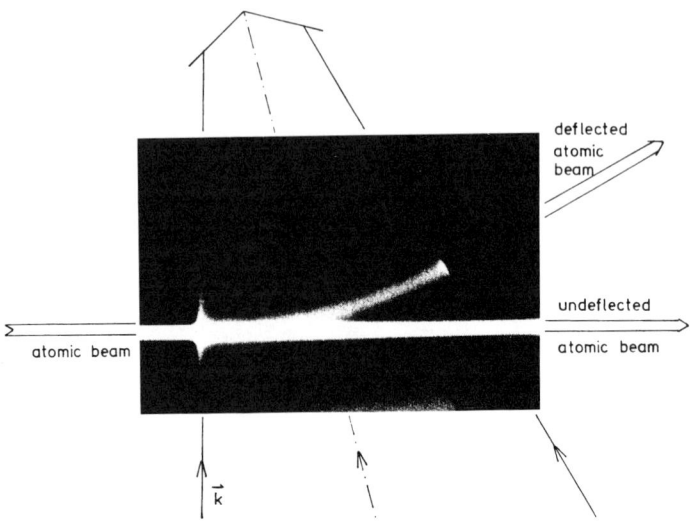

FIG. 7. Photograph of the deflection region with the fluorescence light from atoms, following a circle. The strong fluorescence from the undeflected beam is induced by the decelerating laser beam counter-propagating the atoms.

limit. In an experiment with sodium atoms the cold part with a velocity of 100 m/s could be deflected for angles up to 30°.

Figure 7 is a photograph (top view) of the deflection zone. The resonance fluorescence of the deflected atoms induced by the deflecting laser field shows the circular path of the atoms nicely. Faster atoms still present in the original beam (see also Fig. 3) lead to the fluorescence of the main stream.

With a more extended light field higher deflection angles are possible. Ashkin (1970b) proposed an atom storage ring based on the spontaneous force.

For infinite interaction time in a single laser beam, atoms are still pushed out of resonance; thus mechanisms damping to a definite velocity, for example, standing wave molasses in combination with trapping or focusing, are favored for fully controlled atom manipulation. We will now focus on those configurations.

IV. Trapping and Cooling of Atoms

Confinement of free particles has been an experimental goal for a long time. The successful development of ion traps by Paul *et al.* (1958) in the 1950s soon raised the question for neutral atom traps. Since strong first-order coulomb interaction cancels for neutral particles, only the interaction of the trapping fields with higher order moments of the charge and current distribution of the particle can be utilized to trap neutral atoms.

Since then several types of traps using combinations of static magnetic and time-dependent electromagnetic fields, for example, RF fields or light fields, have been developed. Typical trap depths range from several millikelvins for magnetostatic traps and dipole traps up to some kelvins for magnetooptical traps.

Here we will focus mainly on magnetooptical traps. An excellent overview about other trap configurations can be found in Phillips (1992) and Chu (1992b).

Stable trapping of particles requires a position-dependent force directed to a center. Configurations with light beams directed to a common center and resonant with an optical two-level transition of the atom are excluded by the so-called optical Earnshaw theorem (Ashkin and Gordon, 1979; Ashkin and Gordon, 1983; Ashkin, 1984). If the light scattering force is proportional to the local Poynting vector it is impossible to achieve an overall inwardly directed force. Since there is no sink for photons (e.g., a black absorber) in free space the equation $\nabla F_{scat} = 0$ holds, forbidding a stable equilibrium for an atom at rest. This corresponds to Earnshaw's theorem of electrostatics which states that it is impossible to arrange a set of charges in space to generate a point of stable equilibrium for a test charge. A thorough discussion with exact proofs of several "no-trapping" theorems is given in Chu (1992a).

For a couple of years discussions to circumvent the optical Earnshaw theorem concentrated on purely optical configurations to overcome the global proportionality of the spontaneous force to the Poynting vector. Sophisticated intensity variations in combination with light field gradients were suggested.

The important idea which led to the development of the magnetooptical trap (MOT) came from Pritchard *et al.* (1986), who realized that the spontaneous force need not be proportional to the intensity with the same factor of proportionality for the entire trapping region, if additional spatially varying magnetic or electric fields shift the atomic transition frequency and an at least four-level structure of atomic dipole transitions is considered. Dalibard (1987, in closing remarks in Raab *et al.*, 1987) then suggested the now standard configuration of three pairs of counter-propagating laser beams with mutual opposite circular polarizations and a frequency tuned slightly below the center of the atomic resonance and superimposed by a magnetic quadrupole field (Fig. 8). This configuration was successfully realized for the first time by Raab *et al.* (1987).

The basic principle of trapping in an MOT can be understood from a simplified 1-D configuration (Fig. 9). The Zeeman sublevels of an atom at distance x_1 from the trap center are shifted by the local magnetic field in such a way that due to the selection rules the atom tunes into resonance with the laser directed toward the origin, while for the opposite laser

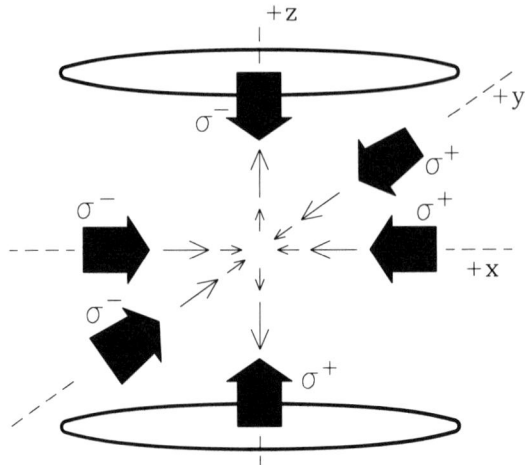

FIG. 8. Principle setup of the magnetooptical molasses.

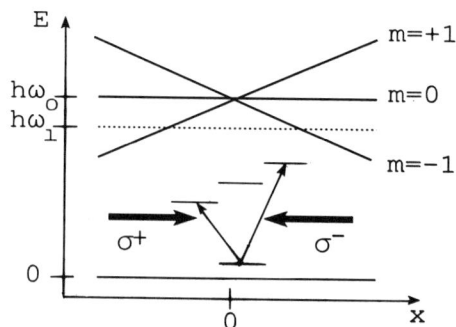

FIG. 9. Energy levels of a $J_g = 0$ to $J_e = 1$ transition in the spatially varying field of a one-dimensional MOT configuration.

beam the atomic transition frequency is shifted away from the laser frequency.

The force acting on an atom in the trap is the sum of the forces exerted by the incident beams in the limit of low saturation. For the 1-D case with the additional restriction to a $J = 0 \to J' = 1$ transition one gets

$$F = F_{\sigma^+} + F_{\sigma^-} = \hbar k \frac{\Gamma}{2} \left[\frac{\frac{I_{\sigma^+}}{I_0}}{1 + \frac{I_{\sigma^+}}{I_0} + \frac{(2\delta - \mathbf{k v} - \zeta x)^2}{\Gamma^2}} \right.$$

$$\left. - \frac{\frac{I_{\sigma^-}}{I_0}}{1 + \frac{I_{\sigma^-}}{I_0} + \frac{(2\delta + \mathbf{k v} + \zeta x)^2}{\Gamma^2}} \right], \quad (32)$$

where ζx denotes the Zeeman shift due to the linearly increasing magnetic field. The important point, making this configuration a standard cold atom source, is the resemblance of the damping mechanism of the standard optical optical molasses to the additional restoring action due to the spatially increasing Zeeman shift.

For low velocities and small deviations of the atoms from the trap center, the dependence of the force on velocity and spatial deviation from the trap center with $B = 0$ is well approximated by a linear relation leading to an equation of motion for a damped harmonic oscillator,

$$x'' + \varepsilon x' + \omega_{osc}^2 x = 0, \tag{33}$$

with the friction characterized by ε,

$$\varepsilon = \frac{4\hbar k^2 \dfrac{I}{I_0} \dfrac{2\delta}{\Gamma}}{m\left(1 + \left(\dfrac{2\delta}{\Gamma}\right)^2\right)^2}, \tag{34}$$

and a natural frequency, ω_{osc}, given by

$$\omega_{osc}^2 = \frac{4\hbar k \dfrac{I}{I_0} \dfrac{2\delta}{\Gamma} \zeta}{m\left(1 + \left(\dfrac{2\delta}{\Gamma}\right)^2\right)^2}. \tag{35}$$

Typically the parameters detuning δ and magnetic field gradient which determine ζ are chosen such that the motion of the atoms is strongly overdamped.

The basic properties characterizing the trapped atomic ensemble are its temperature, its density, which is determined by the balance between capture and loss processes, and the dynamics of the trapped atom cloud. These properties depend strongly on the parameters of the laser beams. Experimentally a variety of interesting observations, for example, laser-catalyzed interaction of the trapped atoms (see, e.g., Lett et al., 1991; Julienne and Heather, 1991, and references therein), density dependence of the temperature (Cooper et al., 1994), sensitivity to misalignment (Chu et al., 1987), and vortex motion of the atoms (Sesko et al., 1991), to name only a few, have stimulated investigations of the rich variety of effects detectable with this particle trap (see also Special Issue, 1989). In addition lots of applications were implemented, making use of the power and simplicity of this atom trap.

Naturally, soon after the observation of sub-Doppler temperatures in optical molasses in 1988 the question arose whether the mechanisms leading to those low temperatures also work in MOTs. The first experiments analyzing the temperature in an MOT showed strong evidence for temperatures below the Doppler limit (Steane and Foot, 1991).

Detailed theoretical analysis demonstrated (Werner et al., 1993) that sub-Doppler cooling mechanisms can act in an MOT as long as the Zeeman energy shift is small compared with the light shift interaction energy. In a standard MOT this condition is fulfilled near the center of the trap, where the magnetic field vanishes, although the so-far-observed temperatures for MOTs are slightly higher than those in pure molasses (Cooper et al., 1994). A fully consistent theoretical explanation is still necessary. To achieve high-density, low-temperature samples the "standard" procedure is therefore loading an optical molasses from a high-density MOT, by switching off the trap magnetic field.

The temperature measurement of a three-dimensional atom cloud is therefore not that easy. We shortly review some methods at this point, before turning back to trap dynamics. A measurement of the velocity distribution (e.g., via the Doppler shift of an additional probe laser) must not disturb the sensitive sub-Doppler cooling processes or heat up the ensemble during the measurement.

If we first look at a molasses configuration, the standard technique today — also developed at NIST during the first sub-Doppler discovery in optical molasses (Lett et al., 1988) – is a time-of-flight method. The molasses laser beams are suddenly switched off and the atoms fall, due to gravity, some centimeters downward through an additional light sheet resonant with the atoms. The fluorescence from the atoms falling through the lower light field is measured as a function of time after the switching off of the molasses. It serves as a measure of the initial momentum distribution in the direction parallel to the gravitational vector.

The method works as long as the initial phase space distribution is small compared with the increase in phase space during the falling time — a condition normally fulfilled for sub-Doppler temperatures and typical experimental conditions. The disadvantage of the method is that the molasses cooling process has to be stopped.

Another method measures the absorption or the gain (!) of a weak probe laser beam sent through the molasses (Verkerk et al., 1992). However, this method still influences the cooling process. Two essentially nondestructive methods use the scattered light from the cold atoms to beat it with a local oscillator (Jessen et al., 1992) or to analyze it in frequency space (Jurczak et al., 1995) and thereby obtain the information about the momentum distribution. The scattered light is frequency shifted by the Doppler effect of the atomic motion. The comparison with the driving molasses beam frequency therefore gives the information about the velocity distribution directly. With these nondestructive *in situ* measurement methods the localization of cold atoms in the antinodes of the standing light fields was investigated (Verkerk et al., 1992; Jessen et al., 1992).

Atoms are localized in the quantized oscillation levels on the scale of wavelengths and further cooled due to Raman transitions to lower oscillator levels. Those "optical lattices," (see, e.g., Courtois and Grynberg, 1992; Lounis et al., 1992, 1993; Grynberg et al., 1993; Hemmerich and Hänsch, 1993; Hemmerich et al., 1994; Verkerk et al., 1994; Petsas et al., 1994) where the atoms are thus confined in the Lamb–Dicke regime (Dicke, 1953), are currently the subject of stimulating investigations, like the scattering of light or X rays on the lattice or the search for collective quantum effects in the ensemble where the de Broglie wavelength reaches the optical wavelength, thus eventually leading to a long-range "contact" of neighborhood lattice places.

If we turn to magnetooptical traps the standard time-of-flight method to measure temperatures often cannot be applied, mainly because of the technical limitations of synchronizing the switching of magnetic and trapping laser fields. New methods derive the damping and the spring constant of the harmonic oscillator model either by the step response due to an additional pushing laser beam (Steane and Foot, 1991) or by observing the forced oscillation of the trapped ensemble due to an oscillating magnetic field (Kohns et al., 1993). The derived coefficients prave the action of sub-Doppler cooling mechanisms, and with additional information about the size of the trapped cloud from the equipartition theorem the temperature of the trapped atoms can be calculated.

If we return to the trap dynamics, in addition to the final temperature, the density and number of trapped atoms turn out to be most important in, for example, frequency standard applications and the search for collective quantum effects. The number N of trapped atoms is determined by the balance between the filling rate R and the loss of atoms from the trap volume. The temporal behavior of the number of trapped atoms is modeled by the differential equation

$$\partial N_{(t)}/\partial t = R - \eta N_{(t)} - \kappa N_{(t)}^2, \tag{36}$$

where the coefficient η describes the loss of individual atoms, for example, due to collisions with residual gas molecules or due to photoionization. The coefficient κ accounts for loss mechanisms due to binary collisions among trapped atoms, like radiative escape processes or associative ionization.

The coefficients η and κ vary over several orders of magnitude depending strongly on the level structure of the trapped atoms and the details of the collision processes. The determination of the coefficients for different trap parameters is an excellent and now intensively used tool to investigate collision processes in a temperature range previously not available (see, e.g., Gould et al., 1988; Bjorkholm, 1988; Julienne, 1988; Prentiss et al., 1988; Sesko et al., 1989; Gallagher and Pritchard, 1989; Julienne and Mies, 1989; Walker et al., 1990; Thorsheim et al., 1990; Bardou et al., 1992).

The filling rate R can be varied independently from the trap parameters by changing the number of atoms in the phase space volume from which atoms are captured. Filling rates ranging from a few atoms/s (Section 6) up to 10^9 atoms/s have been achieved, leading to maximum numbers of more than 3×10^{10} stored atoms (Gibble et al., 1992).

The density distribution of the trapped atoms is determined by the equilibrium between attractive and repulsive forces among the atoms. The density distribution is Gaussian as long as the interaction of the trapped atoms can be considered to be small compared with the restoring trap force; in this case the harmonic oscillator model holds. In contrast, the maximum density achievable in traps is limited by repulsive forces among the atoms. A strong and long-range repulsive force arises from radiation trapping effects by which light scattered from atoms at the trap center repels atoms in the outer regions of the trap. These radiation trapping forces strongly influence the density distribution and can even lead to the formation of stable ring patterns similar to the formation of interstellar clouds (Sesko et al., 1991).

The challenge to increase the number of trapped atoms and to overcome the density limitation of the standard magnetooptical trap has led to numerous refinements and new developments, for example, MOTs directly filled from background pressure gas in simple glass cells (Monroe et al., 1990), the dark spot trap (Ketterle et al., 1993), and the application of additional laser frequencies to influence the collisional losses (Walhout et al., 1995) and to increase the filling rate (Sinclair et al., 1994). However, a complete review is beyond the scope of our chapter.

A lot of measurements with laser-cooled atoms make use of magnetooptically confined ensembles as an atom source, a few examples of which will be discussed in Section VII. As discussed before, the techniques of laser manipulation of atoms developed up to now operate in two different regimes: the confinement of atoms by radiation fields in traps and the manipulation of atomic beams.

For a lot of experiments with cold atoms, for example, interactions with surfaces or cold collisions, a cold, dense, bright and well-collimated beam may be better suited, or is even the ony possible source, compared with a trapped ensemble. One of the aims in atomic beam preparation is the realization of a cold atom source, with a beam diameter of only a few micrometers and a tunable "energy," which can be scanned, for example, over a surface for atom deposition.

In the following we describe techniques suitable for atomic beam manipulation. There mainly exist two methods to produce slow atomic beams. The first one launches atoms from a trapped ensemble with an additional laser beam or running molasses to form an atomic fountain (Kasevich et al., 1990), the so-called "Zacharias fountain" (Ramsey, 1953). Typically the atoms have an average velocity of less than a few m/s, and the temperature—depending on the cooling during the trap period—corresponds to a few recoil velocities. But this method leads to a pulsed experiment due to the different steps of trapping, cooling, and launching, and hence the density of the fountain is typically fairly low.

Fountains were used for ultrahigh Ramsey spectroscopy in the microwave region, for example, to detect a Cs clock transition at 9.1 Ghz with extremely high accuracy and resolution (Clairon et al., 1991). To prepare a beam with an average longitudinal velocity of a few m/s up to 10 m/s the atomic "funnel" (Nellessen et al., 1990), a two-dimensional MOT filled with a

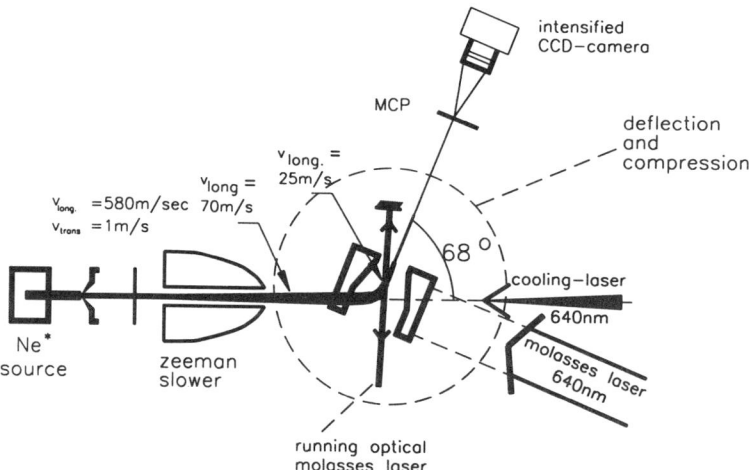

FIG. 10. Experimental setup for deceleration, deflection, and compression of a metastable Ne atomic beam (see the text).

laser-decelerated, precooled beam or filled from a trap (Shimizu et al., 1990; Riis et al., 1990), is well adapted. For additional longitudinal cooling a running molasses can be superimposed (Scholz et al., 1994). A typical experiment, shown in Fig. 10, starts with laser deceleration of an effusive beam from an atomic oven and demonstrates part of the area of different laser manipulation methods.

The slow and cold part of the decelerated atomic beam is directly captured in a 2-D MOT which is placed under a large angle into the beam (Scholz et al., 1994; see Fig. 10), or, alternatively, the cold atoms are first selectively deflected out of the main stream and then captured a few centimeters downstream by a 2-D MOT (Nellessen et al., 1992). The principle of operation in the 2-D MOT is thereby analogous to that of the 3-D setup.

Two counter-propagating pairs of laser beams of different circular polarization are directed toward the symmetry axis z. A linear magnetic quadrupole field in the x–y- plane with zero magnetic field strength on the z axis is superimposed. In the case of negative laser detuning, an atom at rest at the distance x_1 from the axis will absorb more photons from the laser which is directed toward the axis than from the counter-propagating one. Like in the 3-D MOT, for each position x from the center—in this case from the axis—there exists a velocity,

$$v = -\lambda \left(\frac{\mu_B}{h}\right) \frac{\delta B}{\delta x} x \qquad (37)$$

toward the axis for which the forces are balanced.

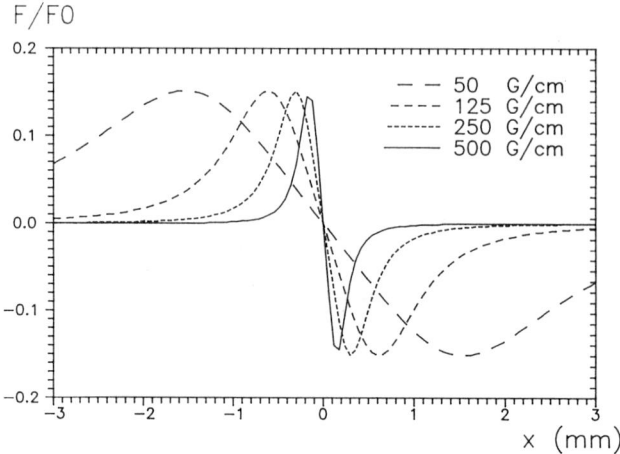

FIG. 11. Radiation pressure force versus distance x from the symmetry axis of a two-dimensional MOT for different values of the magnetic field gradient.

As a consequence the atom will approach the axis like an overdamped harmonic oscillator, as described in Eq. (33). For small displacements and velocities the sum of the radiation pressure forces may be written as

$$F = -\varepsilon v - \omega_{osc}^2 x, \tag{38}$$

where $\varepsilon = \partial F/\partial v$ and $\omega_{osc}^2 = 1/m \, \partial F/\partial x$.

For sodium atoms and a typical experimental situation ($\partial B/\partial x = 50$ G/cm, $\delta_L = -\Gamma/2$, and $S_0 = 0.75$) ε/m is $7.4 \times 10^4 \, \text{s}^{-1}$ and ω_{osc} is $1.75 \times 10^4 \, \text{s}^{-1}$. The force profiles (for zero transverse velocity) along x are shown in Fig. 11, emphasizing the influence of the field gradient $\partial B/\partial x$.

It reveals that for a fixed gradient a compromise between a large "capture range" (for a small gradient, $\partial B/\partial x$) and a strong restoring force near the axis (for a high gradient) is necessary. A solution to overcome this problem is a magnetic field with increasing transverse gradient along the z axis $(\partial B/\partial x)(z)$.

For slow atomic beams with mean velocities below 100 m/s a field gradient increasing from, for example, 50 to 500 G/cm along a distance, Z, of 40 mm allows for a capture range of about 3 mm at the beginning and a compression to final diameters of a few tens of micrometers for sodium atoms (Nellessen et al., 1990). Typical beam diameters at the end of the 2-D MOT as a function of the laser detuning are shown in Fig. 12. Figure 13 is a picture of the fluorescence during the MOT process, which clearly shows the enormous decrease in the beam diameter and increase in density.

The final transverse temperature and the beam diameter depend—as in case of the 3-D trap—on either "simple" radiation pressure, the sub-

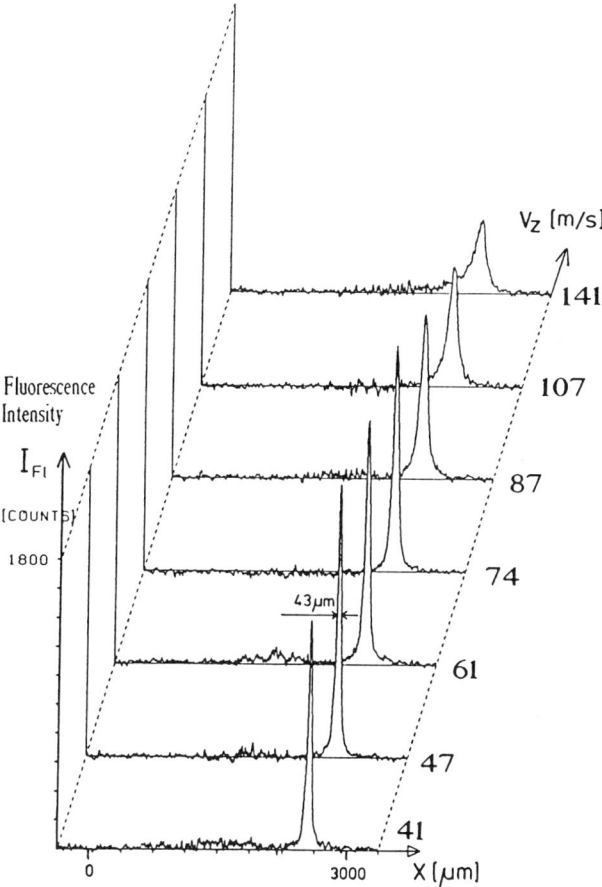

FIG. 12. One-dimensional cut of the transverse fluorescence intensity profile (see Fig. 10) at the end of a 2-D MOT for different longitudinal velocities, v_z, of atoms focused by the MOT.

Doppler mechanism, or both. This, of course, depends on the atomic level scheme involved and the relative parameter strength. Sub-Doppler tori have been observed in 2-D MOTs (Riis et al., 1990; Scholz et al., 1994).

The special atomic funnel configuration of Fig. 10 achieves, beside the transverse compression and cooling, further slowing of the longitudinal velocity and a strong deflection of, for example, 70°. This 2-D MOT can be subdivided into two sections. In a first step atoms with velocities of typically $v_{in} = 70 \, \text{m/s}$ coming from the Zeeman slower are further slowed down and deflected to velocity $v_{long} = v_{in}\cos(\theta) = 24 \, \text{m/s}$ in the z direction (for $\theta = 70°$) until they reach the quadrupole axis. This happens typically within the first 20 mm. In the second part of the MOT the atoms are then captured and bound to the quadrupole axis. Thus in the first part the setup works like a

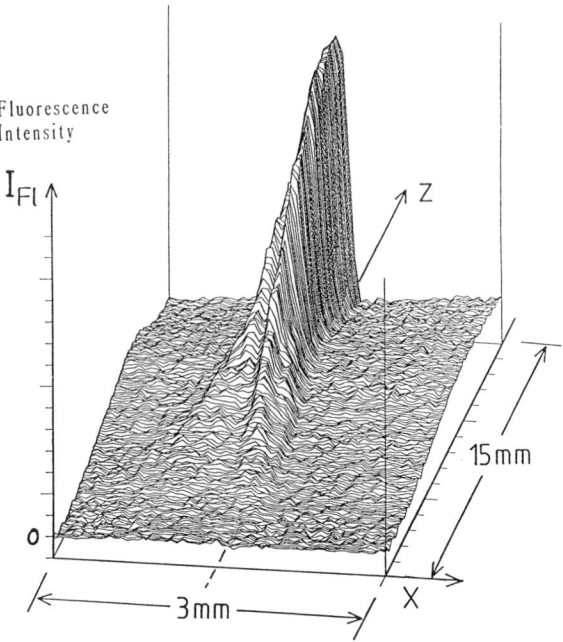

FIG. 13. Digitized photograph of the compression of a sodium atomic beam with a 2-D MOT. The broad transverse distribution of atoms is focused to the MOT axis z within the last 15 mm of the MOT.

Zeeman slower for the x component of the initial motion while the atoms are drifting with a constant velocity parallel to the z axis.

An additional running optical molasses (Faulstich et al., 1992) improves the longitudinal temperature and allows for the fine tuning of the final kinetic energy. In the described experiment the magnetic field gradient is increasing from 65 to 350 G/cm which still allows sub-Doppler cooling of a metastable Ne beam at the end of the MOT for a proper choice of the laser detuning and intensity. The resulting two-dimensional beam shape is monitored with a microchannel plate (MCP) 20 cm behind the MOT region, thus allowing calculation of the transverse beam temperature. The longitudinal velocity distribution can be measured with time-of-flight techniques. With this setup final temperatures down to 15 μK, corresponding to only 3 $\hbar k$, have been measured. These extremely slow, cold, and bright beams are excellent sources to study cold collisions with either surfaces or other beams as well as atom deposition experiments and coherent beam splitters which demand "monochromatic" atomic beams, as described in the next section.

V. Manipulation Schemes Based on the Dipole Force

The possible high strength of the dipole force predestinates it for all manipulation schemes for which a strong change of momentum of atoms within a short distance is necessary, for example, the reflection of atoms from sharp gradient light fields, the channeling in standing wave potential walls, or the strong confinement of atoms in micrometer-sized focused beams. The last point was discussed in the early proposals for manipulation of atoms and micrometer-sized particles by Ashkin (1970a) and Ashkin (1978). It is necessary to keep in mind that the dipole force cannot absorb energy out of the atomic system due to its reactive character.

If we treat particle trapping first, the simplest dipole force trap is a red-detuned focused light beam (see Fig. 14). For red detuning the light shift potential attracts the atoms in the region of high intensity (Chu et al., 1986; Ashkin et al., 1986). The dipole force in the mean can be derived from a potential and is conservative, but due to the diffusion the force fluctuations heat the atoms (for high intensity the heating can dominate the attractive part by orders of magnitude (Gordon and Ashkin, 1980; Cohen-Tannoudji, 1992)), "boiling" them out of the focused light beams.

A possibility to overcome this problem is the superposition of the dipole trap with a standard molasses beam configuration and the time sharing between the light fields (Dalibard et al., 1983). This arrangement provides slow atoms in the molasses to primarily "fill" the dipole trap and later increases the density by the combination of trapping and compensating for dipole force heating from time to time with the molasses (Chu et al., 1986).

The density in this kind of trap is limited again by atomic collisions, but the collisional cross-section σ_{col} is typically smaller in far-off resonance dipole traps than in MOTs (Special Issue, 1989) due to the lower excitation probability in the excited state, typically increasing σ_{col} in MOTs. A dipole force trap, detuned by 65 nm (!), reaching a density of more than 10^{11} atoms/cm^3, was demonstrated (Miller et al., 1993). Also, a promising arrangement of intense gradient fields forming a corner cube was experimen-

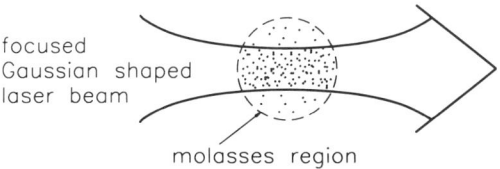

FIG. 14. Dipole force trap operated with a single-focused laser beam, superposed to a molasses region.

tally demonstrated (Davidson et al., 1994), "trapping" the atoms by several thousand bounces between the walls of the corner cube, assisted by gravitation.

Besides atom trapping the early proposals pointed out that small particles in a focused intense light beam will follow laser beam movements (Ashkin, 1970a). This technique is widely carried out with particles such as bacteria or living cells, for example, in water, is known as "optical tweezers," and is currently a fast growing field in cell biology. A review for this field is given in Chu (1991, 1992b).

Another fascinating effect based on the dipole trap is the "channeling" of atoms in the antinodes of a blue-detuned strong standing wave. The channeling of atoms was first demonstrated in Paris, where an atomic beam of Cs atoms was crossed perpendicularly by a standing light wave (Salomon et al., 1987). The atoms channeled into the antinode valleys of the blue-detuned light field. The detection was carried out with a weak probe beam measuring the light shift.

Due to the possible suppression of spontaneous emission in strong gradient far-off resonance fields, the dipole force has generated interest in atom optics experiments. Atom optics is concerned with the coherent propagation of atoms among "optical" devices like mirrors, beam splitters, lenses, and cavities. A special interest lies in "thin" optical elements to reduce chromatic and abberative errors like in real optics (mirrors and thin lenses) and to allow for long regions of free propagation. Thus light fields—which can nowadays realize all of the aforementioned atom optics elements—are predominated for those elements as long as spontaneous emission can be neglected, for example, with far-off resonance dipole forces in addition to short interaction times. Interesting configurations of mirrors and beam splitters for neutral atoms—or more precisely neutral atom wave functions—were realized with evanescent light fields and the strong field gradients existent in the evanescent field.

The evanescent wave is created by the total reflection of a laser beam at a glass vacuum interface (see Fig. 15). The field intensity drops from its maximum at the surface of the glass into the vacuum within one wavelength, forming high-intensity gradients. For blue detuning the dipole force is repulsive and the surface forms a mirror for atoms, as proposed by Cook and Hill (1982) and demonstrated experimentally in bouncing experiments (Kasevich et al., 1990; Aminoff et al., 1993) and for configurations reflecting atoms (Balykin et al., 1987, 1988; Hajnal et al., 1989).

An important step in matter wave optics will be the demonstration of matter wave cavity modes, currently in progress. For atoms as material waves the realization of a matter wave cavity formed by only one mirror becomes possible, if the surface of the mirror is placed horizontally;

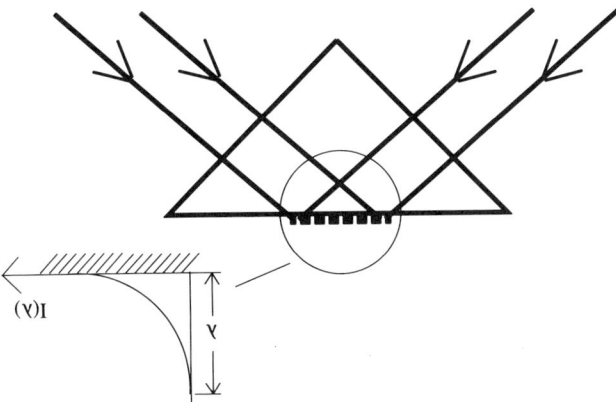

FIG. 15. Evanescent light field on a prism surface. The strong intensity gradient arises from the decrease of the light intensity within approximately one wavelength (see the inset).

gravitation takes over the function of the second "mirror" normally necessary for light cavities. This configuration was proposed (Wallis et al., 1992) and is currently the subject of the theoretical and experimental efforts of several groups. The investigation of the filling of the quantized cavity modes with single particle waves, either bosons or fermions, will lead to a key understanding of collective quantum statistical effects with matter waves. However, not only the reduction of particle losses from the cavities, still present in bouncing experiments, but also the reduction of coherence losses during the interaction and the filling of the modes, for example, with more than one bosonic atom, demands sophisticated experimental schemes.

For example, strong efforts were undertaken to enhance the intensity of the evanescent light field, thus allowing higher detunings during reflection to reduce decoherence by spontaneous emission. Additional thin layers on the glass surface use either dielectric material layers (Esslinger et al., 1993) or multilayer waveguide structures (Kaiser et al., 1993) which enhance the intensity by factors of up to several hundred in the case of waveguides.

Besides acting as mirrors evanescent waves can also form coherent atomic beam splitters if a standing light field is used to form the evanescent wave (Hajnal and Opat, 1989), acting as a grating for the matter wave. This was demonstrated using laser-cooled metastable neon atoms (Christ et al., 1994). A transversely cooled, slow neon atomic beam leaving a two-dimensional MOT configuration, as discussed in Section IV, was sent in grazing incidence on either a running wave or a standing evanescent wave, on a glass prism surface. The running evanescent field led to atom reflection with angles up to 80 mrad for slow atoms with a longitudinal velocity of 25 m/s.

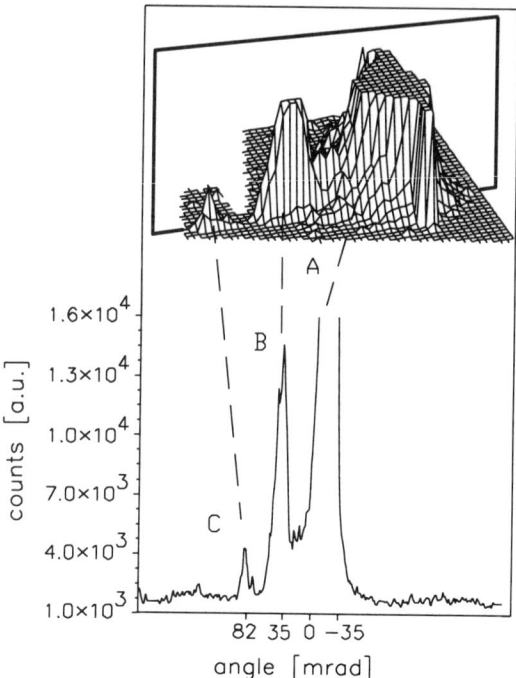

FIG. 16. Deflection and diffusion of metastable neon atoms on an evanescent light field (see the text).

Figure 16 shows a typical result for the diffraction of the evanescent grating. The huge peak A arises from atoms which miss the prism surface and pass straight over the prism toward the CCD detector, placed at a distance of 20 cm to the prism to monitor the reflected or deflected atoms. The specularly reflected atoms form peak B while the first visible diffraction order leads to peak C, separated by 84 mrad from the undeflected part of the atomic beam.

Normalized to the flux of atoms reaching the mirror surface, up to 70% of the atoms are reflected and up to 3% form the diffraction pattern. The results agree with theoretical calculations for the complex diffraction process (Deutschmann et al., 1993a). Higher diffraction efficiency should be reached by applying an additional longitudinal magnetic field, thus optimizing the quantum mechanical amplitude splitting during the complex diffraction process (Deutschmann et al., 1993b). A rigorous proof of the coherence of the outgoing beams remain to be performed, but the large deflection angles of those beam splitters promise to stimulate experimental and theoretical investigations in the field of large-area interferometers.

VI. Experiments with Trapped Atoms

A. Optical Ramsey Spectroscopy on Mg Atoms

High-resolution spectroscopy has evolved into a valuable tool in many areas of modern physics, as, for example, the test of weak interaction in parity violation experiments (see, e.g., Bouchiat, 1991), tests of quantum electrodynamics with hydrogen atoms (see, e.g., Basini et al., 1989) and positronium (Fee et al., 1993), or the test of general relativity, to name only a few. For precise determination of optical frequencies or precise synchronization of time at different places around the world, the current cesium frequency standard already shows deficits, so that the establishment of a new optical frequency standard is currently under way.

With the advent of laser cooling and trapping, great progress has been made in the high-resolution spectroscopy of neutral atoms (see, e.g., Strumia, 1988; Helmke et al., 1992; Riehle et al., 1992). Several laboratories have demonstrated precise measurements in the microwave regime (Kasevich et al., 1989; Clairon et al., 1991), and optical high-resolution spectroscopy has been performed with laser-trapped atoms (Sengstock et al., 1993; Special Issue, 1992). The major improvement in accuracy arises from the dramatic reduction of the relativistic second-order Doppler shift from values of

$$\frac{\delta v}{v} = \frac{1}{2}\frac{v^2}{c^2} = 10^{-12} \tag{41}$$

(for $v = 300$ m/s) to values of 10^{-18} (for $v = 0.3$ m/s) and from the increase of the maximum possible interaction time of the atoms with the spectroscopic fields which determines, according to $\Delta E \cdot \Delta t \simeq \hbar$, the maximum available frequency resolution.

In addition new excitation geometries, possible with trapped atoms, allow compensation for further systematic errors, for example, phase errors in the exciting fields in Ramsey spectroscopy. In traps, atoms are accessible for several seconds compared with approximately 10 ms for the movement of atoms in a beam ($v = 500$ m/s) through an interaction zone of 50 cm.

A successful integration of the techniques from laser manipulation and high-resolution spectroscopy requires the careful choice of a suitable atomic level scheme. The alkaline earth elements offer the interesting connection of fast transitions suitable for cooling and manipulation in the singlet system and optical intercombination transitions from the singlet to the triplet system with line qualities better than 10^{12}. Only for magnesium is the principal cooling transition ($^1S_0 \rightarrow {}^1P_1$, $\lambda = 285$ nm) a closed two-level system, and in addition magnesium offers the narrowest intercombination transition ($^1S_0 \rightarrow {}^3P_1$, $\lambda = 457$ nm) with a linewidth as small as 31 Hz.

The possibilities offered by the magnesium level scheme were demonstrated by a pulsed version of optical Ramse–Bordé interferometry (Bordé, 1989) applied to laser-trapped atoms (Sengstock et al., 1993). A sequence of four light pulses from a highly stable dye laser at 457 nm is irradiated to the cold atoms for a time during which the trapping fields are turned off. Thus free atoms nearly at rest relative to the laboratory frame and undisturbed by other light fields are probed "in the dark" by the Ramsey–Bordé method. Part of the atoms are excited to the metastable $^3P_1$ level and therefore not recaptured when the trapping laser fields are switched on again. The decrease of the trap fluorescence is detected during the fast $^1S_0 \to {}^1P_1$ transition, thus giving a quantum-amplified measure of the electron shelving to the $^3P_1$ level by the spectroscopic Ramsey excitation.

Due to the interferometric character of the optical four-zone (four-pulse) Ramsey excitation (Bordé, 1989), this scheme offers highest sensitivity and the best signal strength for a given interaction time compared with other spectroscopic methods (Hall et al., 1976; Bordé et al., 1984). The period of the Ramsey fringes in frequency space is determined only by the dark interval between the Ramsey laser pulses in which the atomic wave function develops freely. Illustrative examples of measured fringes on trapped ^{24}Mg atoms are presented in Fig. 17. The periodicity of the fringes is only 10 kHz. In the described experiment the possible accuracy in the determination of the line center relative to the "true" transition frequency was estimated to

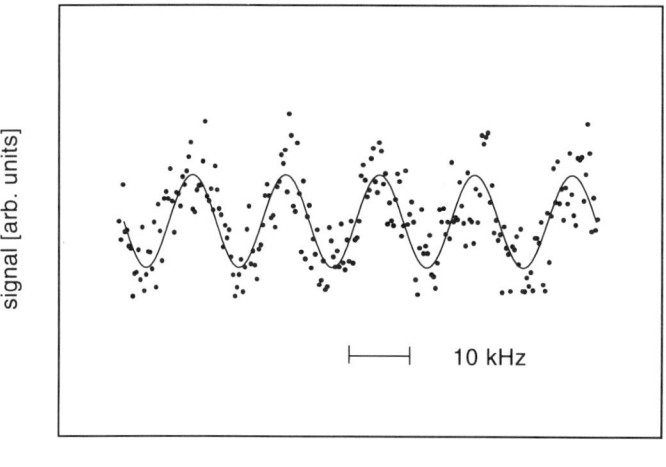

Fig. 17. High-resolution Ramsey fringes from trapped Mg atoms with a periodicity of 10 kHz.

be 2×10^{-15} (Sengstock et al., 1994) which is by a factor of 300 better than the possible accuracy of a thermal atomic beam.

B. TRAPPING OF SINGLE ATOMS

The possibility of observing only a few or even single trapped particles has opened up new dimensions in atomic measurements, as ion traps and neutron traps have already demonstrated and atom traps will continue to demonstrate. Due to the complexity of nature it is a goal to arrange setups with which only a few constituents governed by as few as possible fundamental forces are investigated as precisely as possible. The improvement in precision led and will lead to new phenomena, inspire new ideas, and falsify established theories.

Compared with ions, neutral atoms offer the advantage that they do not interact via the strong coulomb interaction; thus the interesting weak interatomic *higher order* forces between neutral atoms can be studied. Furthermore the center of mass motion is not sensitive to electromagnetic stray fields. As an important point, compared with neutrons, atoms offer a wide range of interactions with their complex internal degrees of freedom, allowing for interactions in higher order topologies (Special Issues, 1992).

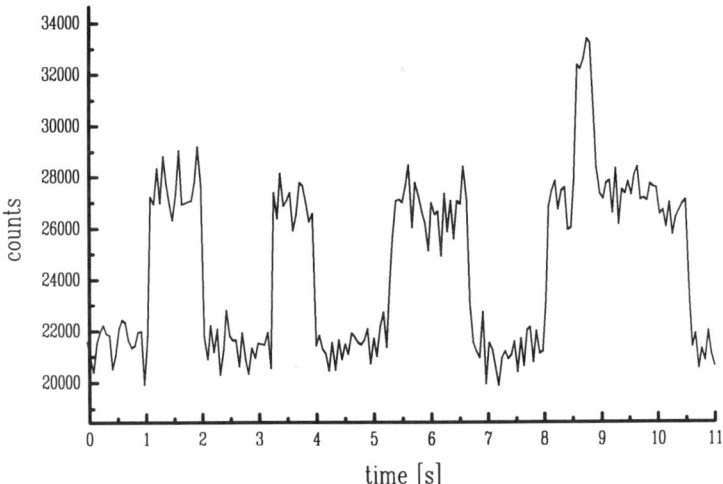

FIG. 18. Time dependence of the fluorescence signal from a few-atom MOT. The jumps in the fluorescence intensity indicate the capture or loss of individual atoms in the trap. The count rate level of approximately 21,000 corresponds to the stray light level.

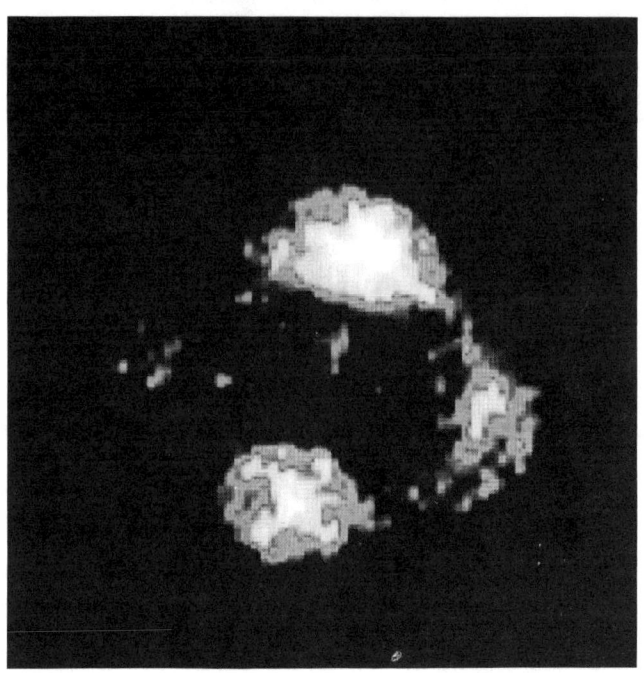

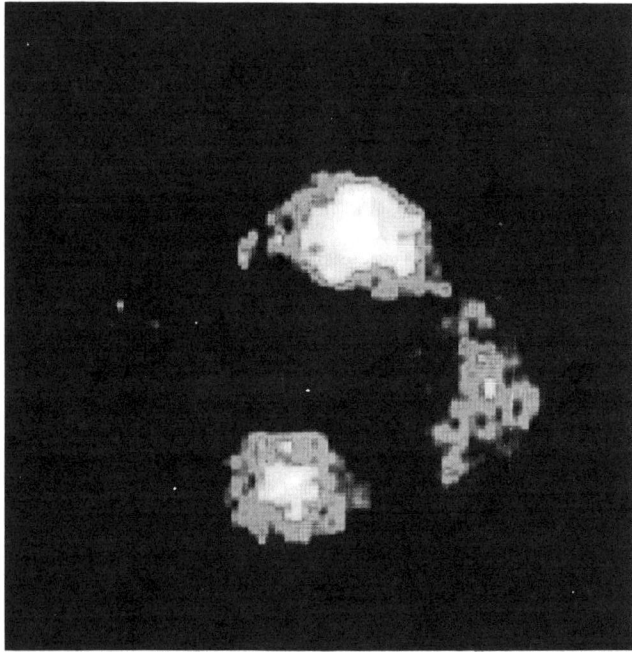

FIG. 19. Single-trapped Mg atom (top). The fluorescence from the atom is seen inside a ring of stray background light. For comparison, the image of the trap zone without trapped atoms is presented (bottom).

The first detection of a single trapped Cs atom was reported by Hu and Kimble (1994). Single atom trapping for storage times up to 0.9 s was achieved by carefully reducing the filling rate of a Cs-background-gas MOT. The storage time was thereby determined by collisions with the background gas in the vacuum cell. The capture or loss of individual atoms is monitored by discrete jumps in the fluorescence light, induced by the trapping laser beams.

Examples for fluorescence jumps coming from single Mg atoms trapped in a standard MOT configuration are shown in Fig. 18 (Bettermann et al., 1995). The long-time statistics of the time intervals of those jumps give then information about the mean storage time and loss mechanisms. Due to the short lifetime of the upper $^1P_1$ Mg level ($\tau = 2$ ns), 10^7 photons stray from a single atom for an overall saturation of $S = 0.1$ of the trapping laser beams. This allows for CCD images of single atoms within integration times of 50 ms. Figure 19 gives an example of the image of a single trapped magnesium compared with the stray light image coming from the trap. These pictures give important information about trap dimensions and asymmetries arising from special MOT geometries. Without changing trap parameters the setup allows for a direct comparison of the trap dynamics, storage times, and trap dimensions for single toms and dense atom clouds with up to 10^8 atoms, simply by tuning the trap filling rate. Currently experiments investigate atom–atom interactions with only a few atoms trapped. Furthermore, single atom spectroscopy will lead to ultimative accuracy for future time standards.

VII. Final Remarks

In conclusion many laser manipulation and cooling schemes have been demonstrated, and it is likely that the field will grow in the near future. After a first decade of fast development, a broad variety of techniques is now readily at hand for further extensions of experiments with cold atoms. Laser cooling allows for a completely new regime in atomic physics, in which the quantum effects of the internal as well as external degrees of freedom dominate experiments, for example, with atoms prepared in the Lamb–Dicke regime.

Furthermore atom cooling works better than hoped for in the first proposals. The temperatures achievable by a combination of Doppler, sub-Doppler, and sub-recoil cooling allow for the coldest atomic samples ever realized. This offers the possibility of connecting atomic physics with multiple particle quantum physics and thermodynamics. The demonstration of an atomic Bose condensate will be an important step for this connection. The field of laser cooling is at the same time a theoretical field, with deep questions in fundamental physics and a domain of applications and connections to many fascinating developments in other fields of physics, as partly discussed in this chapter.

Much about laser cooling is not yet completely understood, and the fully quantized description of multiple atom interactions in the presence of resonant or nearly resonant light fields is just beginning. Extensions of and solutions to the fundamental questions, techniques, and applications in laser manipulation and cooling will likely inspire a broad area of investigation in physics in the future.

References

Adams, C., Sigel, M., and Mlynek, J. (1994). *Phys. Rep.* **240**, 143.
Aminoff, C. G., Steane, A. M., Bouyer, P., Desbiolles, P., Dalibard, J., and Cohen-Tannoudji, C. (1993). *Phys. Rev. Lett.* **71**, 3083.
Andreev, S. V., Balykin, V. L., Letokhov, V. S., and Minogin, V. G. (1981). *JETP Lett. (Engl. Transl.)* **34**, 442.
Andreev, S. V., Balykin, V. L., Letokhov, V. S., and Minogin, V. G. (1982). *Sov. Phys. — JETP (Engl. Transl.)* **55**, 826.
Arimondo, E., Phillips, W. D., and Strumia, F., eds. (1992). *Proc. Int. Sch. Phys. "Enrico Fermi," 1991* **118**.
Ashkin, A. (1970a). *Phys. Rev. Lett.* **24**, 156.
Ashkin, A. (1970b). *Phys. Rev. Lett.* **25**, 1321.
Ashkin, A. (1978). *Phys. Rev. Lett.* **40**, 729.
Ashkin, A. (1984). *Opt. Lett.* **9**, 454.
Ashkin, A., and Gordon, J. (1979). *Opt. Lett.* **4**, 161.
Ashkin, A., and Gordon, J. (1983). *Opt. Lett.* **8**, 511.
Ashkin, A., Dziedzic, J. M., Bjorkholm, J. E., and Chu, S. (1986). *Opt. Lett.* **11**, 288.
Aspect, A., Arimondo, E., Kaiser, R., Vansteenkiste, N., and Cohen-Tannoudji, C. N. (1988). *Phys. Rev. Lett.* **61**, 826.
Aspect, A., Arimondo, E., Kaiser, R., Vansteenkiste, N., and Cohen-Tannoudji, C. N. (1989). *J. Opt. Soc. Am.* **B6**, 2112.
Balykin, V. I., Letokhov, V. S., and Mishin, V. I. (1980). *Sov. Phys. — JETP (Engl. Transl.)* **52**, 000.
Balykin, V. I., Letokhov, V. S., and Didorov, A. I. (1984). *Sov. Phys. — JETP (Engl. Transl.)* **59**, 1174.
Balykin, V. I., Letokhov, V. S., Ovchinnikov, Y. B., and Sidorov, A. I. (1987). *JETP Lett. (Engl. Transl.)* **45**, 282.
Balykin, V. I., Letokhov, V. S., Ovchinnikov, Y. B., and Sidorov, A. I. (1988). *Phys. Rev. Lett.* **60**, 2137.
Bardou, F., Emile, O., Courty, J.-M., Westbrook, C. I., and Aspect, A. (1992). *Europhys. Lett.* **20**, 681.
Basini, G. F., Inguscio, M., and Hänsch, T. W., eds (1989). *The Hydrogen Atom.* Springer-Verlag, Berlin.
Berman, P. (1991). *Phys. Rev. A* **43**, 1470.
Bettermann, D., Ruschewitz, F., Peng, J. L., and Ertmer, W. (1995). To be published.
Bjorkholm, J. E. (1988). *Phys. Rev. A* **38**, 1599.
Bordé, C. J. (1989). *Phys. Lett. A* **140**, 10.
Bordé, C. J., Salomon, C., Avrillier, S., van Lerberghe, A., Bréant, C., Bassi, D., and Scoles, G. (1984). *Phys. Rev. A* **30**, 1836.
Bouchiat, M. A., (1991). *At. Phys.* **12**.

Christ, M., Scholz, A., Schiffer, M., Deutschmann, R., and Ertmer, W. (1994). *Opt. Commun.* **107**, 211.
Chu, S. (1991). *Science* **253**, 861.
Chu, S. (1992a). *Proc. Int. Sch. Phys. "Enrico Fermi," 1991* **118**.
Chu, S. (1992b). *Sci. Am.* **266**, 79.
Chu, S., Hollberg, L., Bjorkholm, J., Cable, A., and Ashkin, A. (1985). *Phys. Rev. Lett.* **55**, 48.
Chu, S., Bjorkholm, J., Ashkin, A., and Cable, A. (1986). *Phys. Rev. Lett.* **57**, 314.
Chu, S., Prentiss, M., Cable, A., and Bjorkholm, J. (1987). In *Laser Spectroscopy VIII* (Persson, W., and Svangerg, S., eds.).
Clairon, A., Salomon, C., Guellati, S., and Phillips, W. (1991). *Europhys. Lett.* **16**, 165.
Cohen-Tannoudji, C. (1992). In *Fundamental Systems in Quantum Optics* (Dalibard, J., Raimond, J. M., and Zinn-Justin, J., eds.), Les Houches, 1990, Vol. 53, Elsevier, Amsterdam.
Cohen-Tannoudji, C., and Reynaud, S. (1977). *J. Phys. B* **10**, 345.
Cohen-Tannoudji, C., Bardou, F., and Aspect, A. (1992). In *Laser Spectroscopy X* (Ducloy, M., Giacobino, E., and Camy, G., eds.), World Sci. Publ. Co., Singapore.
Cook, R. J. (1980). *Phys. Rev. A* **22**, 1078.
Cook, R. J., and Hill, R. (1982). *Opt. Commun.* **43**, 258.
Cooper, C. J., Hillenbrand, G., Rink, J., Townsend, C. G., Zetie, K., and Foot, C. J. (1994). *Europhys. Lett.* **28**, 397.
Courtois, J. Y., and Grynberg, G. (1992). *Phys. Rev.* **46**, 7060.
Dalibard, J., and Cohen-Tannoudji, C. (1985). *J. Opt. Soc. Am. B* **2**, 1707.
Dalibard, J., and Cohen-Tannoudji, C. (1989). *J. Opt. Soc. Am. B* **6**, 2023.
Dalibard, J., Reynaud, S., and Cohen-Tannoudji, C. (1983). *Opt. Commun.* **47**, 395.
Dalibard, J., Raimond, J. M., and Zinn-Justin, J. (1992). *Fundamental Systems in Quantum Optics*, les Houches, 1990, Vol. 53, Elsevier, Amsterdam.
Dalibard, J., Castin, Y., Molmer, K. (1992). *Phys. Rev. Lett.* **68**, 580.
Davidson, N., Lee, H. J., Adams, C. S., Kasevich, M., Chu, S. (1995). *Phys. Rev. Lett.* **74**, 1311.
Davidson, N., Lee, H. J., Kasevich, M., and Chu, S. (1994). *Phys. Rev. Lett.* **72**, 3158.
Deutschmann, R., Ertmer, W., and Wallis, H. (1993a). *Phys. Rev. A* **47**, 2169.
Deutschmann, R., Ertmer, W., and Wallis, H. (1993b). *Phys. Rev. A* **48**, R4023.
Dicke, R. (1953). *Phys. Rev.* **89**, 472.
Drewsen, M., Laurent, P., Nadir, A., Santarelli, G., Clairon, A., Castin, Y., Grison, D., and Salomon, C. (1994). *Appl. Phys. B* to be published.
Einstein, A. (1917). *Phys. Z.* **XVIII**, 121.
Ertmer, W., and Penselin, S. (1986). *Metrologia* **22**, 195.
Ertmer, W., Blatt, R., and Hall, J. L. (1985). *Phys. Rev. Lett.* **54**, 996.
Esslinger, T., Weidemüller, M., Hemmerich, A., and Hänsch, T. W. (1993). *Opt. Lett.* **18**, 450.
Faulstich, A., Schnetz, A., Sigel, M., Sleator, T., Carnal, O., Balykin, V., Takuma, H., and Mlynek, J. (1992). *Euophys. Lett.* **17**, 393.
Fee, M. S., Chu, S., Mills, A. P., Chickester, R. J., Zuchermann, D. M., Shaw, E. D., and Danzmann, U. (1993). *Phys. Rev. A* **48**, 192.
Foot, C. J. (1991). *Contemp. Phys.* **32**, 369.
Frisch, R. (1933). *Z. Phys.* **86**, 42.
Gallagher, A., and Pritchard, D. E. (1989). *Phys. Rev. Lett.* **63**, 957.
Gordon, J. P., and Ashkin, A. (1980). *Phys. Rev. A* **21**, 1606.
Gould, P. L., Lett, P. D., Julienne, P. S., Phillips, W. D., Thorsheim, H. R., and Weiner, J. (1988). *Phys. Rev. Lett.* **60**, 788.
Gibble, K. E., Kasapi, S., and Chu, S. (1992). *Opt. Lett.* **17**, 526.
Grynberg, G., Lounis, B., Verkerk, P., Courtois, J. Y., and Solomon, C. (1993). *Phys. Rev. Lett.* **70**, 2249.
Hänsch, T. W., and Schawlow, A. L. (1975). *Opt. Commun.* **13**, 68.
Hajnal, J. V., and Opat, G. I. (1989). *Opt. Commun.* **71**, 119.
Hajnal, J. V., Baldwin, K. G. H., Fisk, P. T. H., Bachor, H. A., and Opat, G. I. (1989). *Opt. Commun.* **73**, 331.

Hall, J. L., Bordé, C. J., and Uehara, K. (1976). *Phys. Rev. Lett.* **37**, 1339.
Helmcke, J., Riehle, F., Witte, A., and Kisters, T. (1992). *Phys. Scr.* **T40**, 32.
Hemmerich, A., and Hänsch, T. W. (1993). *Phys. Rev. Lett.* **70**, 410.
Hemmerich, A., Zimmermann, C., and Hänsch, T. W. (1994). *Europhys. Lett.* **22**, 89.
Hoffnagle, J. (1988). *Opt. Lett.* **13**, 102.
Hu, Z., and Kimble, H. J. (1994). *Opt. Lett.* **19**, 1888.
Javanainen, J. (1991). *Phys. Rev. A* **44**, 5857.
Javanainen, J. (1992). *Phys. Rev. A* **46**, 5819.
Jessen, P., Gerz, C., Lett, P. D., Phillips, W. D., Rolston, S. L., Spreeuw, R. J. C., and Westbrook, C. I. (1992). *Phys. Rev. Lett.* **69**, 49.
Julienne, P. (1988). *Phys. Rev. Lett.* **61**, 698.
Julienne, P., and Heather, R. (1991). *Phys. Rev. Lett.* **67**, 2135.
Julienne, P., and Mies, F. H. (1989). *J. Opt. Soc. Am. B* **6**, 2257.
Jurczak, C., Sengstock, K., Kaiser, R., Vansteenkiste, N., Westbrook, C. I., and Aspect, A. (1995). *Opt. Commun.*, **115**, 480.
Kaiser, R., Levy, Y., Vansteenkiste, N., Aspect, A., Sciferd, W., Leipold, D., Mlynek, J. (1994). *Opt. Com.* **104**, 234.
Kasevich, M., and Chu, S. (1992). *Phys. Rev. Lett.* **69**, 1741.
Kasevich, M., Riis, E., and Chu, S. (1989). *Phys. Rev. Lett.* **63**, 612.
Kasevich, M., Weiss, O., and Chu, S. (1990). *Opt. Lett.* **15**, 607.
Kazantsev, A. P., Ryabenko, G., Surdutovich, Q., and Yakoulov, V. P. (1985). *Phys. Rep.* **129**, 76.
Kazantsev, A. P., Surdutovich, G. I., and Yakovlev, V. P. (1990). *Mechanical Action of Light on Atoms*. World Sci, Singapore.
Ketterle, W., Davis, K. B., Joffe, M. A., Martin, A., and Pritchard, D. E. (1993). *Phys. Rev. Lett.* **70**, 2253.
Kohns, P., Buch, P., Süptitz, W., Csambal, C., and Ertmer, W. (1993). *Europhys. Lett.* **22**, 517.
Lawall, J., Bardou, F., Saubamea, B., Shimizu, K., Leduc, M., Aspect, A., and Cohen-Tannoudji, C. (1994). *Phys. Rev. Lett.* **73**, 1915.
Letokhov, V. S., and Minogin, V. G. (1981). *Phys. Rep.* **73**, 1.
Lett, P., Watts, R., Westbrook, C., Phillips, W., Gould, P., and Metcalf, H. (1988). *Phys. Rev. Lett.* **61**, 169.
Lett, P., Phillips, W., Rolston, S., Westbrook, C., and Gould, P. (1991). *Phys. Rev. Lett.* **76**, 2139.
Lewenstein, M., You, L., Cooper, J., and Burnett, K. (1994). *Phys. Rev. A* **50**, 2207.
Liang, J., and Fabre, C. (1986). *Opt. Commun.* **59**, 31.
Lounis, B., Courtois, J. Y., Verkerk, P., Salomon, C., and Grynberg, G. (1992). *Phys. Rev. Lett.* **69**, 3029.
Lounis, B., Verkerk, P., Courtois, J. Y., Salomon, C., and Grynberg, G. (1993). *Europhys. Lett.* **21**, 12.
Metcalf, H., and van der Straten, P. (1994). *Phys. Rep.* **244**.
Migdall, A. L., Prodan, J. V., Phillips, W. D., Bergmann, T. H., and Metcalf, H. J. (1985). *Phys. Rev. Lett.* **54**, 2596.
Miller, J. D., Cline, R. A., and Heinzen, D. J. (1993). *Phys. Rev. A* **47**, R4567.
Minogin, V. G. (1980). *Sov. Phys.—JETP (Engl. Transl.)* **52**, 1032.
Minogin, G. V., and Letokhov, V. S. (1987). *Laser Light Pressure on Atoms*. Gordon & Breach, New York.
Minogin, G. V., Letokhov, V. S., and Zueva, T. V. (1981). *Opt. Commun.* **38**, 225.
Moi, L. (1984). *Opt. Commun.* **50**, 349.
Molmer, K. (1991). *Phys. Rev. A* **44**, 5820.
Molmer, K., Berg-Sorensen, K., and Bonderup, E. (1991). *J. Phys. B* **24**, 2327.
Monroe, C., Swann, W., Robinson, H., and Wieman, C. (1990). *Phys. Rev. Lett.* **65**, 1571.
Nellessen, J., Sengstock, K., Müller, J. H., and Ertmer, W. (1989a). *Europhys. Lett.* **9**, 133.
Nellessen, J., Müller, J. H., Sengstock, K., and Ertmer, W. (1989b). *J. Opt. Soc. Am. B* **5**, 2149.

Nellessen, J., Werner, J., and Ertmer, W. (1990). *Opt. Commun.* **78**, 300.
Nienhuis, D., van der Straten, P., and Shang, S.-Q. (1991). *Phys. Rev. A* **44**, 462.
Ol'shanii, M. A. (1991). *J. Phys. B* **24**, L583.
Paul, W., Osberghaus, H. O., and Fisher, E. (1958). *Forschungsber. Wirtsch.- Verkehrminist. Nordrhein-Westfalen* No. 415.
Petsas, K. I., Coates, A. B., and Grynberg, G. (1994). *Phys. Rev. A* **50**, 5173.
Phillips, W. D. (1992). In *Fundamental Systems in Quantum Optics* Les Houches, (Dalibard, J., Raimond, J. M., and Zinn-Justin, J., eds.), Les Houches, 1990, Vol. 53, p. 165, Elsevier, Amsterdam.
Phillips, W. D., Westbrook, C., Lett, P., Watt, R., Gould, P., and Metcalf, H. (1989). *At. Phys.* **11**, 633.
Prentiss, M., Cable, A., Bjorkholm, J., Chu, S., Raab, E., and Pritchard, D. (1988). *Opt. Lett.* **13**, 452.
Pritchard, D. E., Raab, E. L., Bagnato, V. S., Wieman, C. E., and Watts, R. N. (1986). *Phys. Rev. Lett.* **57**, 310.
Prodan, J. V., Migdall, A., Phillips, W. D., So, I., Metcalf, H., and Dalibard, J. (1985). *Phys. Rev. Lett.* **54**, 992.
Raab, E., Prentiss, M., Cable, A., Chu, S., and Pritchard, D. (1987). *Phys. Rev. Lett.* **59**, 2631.
Ramsey, N. F. (1985). *Molecular Beams*, pp. 138, 285, Oxford Univ. Press, London.
Riehle, F., Witte, A., Kisters, T., and Helmcke, J. (1992). *Appl. Phys., Part B* **B54**, 333.
Riis, E., Weiss, D., Moler, K., and Chu, S. (1990). *Phys. Rev. Lett.* **64**, 1658.
Salomon, C., and Dalibard, J. (1988). *C. R. Acad. Sci. Paris* **306**, 1319.
Salomon, C., Dalibard, J., Aspect, A., Metcalf, H., and Cohen-Tannoudji, C. (1987). *Phys. Rev. Lett.* **59**, 1659.
Salomon, C., Dalibard, J., Phillips, W., Clairon, A., and Guellati, S. (1990). *Europhys. Lett.* **12**, 683.
Schieder, R., Walther, H., and Wöste, L. (1972). *Opt. Commun.* **5**, 337.
Scholz, A., Christ, M., Doll, D., Ludwig, J., and Ertmer, W. (1994). *Opt. Commun.* **111**, 155.
Sengstock, K., Sterr, U., Hennig, G., Bettermann, D., Müller, J. H., and Ertmer, W. (1993). *Opt. Commun.* **103**, 73.
Sengstock, K., Sterr, U., Müller, J. H., Rieger, V., Bettermann, D., and Ertmer, W. (1994). *Appl. Phys.,* **B59**, 99.
Sesko, D., Walker, T., Monroe, C., Gallagher, A., and Wieman, C. (1989). *Phys. Rev. Lett.* **63**, 961.
Sesko, D. W., Walker, T. G., and Wieman, C. E. (1991). *J. Opt. Soc. Am. B* **8**, 946.
Shimizu, F., Shimizu, K., and Takuma, H. (1990). *Chem. Phys.* **145**, 327.
Sinclair, A. G., Riis, E., and Snadden, M. J. (1994). *J. Opt. Soc. Am. B* **11**, 2333.
Special Issue (1985). *J. Opt. Soc. Am. B* **2**.
Special Issue (1989). *J. Opt. Soc. Am. B* **6**.
Special Issue (1992). *Appl. Phys. B* **54**, 319.
Steane, A. M., and Foot, C. F. (1991). *Europhys. Lett.* **14**, 231.
Stenholm, S. (1983). *Phys. Rev. A* **27**, 2513.
Stenholm, S. (1986). *Rev. Mod. Phys.* **58**, 699.
Sterr, U., Sengstock, K., Müller, J. H., Bettermann, D., and Ertmer, W. (1992). *Appl. Phys. B* **54**, 341.
Strumia, F. (1988). In *Laser Science and Technology* (Chester, A., Letokhov, V. S., and Mavtelucci, S., eds.), p. 367, Plenum, New York.
Thorsheim, H. R., Wang, Y., and Weiner, J. (1990). *Phys. Rev. A* **41**, 2873.
Ungar, P., Weiss, D., Riis, E., and Chu, S. (1989). *J. Opt. Soc. Am. B* **6**, 2058.
Verkerk, P., Lounis, B., Salomon, C., Cohen-Tannoudji, C., Courtois, J. Y., and Grynberg, G. (1992). *Phys. Rev. Lett.* **68**, 3861.
Verkerk, P., Meacher, D. R., Coates, A. B., Courtois, J. Y., Lounis, B., Guibal, S., Salomon, C., and Grynberg, G. (1994). To be published.

Walhout, M., Sterr, U., Orzel, C., Hoogerland, M., and Rolston, S. L. (1995). *Phys. Rev. Lett.* to be published.

Walker, T., Sesko, D., and Wieman, C. (1990). *Phys. Rev. Lett.* **64**, 408.

Wallis, H. (1995). *Phys. Rep.* to be published.

Wallis, H., and Ertmer, W. (1989). *J. Opt. Soc. Am. B* **6**, 2211.

Wallis, H., Dalibard, J., and Cohen-Tannoudji, C. (1992). *Appl. Phys. B* **54**, 407.

Wallis, H., Werner, J., and Ertmer, W. (1993). *Comments At. Mol. Phys.* **28**, 275.

Walther, H. (1992). In *Fundamental Systems in Quantum Optics* (Dalibard, J., Raimond, J. M., and Zinn-Justin, J., eds.), Les Houches, 1990, Vol. 53, Elsevier, Amsterdam.

Walther, H. (1994). In *Advances in Atomic, Molecular, and Optical Physics*. Academic Press, San Diego.

Weiss, D. S., Riis, E., Shevy, Y., Ungar, P. J., and Chu, S. (1989). *J. Opt. Soc. Am. B* **6**, 2072.

Werner, J., Wallis, H., and Ertmer, W. (1992). *Opt. Commun.* **94**, 525.

Werner, J., Wallis, H., Hildebrandt, P., and Steane, A. (1993). *J. Phys. B* **26**, 3063.

Wineland, D. J., and Dehmelt, H. (1975). *Bull. Am. Phys. Soc.* **20**, 637.

Zoller, P., Marte, M., Walls, D. F. (1987). *Phys. Rev. A* **35**, 198.

ADVANCES IN ULTRACOLD COLLISIONS: EXPERIMENTATION AND THEORY

JOHN WEINER

Department of Chemistry and Biochemistry, University of Maryland, College Park, Maryland

I. Introduction 45
II. Scattering Length 46
   A. Introduction 46
   B. Determination of the Scattering Length in Alkali Systems 47
III. Optical Control of Inelastic Collisions 48
   A. Introduction 48
   B. Photoassociation 49
   C. Optical Suppression and Shielding 57
IV. Trap-Loss Collisional Processes 65
   A. Trap Loss in Rubidium 65
   B. Trap Loss in Lithium 68
   C. Trap Loss in Sodium 69
V. Developments in Theory 70
   A. Complex Potentials 71
   B. Quantum Monte Carlo Wave Functions and the Optical Bloch Equations 73
VI. Future Directions 76
   References 76

I. Introduction

Collisions between optically cooled and trapped atoms have been the subject of intensive investigation since early proposals [1] discussed their novel features and key importance to the achievement of a gaseous ensemble in a single quantum state [2]. Progress in both experimentation and theory has accelerated rapidly over the last three years, and two reviews, one emphasizing theory [3] and the other, experiments [4], recount the state of the art published up to about the midpoint of 1993. The purpose of this chapter is to update continuing lines of research set forth in these and earlier works [5,6] and to relate new results, establishing novel directions for investigation that have appeared in the literature.

Two principal questions motivate research into the nature of ultracold collisions: (1) what new phenomena arise when collisionally interacting particles also exchange photons with modes of the radiation field and (2) what are the important two-body collisional heating mechanisms and how can they be overcome in order to achieve the temperature and density conditions appropriate for Bose–Einstein condensation (BEC)? In fact these two questions are not mutually exclusive, and one of the most notable developments in the past year, relevant to both, has been the demonstration of optical control of two-body ultracold collisional processes. Other important issues touching, on one or both of these questions are the magnitude and sign of the scattering length in s-wave collisions between species in various well-defined quantum states, progress in high-resolution trap loss and photoassociation spectroscopy, and application of optical cooling and compression to atomic beams.

II. Scattering Length

A. INTRODUCTION

As the kinetic energy of a two-body collision approaches zero, the number of partial waves contributing to the elastic collision reduces to the s-wave, and the information inherent in the scattering event can be characterized by the *scattering length*, $a = -\lim_{k \to 0} \delta_0(k)/k$, where k is the wave vector equaling $2\pi/\lambda_{db}$ (λ_{db} is the deBroglie wavelength of the reduced particle) and δ_0 is the s-wave phase shift due to the scattering potential. The sign of the scattering length is of crucial importance to the attainment of BEC. A positive scattering length implies a repulsive scattering potential and stable region of temperature and density for the Bose condensate. A negative scattering length, on the contrary, implies an attractive potential and leads to a more complex situation with a possible phase separation from spinodal decomposition [7]. The phase separation in turn will sharply raise density in the center of the trap, leading to a dramatic increase in the rate of collisional trap loss processes. Although the spinodal decomposition is itself an interesting phenomenon, it may be difficult to distinguish its signature from other trap destruction processes. In contrast the predicted stability of BEC should render it more suitable to unambiguous experimental verification. Furthermore, the weak interactions of a dilute boson gas can be treated as a perturbation series in $(na^3)^{1/2}$, and therefore the properties of the condensate are more tractable to calculate [8]. The currently most promising strategy for the experimental realization of BEC conditions appears to be optical cooling in a magneto optical trap (MOT) [9], followed by transfer to a purely magnetic trap, where BEC conditions can be attained by evaporative cooling [10]. The efficiency of this evaporation process depends on the s-wave elastic collision cross-section, $\sigma = 4\pi a^2$. Thus the determination of signs and magnitudes of the scattering length in trapped atom collisions must necesarily preface any serious attempt to realize a gaseous ensemble interacting in a single quantum state.

B. Determination of the Scattering Length in Alkali Systems

Although collisional frequency shifts in an atomic fountain experiment [11] have led to the determination of negative scattering lengths in Cs collisions [12], the most extensive theoretical and experimental work has been carried out on trapped ^{6}Li and ^{7}Li collisions for two principal reasons. First, the two isotopes obey different statistics: ^{6}Li is a fermion and ^{7}Li is a boson, and their distinctive statistical behavior may be compared under nearly identical experimental conditions. Second, the singlet and triplet ground state potential wells are not very deep and therefore support fewer, more widely spaced bound states than the heavier alkalis. The relative simplicity of the bound state spectroscopy makes it easier to extract reliable scattering lengths from the spectroscopic data. A new analysis of the spectroscopic data for Li$_2$ $^3\Sigma_u^+$ [13] enabled Moerdijk et al. [14] to extract a reliable value for the scattering length of two colliding ^{7}Li atoms in the $F = 2$, $m_F = 2$ ground state. Taking into account uncertainties in the exchange and dispersion terms of the potential, they found the triplet scattering length a_T varied over the range from about $-7a_0$ to $-28a_0$ (a_0 is the atomic unit of length), and therefore triplet collisions of ^{7}Li$_2$ are not suitable for the formation of the Bose condensate. This conclusion has been confirmed by an experiment of Abraham et al. [15], who have directly measured with high precision the binding energy of the last bound energy level in the Li$_2$ $a^3\Sigma_u^+$ triplet ground state. The position of this level determines the molecular potential with sufficient accuracy that the scattering length can be calculated by comparing the limiting phase shift of the continuum solutions to the Schroedinger equation just above threshold with the free-particle solutions in the asymptotic region. Abraham et al. thus determined the triplet scattering length to be $a_T = -27.3 \pm 0.8\, a_0$. Armed with a reliable value for the triplet scattering length and again using the spectrosopic values of Zemke and Stwalley [13], Moerdijk and Verhaar [8] were able to calculate not only the ^{7}Li$_2$ $^1\Sigma_g^+$ scattering length for the molecular potential ($34a_0 < a_S < 36a_0$), but more importantly extract the scattering length for collision between atoms in the $F = 1$, $m_F = -1$ hyperfine state. They find the scattering length positive ($37a_0 < a_{1,-1} < 152a_0$). This result is encouraging because the $F = 1$, $m_F = -1$ can be trapped and therefore is a viable candidate for BEC.

III. Optical Control of Inelastic Collisions

A. INTRODUCTION

The optical control of inelastic and reactive collision processes has captured the imagination of researchers in AMO and chemical physics ever since the first appearance of the CO_2 laser as a practical laboratory instrument [16]. Early attempts to enhance reactivity by exciting localized molecular sites such as double bonds or functional groups were not generally successful because the initial, well-defined optical excitation diffuses rapidly into rovibrational modes on a time scale of the internal molecular motion $\sim 10^{-15}$ s. The development of the pulsed, tunable dye laser together with the proposal for "radiative collisions" [17] revived interest in optical manipulation of inelastic processes. Several experiments [18] reported success in optically transmuting reactant excitation to final products, but in general the high peak powers necessary to significantly enhance the inelastic channel probability masked the optical switching effect by inducing atomic multiphoton excitation and ionization. The necessity for high peak power in turn arose from two important factors: (1) the need to have optical and collisional coupling terms of roughly equal standing in the Hamiltonian describing the collision and (2) the wide dispersion of populated continuum states in the collisional entrance channel due to the Maxwell–Boltzmann distribution at ambient or elevated temperatures typical of atomic-vapor sources. The large Maxwell–Boltzmann width implies that the applied optical field can couple only a small fraction of the available atomic population to excited states of the quasi molecular collision intermediate.

The ultracold regime does not suffer from these drawbacks. Since the collisional interaction is initially at very long range ($\sim 1000 a_0$), the term in the Hamiltonian describing it (usually of the form $V_{\text{collision}} = C_n/R^n$) is quite small. The term describing the optical interaction ($\Omega/2\pi = \vec{\mu} \cdot \vec{E}/h$) therefore requires only very modest optical field intensities to be of the order of the collisional term. For example the long-range collisional interaction between a ground- and an excited-state atom is $V_{\text{collision}} = C_3/R^3$, and for an Na pair interacting at $1200 a_0$, $V_{\text{collision}} \simeq 25$ MHz. To make the corresponding optical term, $\Omega/2\pi$ (MHz) $\simeq (44.4) \sqrt{I \text{ (W cm}^{-2})}$, of equivalent magnitude requires only the laser intensity $I \simeq 320$ mW cm^{-2}. Furthermore, the distribution of populated entrance-channel continuum states is quite narrow (the width of the Maxwell–Boltzmann distribution is comparable to the natural line width of the cooling transition), and therefore almost all the cooled atoms participate in the initial free–bound transition. These conditions lead to favorable signal rates and signal-to-noise ratios without complicating competition from multiphoton effects.

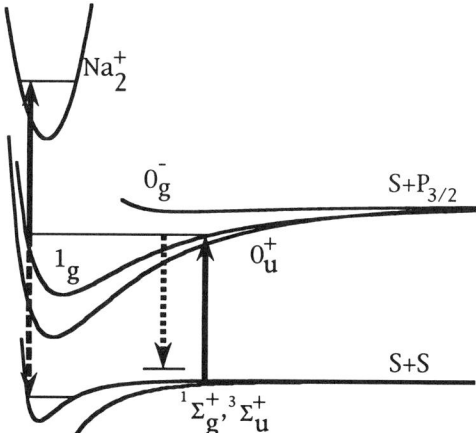

FIG. 1. Schematic showing a two-step pathway to photoassociation products. The end result may be a molecular ion via photoassociative ionization (solid arrows) or a neutral molecule via photoassociation (dashed arrows).

B. Photoassociation

Photoassociation refers to a process in which incoming scattering flux on a relatively flat ground state absorbs a photon and populates a long-range attractive, excited molecular state. The force derived from the attractive potential curve accelerates the particles together, thereby "associating" them. The ultimate fate of the associated flux may be either spontaneous emission relaxation back to bound or continuum levels of the ground state or a second step of photon absorption, resulting finally in ionization. The second photon absorption may come from the same laser source as the first or from a second, independently tunable source. The two cases are distinguished as "one- color" or "two color" experiments. When the ultimate product of the photoassociation is an ion–molecule the process is called photoassociative ionization (PAI). Figure 1 illustrates the various collision pathways for the specific case of sodium. A review [4] well documents the series of experiments dating from the first detection of PAI in an optical dipole trap to the first measurements of extensive molecular spectral features. Here we will confine ourselves to updating this previous review with subsequent developments.

1. Dynamics from Line Shapes

A series of experiments carried out at NIST [19] has established with unprecedented precision the spectroscopy of the 1_g, 0_u^+, and 0_g^- states of

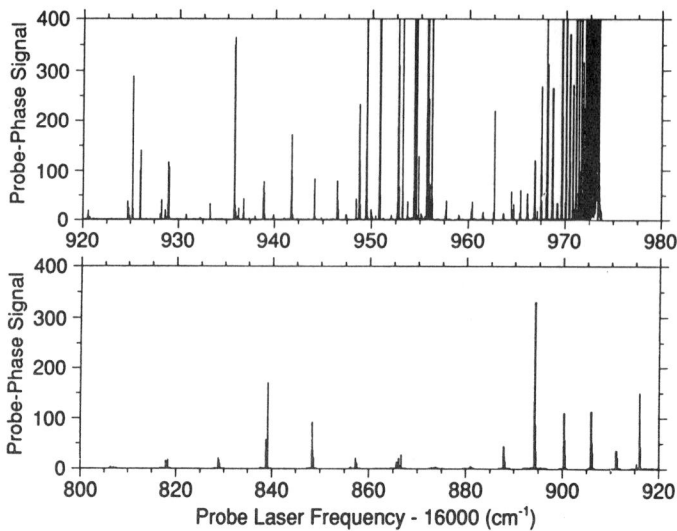

FIG. 2. An example of the PAI spectrum measured by the NIST group, showing free-bound transitions, $1_u(^3\Sigma_u) \to 1_g(^1\Pi_g)$, in first step of the PAI process.

Na_2, near the molecular dissociation limit. The experiments were performed in a "dark spot" magneto optical trap [20], and the spectroscopic signal was derived from a two-step, one-color PAI process (Fig. 1). The experiment itself was divided into two parts: a "trap phase" when the MOT light cooled and confined the atom sample and a "probe phase" when the MOT was turned off and the probe light was turned on. The two phases oscillated at 50 KHz. The 1_g state is of particular interest because good signals were obtained over a wide range of probe detuning (170 cm^{-1}), corresponding to a change in angular momentum coupling from Hunds's case (c) at long range to Hund's case (a) at short range. The natural width of the PAI lines over this spectral range reflect the change in angular momentum coupling. Scattering flux enters on the ground state $^3\Sigma_u$, and at small probe detuning strongly allowed free–bound transitions populate high vibrational levels of the 1_g. As the probe frequency detunes to the red, however, the 1_g state recouples to $^1\Pi_g$, the $^3\Sigma_u^+ \to {}^1\Pi_g$ rovibronic transitions become spin forbidded, and the natural line widths become narrower than the probe laser line width of about 1 MHz. A sample of this precision PAI spectrum is shown in Fig. 2. The excellent signal-to-noise ratio typical of PAI experiments, together with MHz resolution of the probe, permits a detailed analysis of the line shape. Such an analysis has been carried out for transitions to the $J = 1, 2, 3$ and 4 rotational levels of the Na_2 $^1\Pi_g$ ($v = 48$) by Napolitano et al. [21]. The lines shapes show structure and asymmetry due to the presence of excited-state hyperfine levels, Wigner threshold-law behavior for con-

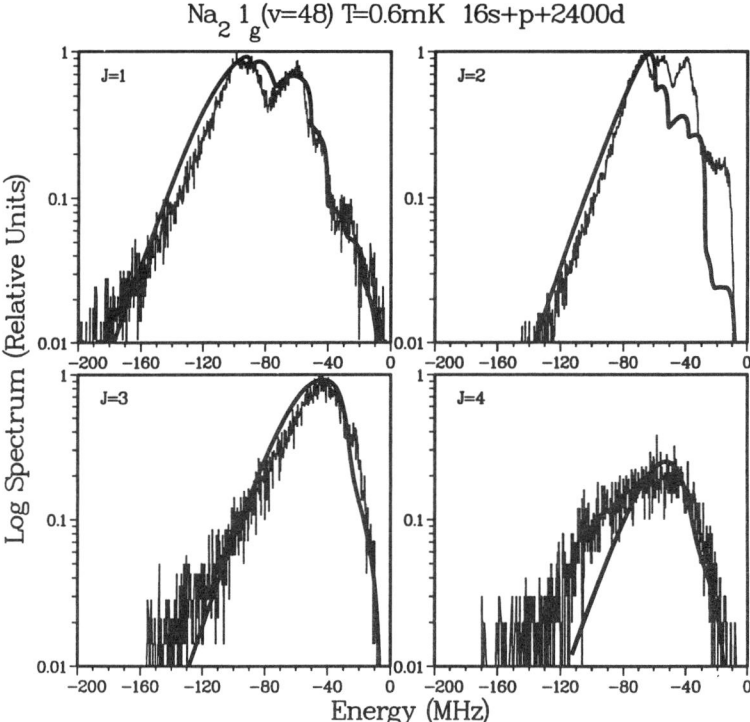

FIG. 3. Experimental high-resolution spectrum of the Na_2 1_g, $v = 48$, $J = 1, 2, 3, 4$, levels together with theoretical line profile calculations (thick solid lines). Note especially that the blue (right) side of the profiles reflects the Wigner threshold law for the contribution partial waves, whereas the red (left) side is dominated by the Maxwell–Boltzmann exponential tail.

tributing partial waves, and the finite kinetic energy distribution of the cooled atoms. Figure 3 shows the experimental line shape from [19] together with the results of the analysis. Evidence of hyperfine contributions can be seen in the incompletely resolved structure in the $J = 1$ and $J = 2$ profiles. The intensity of the photoassociation spectrum is proportional to the rate coefficient $K_p(T, \omega)$ for the product p of the collision. The thermally averaged rate coefficient is given by

$$K_p(T, \omega) = \frac{k_B T}{h Q_T} \sum_{\ell=0}^{\infty} (2\ell + 1) \int_0^{\infty} |S_p(\varepsilon, \ell, \omega)|^2 \, e^{-\varepsilon/kT} \frac{d\varepsilon}{k_B T}, \qquad (1)$$

where ε is the asymptotic kinetic energy; ℓ, the relative angular momentum quantum number; Q_T, the translational partition function; and $S_p(\varepsilon, \ell, \omega)$, the S-matrix element for the process that forms product p from the initial ground state entrance channel. The other terms in this expression have their usual meanings. The form of $S_p(\varepsilon, \ell, \omega)$ around an isolated resonance can be taken as

$$|S_p(\varepsilon, \ell, \omega)|^2 = \frac{\gamma_p \gamma_s(\varepsilon, \ell)}{(\varepsilon - \Delta_b)^2 + (\gamma/2)^2}, \quad (2)$$

where $\Delta_b(\omega)$ is the probe laser detuning from the bound rovibronic state of the excited member of the free-bound transition, γ_p is the decay width to the collision product, and $\gamma_s(\varepsilon, \ell)$ is the stimulated rate connecting the ground and excited levels. The γ term in the denominator is the sum of γ_p and $\gamma_s(\varepsilon, \ell)$ and any other (undetected) decay channel. Now the Wigner threshold behavior intervenes in the form of $\gamma_s \simeq A_\ell \varepsilon^{(2\ell+1)/2}$ as $T \to 0$. Three partial waves, s, p, and d, contribute in varying amounts to the scattering amplitude of each of the four J levels, and their influence can most easily be seen on the blue slope of each line. Note especially in the $J = 2$ panel of Fig. 3 the very sharp onset to the profile, illustrating the influence of the Wigner threshold law for the s-wave contribution (close to infinite slope as $T \to 0$). On the red side of the profile the intensity falls as $e^{-\Delta_b/k_B T}$, and at first thought one might expect to extract a collision temperature from the slope of a semilog plot. However, the s, p, and d contributions to the line shapes all yield somewhat different effective temperatures, again due to the influence of the Wigner threshold laws operating on γ_s. In order to extract a meaningful temperature from the experimental line shape, therefore, the various partial wave contributions to the profile must be known accurately. These contributions, in turn, are extremely sensitive to the choice of potential on which scattering takes place.

Two-color photoassociation spectroscopy has also been studied in Na [22], Rb [23], and Li [24] collisions. The two-color Na experiments explore PAI closer to the dissociation limit than the one-color spectroscopy of the NIST group. In this spectral region $C_3/R^3 \leqslant D_s$, where D_s is the atomic ^2S ground state hyperfine splitting, and the hyperfine interaction mixes the molecular states even to the extent of breaking the gerade/ungerade symmetry. Although this angular momentum recoupling to the nuclear spin precludes labeling curves with Hund's case (c) notation such as 1_u or 0_g^-, the authors were able to identify a series of peaks with a manifold of excited-state hyperfine levels which they termed a "0_g^- bundle" because they have a significant amount of 0_g^- character even for internuclear distances $> 300 a_0$.

Both the one- and two-color PAI spectroscopies interpret the observed structure as free–bound transitions in the first step of photon absorption. However, if the frequency of the first step is held constant while the second photon sweeps to the *blue* of the dissociation limit, a qualitatively different spectrum appears as shown in Fig. 4a. The two-step excitation is diagrammed in Fig. 4b. The fixed MOT laser frequency initiates the first step. The probe laser scans to the blue of atomic resonance, populating a doubly

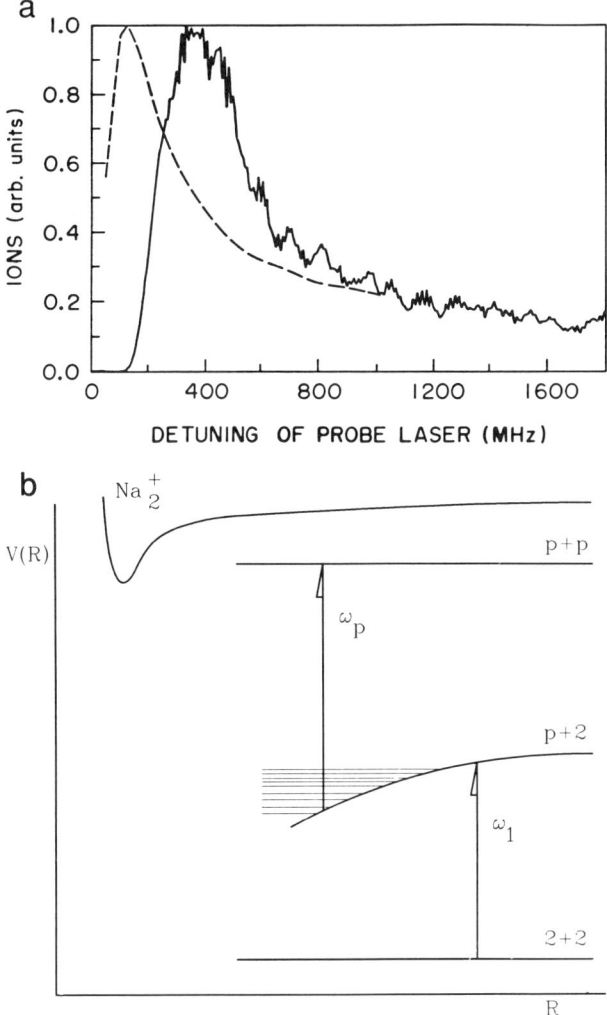

FIG. 4. Two-color photoassociative ionization. (a) Asymmetric line shape measured in two-color PAI with the variable probe laser sweeping to the blue of the trapping transition. (b) Two-step route to photoassociative ionization with ω_p tunable and ω_1 fixed.

excited $p + p$ level from the intermediate $s + p$ level, which subsequently undergoes autoionization at short range. The asymmetric line shape exhibits a clearly observable peak about 375 MHz to the blue of the cooling transition with a long tail degrading to the blue. This line shape resembles the "quasistatic wing" in the theory of collision broadening [25] and is characteristic of optical or radiative collisions [17] In the ultracold regime this frequency dependence of the PAI probability can be roughly interpreted in terms of the R dependence of the product of excitation and survival probabilities, $P_{PAI} = P_{excit} \cdot P_{surv}$. At one extreme, when the atoms are far apart and the probe laser is not far detuned from the cooling laser frequency, the atoms pass slowly through the Condon point, where ω_p, the probe laser frequency, resonantly couples the singly and doubly excited states. The probability of excitation is high. However, after excitation the probability of survival against radiative decay back to the ground state before the colliding partners reach the autoionization region is low, and therefore the overall probability for PAI remains low. At the other extreme, when the atoms are closer together and the probe laser tuned far to the blue, the atoms pass quickly through the Condon point, and the probability of excitation to the doubly excited state is low. Once excited, however, the survival probability is high, but the product of the two probabilities will inevitably tend to zero with increased blue detuning of ω_p. Somewhere between these two extremes the product of excitation and survival probabilities will maximize. Quantitative comparison of the measured profile with predictions from a simple two-state semiclassical mode [26] reveals an unexpected 200-MHz blue shift in the peak position. Considering that this model does not take account of all the hyperfine manifolds present, the lack of quantitative agreement is not unexpected.

Although atom traps most commonly used for collision studies have been variations on the MOT, the research group at the University of Texas lead by Heinzen has pioneered the use of the far-off resonance trap (FORT) [27] to study photoassociation in collisions between Rb atoms. The advantages of the FORT over the MOT are: (1) relative simplicity (no magnetic fields), (2) very low excited state population and therefore negligible diffusional heating, (3) high density ($\sim 10^{12}-10^{13}$ cm^{-3} compared with a MOT density of $10^{10}-10^{11}$ cm^{-3}), and (4) a well defined polarization axis along which atoms can be aligned or oriented (spin polarized). The principal disadvantage is the low *number* of trapped atoms due to the small FORT volume. Miller et al. [27] estimate a density of 8×10^{11} cm^{-3} but only 1300 atoms in the trap (later improvements in technique increased this number to $\sim 10^4$ [28]). The FORT itself is loaded from a MOT by spatially superposing the focused FORT laser on the MOT and alternately chopping the two kinds of traps. This FORT loading sequence continues for 5000 cycles. At the end of the load sequence flourescence detection probes the number of atoms in the FORT. In the first version of this experiment [28] the FORT laser itself was swept in frequency from 50 to 980 cm^{-1} below the Rb ($^2S_{1/2}$) + RB ($^2P_{1/2}$) dissociation limit. When the cold, colliding atoms absorb a photon from the FORT they undergo the first step of the photoassociation process depicted in Fig. 1. The ultimate fate of the collision partners may be formation of a bound molecule, relaxation to a "hot" continuum, or ionization. In any case the pair will excape the trap with a consequent loss of fluorescence signal. These trap-loss dips in the fluorescence intensity

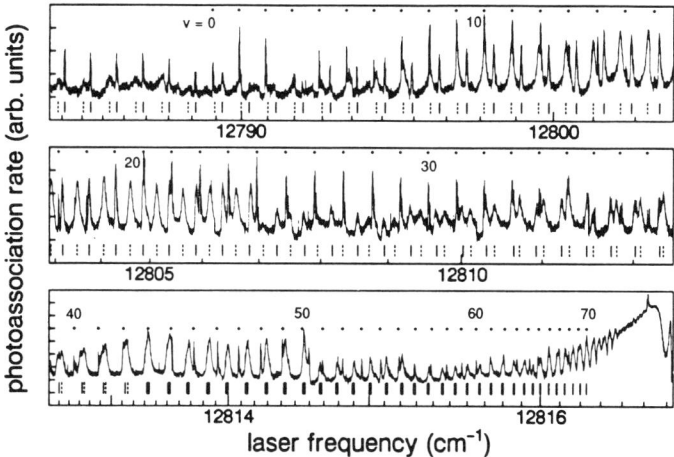

FIG. 5. Photoassociation trap-loss spectrum from a FORT trap.

generate the photoassociation spectra of free–bound transitions to excited states of the Rb_2 molecule. In a second, two-color, version of this experiment [29] a high-resolution probe laser is introduced to sweep the PA spectrum while the FORT laser remains at a fixed frequency. The two lasers, probe and FORT, are alternately chopped in trap and probe cycles to avoid spectral distortion from AC-Stark shifts and power broadening due to the FORT. At the end of the trap-probe cycling (20,000 periods) both lasers are turned off and the remaining atoms detected by laser-induced fluorescence. In contrast to the earlier one-color work, this experiment concentrated on molecular spectra converging on the Rb ($^2S_{1/2}$) + Rb ($^2P_{3/2}$) dissociation limit. The two-color experiment enabled a greatly increased spectral range, from $\sim 0.3 \, cm^{-1}$ to more than $1000 \, cm^{-1}$ below the dissociation limit and with resolution better than $0.002 \, cm^{-1}$. Figure 5 shows the Rb_2 spectrum generated from the FORT experiment. The data show rotationally resolved spectra of the 0_g^- pure long-range state as well as features belonging to the 1_g and 0_u^+ states. The resolution is sufficient to show asymmetry in an individual rotational line due to the Maxwell–Boltzmann distribution as well as broadening in the 0_u^+ state due to predissociation to Rb ($^2S_{1/2}$) + Rb ($^2P_{1/2}$) continuum.

2. Atomic Lifetimes from Molecular Spectroscopy

The detailed, precise vibration–rotation spectra obtained from ultracold photoassociation provide the data necessary for accurate determination of molecular potentials. Accurate potentials in turn yield new determinations of atomic radiative lifetimes at the few-percent level which are important for checking atomic structure calculations. Not far from the dissociation limit the molecular potential of attractive excited states of homonuclear molecules, dissociating to one ^2S atom and one ^2P atom, exhibits the form

$$V \simeq -C_3/R^3 - C_6/R^6 - C_8/R^8, \qquad (3)$$

where the C_3/R^3 term is dominant. Near the dissociation limit, the excited vibrational energies obey a well-known spectoscopic expression [31, 32],

$$v_D - v = a_3 \varepsilon_v^{1/6}, \qquad (4)$$

where v_D is the "effective" vibrational quantum number at the limit of dissocation and v is the vibrational quantum number of a state with binding energy ε_v. The constant a_3 is related to C_3 by

$$C_3 = \frac{(3\hbar)^3}{(60)^3 h^{1/2}} \left(\frac{2\pi}{\mu}\right)^{3/2} \left(\frac{\Gamma(4/3)}{\Gamma(5/6)}\right)^3 a_3^3, \qquad (5)$$

where μ is the reduced mass. Then C_3 is related to the atomic lifetime τ_A through an expression of the form

$$C_3 = \kappa \frac{3\hbar}{4k^3} \cdot \frac{1}{\tau_A}, \qquad (6)$$

where κ is a numerical constant characteristic of the molecular potential and $k = 2\pi/\lambda$. A convenient table of C_3 constants for the attractive molecular states dissociating to $^2P_{3/2} + {}^2S_{1/2}$ has been given by Julienne and Vigué [30]. The strategy therefore is to determine C_3 by fitting a_3 to the measured vibrational progressions using Eq. (4) and then calculating the atomic lifetime from Eq. (6). This program has been carried out by the NIST group [33] to the point of determining C_3 constants for vibrational progressions and, by comparing the measured C_3 to calculated values, identifying the 1_g state as the principal intermediate state in the two-step PAI process. The Texas group [29] has also followed this procedure to determine C_3s for the 1_g and 0_u^+ states in Rb_2, confirming their assignments by comparison to values calculated from Eq. 6. While fitting vibrational progressions only from the long-range part of the potential yields C_3 constants sufficiently accurate to distinguish and identify molecular electronic states, systematic errors due to rotational populations and hyperfine effects limit the accuracy of the "apparent" C_3. The Rice group [24] has extended this approach and improved precision by fitting photoassociation data from $^6Li_2$ experiments to the $1^3\Sigma_g^+$ potential constructed over a much wider range of internuclear separation. By combining results from an RKR fit and *ab initio* calculations for the short and intermediate ranges with the long-range C_3 results, they have been able to determine the Li(2p) lifetime at the 1% level of uncertainty. Improvements in the accuracy of the *ab initio* calculations could lead to an improvement in accuracy to the 0.1% level. The determination of a radiative lifetime in such a simple system as Li at the 0.1% level of accuracy would provide an important new benchmark for the theory of atomic structure.

C. OPTICAL SUPPRESSION AND SHIELDING

1. Introduction

Optical control of collisions implies the ability to turn off inelastic channels as well as turn them on. Within the past year several groups have demonstrated that the normal course of scattering flux can be interrupted by imposing an optical field on the ultracold sample. This "suppressor" field significantly reduces reaction rate constants for inelastic processes such as photoassociative ionization or trap-loss collisions. The suppression effect has been interpreted as the rerouting of ground-state entrance channel scattering flux to excited repulsive states at an internuclear distance localized around a "Condon point," at which the suppressor optical field resonantly couples them. In the ultracold regime particles approaching on flat ground states have very little kinetic energy, and the transfer to repulsive states effectively halts their approach in the immediate vicinity of the Condon point. At weak suppressor field intensity the scattering flux then exits on the repulsive state. As suppressor intensity increases the collision becomes "adiabatic" on the dressed-state potentials, the flux effectively enters and exits on the ground state, and the collisions become effectively elastic. Since the overall effect of the repulsion is to prevent colliding atoms from approaching to close range at which trap-loss processes such as photoassociation or fine-structure changing collisions take place, the ultracold sample is said to be optically "shielded" from reactive or collisional heating processes.

2. Optical Suppression of Trap Loss

Evidence for optical manipulation of trap-loss processes was first reported by Bali et al. [34] in ^{85}Rb collisions in an MOT. In this experiment scattering flux entering on the ground state transfers to excited repulsive molecular states by means of a control or "catalysis" laser tuned 5–20 GHz to the blue of the cooling frequency. When the blue detuning of the control laser becomes greater than about 10 GHz, the scattering partners receding from each other along the repulsive potential gather sufficient kinetic energy to escape the trap. The control laser thus produces a "trap-loss" channel, the rate of which is characterized by the constant, $\beta(I)$, a function of the control laser intensity, I. Figure 6 shows the increase of β up to about 6 W/cm^2. Above this point the trap loss rate constant begins to *decrease* indicating a *suppression* of the loss channel with increasing control laser intensity. Bali et al. interpret this suppression in terms of the dressed-state Landau–Zener crossing model depicted in Fig. 7. The suppression of β results from the increasingly avoided crossing between the ground atom-field state, $|g; N\rangle$ and the excited atom-field state, $|e; N-1\rangle$. As the control laser intensity increases, incoming flux on $|g; N\rangle$ begins to propagate

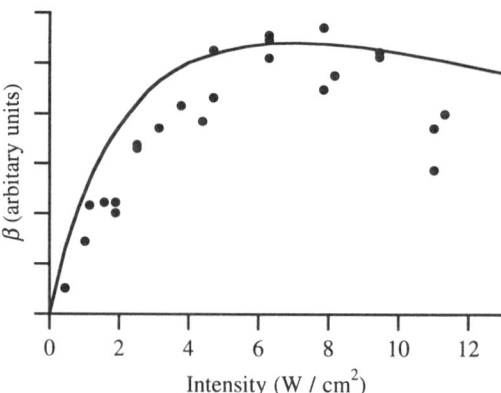

FIG. 6. Rate constant β as a function of catalysis laser intensity. Note that β begins to decrease after about 8 W cm^{-2}.

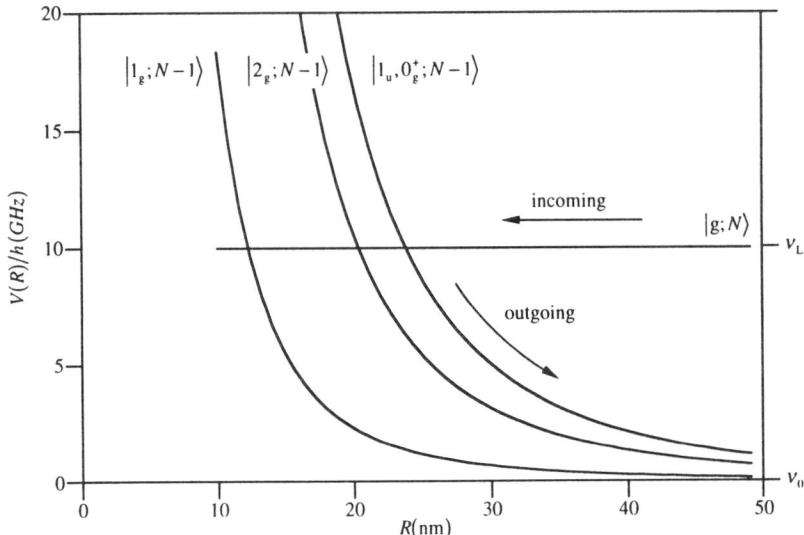

FIG. 7. Dressed-state potential curves for repulsive trap-loss collisions. Incoming flux on the ground state can transfer to outgoing channels.

adiabatically near the crossing and recedes along the same channel from which it entered. Despite the appeal of such a simple two-state model, Fig. 6 clearly shows it underestimates the measured suppression effect.

3. Optical Suppression and Shielding in Photoassociative Ionization

Although collisional trap loss has played a major role in revealing the nature of the ultracold regime, to date only PAI has permitted direct measure of the reaction product as a function of various probe fields and trap parameters. As discussed in Section III.B, the PAI process takes place in two steps (see Fig. 1). Reactant partners first approach on the ground state potential. Near the outer Condon point, at which an optical field, $\hbar\omega_1$, connects the ground and excited attractive states, population transfers to a long-range attractive state, drawing the two atoms together. The second step either photoionizes the excited quasimolecule directly or brings the population to a doubly excited state at a second Condon point from which the collision partners associatively ionize.

The notion of optical suppression intervenes at the first step. Rather than transferring population to an *attractive* potential at the first Condon point, an optical field, $\hbar\omega_3$, tuned to the blue of the excited state asymptote, transfers population to a *repulsive* curve. The collision partners are effectively stopped from approaching further and exit the collision. This process is analogous to the weak-field, trap-loss mechanism described by Bali et al. [34]. As the intensity of the suppressor field increases, scattering flux exits on the ground state, and the suppressed PAI becomes elastically shielded.

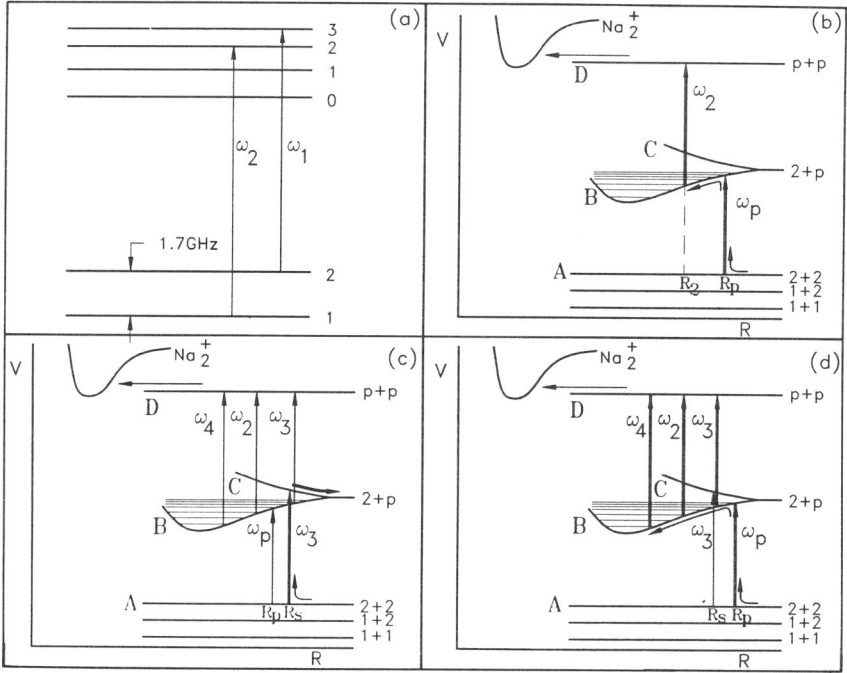

FIG. 8. (a) Relation between trap frequency ω_1 and repumper frequency ω_2. (b) Normal two-step PAI pathway. (c) Probe frequency ω_p scanning to the red (left) of ω_3. The normal pathway is suppressed by $\hbar\omega_3$ coupling the $2+2$ entrance channel to the $2+p$ repulsive level. (d) Probe frequency ω_p scanning to the blue (right) of ω_3. The normal pathway is enhanced by $\hbar\omega_3$ and $\hbar\omega_4$ coupling the $2+p$ attractive level to the $p+p$ doubly excited level.

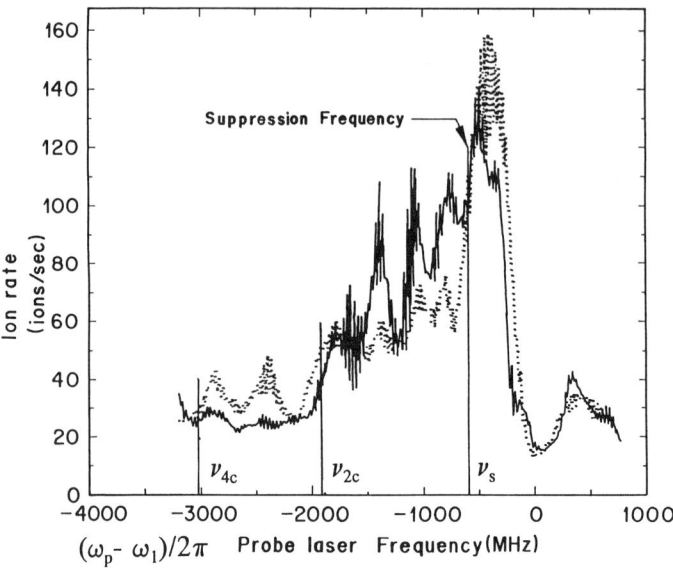

FIG. 9. The solid line is the normal two-color photoassociative ionization spectrum without the suppressor beam. The dotted line shows the same spectra region with the suppressor beam (0.4 W cm^{-2}) applied to the MOT.

Optical suppression of photoassociative ionization was first reported by Marcassa et al. [35] in a MOT setup in which four different optical frequencies were used to: (1) produce the MOT (ω_1, ω_2), (2) generate the suppressor frequency (ω_3), and (3) probe the PAI inelastic channel, (ω_p). Electrooptic modulation of the repumper ω_2 produced the suppressor frequency as the red sideband and another frequency ω_4 as the blue sideband. Although ω_4 was not necessary to observe the suppression effect, its presence added new features consistent with the interpretation of suppression and shielding described here. Figure 8 shows the coupling scheme and routing of the incoming flux. Figure 8b shows the familar two-step PAI process with no suppression. After excitation to the attractive $2 + p$ level, the frequency ω_2 transfers scattering flux in a second step to the $p + p$ level from which associative ionization takes place. As ω_p tunes to the red of the Condon point, R_2, it can no longer, in combination with ω_2, couple to the doubly excited level, and an abrupt cutoff of the PAI signal is predicted at that frequency. Figure 8c shows what happens when a suppressor frequency ω_3 is imposed on the trap. With ω_p tuned to the *red* of ω_3, scattering flux on the $F = 2$, $F' = 2$ entrance channel diverts to the *repulsive* $2 + p$ curve around the first Condon point, R_s and exits via the excited or ground state, depending the intensity of the suppressor field. Figure 8d depicts what happens when ω_p is tuned to the *blue* of ω_3. In this case the incoming flux transfers to the *attractive* $2 + p$ curve around the Condon point R_p. However, now ω_3 and ω_4 add to ω_2 to enhance the probability of second-step excitation to the $p + p$ level. Therefore we expect PAI to be *enhanced* with respect to the normal PAI rate represented in Figure 8b. Futhermore the cutoff for PAI should be *extended* beyond R_2 because ω_4 is the sideband of ω_2, appearing 1.1 GHz farther to the blue. Figure 9 shows a plot of PAI count rate vs ω_p detuning with and without ω_3 and ω_4. All three predicted features, suppression, enhancement, and extension of the cutoff, are evident and appear where expected.

The dependence of any shielding or suppression effect on the intensity of the optical field is obviously of great interest and importance. Marcassa et al. [35] carried out model calculations at three levels of complexity: (1) time-dependent Monte Carlo wave function simulations of wave-packet dynamics, (2) quantum close-coupling calculations without excited-state decay, and (3) a very simple estimate based on a dressed-state Landau–Zener avoided crossing model. The results of the calculations were in close agreement among themselves and within about 15% of the experiment. However, instrumental limitations precluded intensity measurements over ~ 1 W cm^{-2}. In a follow-up article Marcassa et al. [36] were able to measure the suppression power dependence up to 8 W cm^{-2}. Figure 10 plots the fractional intensity of the PAI signal as a function of suppressor power density. The experimental results are shown together with calculations from quantum close coupling and the Landau–Zener semiclassical model. Agreement is good at low field, but above about 5 W cm^{-2} the data tend to lie above the theoretical trace, suggesting that suppression "saturates" at an efficiency less than 100%. Calculated values do not lie outside the statistical error bars on the data points, however, so these data do not establish a saturation effect at high field beyond question.

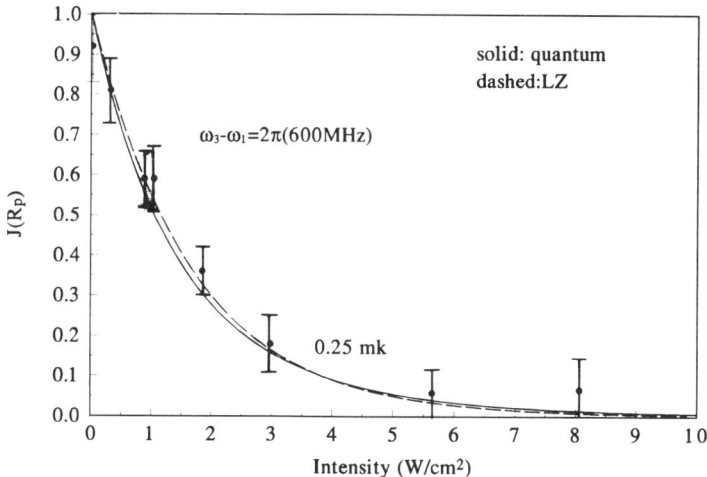

FIG. 10. Probability of flux penetration $J(R_p)$ to the PAI channel as a function of suppression intensity. Points are measured data with statistical error bars; curves are theoretical models. Note the tendency of data to lie above model results at a higher field.

4. Optical suppression and Shielding in Xenon and Krypton Collisional Ionization

Another example of optical shielding in metastable xenon collisions has been reported by the NIST group [37]. The experiment takes place in an MOT with the cooling transition between the metastable "ground state" and a dipole-allowed excited state, $6s[\frac{3}{2}]_2 \leftrightarrow 6p[\frac{5}{2}]_3$. The MOT is time chopped with a 150 μ period and an "on" duty cycle of $\frac{2}{3}$. Measurements were carried out in the probe cycle (MOT off) and the trap cycle (MOT on). In the probe cycle experiments a control laser, starting from the cooling transition, sweeps over a range of detuning from about 1.5 GHz red to 500 MHz blue. The control laser induces a strong enhancement of the ionization rate over a red detuning range of about 500 MHz, peaking at 20 MHz, and produces suppression at blue detunings over several hundred megahertz. The maximum suppression factor (~ 5), occurring at 200 MHz blue, appears to "saturate" at the highest control laser power of 0.5 W cm^{-2}. With the MOT beams illuminating the sample the experiments show an even more dramatic suppression effect. Figure 11 shows the suppression factor, the ratio of ionization rate constant with and without the control laser, plotted against control laser detuning. At about 250 MHz blue detuning the suppression factor reaches a maximum of greater than 30. The interpretation is similar to that of the sodium case [35]. Without the control laser field present, reactant scattering flux, approaching on the metastable ground state, is transferred to an attractive excited state at far internuclear separation by the trapping laser. The colliding partners start to accelerate toward each other and during their inward journey radiatively relax back to the metastable ground state. The consequent increase in kinetic energy allows the colliding partners to surmount centrifugal barriers that would

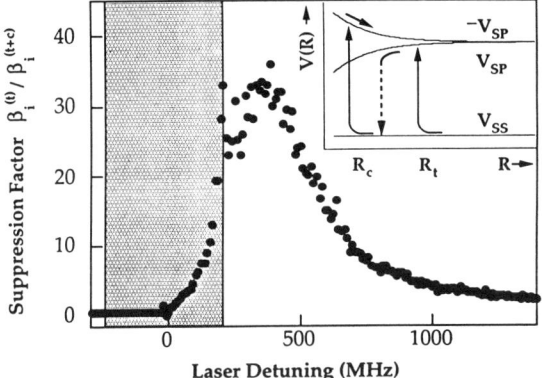

FIG. 11. Suppression factor vs detuning in Xe collisions with the MOT light on. The inset shows the shielding mechanism.

otherwise have prevented higher partial waves from penetrating to near internuclear separation where Penning and associative ionization takes place. The net result is an increase in the ionization rate constant during the MOT on cycle. With the control laser present and tuned to the blue of the trapping transition, the previously accelerated incoming scattering flux is diverted to a repulsive excited state before reaching the ionization region. The colliding atoms are prevented from approaching further, and the result is the observed marked reduction in ionization rate constant.

An experiment similar to the xenon work has been carried out in a krypton MOT by Katori and Shimizu [38]. Analogous to the xenon case, the cooling transition cycles between $5s[\frac{3}{2}]_2 \leftrightarrow 5p[\frac{5}{2}]_3$ and a control laser sweeps from red to blue over the cooling transition from about -600 MHz to $+100$ MHz. Suppression of the ionization rate was again observed with the control laser tuned to the blue. The power in the control laser was only about 4 mW cm^{-2}, and the suppression factor peaked at a blue detuning of about 20 MHz. Although this detuning is very close to the trapping transition, apparently the low power in the suppression laser and time chopping of the MOT cycle and probe cycle permitted ionization rate constant measurements without disruption of the MOT itself. Katori and Shimizu also determined the branching ratio between associative and Penning ionization and found it to be 10%. The power dependence of the suppression effect was not investigated over a very wide range of control laser power. The maximum power density was only 25 mW cm^{-2}, so no firm conclusions concerning strong field effects can be drawn from this report.

Application of weak-field theories to either the Na or Xe experiments, however, does not produce the correct dependence of ionization rate on suppression power density. In particular both quantum and semiclassical theories predict that scattering flux penetration to the ionization region should diminish to zero with increasing suppression or control laser power [39]. Instead the suppression effect saturates, and in the Xe case the suppression factor never increases above about a factor of 5 or 8, depending on polarization of the control laser. Explanation of the strong-field behavior must await development of a strong-field theory, and work in this area is proceeding apace [40].

5. Temperature-Controlled Suppression of Trap Loss

In the foregoing examples of optical suppression, colliding atoms are shielded from each other by interposing an external optical field which couples the atoms to a repulsive potential. Another kind of optical suppression, in which trap-loss processes such as radiative escape (RE) and fine-structure-changing (FS) collisions are avoided by lowering the MOT temperature, has been reported by Wallace et al. [41]. These studies were carried out in an ^{85}Rb MOT. The basic idea is that RE and FS trap–losses

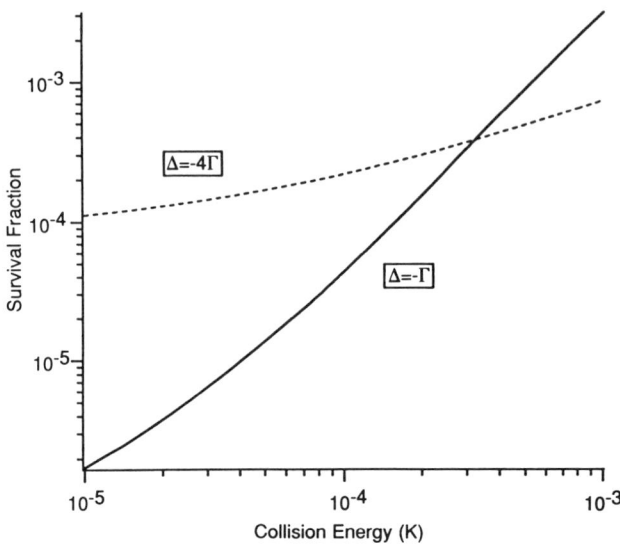

FIG. 12. Probability for excitation to survive to short range ($R = 5$ nm) vs initial collision energy on the 0_u^+ potential curve. The detuning of the initial excitation is $\Delta = -\Gamma$, equivalent to the Condon point distance $R_C = 141$ nm.

always start with the collision partners approaching on a long-range attractive, C_3/R^3, electronically excited molecular state and that both these mechanisms result in a conversion of this electronic energy to nuclear kinetic energy. When this conversion yields kinetic energy equal to twice the MOT trap depth, the atoms excape. The strategy of the Connecticut group is to lower the temperature of the trap, thereby *decreasing* the starting relative collision velocity of the approaching partners and *increasing* the probability of radiative relaxation to the ground state *before* the colliding partners begin to accelerate on the attractive potential and reach the range of internuclear separation in which RE and FS can take place. Figure 12 shows the strong sensitivity of the "survival fraction" (the fraction of population surviving to short range on the excited state) to detuning and to temperature. Lowering the intensity of the MOT lasers lowers the temperature of the trapped atoms. Unfortunately it also lowers the trap depth and the excited state fraction at the same time. However, by normalizing the measured trap loss rate constant β to a "reduced intensity," $I^* = I_t/(10 \, \text{mW/cm}^2)$, the expected linear dependence of β due to simple excited-state population can be divided away to reveal the intensity dependence due to changing trap temperature. Wallace *et al.* therefore defined a "reduced" trap-loss rate constant, $\beta^* = \beta/I^*$, and plotted β^* against trap temperature. The result, shown in Fig. 13, clearly shows a temperature-controlled suppression of β^* at temperatures lower than about $70 \, \mu$K. The use of β^* makes sense only if the dominant trap-loss process is essentially independent of trap depth once the loss

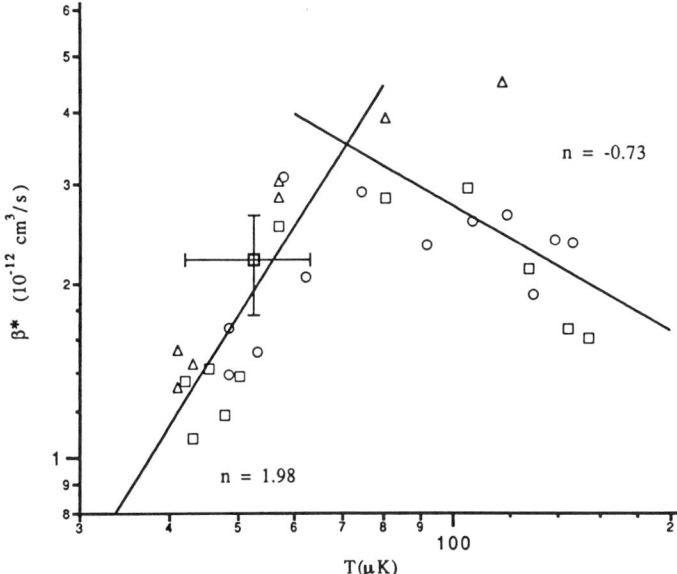

FIG. 13. Temperature dependence of β^*. Note that both axes are logarithmic. The solid curves are best-fit lines to data below and above $T = 70\,\mu K$.

threshold has been reached. This attribute is characteristic of FS, but not of RE. In fact radiative escape would tend to increase β^* with decreasing intensity because as the trap depth decreases more atoms would escape due to this mechanism. The Connecticut group interpret this behavior of β^* as indirect evidence that fine-structure changing collisions dominate the collisional trap-loss phenomena.

IV. Trap-Loss Collisional Processes

Trap-loss collisional studies continue to provide important new insight into various inelastic processes occurring in traps and contributing to limitation of atom densities in traps. A review of ultracold collisions [4] discussed trap-loss processes extensively. This chapter, therefore, restricts discussion to an update of advances since the appearance of the earlier review.

A. Trap Loss in Rubidium

In addition to the temperature-controlled optical suppression study of trap loss in Rb [41], summarized in the preceding section, the Connecticut

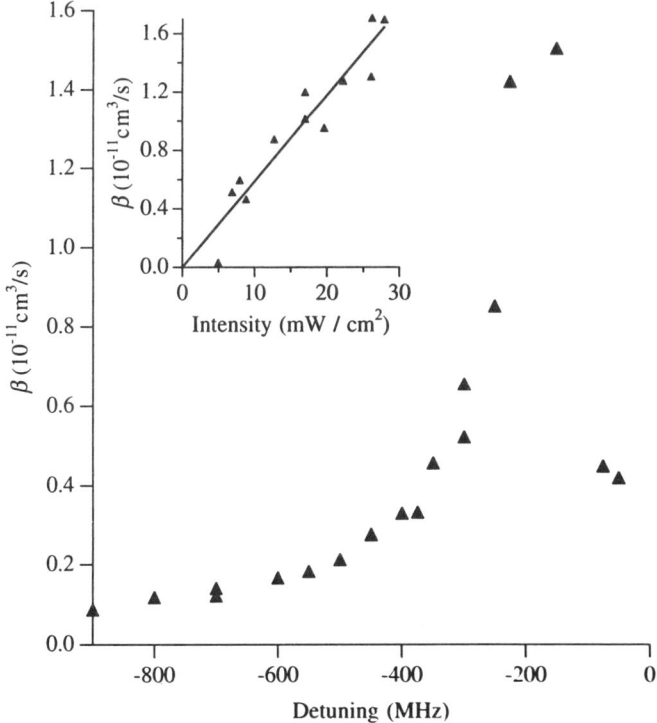

FIG. 14. Trap-loss rate coefficient β as a function of catalysis laser detuning from the ^{85}Rb trapping transition.

group, in collaboration with groups at NIST and Århaus University, has carried out a detailed study of the influence of hyperfine stucture on trap-loss rate constants [42]. The basic idea behind this study is to examine carefully the frequency profile of the collisional trap-loss rate constant β in an Rb MOT as a function of an auxiliary or "catalysis" laser, red tuned over a range of about 500 MHz from the trapping transition in ^{85}Rb. Earlier trap-loss measurements [43] had delineated the overall shape of this profile and interpreted it in terms of the semiclassical Gallagher–Pritchard model [26, 44]. As shown in Fig. 14 β rises sharply to a peak around 170 MHz to the red of the trapping transition and then decreases smoothly out to about 1 GHz detuning. The conventional interpretation of the "Gallagher–Pritchard profile" is that it reflects the trap-loss probability by collisions between atoms, one of which is in the ground state and the other in an excited state coupled to the ground state by an allowed dipole transition moment. At small red detunings the loss probability is small because the excited atom radiatively relaxes to the ground state before FS or RE processes can take place. At large detunings the probability is small because

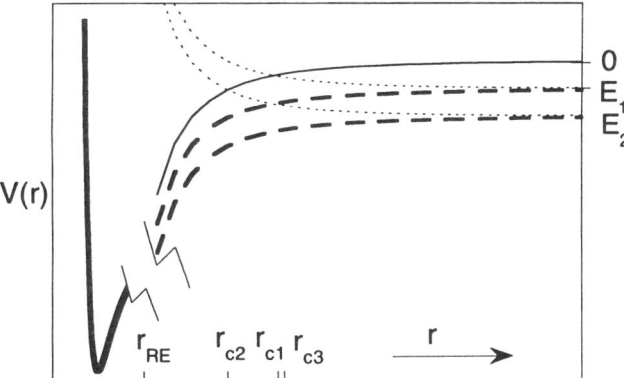

FIG. 15. Hyperfine model of trap loss. Five potentials having a total of three curve crossings at long range are included.

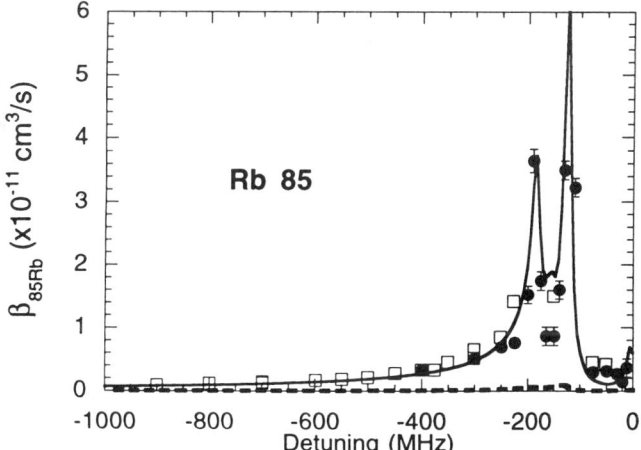

FIG. 16. The heavy solid curve shows the model calculation using 2_u molecular potential parameters and ^{85}Rb parameters where needed. The open squares are taken from reference [43], and the closed circles are from reference [42].

the pair distribution function falls off at short internuclear separation. At some intermediate range, therefore, β peaks where the joint probability between survival and excitation is at a maximum. This simple model, however, ignores any effects due to molecular hyperfine structure, and the contribution of Lett et al. [42] is to demonstrate that hyperfine effects lead to structure in the spectrum of β vs detuning. Figure 15 shows the essential physical picture. The potential curve manifold shows five excited molecular hyperfine levels, three of which are attractive and two repulsive. These attractive and repulsive curves interact at localized Landau–Zener avoided crossings at three different points. The auxiliary scanning laser connects the ground state to the excited molecular state labeled 0, and, as this laser scans through the Landau–Zener regions, the coupling to the repulsive curves perturbs the normal Gallagher–Pritchard profile. Figure 16 shows the model spectrum together with the data from [43] and the new measurements reported by Lett et al. [42]. Close scrutiny of Fig. 14 reveals gaps in the β spectrum at precisely those detunings at which the hyperfine model predicts and the new measurements confirm structure. Lett et al. conclude that, in order to achieve this good agreement between the hyperfine model calculations and the measured β spectrum close to the trapping resonance, the principal excited molecular state must be the 2_u and that, not suprisingly, radiative escape is the principal mechanism. The identification of the 2_u state as the principal actor, however, is remarkable since this state, absent retardation effects, has no dipole transition moment to the ground state. It must be borne in mind that in this region of strong hyperfine mixing the use of Hund's case c coupling to label molecular states breaks down and one expects a strong sensitivity of transition moments to internuclear separation.

B. TRAP LOSS IN LITHIUM

Kawanaka et al. [45] and Ritchie et al. [46] have measured β in an ^{7}Li MOT as a function of trap laser intensity and at several red detunings (2.3Γ to 4.1Γ) from the trapping transition. As usual, trap loss from cold collisions is believed to arise from FS and RE so that $\beta_{\text{total}} = \beta_{\text{FS}} + \beta_{\text{RE}}$. The novel feature of these experiments is that, due to the very small fine structure splitting between the excited $^2P_{3/2}$ and the $^2P_{1/2}$ ^{7}Li levels (only $0.34\,\text{cm}^{-1}$), the kinetic energy release from FS collisions is comparable to the trap depth ($T_{\text{FS}} = E_{\text{FS}}/k = 0.48\text{K}$). The trap depth increases with MOT intensity so that at some critical intensity I_c, FS collisions no longer release enough kinetic energy to escape the trap and therefore no longer contribute to trap loss. Above I_c the trap-loss rate comes principally from radiative escape. Figure 17 shows a typical plot of β vs MOT intensity. The rapid decrease of β with increasing intensity reflects the "freezing out" of the FS collisional loss. The dashed vertical line indicates I_c calculated from a three-dimensional trajectory model, and measurements of β above I_c are believed to arise primarily from RE collisions. Although semiclassical trajectory optical Bloch equation (OBE) calculations predict an increasing β_{RE} with intensity, the Rice group finds that trap-loss rates above I_c vary little. The discrepancy is not well understood, and better insight may have to await the development of a full three-dimensional high-intensity quantum close coupling theory.

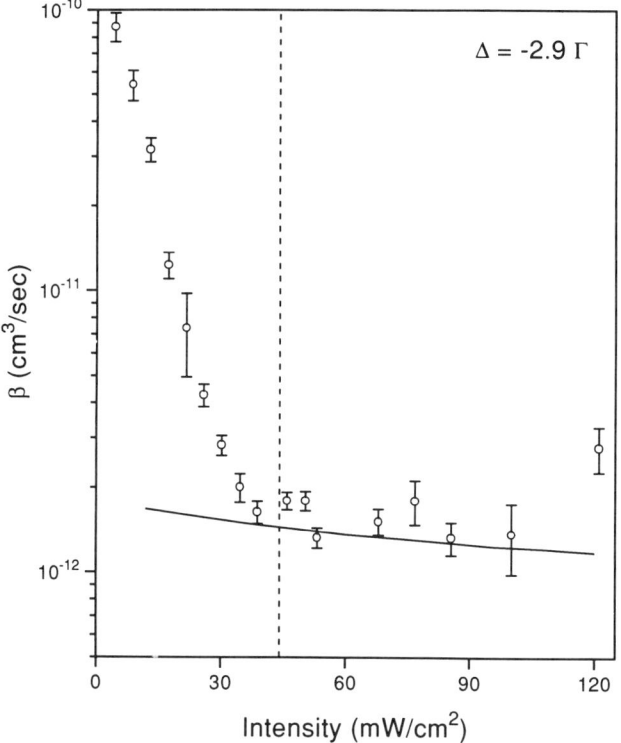

FIG. 17. Two-body trap-loss collision rates for ^{7}Li as a function of total trap laser intensity. The curve shown is for a trap detuning of $\Delta = -2.9\Gamma$. The minimum in the observed loss rate corresponds to the intensity at which the trap depth is sufficient to capture the products of fine structure changing collisions. Above this intensity only radiative escape can contribute to loss. The dotted vertical line in the plot is the calculated model intensity at which FS confinement occurs.

C. Trap Loss in Sodium

The dependence of β on trap light intensity has also been of interest in sodium MOTs. Early work inferred that collisional loss rates are independent of intensity [47], but subsequent work [48] showed β increases with MOT intensity as expected from loss processes arising from collisions between ground and excited atoms. The latest contribution to this line of investigation confirms that intensity dependence and shows the important role played by hyperfine-changing collisions under well-defined experimental conditions [49]. Shang et al. carried out measurements of β on two different MOTs (type I and type II) corresponding to trapping transitions on the $F = 2 \to F' = 3$ and $F = 1 \to F' = 1$ hyperfine levels, respectively.

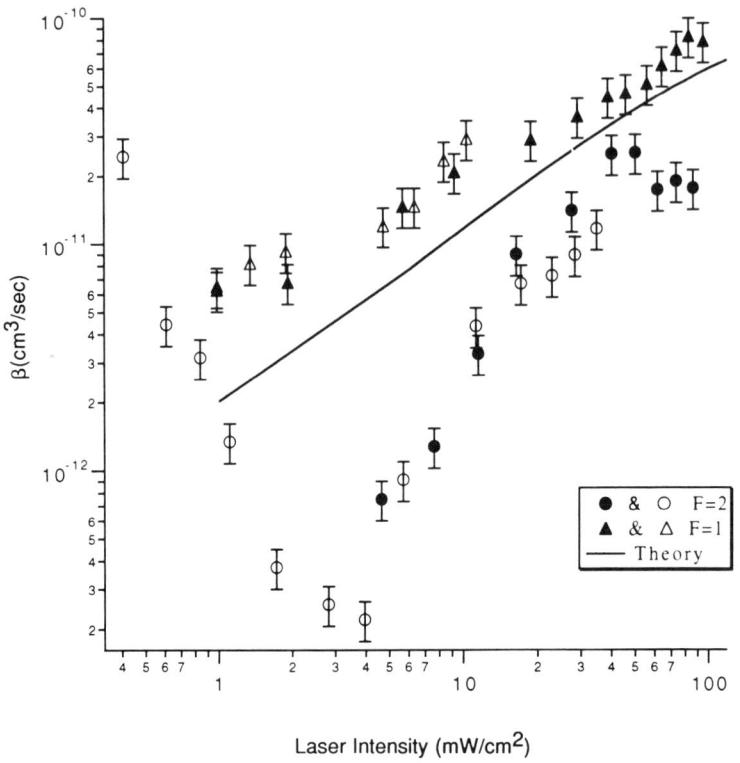

FIG. 18. The cold-collision loss-rate constant β as a function of trapping laser intensity in Na. The closed and open circles are from the $F = 2$ ground state hyperfine level, and they show the marked onset of hyperfine changing collisions at about $4\,\text{mW cm}^{-2}$. The closed and open triangles are for $F = 1$ collisions and exhibit no hyperfine changing effect.

Figure 18 shows β_I and β_II as a function of MOT laser intensity over a wide range from 0.1 to 100 mW cm^{-2}. The rapid increase in β_I below 4 mW cm^{-2} is due to hyperfine changing collisions $F = 2 \to F = 1$ with consequent release of kinetic energy. As expected, β_II shows no evidence of this abrupt, low-intensity increase.

V. Developments in Theory

Both fundamental theoretical approaches and computational techniques have evolved rapidly so as to keep pace with new developments in the laboratory. We summarize here and update the principal results of research in the theory of ultracold collisions since the appearance of an excellent and thorough review [3].

A. COMPLEX POTENTIALS

Julienne et al. [50] and Boesten et al. [51] have each developed quantum mechanical complex potential models of ultracold collisions in order to test semiclassical theories of collisions between excited- and ground-state atoms. The motivation was to calculate accurately measured rates of collisional trap-loss in alkali MOTS and to investigate the possibility of unexpectedly low collision rates at temperatures below 1 mK, a phenomenon termed "quantum suppression" [51]. The basic approach, common to many theoretical treatments of cold collisions, is to factor the probability for trap loss into two parts: one factor $J(E,\ell,\Delta,I;\gamma)$, corresponds to long-range excitation and subsequent survival against spontaneous emission, while the other factor, P_X, reflects the short-range collisional excitation itself. The subscript X denotes either RE or FS. The excitation/survival term is a function of E, the total collision energy; ℓ, the angular momentum; I, the optical field intensity detuned from atomic resonance by Δ; and the excited state decay rate, γ. The overall probability for the trap-loss process is, therefore,

$$P(E,\ell,\Delta,I;\gamma) = P_X(E,\ell) \cdot J(E,\ell,\Delta,I;\gamma). \tag{7}$$

In principle the inclusion of a dissipative term like spontaneous decay means that setting up the problem in terms of a conservative Hamiltonian becomes inappropriate, and an approach such as OBE or quantum Monte Carlo techniques, which take dissipation into account naturally, should be the chosen. However, in [50] the decay rate was kept very small (weak coupling case) so that a three-state quantum close-coupling model could be brought to bear on the problem. The three states are the entrance channel ground state, the molecular state optically excited at long range, and the *probe state* which simulates the final hot-atom channel populated by the RE or FS processes. The trap-loss probability is then written in terms of the S matrix connecting the ground state to the probe state,

$$P(E,\ell,\Delta,I;\gamma) = |S_{gp}(E,\ell,\Delta,I;\gamma)|^2 = P_{XQ}(E,\ell) \cdot J_Q(E,\ell,\Delta,I;\gamma). \tag{8}$$

The subscript Q denotes quantum close coupling, and Julienne et al. [50] compare J_Q to four approximations for J, proposed by Julienne and Vigué [30] J_{JV}; Band and Julienne [52] J_{BJ}; Boesten and Verhaar [53], J_{BV}; and Julienne et al. [50] J_{LZ}. Both J_{BV} and J_{LZ} are variants on the celebrated Landau–Zener formula,

$$J_{LZ} \simeq S_a e^{-A},$$

where S_a is the survival factor and A has the familar form

$$A = \frac{2\pi|V_{ag}(I)|^2}{\hbar\left[\left(\frac{dVa}{dR}\right)_{RC} - \left(\frac{dVg}{dR}\right)_{RC}\right]v_a}. \quad (9)$$

In Eq. (9) $V_{ag}(I)$ is the long-range optical coupling, the bracketed term in the denominator is the difference in slopes between the excited and the ground states, evaluated at the Condon point R_C, and v_a is the local collision velocity at the Condon point. The inner-region collisional excitation $P_{XQ}(E,\ell)$ is insensitive to the intensity and detuning of the outer-region optical excitation and to the decay rate γ or the collision energy E. Therefore, the overall trap-loss probability [Eqs. (7, and 8)] is controlled by the behavior of J. Figure 19 shows the results for the various Js as a function of collision temperature. Two important features of these curves stand out: (1) the close coupling calculation J_Q confirms the quantal calculation J_{BV} and (2) a "quantum suppression" effect indeed exists. With a molecular decay rate, $\gamma = \frac{4}{3}\gamma_A$ (γ_A is the atomic radiative rate), the

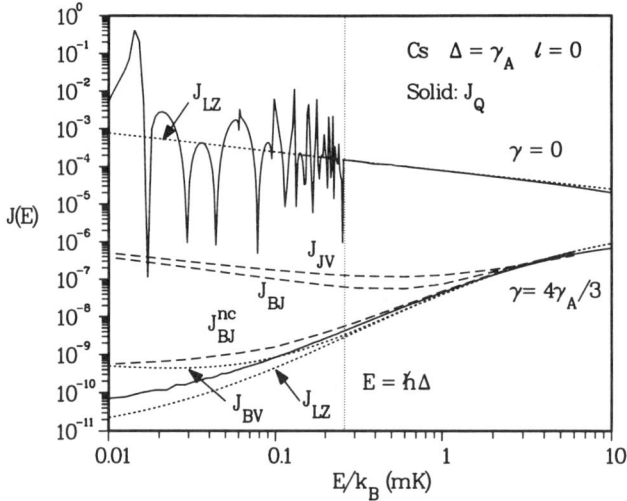

FIG. 19. Comparison of J calculated by quantal and semiclassical methods. The figure shows that for the case of no dissipation ($\gamma = 0$) J_Q and J_{LZ} are many orders of magnitude greater than those resulting when dissipation ($\gamma = 4\gamma_A/3$) is included. The plot also shows two variants of the semiclassical, diabatic OBE calculation, J_{BJ} and J_{BJ}^{nc}. Both J_{BJ} and J_{JV} fail at low temperature, but J_{BJ}^{nc} shows the temperature suppression effect much better. The difference between the two OBE calculations is in the deterimination of the semiclassical trajectories on which the density matrices were propagated. The superscript nc denotes "not corrected" which appears to give better results than the trajectories "corrected" relative to a reference trajectory to take account of the average propagation paths of the coherence terms in the density matrix.

trap-loss rate drops by about one order of magnitude between 1 and 0.1 mK. This effect is completely missed by the semiclassical calculation of Julienne and Vigué [30] and, more surprisingly, by the OBE calculation of Band and Julienne [52]. What is most surprising of all is the *excellent* agreement between the quantal calculations and the humble Landau–Zener formula. This finding indicates that there is nothing particularly exotic about quantum suppression, and it is not neccessary to invoke destructive interference of radial waves [51] to explain it. The dramatic reduction in trap-loss probability below 1 mK can simply be interpreted as poor survival against spontaneous emission when the approaching partners are excited at very long range ($\Delta = \gamma_A$ in Fig. 19) and when they are massive and moving very slowly. To check this interpretation, calculations were carried out at detuning farther to the red and with ligher alkalis. In both cases the temperature-dependent reduction in trap-loss probability was greatly attenuated, consistent with the more prosaic, semiclassical interpretation of quantum suppression. As discussed in Section III.C.5, the careful trap-loss measurements of Refs. [41] in an Rb MOT establish convincing experimental evidence that temperature-controlled suppression of trap-loss collisional rates actually takes place.

B. QUANTUM MONTE CARLO WAVE FUNCTIONS AND THE OPTICAL BLOCH EQUATIONS

Suominen *et al.* [54] have carried out a two-state model study of survival in the very low temperature regime in which the quantal and Landau–Zener results show a dramatic reduction in trap-loss probabilities. Their approach is to apply the Monte Carlo state vector method [55] to wave packet dynamics [56] to obtain a statistical sampling of the density matrix. This approach treats the dissipative spontaneous emission of the problem rigorously while maintaining the calculational burden just within manageable limits. The motivation for the study was to provide a definitive benchmark against which all approximate (but numerically more manageable) theories could be compared and to verify the apparent discovery [50] that the OBE calculations resulted in trap-loss probabilities too high by orders of magnitude at collision energies below 1 mK (at least for the heavy alkalis). Further motivation was to investigate from an entirely different approach the surprising resilience and robustness of the Landau–Zener formula in an energy regime in which a semiclassical formula would be *a priori* suspect.

Although the optical Bloch equations treat spontaneous emission properly in principle, problems arise when the time dependence of these equations maps to a spatial dependence. This mapping converts the time coordinate to a semiclassical reference trajectory. Because the potentials of the ground and excited states are quite different, the state vectors evolving

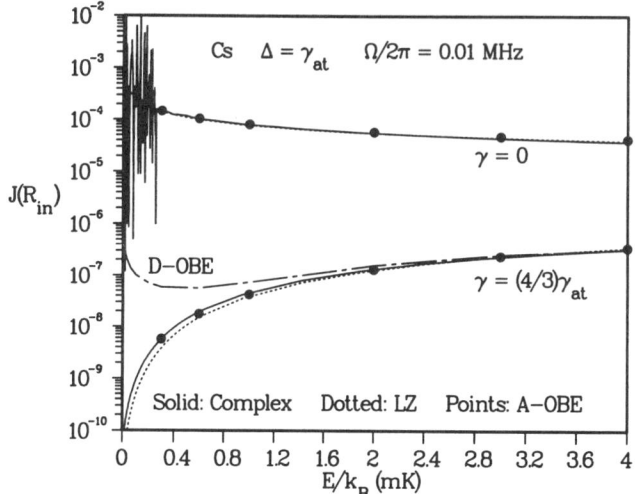

FIG. 20. Comparison of diabatic OBE (D-OBE, dotted–dashed line), complex potential method (solid line), Landau–Zener (dotted line), and adiabatic OBE (A-OBE, heavy black points). Excellent agreement is evident among the last three in the low-temperature region. These model calculations are carried out with Cs parameters, a detuning of $\Delta = \gamma_{at}$, and a molecular spontaneous decay rate, $\gamma = 4\gamma_A/3$.

on these two potentials, in the diabatic representation, actually follow very different trajectories. Correction factors must be applied to the initial reference trajectory to accurately reflect the actual semiclassical path followed by each state vector component. The problem appears to be centered on the way the reference trajectory is corrected and in particular how the correction factors are applied to the field-induced off-diagonal density matrix "coherences." Figure 19 shows that the uncorrected OBE results yield much better results at low energies. Band et al. [57] have recast the OBE calculations on an adiabatic basis in which the field interaction is diagonalized and the interaction coherences are therefore eliminated. This approach gives much better results at low temperature, as can be seen in Fig. 20.

In contrast to the semiclassical OBE approach, the complex-potential, time-independent quantal close coupling formalism appears to give more accurate results at low energy. However, Suominen et al. [54] point out that the complex term in the Hamiltonian may not remain perturbative at very low collision energy which could lead to spurious results. Furthermore, the complex potential formalism is restricted to very weak optical fields and cannot handle spontaneous emission with the possibility of population recycling between ground and excited states. Extension of the time-independent complex potential formalism to the intense field regime is,

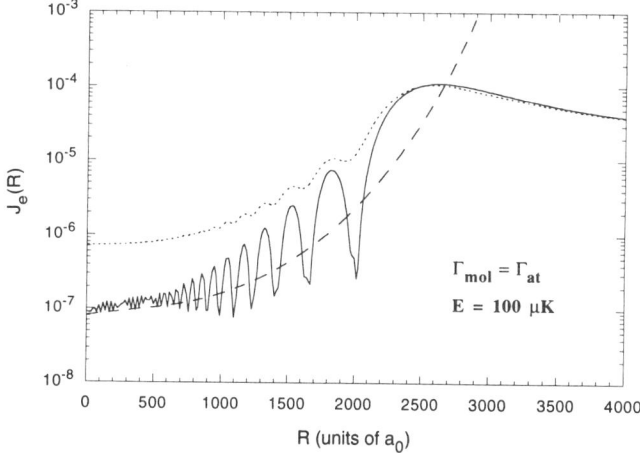

FIG. 21. Comparison of the diabatic OBE (dotted line), wave-packet (solid line), and Landau–Zener (dashed line) results for the weak-field, two-state model. Note that the collision energy is 100 μK and the molecular spontaneous emission rate is taken to be equal to the atomic rate.

however, in active development [40] for an important subset of problems for which these restrictions do not apply. Although computationally more tedious, the wave-packet, quantum Monte Carlo approach is unencumbered by semiclassical or perturbative assumptions and can thus act as a test of their validity.

Figure 21 shows the results of a diabatic OBE, a wave packet, and a Landau–Zener calculation of the excited-state flux survival at a collision energy corresponding to 100 μK. The diabatic OBE clearly fails, and the Landau–Zener appears as the semiclassical average of the wave-packet result. The conclusions of this study are therefore that (1) semiclassical, diabatic OBE calculations cannot be trusted below about 1 mK, (2) the quantal wave-packet calculations confirm previous findings [50, 51] that trap-loss collision probabilities drop off dramatically with temperature, and (3) the Landau–Zener calculations are in excellent accord with the oscillation-averaged fully quantal results. This latter conclusion has also been reached by Boesten and Vehaar [57].

It should be borne in mind, however, that none of these theoretical studies take into account the complications of realistic molecular hyperfine states or go beyond weak-field optical coupling. These areas are the focus of continuing efforts. In particular Holland et al. [58] have extended the wave-packet quantum Monte Carlo technique to consider radiative heating of an ensemble of two-level Cs atoms; i.e., the reduced mass of the two-body collision was chosen to be that of Cs and the excited state was chosen to have the C_3 parameter of the attractive 0_u^+ state. An important advantage of wave-packet quantum Monte Carlo is that the results can reveal the difference between initial and final momentum distributions even if the atoms are not lost from the trap. In fact a major finding of this study is that trapped atom ensembles subject to optical excitation, at densities typical of MOTs, pick up significant translational heating through bimolecular collisions, *even though trap loss does not occur*. Although it is too early to draw definite conclusions, significant heating without trap loss may have an important bearing on the limiting temperatures and densities attainable in optical traps.

VI. Future Directions

The study of ultracold collisions continues to evolve rapidly. New results in optical shielding and suppression reveal a polarization dependence increasing with control-field intensity. Full three-dimensional close-coupling calculations on spinless model potentials also show marked polarization effects. Spatial anisotropy of optical supression will soon be investigated in cold, compressed atomic beams. Modification of inelastic processes through manipulation of polarization and intensity appears to offer new avenues to "coherent control." Proposals for new routes to molecule formation *via* photoassociation with internal state specificity will be explored, and the race to cool and confine molecules is on. If the recent past is any guide to the near future, we may anticipate novel, exciting, and unexpected developments in the continuing exploration of ultracold collision dynamics.

References

[1] H. R. Thorsheim, J. Weiner, and P. S. Julienne, *Phys. Rev. Lett.* **58**, 2420 (1987).
[2] J. Vigué, *Phys. Rev. A* **34**, 4476 (1986).
[3] P. S. Julienne, A. Smith, and K. Burnett, in *Advances in Atomic, Molecular, and Optical Physics* (D. R. Bates and B. Bederson, eds.), Vol. 30, p. 141, Academic Press, San Diego, 1993.
[4] T. Walker and P. Feng, in *Advances in Atomic, Molecular, and Optical Physics* (B. Bederson and H. Walther, eds.) Vol. 34, Academic Press, San Diego, 1994.
[5] See, e.g., Special Issue on Laser Cooling and Trapping, *J. Opt. Soc. Am. B* **6**, 2020 (1989).
[6] C. J. Foot, *Contemp. Phys.* **32**, 369 (1991).
[7] H. T. C. Stoof, *Phys. Rev. A* **49**, 3824 (1994)
[8] A. J. Moerdijk and B. J. Verhaar, *Phys. Rev. Lett.* **73**, 518 (1994).
[9] E. Raab, M. Prentiss, A. Cable, S. Chu, and D. Pritchard, *Phys. Rev. Lett.* **59**, 263 (1987); C. Monroe, W. Swann, H. Robinson, and C. Wieman, *Phys. Rev. Lett.* **65**, 1571 (1990).
[10] J. M. Doyle, J. C. Sandberg, I. A Yu, C. L. Cesar, D. Kleppner, and T. J. Greytak, *Phys. Rev. Lett.* **67**, 603 (1991), and references therein.
[11] K. Gibble and S. Chu, *Phys. Rev. Lett.* **70**, 1771 (1993).

[12] B. J. Verhaar, K. Gibble, and S. Chu, *Phys. Rev. A* **48**, R3429 (1993).
[13] W. T. Zemke and W. C. Stwalley, *J. Phys. Chem.* **93**, 2053 (1993).
[14] A. J. Moerdijk, W. C Stwalley, R. G. Hulet, and B. J. Verhaar, *Phys. Rev. Lett.* **72**, 40 (1994).
[15] E. R. I. Abraham, W. I. McAlexander, C. A. Sackett, and R. G. Hulet, *Phys. Rev. Lett.* **74**, 1315 (1995).
[16] See, for example, *Laser Induced Processes in Molecules*, Vol. 6, Springer Series in Chemical Physics, edited by K. L. Kompa and S. D. Smith (Springer-Verlag, Berlin, 1979).
[17] For early theory see, e.g., L. I. Gudzenko and S. I. Yakovlenko, *Zh. Eksp. Teor. Fiz.* **62**, 1686 (1972) [*Sov. Phys.–JETP (Engl. Transl.)* **35**, 877 (1972)] and V. S. Lisitsa and S. I. Yakovlenko, *Zh. Eksp. Teor. Fiz.* **66**, 1550 (1974) [*Sov. Phys.–JETP (Engl. Trans.)* **39**, 759 (1974)].
[18] See, e.g., R. W. Falcone, W. R. Green, J. C. White, J. F. Young, and S. E. Harris, *Phys. Rev. A* **15**, 1333 (1977); P. L. Cahuzac and P. E. Toschek, *Phys. Rev. Lett.* **40**, 1087 (1978); P. Polak-Dingels, R. Bonanno, J. Keller, J. Weiner, and J.-C. Gauthier, *Phys. Rev. A* **25**, 2539 (1982), and references therein.
[19] L. P. Ratliff, M. E. Wagshul, P. D. Lett, S. L. Rolston, and W. D. Phillips, *J. Phys. Chem.* **101**, 2638 (1994).
[20] W. Ketterle, K. B. Davis, J. A. Joffe, A. Martin, and D. Pritchard, *Phys. Rev. Lett.* **70**, 2253 (1993).
[21] R. Napolitano, J. Weiner, C. J. Williams, and P. S. Julienne, *Phys. Rev. Lett.* **73**, 1352 (1994).
[22] V. S. Bagnato, J. Weiner, P. S. Julienne, and C. J. Williams, *Laser Phys.* **4**, 1062 (1994), and references therein.
[23] R. A. Cline, J. D. Miller, and D. J. Heinzen, *Phys. Rev. Lett.* **73**, 632 (1994).
[24] W. I. McAlexander, E. R. I. Abraham, N. W. M. Ritchie, C. J. Williams H. T. C. Stoof, and R. G. Hulet, *Phys. Rev. A* **51**, R871 (1995).
[25] A. Gallagher and T. Holstein, *Phys. Rev. A* **16**, 2413 (1977), and references therein.
[26] A. Gallagher, *Phys. Rev. A* **44**, 4249 (1991).
[27] J. D. Miller, R. A. Cline, and D. J. Heinzen, *Phys. Rev. A* **47**, R4567 (1993).
[28] R. A. Cline, J. D. Miller, and D. J. Heinzen, *Phys. Rev. Lett.* **71**, 2204 (1993).
[29] R. A. Cline, J. D. Miller, and D. J. Heinzen, *Phys. Rev. Lett.* **73**, 632 (1994).
[30] P. S. Julienne and J. Vigué, *Phys. Rev. A* **44**, 4464 (1991).
[31] R. J. LeRoy and R. B. Bernstein, *J. Phys. Chem.* **52**, 3869 (1970).
[32] W. C. Stwalley, *Chem. Phys. Lett.* **6**, 241 (1970).
[33] P. D. Lett, W. D. Phillips, L. P. Ratliff, S. L. Rolston, and M. E. Wagshull, *Phys. Rev. Lett.* **71**, 2200 (1993).
[34] S. Bali, D. Hoffmann, and T. Walker, *Europhys. Lett.* **27**, 273 (1994).
[35] L. Marcassa, S. Muniz, E. de Queiroz, S. Zilio, V. Bagnato, J. Weiner, P. S. Julienne, and K.-A. Suominen, *Phys. Rev. Lett.* **73**, 1911 (1994).
[36] L. Marcassa, R. Horowicz, S. Zilio, V. Bagnato, and J. Weiner, *Phys. Rev. A* **51**, in press (1995).
[37] M. Walhout, U. Sterr, C. Orzel, M. Hoogerland, and S. L. Rolston, *Phys. Rev. Lett.* **74**, 506 (1995).
[38] H. Katori and F. Shimizu, *Phys. Rev. Lett.* **73**, 2555 (1994).
[39] K.-A. Suominen, M. J. Holland, K. Burnett, and P. Julienne, *Phys. Rev. A* **51**, in press (1995).
[40] R. Napolitano, J. Weiner, and P. S. Julienne, unpublished results.
[41] C. D. Wallace, V. Sanchez-Villicana, T. P. Dinneen, and P. L. Gould, *Phys. Rev. Lett.* **74**, 1087 (1995).
[42] P. D. Lett, K. Mølmer, S. D. Gensemer, K. Y. N. Tan, A. Kumarakrishnan, C. D. Wallace, and P. L. Gould, *J. Phys. B* **28**, 65 (1995).

[43] D. Hoffmann, P. Feng, R. S. Williamson, and T. Walker, *Phys. Rev. Lett.* **69**, 753 (1992).
[44] A. Gallagher and D. Pritchard, *Phys. Rev. Lett.* **63**, 957 (1989).
[45] J. Kawanaka, K. Shimizu, H. Takuma, and F. Shimizu, *Phys. Rev. A* **48**, R883 (1993).
[46] N. W. M. Ritchie, E. R. I. Abraham, Y. Y. Xiao, C. C. Bradley, and R. G. Hulet, *Phys. Rev. A* **51**, R890 (1995).
[47] M. Prentiss *et al.*, *Opt. Lett.* **13**, 452 (1988).
[48] L. Marcassa, V. Bagnato, Y. Wang, C. Tsao, J. Weiner, O. Dulien, Y. B. Band, and P. S. Julienne, *Phys. Rev. A* **47**, R4563 (1993).
[49] S.-Q. Shang, Z.-T. Lu, and J. Freedman, *Phys. Rev. A* **50**, R4449 (1994).
[50] P. S. Julienne, K.-A. Suominen, and Y. Band, *Phys. Rev. A* **49**, 3890 (1995).
[51] H. M. J. M. Boesten, B. J. Verhaar, and E. Tiesinga, *Phys. Rev. A* **48**, 1428 (1993).
[52] Y. B. Band and P. S. Julienne, *Phys. Rev. A* **46**, 330 (1992).
[53] H. M. J. M. Boesten and B. J. Verhaar, *Phys. Rev. A* **49**, 4240 (1994).
[54] K.-A. Suominen, M. J. Holland, K. Burnett, and P. S. Julienne, *Phys. Rev. A* **49**, 3897 (1994).
[55] Y. Castin, K. Mølmer, and J. Dalibard, *J. Opt. Soc. Am. B* **10**, 524 (1993).
[56] W. K. Lai, K.-A. Suominen, B. M. Garraway, and S. Stenholm, *Phys. Rev. A* **47**, 4779 (1993).
[57] Y. B. Band, I. Tuvi, K.-A. Suominen, K. Burnett, and P. S. Julienne, *Phys. Rev. A* **50**, R2826 (1994).
[58] M. J. Holland, K.-A. Suominen, and K. Burnett, *Phys. Rev. A* **50**, 1513 (1994).

IONIZATION DYNAMICS IN STRONG LASER FIELDS

L. F. DiMAURO* and P. AGOSTINI[†]

**Chemistry Department, Brookhaven National Laboratory, Upton, New York;
and [†]Service des Photons, Atomes et Molécules, CE Saclay 91191 Gif Sur Yvette, France*

I. Introduction	79
II. The Bound–Free Step	82
A. Transient Resonances and Excited State Population Trapping	82
B. Evolution from Multiphoton to Tunneling	89
III. The Free–Free Step	92
A. Electron Energy Distributions	92
B. Effects of Rescattering on the Photoelectron Energy and Momentum	97
IV. Strong-Field Double Ionization	108
A. "Direct" Channel in Multiphoton Double Ionization	108
B. "Nonsequential" Channel in Tunneling Double Ionization	111
C. Nonsequential Rate	114
V. Conclusion	116
References	118

I. Introduction

The behavior of atoms submitted to intense electromagnetic fields is a subject of wide interest and active research. Much of the knowledge in this field is provided by studying the ionization dynamics, the energy and momentum of the photoelectrons, and the spectrum of emitted photons. *Multiphoton ionization* (MPI) is the process by which an atom is ionized by simultaneous absorption of several photons. The number of photons absorbed is in general even larger than the minimum required by energy conservation. The excess energy can be transferred to the photoelectron whose energy spectrum is composed of a series of lines separated by the photon energy. This process is known as *above-threshold ionization* (ATI). Alternately, the electron can be recaptured, emitting a series of energetic photons at odd harmonic frequencies of the driving field, dubbed *optical harmonic generation* (OHG).

Much of the progress in this field, over the past 20 years, has been coupled to the advances in intense laser technology. Developments in kilohertz repetition rate laser sources with high pulse energy and excellent beam quality have been made. In addition, techniques capable of producing usable short pulses of soft X rays through either OHG or laser plasma have been improved. These technological advances result in the ability not only to subject atoms to electromagnetic field strengths of the order of an atomic unit[1] but also (1) to make measurements with unprecedented dynamic range and precision and (2) to explore entirely new experimental situations. As a consequence, unsuspected aspects of MPI have been uncovered; new processes, observed; and critical tests of theoretical models, achieved during the last two years. These advances have provided the impetus for viewing all strong-field phenomena in a consistent and unified manner. It is this aspect which will be the focus of this chapter.

During the course of the ionization process, the electron gets displaced from the nucleus but still interacts strongly with the field (Kupersztych, 1987). This rational provides a view of strong-field ionization as resulting from two steps: (1) a bound–free step whose characteristics evolve with the field strength from multiphoton to tunneling and (2) a free–free step in which the interaction with the electromagnetic field dominates. We will successively deal with these two aspects.

During the first step, at intermediate intensities, the discrete spectrum of the atom plays a very important role even in *a priori* nonresonant situations. The reason is that the atomic spectrum is strongly distorted by AC-Stark shifts during the light pulse, and transient resonances virtually always occur. Simultaneous Stark shift of the continuum limit induces a change in the photoelectron energy, and the spectrum actually mimics the intensity dependence of the ionization rate (Freeman *et al.*, 1992). During the last two years, the dynamical behavior of transient resonances has been investigated in more details by several groups. These studies were triggered by the surprising demonstration that real population was transferred to the excited state during the transient resonance and could actually survive the whole exciting pulse (deBoer and Muller, 1992). Subsequent experimental and theoretical studies have addressed a number of questions regarding intensity-tuned multiphoton resonances, including the adiabaticity of the crossing, the degree of trapping, and the space–time dependence of ionization. This work is reviewed in Section II.A.

For laser field strengths approaching the coulomb field binding of the electron, the sharp ATI structures of the electron spectrum progressively disappear, becoming a continuous distribution. This evolution (Mevel *et al.*, 1993) is loosely correlated to the values of the adiabaticity parameter (Keldysh, 1964), signifying a transition from the multiphoton to the tunneling regime of ionization (Section II.B). However, no clear-cut separation exists between the two regimes, and in most experiments the ionization rate

---

[1]In general, atomic units will be used in this chapter. We recall that the atomic unit of intensity is 3.5×10^{16} W/cm^2.

is best characterized as an admixture of multiphoton and tunneling components. A major breakthrough in the theoretical approaches has been the development of powerful numerical methods to directly solve the time-dependent Schrödinger equation (TDSE) using a single-active electron (SAE) approximation. Such methods have proven to be very efficient in accurately describing strong field ionization of rare gas atoms (Schafer et al., 1993; Yang et al., 1993). SAE analysis of high-precision measurements of the intensity dependence of ion yields allows (Walker et al., 1994) determination of the fraction of tunneling versus multiphoton and is discussed in Section II.B.

The simplest treatment of the free–free part of the problem (Section III.A.1) is to neglect the coulomb interaction and treat the electron motion by classical mechanics (van Linden van de Heuvell and Muller, 1988). It is actually this physics that underlies KFR (Keldysh, 1964; Faisal, 1973; Reiss, 1980, 1987) quantum theories, in which the final state is described by a Volkov state (Becker et al., 1994). It is justified in the tunneling regime, since the electron wave packet is created in the first step, displaced by a few Bohr radii from the nucleus, and consequently interacts essentially with the electromagnetic field.

The classical motion in the field is separable into an oscillatory motion, with an associated average kinetic energy called the ponderomotive energy[2] and henceforth denoted U_p, and a drift motion which depends upon the phase of the field at the instant the electron is placed in the continuum. The subsequent dynamics in the continuum are intimately connected to the initial conditions produced in the first step. We first examine a situation in which electrons are dropped at randomly distributed phases with a large initial velocity (Section III.A.2). Such conditions exist in a new process called laser-assisted Auger decay (Schins et al., 1994). The measured electron energy distribution gives excellent agreement with the classical prediction. A second scenario is presented for the strong-field ionization of helium, in which the initial conditions result in a narrow phase distribution and small initial velocities (Section III.A.3), characteristic of tunnel ionization (TI) which was first investigated by Corkum et al. (1989). It can be shown that under such circumstances, the maximum cycle averaged kinetic energy is $2U_p$. Again, the qualitative agreement with the classical model is excellent, with 99% of the electrons bound between zero and $2U_p$ energy.

Despite its success in predicting some features of the electron energy spectrum, this simple model cannot account for all the experimental facts. For instance, high-sensitivity measurements (Section III.B.2) show that some fraction of the electron distribution extends beyond $2U_p$, and the degree depends upon the ionization regime. This was observed in MPI of rare gases (Schafer et al., 1993; Paulus et al., 1994a; Walker et al., 1994). The

---

[2]The ponderomotive energy U_p is defined in atomic units as $I/4\omega^2$, where I and ω are the laser intensity and frequency, respectively. For 780-nm photons, $U_p = 5.7$ eV at 10^{14} W/cm^2.

existence of fast electrons can be explained by the classical model if the oscillating electron *rescatters* with the ionic core (Section III.B.1) (Schafer *et al.*, 1993; Corkum, 1993). Quantum mechanically, this corresponds to a multiphoton inverse Bremsstrahlung process. Numerical calculations using realistic 3-D (Schafer *et al.*, 1994) or 1-D model potentials (Paulus *et al.*, 1994) and analytical (Lewenstein *et al.*, 1995; Becker *et al.*, 1994) solutions have confirmed the rescattering mechanism. However, experiments (Walker *et al.*, 1994) show that the fraction of electrons with energies in excess of $2U_p$ decreases as ionization approaches the "pure" tunneling regime, suggesting that TI results in ineffective rescattering (Section III.B.3). Dramatic evidence for the rescattering mechanism (Section III.B.4) is observed as *scattering rings* in the angular distribution of high-order ATI electrons (Yang *et al.*, 1993; Paulus *et al.*, 1994b). This strongly intensity-dependent off-axis electron emission appears in a narrow energy range around $8U_p$ and suggests the importance of backscattering and quantum interference.

Two-electron ejection (double ionization) can be viewed as another consequence of rescattering (Corkum, 1993) via $e-2e$ collisional ionization. Experiments on helium and neon have established the existence of a nonsequential process for double ionization in strong fields. The nonsequential production has a conspicuously strong polarization dependence, easily accounted for by the classical view. However, measurements using *kilohertz* technology shows gross inconsistencies with the rescattering mechanism. They can be attributed to the wave packet character of the electron which is absent from the classical view. Consequently, the free evolution of the wave packet (electron) prior to rescattering results in significant spreading, reducing the effectiveness of the $e-2e$ event. The current status of strong-field double ionization will be addressed in Section IV.

Due to the limited available space, the authors chose to focus on some of the latest strong-field experiments and the basic theoretical approaches. The reader is referred to *Atoms in Intense Laser Fields* (Gavrila, 1992) and *Super-Intense Laser-Atom Physics* (Piraux *et al.*, 1993) for earlier works and in-depth theory. Furthermore, some other aspects of strong-field ionization, like adiabatic stabilization (deBoer *et al.*, 1993a) and multiphoton detachment of negative ions (Davidson *et al.*, 1992; Stapelfeldt *et al.*, 1994), could not be included in this chapter.

II. The Bound–Free Step

A. Transient Resonances and Excited State Population Trapping

If no harmonics of the laser frequency coincide with the energy of an atomic state, MPI is nonresonant. However, under intense irradiation, nonresonant MPI may become resonant at some intensity due to AC-Stark shifts in the atomic energies. Stark-induced transient resonances were revealed by photoelectron spectroscopy using very short laser pulses (Freeman et al., 1987). MPI theory incorporating AC-Stark-shifted atomic levels easily predicts such resonances (Keldysh, 1964; Gontier et al., 1975), but they normally remain unobservable in the total rate. Analogous to normal spectroscopy, in which resonances are observed by tuning the frequency, sharp peaks appear by tuning the *light intensity* around the value which adjusts the shifted energy of a state into resonance with an integer number of photons. So in a sense, *the atom is tuned, not the light source*. For conditions defined by a strongly focused pulse, the resonances are experimentally lost since the signal is averaged over the space–time intensity distribution. However, for short-pulse photoelectron energy spectroscopy (PES) the resonances are observable since (1) the ionization limit is increased by U_p, inducing an equal but opposite shift of the electron energy, and (2) the shift is *not* compensated by the ponderomotive acceleration, as is the case for long pulses. Consequently, the intensity dependence of the ionization rate is mapped onto its energy dependence, i.e., the PES spectrum. Several resonances may be crossed during the pulse, each giving rise to an increase in the number of electrons at the prescribed energy. Thus, an ATI energy spectrum consists of groups of sharp structures (each corresponding to an atomic state) repeated with a period equal to the photon energy (Freeman et al., 1992). The widths of the structures has important implications about the dynamics of the process. Neglecting the small Stark shift of the ground state, the photoelectron energy depends on the intensity I through (in atomic units)

$$E(I) = N\omega - E_0 - U_p(I), \qquad (1)$$

where E_0 is the field free ionization energy. Thus, narrow structures must have been produced over a small intensity range (small spread in U_p). The widely accepted scenario is that ionization must occur *at the resonance intensity* I_r. Introducing the static detuning for an *m*-photon resonance, $\Delta_0 = E_r - E_g - m\omega$ (where E_g and E_r are the unperturbed ground and resonant state, respectively), and the linear AC-Stark shifts coefficients α_g and α_r, the resonance intensity is

$$I_r = \Delta_0/(\alpha_g + \alpha_r). \qquad (2)$$

For a pulse with a peak intensity of $I_0 > I_r$, $I(t) \approx I_r$ for a very *short* time during the rising and falling edges. The uncertainty principle would argue for correspondingly *broad* energy structures, at variance with the experiments. However, by assuming that sharp peaks are produced in spatial regions of the beam in which *the peak intensity of the pulse is equal to I_r*, the resonance time can be on the order of the pulse duration. Since in a gaussian beam this region is a thin shell of resonant ionization, this scenario is referred to as the "shell" model (McIlrath et al., 1989). The model's validity was questioned by the observation that real excited state population does survive the exciting pulse (deBoer and Muller, 1992). The experiment used a low-intensity, large-fluence probe pulse to interrogate the interaction volume after the strong ionizing pulse. The delayed PES showed *new peaks* which were assignable to *populated excited states*. Contrary to the shell model, this observation implied that ionization occurs over a *wide* range of intensity, defined by the time at which I_r is first reached on the rising edge and the remainder of the pulse. An alternate scenario in which the sharp peaks were recovered by recognizing that Rydberg states shift ponderomotively was proposed. More precisely, the electron energy is defined as the difference in shift between the resonant state and the continuum limit and reads (assuming a *p*-photon ionization from this state)

$$E(I) = p\omega - E_0 - U_p(I) + E_r + \alpha_r I. \tag{3}$$

which is *independent* of intensity, if $\alpha_r I \sim U_p(I)$. Another consequence of this explanation is that ionization occurs over the whole ionization volume rather than in shells.

Both theories and experiments reviewed in the following sections were stimulated by these apparently conflicting models. The current status is that the shell model is essentially correct, but it is now known that population trapping is possible. The degree of population trapping will depend on the excited state ionization rate and the temporal shape of the pulse.

1. Dynamics of Transient Stark-Induced Resonances

Resonant MPI is often reduced to a two-level problem treated within the dipole and rotating-wave approximations. The evolution of the system under a *constant* perturbation can be treated by a resolvent operator approach (Gontier and Trahin, 1992) or Floquet theory (Potvliege and Shakeshaft, 1992). Within the dressed-atom picture, the ground state dressed by *m* photons and the resonant state undergo an avoided crossing at the resonance intensity I_r. The minimum distance, V_m, between the crossing levels is proportional to the generalized Rabi frequency $\Omega(t) = \Omega_m I(t)^{m/2}$ coupling the two states. Some insight into the time evolution of the system can result from this static description (Agostini and DiMauro, 1993), but only a full time-dependent treatment can provide the detailed dynamics.

As a first approximation, the Landau–Zener theory of time-dependent level crossings provides a simple method of calculating the amount of population transferred from one state to the other. For instance, the probability of remaining in the ground state is given by (Story *et al.*, 1994b)

$$P_g = \exp\left[\frac{-2\pi V_m^2}{dW/dt}\right], \tag{4}$$

where W is the time-dependent energy difference between the two dressed states. For a very slow crossing dW/dt is small and $P_g \approx 0$; thus all population is adiabatically transferred into the excited state. This is the essence of the spatial shell model (McIlrath et al., 1989) since the intensity at the peak of the pulse is varying slowly in time, causing efficient transfer and ionization. Elsewhere in the beam, the crossing is traversed diabatically and the population remains in the ground state.

A time-dependent description is afforded by a density matrix approach (Lambropoulos and Tang, 1992) or by solution of the time-dependent Schrödinger equation (Gibson et al., 1994):

$$i \frac{d}{dt} \psi(t) = H\psi$$

with

$$H(t) = \begin{pmatrix} 0 & \Omega(t) \\ \Omega(t) & \Delta(t) - i\Gamma(t) \end{pmatrix}$$

and

$$\psi(t) = \begin{pmatrix} g(t) \\ a(t) \end{pmatrix}$$

$$\Omega(t) = \Omega_m |\mathscr{E}(t)|^m$$

$$\Delta(t) = \Delta_0 + \alpha |\mathscr{E}(t)|^2 \tag{5}$$

$$\Gamma(t) = \Gamma_0 |\mathscr{E}(t)|^2,$$

which is numerically integrated for a gaussian pulse, $\mathscr{E}(t) = \mathscr{E}_0 e^{-t^2/\tau^2}$ with a peak field, $\mathscr{E}_0$. The other parameters in the problem are the ionization cross-section of the excited state Γ_0, the intensity-dependent ionization rate $\Gamma(t) = \Gamma_0 I(t)$, and the detuning $\Delta(t) = \Delta_0 - (\alpha_r + \alpha_g)I(t)$.

Typical space–time dependences for the excited and ionized fractions are displayed in Fig. 1 along with a section of the beam showing the $4I_r$, $2I_r$, I_r, and $0.8I_r$ intensity zones. The computations result in the following conclusions for typical atomic parameters (see Table I). (1) No population is transferred until I_r is reached on the rising edge of the pulse. (2) Beyond I_r, the populations evolve under the combined influence of the nonresonant rate, the Rabi coupling (which causes the oscillations), and the ionization damping at a rate $\Gamma(t)$. (3) I_r is again obtained on the falling edge, where *additional* population is again efficiently transferred, unless the ground state is depleted. After this point, the population remains trapped in the excited state and survives the pulse. (4) Except for the zone $I(r, z, 0) = I_r$, *ionization occurs over a wide range of intensity*. However, the influence of each zone

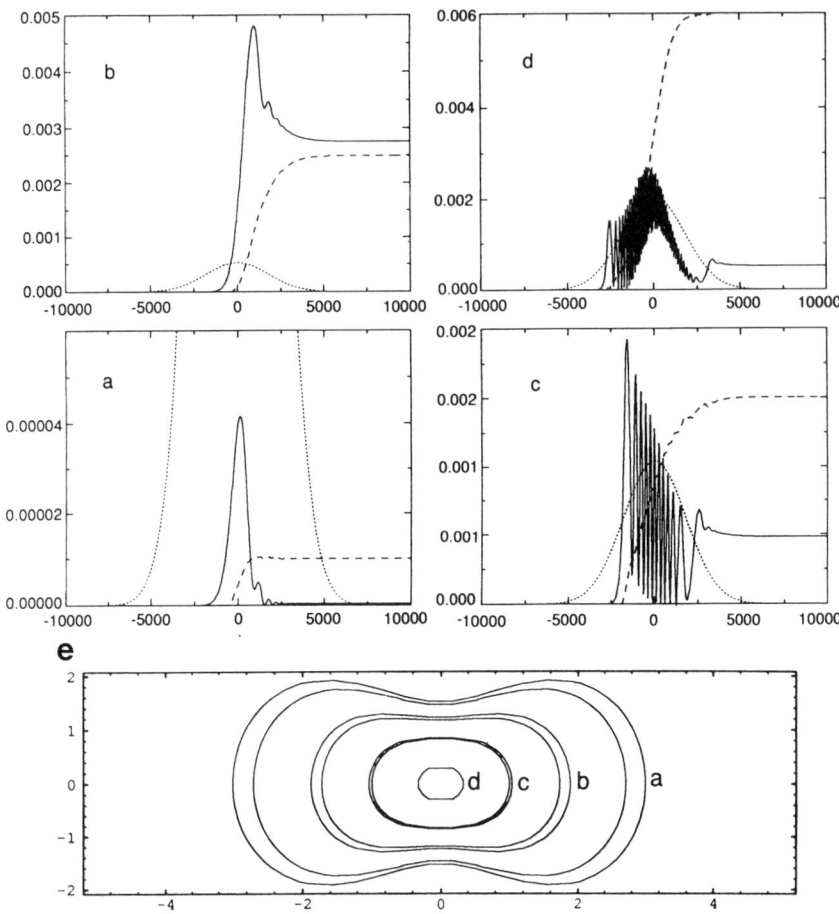

FIG. 1. Typical time dependence of the excited state (solid line) and ionized fractions (dashed line) and intensity (dotted line) for different peak intensities of a gaussian pulse: (a) $0.8I_r$, (b) I_r, (c) $2I_r$, and (d) $4I_r$. The inset (e) shows the corresponding shells in a gaussian beam with $I(0,0,0) = 4I_r$. The atomic parameters are (atomic units): $\Gamma_0 = 0.3$ and $\Omega_0 = 3.85 \times 10^5$.

must be weighted by a volume factor which can be shown to be (Gibson et al., 1994)

$$\frac{2I + I_0}{I^2} \sqrt{\frac{I_0 - I}{I}}. \tag{6}$$

The weighting essentially eliminates contributions from regions in which $I > I_r$. Zones in which $I < I_r$ give negligible contributions since the rate is small. Consequently, the averaged PES distributions are extremely peaked around one intensity (I_r) and energy. (5) This *does not prevent population*

TABLE I

SIX-PHOTON RABI COEFFICIENTS Ω_n AND
IONIZATION CROSS-SECTION Γ_n FOR
HYDROGEN nf STATES

n	Ω_n	Γ_n
4	3.990×10^4	0.966
5	2.899×10^4	0.605
6	1.669×10^4	0.378

trapping since it depends critically on the ionization cross-section Γ_0, as well as on the pulse duration τ. As intuitively expected, it decreases with long pulses or large cross-sections, but, interestingly, both dependencies have a maximum. This behavior is well known in the context of molecular predissociation but was not apparently noticed in the present context.

2. Experiments

The above analysis has been confirmed by a series of experiments described below.

a. Nonponderomotive Shifts. Equation (3) implies that a narrow PES peak must have resulted from ponderomotively shifted state. However, the AC-Stark shifts of deeply bound states do not satisfy this condition. Experimental measurements (Gibson et al., 1992) showed that such states *do give narrow peaks* in the PES, in strong support of the shell model.

b. Large Ionization Cross-Section. The resonance intensity and the corresponding ionization rate depend on the static detuning Δ_0 [Eq. (2)]. For the six-photon resonances in xenon, Muller and co-workers found that by varying the laser wavelength toward larger static detuning less population trapping occurred.

c. Diabatic vs Adiabatic Crossings. Except at the top of the pulse, at which the intensity is constant, the transition is neither purely diabatic or adiabatic. The admixture can be altered by experimentally varying the pulse shape. For instance, under certain conditions a dressed ground state successively crossing several Rydberg states during the rise of the pulse may deplete the ground state population before the pulse reaches its peak intensity, i.e., maximal transfer rate (adiabatic) in the shell model. This effect was demonstrated by Vrijen et al. (1993) by examining the xenon PES for two pulse durations (300 and 600 fs). In the experiment, nf resonances are successively crossed on the rising edge of the pulse with $n = 4$, 5, and 6 clearly resolved in the PES. For the longer pulse the crossings are more

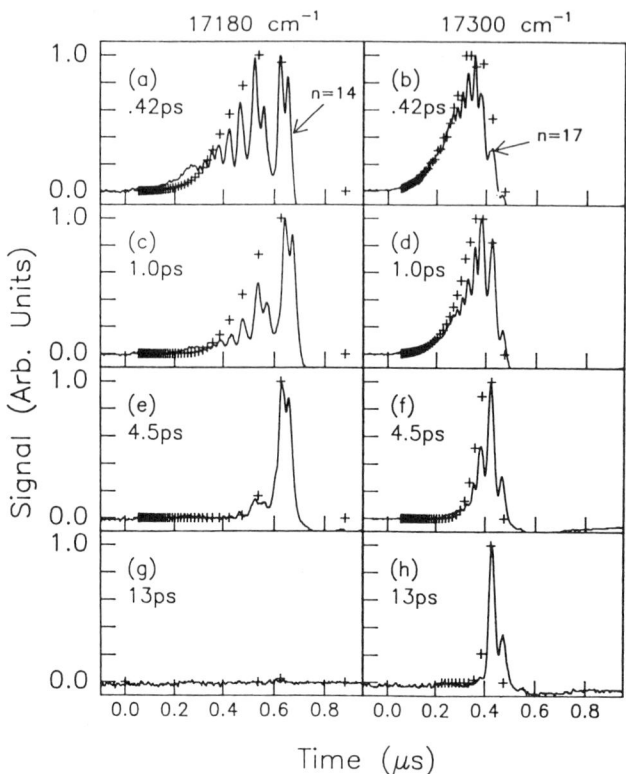

FIG. 2. The residual Rydberg state population detected by field ionization for pulse widths of 0.42, 1.0, 4.5, and 13 ps with laser frequencies of 17,180 and 17,300 cm^{-1}. (From Story et al., 1993.

adiabatic, depleting the ground state before the pulse reaches the $4f$ resonance intensity. Consequently, the $4f$ peak is suppressed in the PES.

A similar case was investigated by Story et al. (1993, 1994a) by examining the residual Rydberg state population for the case of $(2 + 1)$ ionization of potassium. The short excitation wavelengths required for this experiment resulted in negligible ponderomotive shift. The bulk of the $4s$ ground state shift resulted from the strong coupling to the $4p$ excited state. After the laser pulse, the population trapped in the Rydberg states was interrogated using field ionization. Figure 2 shows the Rydberg population as a function of pulse duration (0.42–13 ps) at two laser frequencies. As in the earlier case, the results are in excellent agreement with the Landau–Zener probabilities. For short pulses, the passage through the crossings with nd states is relatively fast and little population is transferred to any individual state, thus producing a large n distribution. As the pulse duration is increased, the Rydberg state population is observed to become concentrated in lower n

states (Figs. 2c, 2e, and 2g) due to the more adiabatic nature of the crossings which quickly depletes the ground state.

In summary, the dynamics of Stark-induced resonances show that ionization occurs over a wide range of intensity. However, > 90% of the electrons originate in the beam region in which $I(r, z) = I_r$, thus asserting the validity of the shell model in describing and assigning the PES of strongly perturbed states (Agostini et al., 1989). Population trapping is nevertheless possible, depending on the circumstances, and is particularly large for circular states ($l = n - 1$) which have small ionization cross-sections (deBoer et al., 1993b). In fact, transient resonances were used to prepare such states for the demonstration of atomic stabilization (deBoer et al., 1993a).

B. Evolution from Multiphoton to Tunneling

The combined potential of an atom in a DC electric field has a nonzero ionization probability since the electron can tunnel through the barrier. Furthermore, if the saddle point is lower than the ionization potential, the electron can escape over the barrier. Keldysh (1964) extended the DC formalism to include AC fields and defined an adiabaticity parameter, γ, as

$$\gamma = \frac{\omega_{\text{laser}}}{\omega_{\text{tunnel}}}, \quad (7)$$

where ω_{laser} and ω_{tunnel} are the laser and tunnel frequencies, respectively. In the limit $\gamma \ll 1$ the ionization mechanism is tunneling, while for $\gamma \gg 1$, for which the field changes rapidly compared with the tunneling time, ionization is multiphoton. Using the width of the barrier, $E_0/\sqrt{I}$, and the electron velocity, $\sqrt{2E_0}$, Keldysh (1964) derived an expression for γ in terms of the ionization potential, E_0, and ponderomotive energy (field strength):

$$\gamma = \sqrt{\frac{E_0}{2U_p}}. \quad (8)$$

This expression implies that γ increases with E_0 which is somewhat misleading since, in practice, the intensity required to observe ionization also increases with E_0. For example, Table II lists experimental appearance and saturation intensities and the resultant γs for different rare gases ionized by 100-fs, 620-nm pulses (Mevel, 1994). It is clear that as Z decreases, (1) γ *decreases* and (2) ionization evolves toward tunneling. Experimentally, the PES and total ion yield will reflect this evolution. In the PES, the most conspicuous feature is the progressive loss of sharp structures as $\gamma < 1$. Due to the limited dynamic range, experiments employing low-repetition-rate lasers could not observe this evolution by studying an individual rare gas atom. However, comparison of the PES from different rare gases (Mevel et al., 1993) resulted in some early insights into this evolution. A study using

TABLE II
EXPERIMENTAL APPEARANCE, SATURATION INTENSITIES, AND ADIABATICITY PARAMETERS FOR MPI OF RARE GASES BY 0.62-μm, 100-fs PULSES

	E_0 (eV)	I_{app}	I_{sat}	γ_{app}	γ_{sat}
Xenon	12.126	4.0×10^{12}	7.0×10^{13}	6.53	1.56
Krypton	13.99	1.6×10^{13}	1.3×10^{14}	3.48	1.23
Argon	15.755	2.0×10^{13}	2×10^{14}	3.33	2.05
Neon	21.56	2.0×10^{14}	8.3×10^{14}	1.23	0.60
Helium	24.58	2.5×10^{14}	1.5×10^{15}	1.18	0.48

a kilohertz titanium–sapphire laser (Walker *et al.*, 1994) provided enough accessible dynamic range to clearly map this evolution in helium following ultra-short-pulse excitation. The intensity range covered was from 0.2 PW/cm^2 ($\gamma = 1.0$) to 0.8 PW/cm^2 ($\gamma = 0.5$). The electron distribution shown in Fig. 3a clearly evolves from a typical ATI into a structureless spectrum. The reason for this change is not obvious. According to the AC-tunneling theory (Amosov *et al.*, 1986), the ionization rate is given in atomic units by

$$W = \frac{4(2E_0)^{5/2}}{\mathscr{E}} \exp\left[-\frac{2(2E_0)^{3/2}}{3\mathscr{E}}\right], \tag{9}$$

where $\mathscr{E}$ is the laser electric field strength. Over the intensity range used in the helium experiment ($E_0 = 0.9$), the rate increases from 1.4×10^{-7} to 2.5×10^{-3}. Thus, the rate remains small compared with the photon frequency (9.31×10^{-3} atomic units), indicating that ionization proceeds over many optical cycles as short electron bursts peaked around the field extrema, which in principle should preserve the ATI (photon) structure.

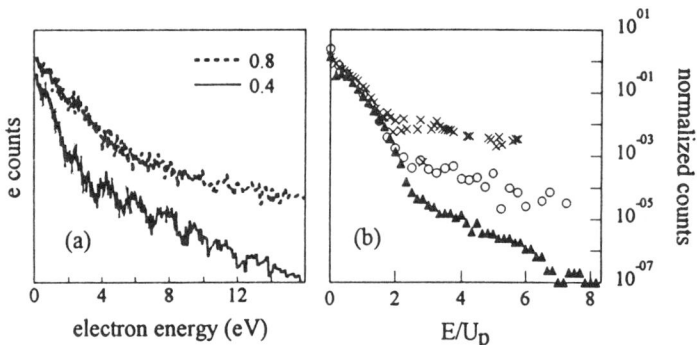

FIG. 3. Helium photoelectron spectra for 0.78-μm excitation. Low-energy electrons plotted at (a) 0.4 and 0.8 PW/cm^2. The ATI structure is evident in the 0.4-PW/cm^2 spectrum. Scaled energy, E/U_p, plotted at (b) 0.2 ($\times$), 0.4 ($\bigcirc$), and 1.0 ($\blacktriangle$) PW/cm^2.

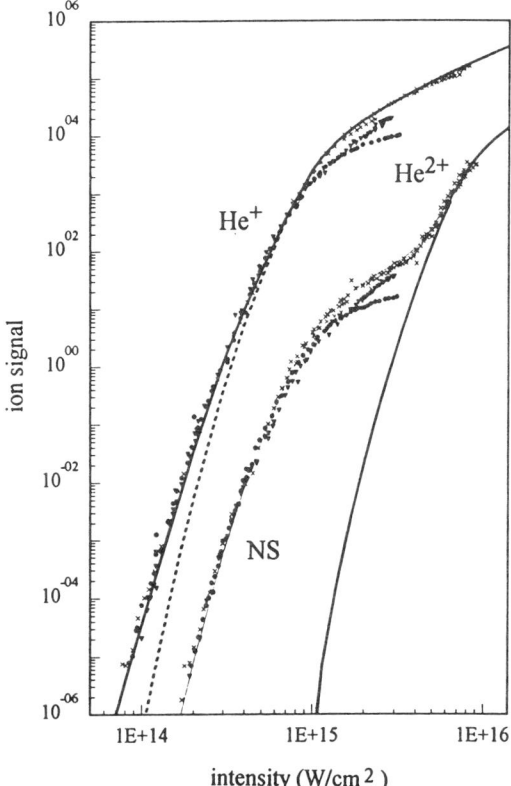

FIG. 4. Measured (symbols) helium ion yields for linearly polarized, 100-fs, 0.78-μm light. Calculations are shown as solid (SAE) and dashed (AC-tunneling) lines. The measured intensities are multiplied by 1.15. The solid curve on the right is the calculated (SAE and tunneling) sequential He^{2+} yield.

However, contrary to the case of Stark-induced resonances discussed above, the relatively weak intensity dependence of the tunnel rate and the large ponderomotive shifts when averaged over the experiment's interaction volume (intensity distribution) smear out the ATI structure which is otherwise present in the "single" intensity calculation.

Total yield measurements studied over a sufficient dynamic range also provide a test of the onset of tunneling. Figure 4 shows a precision measurement (Walker et al., 1994) of the MPI of helium, using 160-fs, 0.78-μm pulses covering 12 orders of magnitude dynamic range. Although the tunneling rate from Eq. (9) (dashed line) fits the data well above 0.6 PW/cm^2, the equation clearly underestimates the experimental rate at "low" intensity and signifies the importance of multiphoton processes. Although the discrepancy is not as important as that in the case of double ionization (discussed in Section IV), it is sufficient to require a more elaborate theoretical modeling. This has been accomplished by numerically solving the TDSE using an SAE approximation (Kulander et al., 1992). This powerful, albeit computer-intensive, method has proven to be extremely successful for accurately describing the strong-field physics of noble gases (Schafer et al., 1993). Its high degree of accuracy is demonstrated by the solid line in Fig. 4 which fits the experimental data extremely well over the entire range.

Based on the precise evidence in helium, the AC-tunneling rates are not very accurate since the majority of experiments performed at near visible wavelengths never achieve a pure tunneling condition. However, they are sufficient to estimate the appearance intensities (Luk et al., 1992; Auguste et al., 1992). The rates are more reliable at the saturation intensity, at which γ is smallest.

III. The Free–Free Step

A. ELECTRON ENERGY DISTRIBUTIONS

Much of our knowledge about strong-field interactions has been derived, over the past decade, from studies which examine the energy and momentum distributions of the photoelectron. In the following sections we will focus on the basic quasiclassical model of strong-field physics which has evolved and been refined with the advent of high-sensitivity photoelectron measurements.

1. Classical ATI

Ionization results from interaction with both the electromagnetic and the coulomb fields. However, once the electron is displaced from the core, interaction with the electromagnetic field dominates. In the first approximation, one can simplify the problem by treating both the electron and the field classically. The success of this amazingly simple idea (van Linden van den Heuvell and Muller, 1988; Muller, 1990; Gallagher, 1988; Corkum et al., 1989) has formed the intuitive basis for understanding strong-field phenomena, including ATI and high-order harmonic generation. This model has been dubbed the "Simpleman" theory. The model assumes that an electron is created in the field, $\mathscr{E} \sin \omega t$, at a time, t_0, with some initial velocity, v_0, and a corresponding kinetic energy, $T_0 = v_0^2/2$. It then acquires through its motion in the field an extra kinetic energy term which is the classical counterpart of the ATI process (see Fig. 5).

Integrating the 1-D equation of motion for a linearly polarized field results in the following solutions for the velocity, position, and cycle-

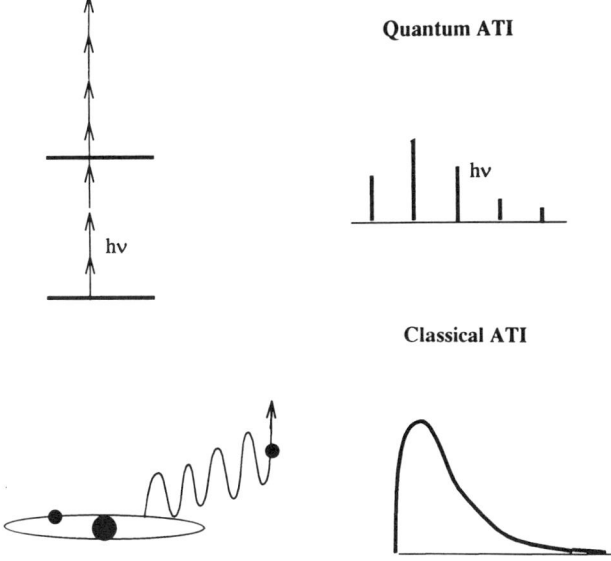

FIG. 5. Schematic quantum and classical ATI.

averaged kinetic energy:

$$v(t) = v_0 + 2\sqrt{U_p}(\cos \omega t - \cos \omega t_0) \tag{10}$$

$$x(t) = x_0 - \frac{\mathscr{E}}{\omega^2}(\sin \omega t - \sin \omega t_0) + (v_0 - 2\sqrt{U_p}\cos \omega t_0)(t - t_0) \tag{11}$$

$$T = T_0 + 2U_p(\tfrac{1}{2} + \cos^2 \omega t_0) - 2\sqrt{U_p}\sqrt{2T_0}\cos \omega t_0, \tag{12}$$

where $\mathscr{E}/\omega = 2\sqrt{U_p}$ has been used. The motion can be visualized as the superposition of an oscillation and a drift. The cycle-average kinetic energy, T, is the sum of the initial energy T_0, the quiver energy U_p, the drift energy $2U_p \cos^2 \omega t_0$, and the cross-term $2\sqrt{U_p}\sqrt{2T_0}\cos \omega t_0$. In a case with zero initial velocity ($T_0 \sim 0$), the upper limit of T is $3U_p$ for electrons born at $\omega t_0 = 0$. However, in short-pulse experiments, the quiver energy, which contributes $1U_p$ to the electron's energy, goes to zero at the end of the pulse and is not detected. The limit[3] is therefore $2U_p$. Thus, the envelope of the ATI amplitudes in the PES between zero and $2U_p$ depends on the statistical distribution of electrons at the initial phase ωt_0. Experiments, discussed below, have demonstrated the reality of the cross-term and the validity of the $2U_p$ limit.

---

[3] Note that the $1U_p$ distinction between short and long pulses should be kept in mind in the proceeding discussions.

Corkum et al. (1989) pointed out that circular polarization *always* results in a drift motion with a velocity, $\mathscr{E}/\omega$, perpendicular to the field direction at $t = t_0$. This constant drift significantly modifies the electron energy spectrum and has repercussions on the subsequent interaction of the electron when the oscillatory motion returns it to the vicinity of the core, as discussed in Section III.B. The quantum mechanical counterpart of this drift is the effect of the centrifugal barrier: if N circularly polarized photons are absorbed to promote the electron into the continuum, a corresponding angular momentum ($L = N$) is transferred to the electron whose wavefunction is repelled from the nucleus.

Naturally all quantum (photons) aspects are lost in the classical model which predicts only a continuous electron energy spectrum determined by the initial phase distribution imposed externally into the model. Thus, it is advantageous to consider the resulting energy distribution as an envelope of the real spectrum which is composed of discrete structure representing the quantum aspects.

2. Large Initial Velocity and Uniform Phase Distribution: Laser-Assisted Auger Decay

Let us assume that in an intense electromagnetic field it is possible to create electrons with large initial velocities at random times during an optical cycle. Specifically, if ωt_0 is uniformly distributed over 2π, then Eq. (12) predicts that the initial spectrum, at a single energy, T_0, is symmetrically broadened by an amount, $2\sqrt{U_p}\sqrt{2T_0}\cos\omega t_0 \cos\alpha$, where α is the angle between the initial velocity and the field direction. Such a scenario is effectively realized in a laser-assisted Auger decay process (Schins et al., 1994). Auger decay results from an inner-shell rearrangement following the creation of a core-hole ion by photoionizing a primary inner-shell electron. Quantum mechanically, the photoeffect (Cionga et al., 1993) and Auger decay in the presence of a laser field has been treated within the S-matrix formalism (Fiordilino et al., 1988; Zangara et al., 1990). Following the KFR spirit, the calculation neglects the dressing of the bound atomic states and uses a Volkov state to describe the continuum dressed by the laser field. The probability as a function of the number of exchanged photons, N, is shown in Fig. 6 and compared with the envelope calculated using the above classical expression, with parameters of the quantum calculation assuming a uniform initial distribution. The probability for exchanging N photons is proportional to the square of the Bessel function of order N, and the argument is the scalar product of the electron momentum $\vec{k}$ and quiver amplitude $\vec{a}$: $J_N^2(\vec{k}\cdot\vec{a})$. The drop for $N > 5$ is a consequence of the well-known property of Bessel functions to decrease exponentially for orders larger than the argument. Clearly, the classical picture captures a good deal of the physics and leads to a simple interpretation of the limit.

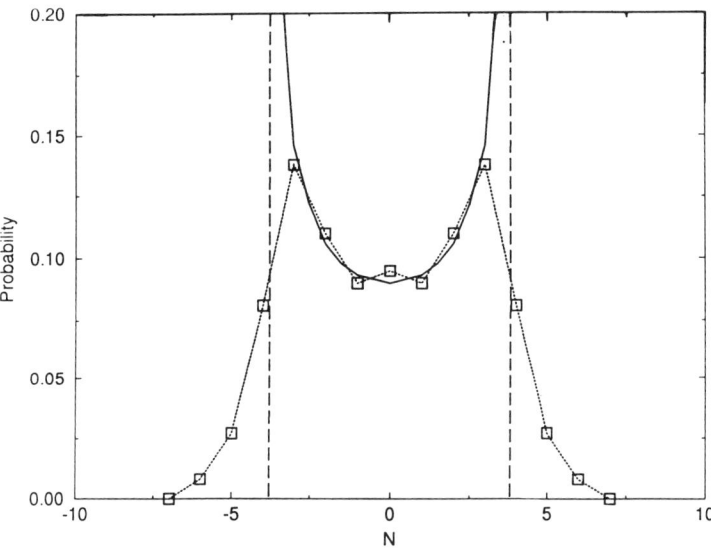

FIG. 6. Spectrum (squares and dotted line) of Auger electrons as a function of the number of exchanged photons (from Zangara et al., 1990). The solid curve is the classical energy distribution, assuming a uniform phase distribution and using the parameters of the quantum calculation ($T_0 = 1074$ eV, $\omega = 1$ eV, and $I = 1.69 \times 10^{10}$ W/cm^2). The position of the dashed lines is $\pm \sqrt{8 U_p T_0}$.

The experiment irradiates argon atoms with a short XUV broadband pulse and a synchronous strong laser field. An L-shell electron is ionized by photons with energies of > 250 eV, producing a core-hole $(A^+)^*$ ion. The excited ion decays into the A^{2+} ground state by interaction of two M-shell electrons: one fills the hole while the other is ejected in the continuum with a kinetic energy characteristic of the Auger transition. In the present study, the L_{23}; M_{23}; M_{23} transition has a mean energy of 207 eV. The XUV pulse is generated by a gallium plasma source excited by an intense 150-fs, 0.8-μm pulse. A fraction of the laser output irradiates the argon atoms, and thus the Auger decay occurs in the presence of a strong 0.8-μm field. The mechanism through which the Auger electron decays ensures a uniform phase distributed over 2π. We briefly stress that (1) the liquid gallium target provides a regenerated surface between laser shots; (2) a 1.5-m magnetic bottle time-of-flight PES spectrometer is used; and (3) the space and time overlap between the laser and the XUV pulses is subjected to careful diagnostics using collimating pinholes on the XUV source and an optical delay line on the laser. The 0.8-μm dressing field intensity was 10^{11-12} W/cm^2 which results in a small value for U_p. Using values of $T_0 = 207$ eV and $U_p = 20$ meV in the above equation, one obtains a broadening of ~ 6 eV on either side of the central Auger transition, which corresponds to four photon sidebands. In

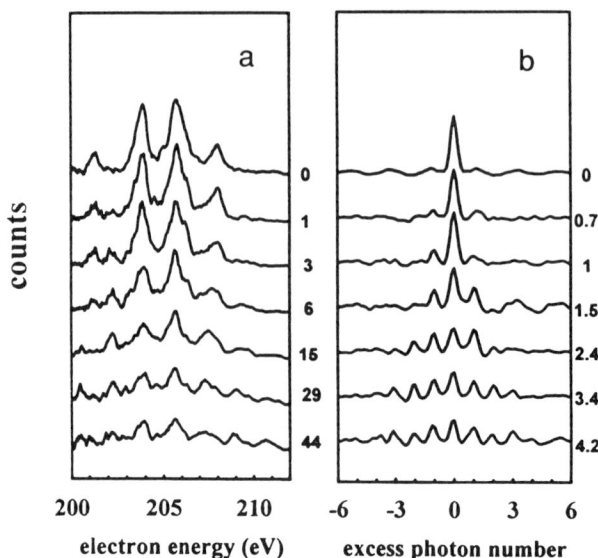

FIG. 7. Laser-assisted Auger decay: raw (a) and deconvoluted (b) spectra for different intensities of the laser field. The figures on the right-hand side are the intensities in units of 10^{11} W/cm² (a) and the theoretical number of sidebands (b).

practice, the observation of the sidebands is complicated by the presence of four main peaks from the "undressed" Auger transition. This problem can be circumvented by deconvoluting the spectrum from the one obtained without the 0.8-μm field. The effect is clearly demonstrated in the PES spectra shown in Fig. 7 at different intensities. Increasing intensity results in additional sidebands which show good agreement with both the quantum and the classical predictions. A quantitative analysis shows that the conservation of the total number of electrons is approximately verified in the experiment, which in the quantum calculation corresponds to a sum rule of the Bessel functions (Kroll and Watson, 1973).

A novel ultra-short XUV cross-correlation technique is a consequence of this experiment since the sidebands' amplitude depends critically on the temporal overlap between the two pulses. In the present study, measured subpicosecond durations were produced in the energy range of 250–400 eV.

3. Small Initial Velocity and Narrow Phase Distribution: Tunnel Ionization and the $2U_p$ Limit

A different scenario arises in the long wavelength (Corkum et al., 1989) or large U_p regime ($\gamma \ll 1$), in which electrons tunnel into the continuum with zero initial velocity, peaked in phase near $\omega t_0 = \pi/2$. The AC-tunneling rate

[Eq. (9)] dependence on the electric field amplitude elucidates the concentrated electron emission around the field extrema. So electrons are produced in short bursts on every half cycle. Defining the initial phase $\omega t_0 = \varphi$, substituting $\mathscr{E} \sin \varphi$ for $\mathscr{E}$ in Eq. (9), and noting that the drift energy can be written as $T = 2U_p \cos^2\varphi$ and therefore $\varphi = \arcsin\sqrt{1 - T/2U_p}$, the energy spectrum follows as (Burnett and Corkum, 1989)

$$\frac{dW}{dT} = \frac{dW}{d\varphi}\frac{d\varphi}{dT} = \frac{2^{3/2}}{\sqrt{1-T/2U_p}\sqrt{T/U_p}} \frac{4(2E_0)^{5/2}}{\mathscr{E}\sqrt{1-T/2U_p}} \exp\left[-\frac{2(2E_0^{3/2})}{3\mathscr{E}\sqrt{1-T/2U_p}}\right]. \quad (13)$$

The energy distribution is bound between zero and $2U_p$ with the maximum emission around zero. The validity of this expression was experimentally demonstrated by observing the energy distribution produced by ionizing xenon atoms with an intense 2.5-ps, 9-μm CO_2 laser source. The results showed good agreement for both linear and circular excitation. However, the quality of the agreement was limited to one order of magnitude dynamic range due to experimental sensitivity. Using femtosecond pulses from a titanium–sapphire laser operating at a kilohertz repetition rate, electron energy spectra from the ionization of helium atoms have been recorded over a very large dynamic range (Walker et al., 1994). Figure 3b shows electron energy spectra for three different laser intensities. Each spectrum is plotted in energy units scaled to U_p associated with the laser peak intensity. One obvious feature is the qualitative agreement with the distribution predicted by Eq. (13) for the three intensities shown. The bulk of the electron emission, ⩾99%, occurs below $2U_p$, but the figure also shows a fraction of the emission extending beyond $2U_p$, a feature which cannot be reproduced with the Simpleman model. It is the interpretation of the high-energy electron emission which forms the basis of a more complete strong-field model.

B. EFFECTS OF RESCATTERING ON THE PHOTOELECTRON ENERGY AND MOMENTUM

1. Theory of Rescattering

Fast electrons, high harmonic emission, and other results, to be discussed hereafter, led to the introduction of the *two-step* rescattering mechanism into the theory: (1) first in the Simpleman's model (Schafer et al., 1993; Corkum, 1993) and (2) then in the quantum KFR theory (Becker et al., 1994) and (iii) finally in a simiclassical model (Lewenstein et al., 1995). Of course, this aspect was *a priori* present in the TDSE–SAE method.

The fundamental feature of the two-step model is revealed by examining the electron trajectories in momentum space shown in Fig. 8 for a linearly polarized field at different initial phases, φ. The figure shows that, for certain

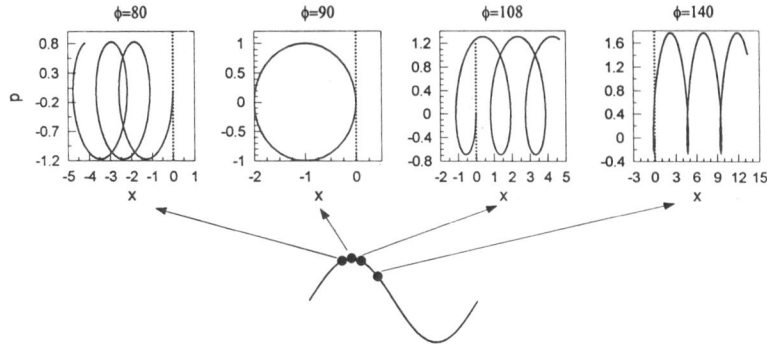

FIG. 8. Electron trajectories in momentum space for different initial phases, ϕ, using Eqs. (10) and (11) and assuming zero initial velocity. The dotted lines show the origin (core) position.

initial phases, the electron trajectories do return to the origin (ionic core) at least once. An equivalent statement in terms of Eq. (10) is that, if the drift velocity is too high, the electron will never return. Analysis of the classical equations of motion show that the electrons which revisit the core have an initial phase bound by $n\pi/2 \leqslant \varphi < (n + 1)\pi/2$, for n odd. Thus emerges a physical picture in which electrons promoted into the continuum near the nucleus at arbitrary times during an optical cycle undergo two types of classical orbits: those that return to the region of the nucleus and those that do not. Those that do not return result in observables consistent with the Simpleman theory, but those that do (approximately half) can elastically or inelastically rescatter at the nucleus.

We first examine the consequences of the rescattering on the *instantaneous* kinetic energy. From Eq. (10) the electron's instantaneous kinetic energy, $\mathcal{T}$, at the return time, t_r, as a function of initial phase, ωt_0, is given by

$$\mathcal{T}_r = 2U_p[\cos^2\omega t_r + \cos^2\omega t_0 - 2\cos\omega t_0 \cos\omega t_r]. \tag{14}$$

The relationship between t_r and t_0 is determined by setting $x(t) = 0$ in Eq. (11). Evaluation of this function shows that the *maximum* return energy is given by $3.17U_p$ for an initial phase of $\sim 108°$.

The effect of the rescattering on the *cycle average* kinetic energy $T = \langle \mathcal{T} \rangle$ has a direct bearing on the electron energy spectrum. Assuming that the electron backscatters at the return time, t_1, then the velocity at subsequent times is given by

$$v(t > t_r) = \frac{E}{\omega}[(\cos\omega t - \cos\omega t_r) - (\cos\omega t_r - \cos\omega t_0)]. \tag{15}$$

The first term in parentheses looks like the normal Simpleman velocity with the *new* phase defined by t_r, and the second term is an initial backscattering velocity. Thus the electron can acquire additional drift energy by rephasing itself in the field via the scattering event. In general, after an elastic collision at an angle θ, at the return time t_r, T is given by (Paulus et al., 1994c)

$$T(t > t_r) = 2U_p[\cos^2 \omega t_0 + 2(1 - \cos\theta)\cos \omega t_r(\cos \omega t_r - \cos \omega t_0)]. \quad (16)$$

Depending on the initial phase, T can reach $10U_p$ ($11U_p$) for short (long) laser pulses, well beyond the $2U_p$ limit.

The effects of rescattering have been investigated in more elaborate theories, most notable in time-dependent methods for describing the single-electron dynamics of rare gas atoms, including electron distributions, harmonic generation, and total yields. Several papers (L'Huillier et al., 1991; Yang et al., 1993; Schafer et al., 1993) have illustrated the success of such methods. The reader is referred to the article by Kulander et al. (Gavrila, 1992) for details. The method solves the TDSE on a three-dimensional numerical grid. An SAE approximation which assumes that the MPI occurs for one valence electron at a time in the mean field of the remaining unexcited electrons is incorporated. The quantum mechanical nature of the calculation includes all aspects of the ionization including both multiphoton and tunneling, plus the time dependence allows for the rescattering under the influence of the laser field and ion core.

The rescattering was introduced in the KFR formalism by Becker et al. (1994) by expanding the electron propagator in terms of the atomic potential. The resulting amplitude is the sum of terms involving *zero, one, etc.* interactions (returns) with the core. Calculations using the first two terms result in a modified energy spectrum which extends toward high energies, showing a marked change in the distribution's slope around $8U_p$. As will be seen in the Section III.B.3 contributions from any higher order returns are negligible due to the spreading of the electron wave packet. Further evidence into the energy dependence of the differential cross-section is provided by semiclassical calculations (Lewenstein et al., 1995) using a quantum mechanical source term.

2. Experiments: The High Harmonic Cutoff

The high harmonic emission (L'Huillier et al., 1991) can be characterized as a flat distribution as a function of increasing harmonic order, dubbed the "plateau," followed by an abrupt decrease in efficiency, referred to as the cutoff. Thus, the highest harmonic order possible is determined by the position of the cutoff which is intensity dependent and further complicated by macroscopic phase matching conditions of the medium. However, it was quickly recognized that the cutoff scaled semiempirically as $\sim 3U_p + E_0$ (L'Huillier et al., 1991). Theoretical treatments incorporating one-electron approximations in a variety of model potentials reconstructed this simple cutoff scaling. This established the importance of the single-electron dynamics and the weak dependence on the atomic potential. The classical model above suggests a simple and clear interpretation of the cutoff: the harmonics are generated when the electron recombines with the core at the return time, and the maximum photon energy that can be generated is therefore $3.17U_p + E_0$ (Schafer *et al.*, 1993; Corkum, 1993).

3. *Experiments: Electron Distributions Beyond* $2U_p$

Precision measurements of PES and angular distributions have been studied both in the MPI and in the tunneling regimes using *kilohertz*-repetition-rate, high-intensity lasers (Yang *et al.*, 1993; Paulus *et al.*, 1994a, b; Walker *et al.*, 1994). The impact of kilohertz laser technology on this problem became evident in a series of experiments (Nicklich *et al.*, 1992; Schafer *et al.*, 1993). Figure 9 shows a photoelectron spectrum of xenon taken with 30 TW/cm², 50-ps, 1-µm radiation. The ordinate is a logarithmic scale spanning six orders of magnitude in electron counts. Simple physical arguments demonstrate the usefulness of this technology. Consider that for photoelectron spectroscopy the resolution is often limited by space–charge effects. A typical laser focus used in strong-field studies is $\sim 5\,\mu m$, placing a limit of 10–20 ions per laser shot. Since a time-of-flight electron spectrometer has $\sim 10\%$ collection efficiency, a 10-Hz laser system is limited to $\sim 10^{-4}$ detection sensitivity for 1 h averaging. Obviously two orders of magnitude are immediately gained by virtue of the kilohertz duty cycle, but the

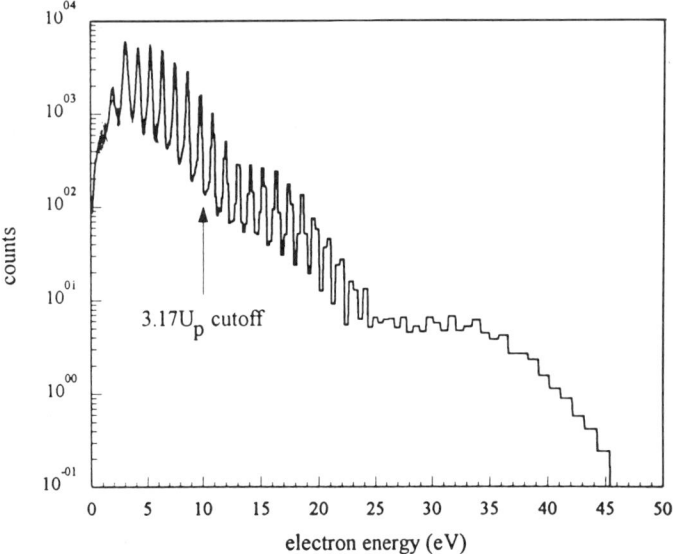

FIG. 9. The photoelectron spectrum of xenon atoms resulting from 30-TW/cm², 50-ps, 1.05-µm excitation. The arrow indicates the $3.17U_p$ high harmonic cutoff. The "Simpleman" limit is $3U_p$ for long pulse excitation.

advantages are more subtle. Other advantages are the application of new detection schemes, such as coincidence measurements (Walker et al., 1993), stable shot-to-shot performance (Saeed et al., 1990), and, most importantly, the ability to perform systematic studies in a manageable time period which for all practice reasons were virtually impossible at low repetition rates.

The PES of helium atoms exposed to 100-fs, titanium–sapphire pulses provides an excellent comparison with the rescattering model since the bound–free step is nearly pure tunneling. An early study (Mohideen et al., 1993) had shown an inconsistency with the measured electron distribution below $2U_p$ with both the Simpleman and the KFR models. However, the experiment could not observe the electrons beyond $2U_p$ and thus was unable to probe the most sensitivity region of the rescattering model. A high-sensitivity, *kilohertz* experiment, already introduced, showed that $>99\%$ of the electrons are in qualitative agreement with the Simpleman model but at variance with the quantitative prediction of Eq. (13). A discerning view of the spectral evolution, which equates to the change in the initial continuum conditions, is illustrated in Fig. 3b at three different intensities. The fraction of electrons with energies $>2U_p$ decreases at a rate $\propto I^{-2.6}$. Clearly the decrease is more rapid than that predicted by the energy dependence of the elastic cross-section (Mott and Massey, 1965). As discussed in Sections II.B and III.A.3, the ionization regime evolves from MPI to pure tunneling at $0.8\,\text{PW/cm}^2$. Thus, the interpretation of the data should be viewed as the importance of MPI at producing high-energy electrons at low intensity and the relative ineffectiveness of rescattering in the tunneling regime (high intensity).

The decrease in high-energy electrons in the tunneling regime could indicate (Walker et al., 1994) the role of transverse wave-packet spread prior to rescattering. Clearly the electron is not a classical object but a quantum mechanical wave packet which propagates outside the influence of the core for almost half of an optical cycle before rescattering. During this time the wave packet should be comparable to a freely spreading gaussian packet whose width is given in atomic units by $\alpha_t = \sqrt{\alpha_0^2 + (2t/\alpha_0)^2}$, where α_0 is the width at $t = 0$. At time t, α_t can be no smaller than $\sqrt{4t}$, corresponding to an initial width of $\sqrt{2t}$. The spread is intensity *independent* but wavelength *dependent*. Figure 10 plots α_t versus α_0 after half of an optical cycle for a 0.8-μm titanium–sapphire field. For a typical initial condition of $(3-4)a_0$ the wave packet can spread an order of magnitude in a half cycle, thus diluting the effectiveness of the rescattering event. This interpretation is consistent with the observed (L'Huillier and Balcou, 1993) decrease in high-order harmonic efficiency as one approaches the pure tunneling regime. The effect of the spreading of the wave packet on the rescattering will be further discussed in Section IV.

Electron distributions were examined (Paulus et al., 1994b) for a number of different inert gases using a 40-fs, 0.63-μm kilohertz laser. Figure 11 shows

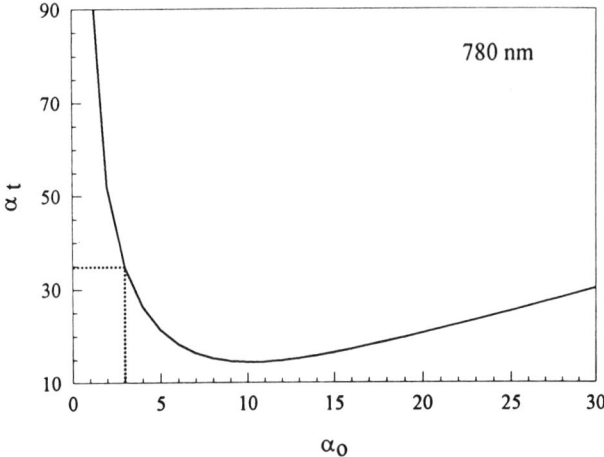

FIG. 10. The wave packet spread, α_t, as a function of initial width, α_0, for 0.78-μm light. The axes are in atomic units. The dashed line shows the spread for a realistic initial condition of $3a_0$.

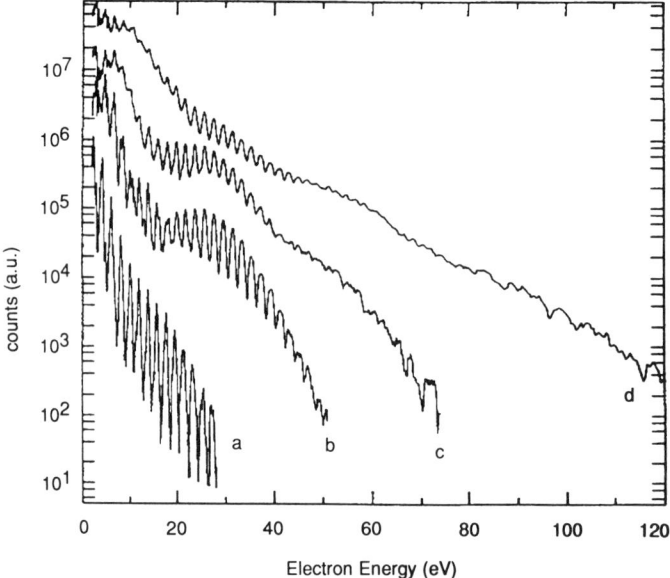

FIG. 11. Argon ATI spectrum recorded with 40-fs, 0.63-μm pulses at intensities of (a) 6×10^{13} W/cm^2, (b) 1.2×10^{14} W/cm^2, (c) 2.4×10^{14} W/cm^2, and (d) 4.4×10^{14} W/cm^2.

the ATI spectra as a function of intensity for argon. The spectra show a marked change in the slope of the distribution at higher intensity which was dubbed a "plateau," analogous to the high harmonic plateau. Note that there exists a number of distinct physical differences between the two, and this will be addressed in Section III.B.5. This study further exemplified the universal nature of the general shape of the distributions beyond $2U_p$, which show little atom or wavelength dependence. Another interesting feature is that the energy difference between the ATI peaks in the plateau and those at lower energy do not correspond to an integral number of photons. The dressed-state picture (Kroll and Watson, 1973) was proposed as a potential resolution to this observation, but the role of the spatial averaging in the experiment may also lead to an apparent shift. Furthermore, the plateaus are absent using circularly polarized light, consistent with the polarization dependence of the rescattering mechanism in the two-step model.

4. Experiments: Scattering "Rings" in Angular Distributions

A significant fraction of our knowledge about strong-field ionization comes from studies of the momentum characteristics of the photoelectron. For light linearly polarized along the z axis, the angular distributions are expected to become more strongly peaked along this axis as the ATI order increased, due to the propensity rule that favors increasingly higher angular momentum states as additional photons are absorbed. This expectation is also compatible with the quasiclassical prediction which views the acceleration (drift) of the tunneled electrons along the laser's electric field direction. Since the faster electrons are emitted before or after the peak of the field, they are more likely to emerge closer to the z axis.

A number of *kilohertz* experiments (Yang et al., 1993; Paulus et al., 1994b) have revealed that these simple ideas are incomplete and provided a window on the underlying physics. Figure 12 shows photoelectron spectra of xenon recorded at two intensities using 50-ps, 1-μm light. The distributions show a propensity for producing higher energy ATI peaks with increasing intensity, as well as the presence of a change in the slope above 25 eV for the highest intensity. The insets are polar plots of the angular dependence for different ATI orders ($S = 5, 10, 15, \ldots$). The data ($\times$s) are fit by a sum of even Legendre polynomials represented by the solid lines. The angular distributions (AD) for the first several peaks are consistent with the aforementioned expectations. However, as the order increases further the ADs for several peaks exhibit off-axis structures which have been dubbed *scattering rings* (Yang et al., 1993). For example, nearly one-half of the ionization signal in the 20th ATI peak ($\sim 24\,\text{eV}$) appears around 45° from the laser polarization axis. Beyond these "rings" the ADs return to the strongly aligned shape. Furthermore, the two plots show that the energy at which the rings appear is strongly intensity dependent, shifting to lower energy as the intensity decreases.

The strong intensity scaling of the rings becomes clear in Fig. 13. The symbols represent the half-flux angle of the ADs for each xenon ATI peak for three different intensities. The rapid narrowing for the lowest-order ATI peaks is due to the well-known (Freeman et al., 1992) macroscopic effect of ponderomotive scattering of low-energy electrons in the long pulse limit

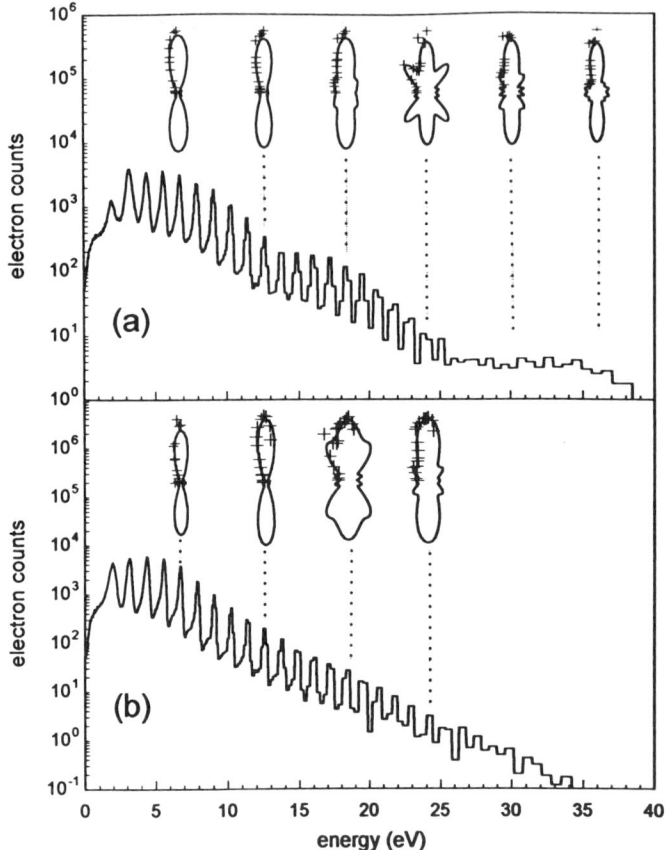

FIG. 12. The photoelectron energy spectrum of xenon with 50-ps, 1.05-μm excitation at (a) 30 TW/cm^2 and (b) 19 TW/cm^2. The spectra are an average of 7.2×10^6 laser shots. The inserts are polar plots of the ADs for the $S = 5, 10, 15, \ldots$ ATI peaks. The solid line results from the Legendre fit, and the raw data are shown as $\times$ s.

($\tau_p > 1$ ps). The presence of the rings in this representation causes the later increase in the half-angle. It is clear that the position and width of the peaks in the half-angle distribution are strongly intensity dependent. The exact intensity dependence is revealed by plotting the half-angle against the normalized electron energy, E/U_p. The plot shows a clear and dramatic *rephasing* of the peaks within a narrow energy range centered around $9U_p$. The $9U_p$ scaling signifies that the rings cannot be due to a mechanism associated with the $1U_p$ shift of the ionization threshold. The degree of manifestation of the rings varied with the inert gas examined and was strongest for xenon.

An excellent example of the appropriateness of the TDSE–SAE method is shown in Fig. 14. This is a half-angle plot prepared in the same manner

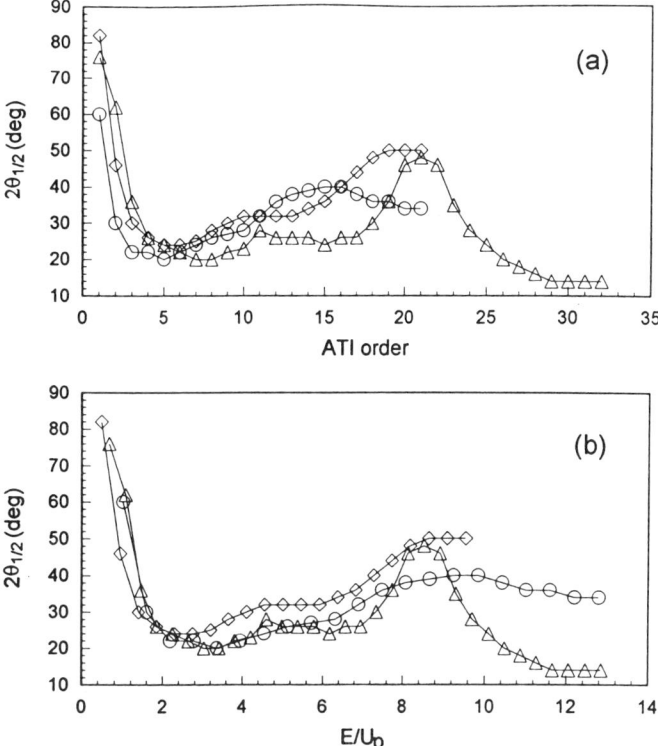

FIG. 13. Xenon half-angles as a function of (a) ATI order and (b) E/U_p, the electron energy over the ponderomotive energy, at 19 (circles), 25 (diamonds), and 30 (triangles) TW/cm². The value of U_p used for each curve is defined by the laser's peak intensity.

as that in Fig. 13, for the three different intensities used in the experiment. The $9U_p$ scaling is clearly reproduced. Furthermore, the polar inserts are representative ADs which agree remarkably well with those observed in the experiment. Similar comparison with the angular resolved photoelectron spectra also show excellent quantitative agreement. Most importantly, the SAE calculation provides a crucial clue as to the origin of the scattering rings. Since this is a single electron approximation, the source of the rings must be in the single-electron dynamics.

Examination of Eq. (16) suggests that electrons with $8U_p$ energy result from backscattering. A quasiclassical calculation (Lewenstein et al., 1995) using a model short-range potential in a generalized strong-field approximation demonstrated this fact by separately analyzing the classical trajectories in the continuum. The result is shown in Fig. 15 for the case of $U_p > E_0$. The plot clearly shows that the fraction of electrons with high energies ($>6U_p$) is solely due to backscattering. Further evidence can be seen in the evolution of the time-dependent wave function (Kulander et al., 1992) which

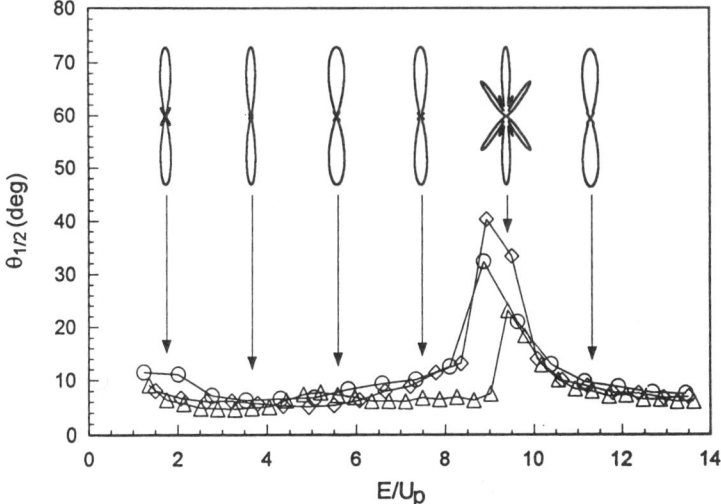

FIG. 14. Calculated TDSE–SAE xenon half-angle as a function of E/U_p for 15 (circles), 20 (diamonds), and 30 (triangles) TW/cm^2, 1-μm radiation. The inserts are polar plots of the calculated ADs for the indicated ATI peaks at 30 TW/cm^2.

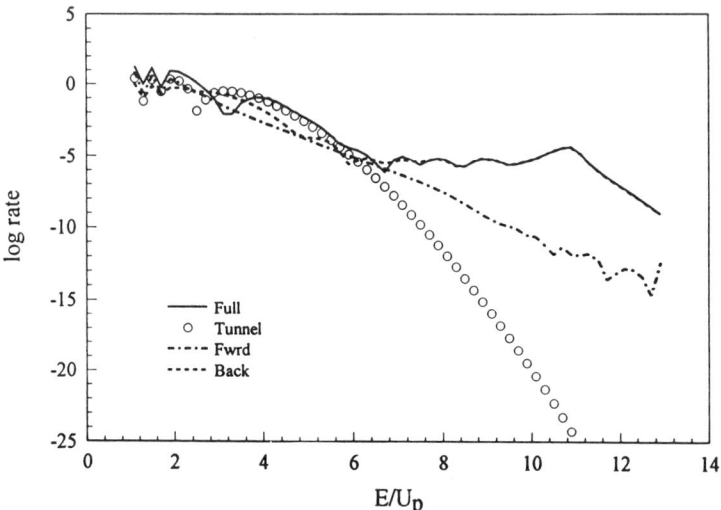

FIG. 15. The calculated curves of Lewenstein *et al.* (1995) for $U_p = 3E_0$. The solid line and circles are the full rate of the quantum mechanical and tunneling source terms, respectively. The dashed and dashed–dotted lines are the backward and forward scattered contributions. Note that beyond $6U_p$ the total and backward rates are indistinguishable, clearly demonstrating the importance of *backscattering*.

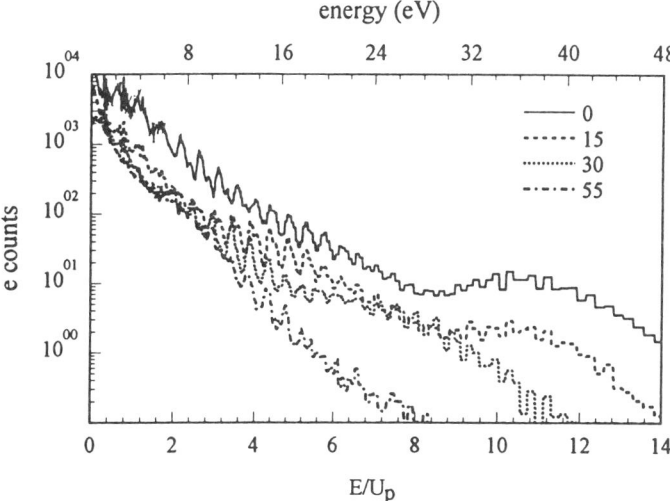

FIG. 16. The photoelectron spectra of xenon at different detector angles (degrees) with respect to the laser's polarization direction. Note the dips and crossover around $8U_p$ energy. The spectrum results from 70-TW/cm², 100-fs, 0.78-μm excitation. The top scale is absolute electron energy, and the bottom scale is in normalized energy units.

reveals that after a half-cycle of free propagation in the field the returning wave packet interacts strongly with the core. Furthermore, some fraction of the packet continues to evolve against the field directions, indicative of backscattering. The rings are also predicted by the two-step model (Paulus et al., 1994c) by assuming a uniform initial distribution of electrons scattered at an angle θ at t_r which emerge from the beam at an angle θ_1, given by

$$\cot\theta_1 = \frac{p_x}{p_y} = \cot\theta - \frac{\cos\omega t_r}{\cos\omega t_r - \cos\omega t_0}\frac{1}{\sin\theta}. \qquad (17)$$

However, the two-step model's prediction fails to explain many of the experimental observations, including the atomic dependence of the ring strength and their appearance in a narrow electron energy window. Such details could indicate the neglect of the quantum nature of the scattering event in the classical treatments.

Another means for examining this effect is shown in Fig. 16 for a plot of the xenon electron distributions at different angles relative to the polarization direction. The experiment uses 100-fs, 0.78-μm excitation at an intensity of 70 TW/cm². The distribution shows that around $8U_p$ the count rates at 30° and 15° become equivalent. This would manifest itself as a weak scattering ring in the polar representation. Other inert gas atoms demonstrate similar behavior near $8U_p$, providing evidence into the universal nature of this effect. Furthermore the shape of the distributions suggest some form of quantum interference.

The xenon rings have been observed by Paulus et al. (1994b) at a different wavelength. The experiment shows some differences regarding the angle at which the rings appear. The cause of this difference is not yet clearly identified but it could reflect a slight dependence on wavelength and pulse duration. Furthermore, they found that the rings appeared at the onset of the plateau, again suggesting a quantum interference interpretation. It would be interesting to investigate the angular distributions from helium in the pure tunneling regime to test the predictions of the classical calculation. This is obviously a difficult experiment considering the small number of rescattered electrons (see Fig. 3b).

5. *Connection between ATI and OHG*

High-sensitivity experiments found their genesis in an attempt to address the connection between ATI and high harmonics (Javanainen et al., 1988; Kulander et al., 1992; Schafer et al., 1993), that is, to uncover a more universal view of strong-field interactions. The spectrum in Fig. 9 shows a well-developed series of ATI peaks extending out to 40 eV with no evidence of any cutoff at $3.17U_p$ (indicated by an arrow). Thus, the harmonic spectra have a cutoff; ATI spectra do not. This statement is consistent with the two-step model which views rescattering as the source of both high-energy harmonics and electrons. The key difference between the PES and the OHG is the distinction between the averaged [Eq. (14)] and the instantaneous [Eq. (16)] energies, respectively. Furthermore, any correlation between the electron and the photon distributions (plateaus) is lost since, in OHG, the rescattering recaptures the electron to the ground state, terminating its evolution in the continuum. However, for ATI, the rescattering rephases the evolution of the quiver motion in the field.

IV. Strong-Field Double Ionization

A. "Direct" Channel in Multiphoton Double Ionization

Multiphoton double ionization was observed for the first time in the mid-1970s (Suran and Zapesochnyi, 1975). It was immediately realized that the rate was several orders of magnitude larger than that predicted by assuming *sequential* nonresonant single-electron dynamics within the context of lowest-order perturbation theory. Delone et al. (1984) proposed a solution to this puzzle in which two electrons are first driven above the first

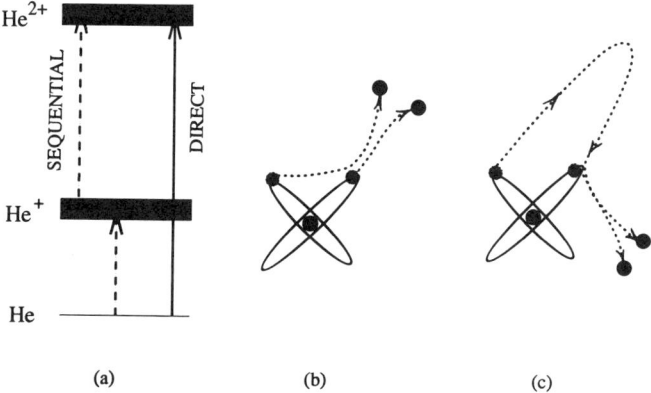

FIG. 17. Schematic representations of the double ionization mechanisms in helium. (a) Multiphoton regime sequential vs direct. Nonsequential strong-field regime: (b) shake-off and (c) $e-2e$. The mechanisms can be generalized for any atom.

ionization threshold through two-electron immediate states followed by rapid excitation above the two-electron limit via a dense manifold of quasicontinuum doubly excited states. This mechanism was dubbed *direct* excitation in sharp contrast to the single-electron sequential dynamics; both mechanisms are depicted in Fig. 17a. Many studies using tunable lasers and electron spectroscopy (DiMauro, 1993), which investigated the MPI of alkaline-earth atoms, followed. The conclusion was that direct ionization was not responsible for the anomalous ratio but instead involved single-electron ejection of excited electronic states of the ion, as well as resonant enhancement. In certain cases, correlated double excitation played a significant role (Tang *et al.*, 1990), but the ionization always proceeded in a sequential manner. The reason is that it is very difficult to keep the two electrons at the same level of excitation and that the decay rate of doubly excited states to a single ionization continuum is faster than the decay to the double continuum. If direct ionization does not occur *a priori*, it may be possible to considerably enhance this channel by carefully choosing the ionization path. One way is to select intermediate doubly excited states in l^2 configurations. This was attempted in the MPI of magnesium (Hou *et al.*, 1990), and some hint of a direct excitation of high doubly excited states were reported. However, a similar experiment (van Druten *et al.*, 1994) did not confirm them. *Planetary* double Rydberg states, on the other hand, have slow autoionization rates. Cohen *et al.* (1993) exploited this fact by choosing an excitation scheme resonant with such states and indeed found a strong suggestion of direct double ionization. These authors have studied the double ionization of Ba $6snl$ Rydberg states through resonant two-photon excitation of $8snl$ doubly excited Rydberg states with $n = 13$ and 14 and different l. The dependence on the intensity of the Ba^{2+} yield changes from I^4 for $l = 8$ to $I^{2.6}$ for $l = 11$. For $l \leqslant 10$, the double ionization proceeds via the usual channel: decay of the doubly excited Rydberg to an ion-excited state followed by the four-photon ionization of this state. For $l > 10$ (planetary states), this decay vanishes and the direct three-photon channel dominates.

By the early 1980s, multiple ionizations of rare gases (L'Huillier et al., 1983) and other atoms (Luk et al., 1992) were well documented. Since the ionization regime in this case is much more tunneling, the above concept of the direct process is meaningless. Actually, except for a small number of cases, it was found that the various charge states were reasonably predicted by assuming *sequential* mechanisms in which the electrons are peeled off successively during the rising edge of the pulse. The corresponding rates are roughly given by scaling the generalized cross-sections (Lambropoulos and Tang, 1987) to the appropriate number of photons and ionization energy or by using the AC-tunneling rates (Luk et al., 1992; Auguste et al., 1992) or over-the-barrier ionization threshold intensities (Augst et al., 1989).

The interesting and intriguing cases are those for which the rates cannot be fit by any of the above models. One long standing case of direct or nonsequential ionization of a rare gas atom involved an MPI study (L'Huillier et al., 1983) of xenon with 0.5-μm radiation. For xenon, under these conditions, the ionization dynamics are predominantly multiphoton in nature. The signature for a nonsequential process is provided by a characteristic hump or knee on the doubly charged ion yield as a function of intensity which corresponds to a premature production of Xe^{2+}. The salient point is that the Xe^{2+} yield *cannot* be fitted by a single kinetic rate. This experiment remained for a decade the best evidence for direct two-electron ionization, but questions persisted as to exact nature of the nonsequential mechanism especially since the alkaline-earth investigation proved to be fruitless. Walker et al. (1993) using a 1-kHz repetition rate, 0.5-μm laser investigated this question by (1) observing the ellipticity dependence of the nonsequential yield and (2) applying electron–electron and electron–ion coincidence techniques. The nonsequential yield showed a much stronger polarization dependence than the *sequential* production, becoming completely extinguished with circular polarized light. Furthermore, the coincidence measurements with linear polarized light showed no evidence for the direct process but instead identified the electrons associated with the sequential process. The authors proposed that the cause of the nonsequential signature was a higher-order sequential process. In this scenario the increased rate giving rise to the knee structure was due to a continuum of two-electron resonances that autoionizes, followed by ionization of the excited ion. The polarization dependence implied that the continuum state must have relatively low angular momentum, which suppresses the resonance in circular polarization by simple selection-rule arguments. Furthermore, this explanation required no unusually large cross-sections or inconsistencies with lowest-order perturbation theory. This high-order sequential mechanism was later clarified (Charalambidis et al., 1994) using a rate equation approach and involved a (5 + 5 + 1)-photon ionization, leaving the Xe$^+$ ion in the excited $5s5p^6$ state which efficiently five-photon ionizes to form Xe^{2+}. The enhanced rate for producing the excited ion is a consequence of an inner subshell excitation of the 5s electron.

B. "Nonsequential" Channel in Tunneling Double Ionization

Helium atoms (Fittinghoff et al., 1992) exposed to an intense pulse of 100-fs, 0.8-μm light exhibit a behavior qualitatively identical to that of xenon yield curves: a premature production of doubly ionized helium which is strongly polarization dependent (Walker et al., 1993). The obvious question is, does such similar behavior imply a common mechanism which in the case of xenon does not lead to direct two-electron ionization. The major difference, as discussed earlier, is that helium ionizes nonperturbatively (tunneling) while xenon is perturbative (multiphoton). This implies that the initial conditions determined at the time the electron is being promoted to the continuum should result in very different nonsequential dynamics. In fact, in the intensity regime used in the experiment, helium has no bound excited states. Thus, the "resonant" scenario used to explain the Xe^{2+} nonsequential production becomes physically nonapplicable to helium. Fittinghoff et al. (1992) suggested that helium then must proceed via direct ionization involving a mechanism analogous to the two-electron shake-off process observed with high-frequency photons (Wehlitz et al., 1991; Mittelman, 1966). In this scenario the sudden loss of screening by the instantaneous removal of the first electron results in some probability of diabatic two-electron ejection. For synchrotron studies using high-energy photons (keV) this condition is met by the high kinetic energy imparted to the first electron and produces $\sim 3\%$ double ionization in helium. In the strong-field scenario the sudden loss of screening is provided by the instantaneous displacement of the electron from the core via tunneling. A calculation (Fittinghoff et al., 1992) using a kinetic description resulted in good agreement with experimental results. However, the polarization dependence of the nonsequential process remained not a priori clear in a strong-field shake-off model (Fig. 17b).

A second scenario (Corkum, 1993) based on the two-step model results in a clear intuitive explanation of the nonsequential production in helium as an e–$2e$ scattering event (Fig. 17c). Since the electron's rescattering after half a cycle is well supported by other evidence (plateaus and rings in ATI), it is natural to envision that this scattering could liberate a second electron. Using the known e–$2e$ cross-sections, Corkum (1993) could fit the observed helium yields with good agreement. Likewise, the sharp dependence of the nonsequential rate on the polarization is immediately interpreted in this model as the perpendicular drift motion of the electron. As mentioned above, in circular polarization the classical motion has a constant $(\mathscr{E}/\omega)$ *transverse* drift velocity which is *independent* of the initial phase. The drift prevents the electron from returning to the core and therefore quenches the nonsequential process, consistent with the experiments.

The existence of the two different models prompted more precise measurements of both the rates and their ellipticity dependence. The ion yield curves (Walker *et al.*, 1994) for linearly polarized excitation are shown in Fig. 4; the resulting nonsequential (NS) He^{2+} production is clearly evident, saturating at $0.8\,PW/cm^2$. This measurement was produced using a *kilohertz* titanium–sapphire laser capable of producing $10\,PW/cm^2$. The laser's high repetition rate made observation of the helium production over an unprecedented 12 orders of magnitude in dynamic range. These data exceed previous measurements by 5 orders of magnitude and provide the first view into the exact nature of the NS process. The relationship between the He^+ and the He^{2+} NS production was experimentally established by systematically varying the confocal spot size of the laser focus with respect to the ion spectrometer's detection image. As the spot size exceeded the detection image a reduction from the $I^{3/2}$ gaussian scaling was observed above saturation. This purely spatial effect clearly demonstrates that the NS He^{2+} production is exactly linked to the saturation of the He^+ yield and verifies a connection with the depletion of neutral helium atoms.

A sensitive measure of the NS dynamics is illustrated in Fig. 18 which plots the $He^{2+}(NS)/He^+$ ratio as a function of intensity. The closed circles

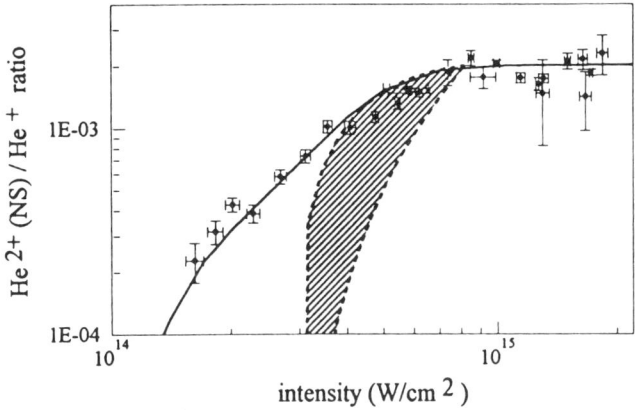

FIG. 18. Intensity dependence of the He^{2+} (NS)/He^+ ratio for 0.78-μm excitation. Error bars indicate one standard deviation. The solid line is calculated; see the text for details. The hatched area is calculated using the two-step model; see the text for details.

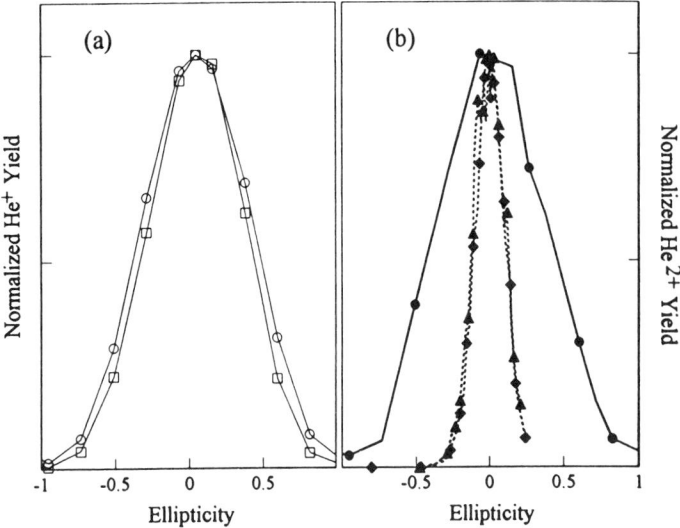

FIG. 19. Ellipticity dependence for (a) He$^+$ and (b) He^{2+} ion yields at different 0.78-μm intensities. The values of 1, 0, and -1 correspond to right, linear, and left circularly polarized light, respectively. The dotted lines correspond to the intensities at which nonsequential He^{2+} production dominates.

are the experimentally measured ratio which at saturation is 0.0020(3). This value is over 10 times smaller than the single photon ratio (Wehlitz et al., 1991). Although the ion curves in Fig. 4 show a strong intensity dependence, changing by seven orders of magnitude between 0.15 and 8.0 PW/cm^2, the ratio exhibits a gentle slope, scaling as $I^{1.3}$. The solid line in Fig. 18 is the ratio between the calculated AC-tunneling and the SAE (total) rates shown in Fig. 4 normalized to the experimentally saturated value. The striking agreement implies that the NS two-electron process depends solely on tunneling, even when the multiphoton process dominates the total rate. Although this does not define the mechanism responsible for NS production, it does provide an important insight into the underlying physics.

Finally, studies of the nonsequential production ellipticity dependence have been carried out for neon (Dietrich et al., 1994) and helium (Walker et al., 1994). It is clear in Fig. 19 that this dependence is much stronger for the nonsequential than for the sequential process. The dependence has no *a priori* explanation in the shake-off model. However, a quantitative comparison (Dietrich et al., 1994) with the prediction of the two-step model yields excellent agreement for the polarization dependence of the high harmonic and nonsequential yields for neon. However, the agreement with the absolute nonsequential yield in neon and helium is poor and will be discussed in the next section.

C. Nonsequential Rate

The two proposed models share the common requirement of tunnel ionization but differ on when and how the second electron escapes. However, the interpretation of the existing experimental evidence is somewhat ambiguous and provides some serious concerns for both mechanisms. Let us begin by examining the two-step rescattering model. In the rescattering model the tunneled electron revisits the core after a free propagation period of at least half an optical cycle with a maximum kinetic energy of $3.2U_p$. The second electron will be freed if the return energy exceeds the He^+ binding energy, that is, $3.2U_p \geqslant 54.4\,\text{eV}$. Consequently a threshold exists at $U_p = 17\,\text{eV}$ or an intensity of $0.3\,\text{PW/cm}^2$. *No such threshold is observed in the 0.78-μm experiment* which extends down to $0.14\,\text{PW/cm}^2$. Additionally, by allowing for the possibilities that NS ionization may result from collisional excitation of the second electron to excited states (41 eV) which rapidly field ionize and that the core potential is suppressed by the field at the rescattering time, this threshold is reduced to $0.23\,\text{PW/cm}^2$. This could be further quantified by expressing the e–$2e$ rescattering rate as $\propto \sigma(\epsilon)F(\epsilon, I)$, where σ is the field-free e–$2e$ inelastic cross-section (Lotz, 1968) and $F(\epsilon, I)$ is the intensity-dependent distribution of returning electrons with energy, ϵ. The result is shown in Fig. 18 by the hatched area which is bound on the left side by assuming that $F(\epsilon, I)$ is a δ function at $3.2U_p$ and on the right by a flat distribution between 0 and $3.2U_p$. Furthermore, the calculated curves are fit to the experimental He^{2+}/He^+ ratio at saturation, the physical meaning of which will be discussed shortly. Clearly, the quasiclassical approximation to the rescattering model fails to predict the low-intensity threshold behavior of the measurement.

Walker et al. (1994) considered the higher intensity behavior of the rescattering model by accounting for the half-cycle evolution of a freely propagating wave packet. The transverse spread, α_t, of a gaussian wave packet at time t was shown earlier to be in atomic units, $\alpha_t = \sqrt{\alpha_0^2 + (2t/\alpha_0)^2}$, where α_0 is the width at $t = 0$. The adjustable parameter used in obtaining the fitted curves in Fig. 18 has its physical origin in the transverse spread. For this fit, α_t *after a half cycle of propagation* in the 0.78-μm field is $3 - 4a_0$, consistent with the value used earlier (Corkum, 1993). However, using a realistic initial condition of $\alpha_0 = 3 - 4a_0$, derived from time-dependent calculations (Kulander et al., 1993), a value of $\alpha_t = 30 - 40a_0$ is obtained at the rescattering time. Analysis of a calculated tunneling wave packet recrossing the core also yields a width of $30a_0$. This width is an order of magnitude larger than that derived from the fit. Knowing α_t, Walker et al. (1994) placed an upper limit on the He^{2+}(NS)/He^+ ratio of 1.5×10^{-4}, more than an order of magnitude smaller than the experimental ratio (0.002). Clearly, the NS ionization due to e–$2e$ rescattering is much smaller than the measured yields over the entire dynamic range of the experiment. One possible source of this discrepancy could be due to the use of field-free $e-2e$ inelastic cross-sections in the calculations. Kulander et al. (1995) examined this issue by including the influence of the 0.78-μm laser field on the impact process using an approximate 1-D two-electron model. The result showed that the field has little effect on the total cross-section since it is the short-range physics which dominate the collisional process.

On the other hand, the shake-off mechanism (Fittinghoff et al., 1992) focuses on the behavior of the second electron at the instant at which the first electron enters the continuum. This *simultaneous* two-electron ejection cannot be tested as thoroughly as the rescattering mechanism since it requires a realistic two-electron calculation. The mechanism is qualitatively consistent with the observation that the NS production scales with the tunneling rate, since the sudden approximation should be satisfied by an *instantaneously displaced* tunneling wave packet which is absent in the MPI regime. However, it lacks a clear explanation of the observed NS polarization dependence. Walker et al. (1994) have suggested that this could reflect the substantial amount of angular momentum gained by the electron in an elliptically polarized field which could inhibit the energy transfer between the two electrons.

In summary, the signature of an NS process when observed in ion yield measurements is not necessarily an adequate proof of direct two-electron ejection, as is the long-standing xenon case. However, the experimental evidence strongly suggests that direct ejection is occurring in the strong-field ionization of helium. In fact, such a mechanism should be universal whenever tunnel ionization starts to become appreciable The lack of quantitative agreement with experiments and theory would indicate that an $e-2e$ rescattering process is not responsible for the NS production in helium. Consequently, the mechanism must be a simultaneous two-electron ejection process through either a shake-off or a threshold mechanism involving some form of electron correlations. Unfortunately, the underlying mechanism for this process remains uncertain and would benefit from a theoretical treatment involving appropriate two-electron wave functions. It is interesting to speculate upon a possible connection with the nonsequential double ionization and the observation of strong X-ray emission in the kilo-electron volt range from laser-excited rare gas clusters (McPherson et al., 1994; Boyer et al., 1994). Intracluster processes involving inelastic electron–ion collisions of the type

$$e^- + X^{q+} \to e^- + e^- + (X^{(q+1)+})^*$$

are invoked to explain the formation of the multicharged ions required to produce the M-shell and L-shell observed transitions. However, (1) analogous to the nonsequential process, the emission starts at laser intensities much lower than those predicted by this model and (2) the number of detected photons is much larger than that estimated on the same basis.

V. Conclusion

We have described a variety of physical effects associated with the dynamics of single and double strong-field ionization. Progress in short-pulse, high-intensity laser has led to measurements with unprecedented precision of the rates, electron energy spectra, and momentum distributions in multiphoton ionization. These experiments have supplied the impetus for a consistent and unified picture of all strong-field interactions.

The emerging picture of strong-field interactions is based on simple classical insights which have evolved with the mounting experimental evidence. The current view is as follows: a bound electron is promoted to the continuum, and, depending upon the intensity regime, the subsequent dynamics are defined by these initial bound–free conditions. Once in the continuum, the electron's motion is dominated by the field interaction (free–free part). This alone is sufficient to predict the amount of free–free transition in a laser-assisted Auger process. However, high-sensitivity photoelectron and harmonic measurements have prescribed the need for an additional electron interaction (rescattering) with the core after a half-cycle of free propagation in the continuum. Furthermore, this free propagation results in significant spreading of the electron wave packet prior to rescattering. Consequently, the effectiveness for tunneling to produce high-energy electrons, OHG, and two-electron emission is greatly diluted.

These physical notions do render an extremely useful and simple intuitive basis for visualizing strong-field interaction. However, quantitative comparison with the experiments is poor and indicates the necessity for applying quantum aspects to the problem. A remarkably effective method for predicting the quantitative behavior of the single-electron dynamics of inert gas atoms is provided by time-dependent quantum mechanical methods (TDSE–SAE).

Strong-field two-electron ejection remains a largely open question and would greatly benefit from both experimental and theoretical contributions. Our current understanding is that double ionization in the multiphoton regime usually occurs through the excitation of autoionizing states which normally decay faster to excited ionic states than to the double continuum. However, the opposite is true for planetary double Rydberg states which decay through the direct channel. In strong fields, multiple ionization generally occurs sequentially, and the corresponding rates are reasonably predicted by tunneling ionization. In some cases, considerable enhancements in double ionization rates and a high degree of sensitivity to polarization are observed. In the multiphoton regime this often corresponds to higher-order sequential processes, but in the tunneling limit two-electron ejection seems likely. The evidence points to a mechanism which involves *simultaneous* two-electron ejection since the rescattering model fails to predict the intensity dependence of the experimentally measured rates. Additional insight into double ionization physics could result from angular distributions which should be very sensitive to electron–electron correlations in the final state (Faisal, 1994). New two-electron effects in the single ionization of complex atoms have been predicted: Stark-tuned resonances with autoionizing states (Latinne et al., 1995) and core-excitation effects in two-electron systems (Grobe and Eberly, 1992).

The laser-assisted Auger decay can be viewed as a first step toward nonlinear X-ray process. Using OHG and special XUV optics, intensities in the 10^{10-11} W/cm^2 range should be obtainable. Consequently, nonlinear optics in the high-frequency regime, i.e., two-photon inner-shell ionization, should be feasible. From this perspective, it is expected that the new generation of high-intensity lasers (1 J, 50 fs, and 10 pulses/s at 0.8 μm) will have a significant impact.

Acknowledgments

Special appreciation is expressed to our colleagues, Ken Kulander and Ken Schafer, whose fruitful discussions over the years have resulted in a productive collaboration. We thank the people who have tirelessly carried out the research: the Brookhaven group (Brian Sheehy, Barry Walker, and Baorui Yang) and the Saclay group (Eric Mevel, Juleon Schins, and Pierre Breger). Very fruitful collaborations with the FOM group [Raluca Constantinescu and Harm Muller (who suggested the Augier idea)] and the LOA, ENSTA, and Ecole Polytechnique group (George Grillon, André Mysyrowicz, and André Antonetti) were all key factors in the success of the laser-assisted Auger experiment. The help of Michel Trahin in providing the H parameters is greatly appreciated. Both authors acknowledge travel support from a NATO Collaborative Research Grant, Contract No. SA.5-2-05(RG910678). The research at Brookhaven National Laboratory was carried out under Contract No. DE-AC02-76CH00016 with the U.S. Department of Energy and was supported by its Division of Chemical Science, Office of Basic Energy Sciences. The research at Saclay and Palaiseau was partially supported by three EC grants under Contract Nos. SC1-0103C, ERB4001GT921553, and ERB4050PL921025.

References

Agostini, P., and DiMauro, L. F. (1993). *Phys. Rev. A* **47**, R4573.
Agostini, P., Breger, P., L'Huillier, A., Muller, H. G., and Petite, G. (1989). *Phys. Rev. Lett.* **63**, 2208.
Amosov, M. V., Delone, N., and Krainov, V. (1986). *Sov. Phys.—JETP (Engl. Transl.)* **64**, 1191.
Augst, S., Strickland, D., Meyerhofer, D. D., Chin, S. L., and Eberly, J. H. (1989). *Phys. Rev. Lett.* **63**, 2212.
Auguste, T., Monot, P., Lompré, L. A., Mainfray, G., and Manus, C. (1992). *J. Phys. B* **25**, 4181.
Becker, W., Lohr, A., and Kleber, M. (1994). *J. Phys. B* **27**, L325.
Boyer, K., Thomson, B. D., McPherson, A., and Rhodes, C. K. (1994). *J. Phys. B* **27**, 4373.
Burnett, N. H., and Corkum, P. B. (1989). *J. Opt. Soc. Am. B* **6**, 1195.
Charalambidis, D., Lambropoulos, P., Schröder, H., Faucher, O., Xu, H., Wagner, M., and Fotakis, C. (1994). *Phys. Rev. A* **50**, R2822.
Cionga, A., Florescu, V., Maquet, A., and Taieb, R. (1993). *Phys. Rev. A* **47**, 1830.
Cohen, S., Camus, P., and Bolovinos, A. (1993). *J. Phys. B* **26**, 3783.
Corkum, P. (1993). *Phys. Rev. Lett.* **71**, 1994.
Corkum, P., Burnett, N., and Brunel, F. (1989). *Phys. Rev. Lett.* **62**, 1259.
Davidson, M. D., Schumacher, D. W., Bucksbaum, P. H., Muller, H. G., and van Linden van den Heuvell, H. B. (1992). *Phys. Rev. Lett.* **69**, 3549.
deBoer, M. P., and Muller, H. G. (1992). *Phys. Rev. Lett.* **68**, 2727.
deBoer, M. P., Hoogenraad, J. H., Vrijen, R. B., Noordam, L. D., and Muller, H. G. (1993a). *Phys. Rev. Lett.* **71**, 3263.
deBoer, M. P., Noordam, L. D., and Muller, H. G. (1993b). *Phys. Rev. A* **47**, 45.
Delone, N. B., Suran, V. V., and Zon, B. A. (1984). In *Multiphoton Ionization of Atoms* (Chin, S. L., and Lambropoulos, P., eds), Academic Press, New York.
Dietrich, P., Burnett, N. H., Ivanov, M., and Corkum, P. B. (1994). *Phys. Rev. A* **50**, R3585.
DiMauro, L. F. (1993). *Many-Body Theory of Atomic Structure and Photoionization*. World Sci. Publ. Co., Teaneck, New Jersey.
Faisal, F. H. M. (1973). *J. Phys. B* **6**, L89.
Faisal, F. H. M. (1994). *Phys. Lett.* **187**, 180.
Fiordilino, E., Zangara, R., and Ferrante, G. (1988). *Phys. Rev. A* **38**, 4369.
Fittinghoff, D. N., Bolton, P. R., Chang, B., and Kulander, K. C. (1992). *Phys. Rev. Lett.* **69**, 2642.
Freeman, R. R., Bucksbaum, P. H., Milchberg, H., Darack, S., Schumacher, D., and Geusic, G. M. (1987). *Phys. Rev. Lett.* **59**, 1092.
Freeman, R. R., Bucksbaum, P. H., Cooke, W. E., Gibson, G., McIlrath, T. J., and van Woerkom, L. D. (1992). In *Atoms in Intense Laser Fields* (Gavrila, M., ed.), Academic Press, San Diego.
Gallagher, T. F. (1988). *Phys. Rev. Lett.* **61**, 2304.
Gavrila, M. (1992). *Atoms in Intense Laser Fields*. Academic Press, San Diego.
Gibson, G. N., Freeman, R. R., and McIlrath, T. J. (1992). *Phys. Rev. Lett.* **69**, 1904.
Gibson, G. N., Freeman, R. R., McIlrath, T. J., and Muller, H. G. (1994). *Phys. Rev. A* **49**, 3870.
Gontier, Y., and Trahin, M. (1992). *Phys. Rev. A* **46**, 1488.
Gontier, Y., Rahman, N. K., and Trahin, M. (1975). *Phys. Rev. Lett.* **34**, 779.
Grobe, R., and Eberly, J. H. (1992). *Phys. Rev. Lett.* **68**, 2905.
Hou, M., Breger, P., Petite, G., and Agostini, P. (1990). *J. Phys. B* **23**, L583.
Javanainen, J., Su, Q., and Eberly, J. H. (1988). *Phys. Rev. A* **38**, 3480.
Keldysh, L. V. (1964). *Zh. Eksp. Teor. Fiz.* **47**, 1945.
Kroll, M. N., and Watson, K. M. (1973). *Phys. Rev. A* **8**, 804.

Kulander, K. C., Schafer, K. J., and Krause, J. L. (1992). In *Atoms in Intense Laser Fields* (Gavrila, M., ed.) Academic Press, San Diego.
Kulander, K. C., Schafer, K. J., and Krause, J. L. (1993). In *Super-Intense Laser-Atom Physics* (Piraux, B., L'Huillier, A., and Rzazewski, K., eds.), Plenum, New York.
Kulander, K. C., Cooper, J., and Schafer, K. J. (1995). *Phys. Rev. A* **51**, 561.
Kupersztych, J. (1987). *Europhys. Lett.* **4**, 23.
Lambropoulos, P., and Tang, X. (1987). *J. Opt. Soc. Am. B.* **4**, 821.
Lambropoulos, P., and Tang, X. (1992). In *Atoms in Intense Laser Fields* (Gavrila, M. ed.), Academic Press, San Diego.
Latinne, O., Klystra, N. J., Dörr, M., Purvis, J., Terao-Dunseath, M., Joachain, C. J., Burke, P. G., and Noble, C. J. (1995). *Phys. Rev. Lett.* **74**, 46.
Lewenstein, M., Kulander, K. C., Schafer, K. J., and Buckbaum, P. H. (1995). *Phys. Rev. A* **51**, 1495.
L'Huillier, A., and Balcou, P. (1993). *Phys. Rev. Lett.* **70**, 774.
L'Huillier, A., Lompré, L. A., Mainfray, G., and Manus, C. (1983). *Phys. Rev. A* **27**, 2503.
L'Huillier, A., Schafer, K. J., and Kulander, K. C. (1991). *Phys. Rev. Lett.* **66**, 2200.
Lotz, W. (1968). *Z. Phys.* **216**, 241.
Luk, T. S., McPherson, A., Boyer, K., and Rhodes, C. K. (1992). In *Atoms in Intense Laser Fields* (Gavrila, M., ed.) Academic Press, San Diego.
McIlrath, T. J., Freeman, F. R., Cooke, W. E., and van Woerkom, L. D. (1989). *Phys. Rev. A* **40**, 2770.
McPherson, A., Luk, T. S., Thomson, B. D., Borisov, A. B., Shyraev, O. B., Chen, X., Boyer, K., and Rhodes, C. K. (1994). *Phys. Rev. Lett.* **72**, 1810.
Mevel, E. (1994). Thèse de doctorat, Paris (unpublished).
Mevel, E., Breger, P., Trainham, R., Petite, G., Agostini, P., Migus, A., Chambaret, J. P., and Antonetti, A. (1993). *Phys. Rev. Lett.* **70**, 406.
Mittelman, M. (1966). *Phys. Rev. Lett.* **16**, 498.
Mohideen, U., Sher, M. H., Tom, H. W. K., Aumiller, G. D., Wood, O. R., II, Freeman, R. R., Bokor, J., and Buckbaum, P. H. (1993). *Phys. Rev. Lett.* **71**, 509.
Mott, N. F., and Massey, H. S. W. (1965). *The Theory of Atomic Collisions.* (Oxford Sci. Publ., New York.
Muller, H. G. (1990). *Comments At. Mol. Phys.* **24**, 355.
Nicklich, W., Kumpfmüller, H., Walther, H., Tang, X., Xu, H., and Lambropoulos, P. (1992). *Phys. Rev. Lett.* **69**, 3455.
Paulus, G. G., Nicklich, W., Huale, X., Lambropoulos, P., and Walther, H. (1994a). *Phys. Rev. Lett.* **72**, 2851.
Paulus, G. G., Nicklich, W., and Walther, H. (1994b). *Europhys. Lett.* **27**, 267.
Paulus, G. G., Becker, W., Nicklich, W., and Walther, H. (1994c). *J. Phys. B* **27**, L1.
Piraux, B., L'Huillier, A., and Rzazewski, K., eds. (1993). *Super-Intense Laser-Atom Physics*. Plenum, New York.
Potvliege, R., and Shakeshaft, R. (1992). In *Atoms in Intense Laser Fields* (Gavrila, M., ed.), Academic Press, San Diego.
Reiss, H. R. (1980). *Phys. Rev. A* **22**, 1786.
Reiss, H. R. (1987). *J. Phys. B* **20**, L79.
Saeed, M., Kim, D., and DiMauro, L. F. (1990). *Appl. Opt.* **29**, 1752.
Schafer, K. J., Yang, B., DiMauro, L. F., and Kulander, K. C. (1993). *Phys. Rev. Lett.* **70**, 1599.
Schins, J., Breger, P., Agostini, P., Constantinescu, R., Muller, H. G., Grillon, G., Antonetti, A., and Mysyrowicz, A. (1994). *Phys. Rev. Lett.* **73**, 2180.
Stapelfeldt, H., Kristensen, P., Ljungblad, U., Andersen, T., and Haugen, H. K. (1994). *Phys. Rev. A* **50**, 1618.
Story, J. G., Ducan, D. I., and Gallagher, T. F. (1993). *Phys. Rev. Lett.* **70**, 3012.
Story, J. G., Ducan, D. I., and Gallagher, T. F. (1994a). *Phys. Rev. A* **49**, 3875.

Story, J. G., Ducan, D. I., and Gallagher, T. F. (1994b). *Phys. Rev. A* **50**, 1607.
Suran, V. V., and Zapesochnyi, I. P. (1975). *Sov. Tech. Phys. Lett.* (*Engl. Transl.*) **1**, 420.
Tang, X., Chang, T. N., Lambropoulos, P., Fournier, S., and Di Mauro, L. F. (1990). *Phys. Rev. A* **41**, 5265.
van Druten, N. J., *et al.* (1994). *Phys. Rev. A* **50**, 1593.
van Linden van den Heuvell, H. B., and Muller, H. G. (1988). In *Multiphoton Processes* (Smith, S. J., and Knight, P. L., eds.), Cambridge Univ. Press, London.
Vrijen, R. B., Hoogenraad, J. H., Muller, H. G., and Noordam, L. D. (1993). *Phys. Rev. Lett.* **70**, 3016.
Walker, B., Mevel, E., Yang, B., Breger, P., Agostini, P., and DiMauro, L. F. (1993). *Phys. Rev. A* **48**, R894.
Walker, B., Sheehy, B., DiMauro, L. F., Agostini, P., Schafer, K. J., and Kulander, K. C. (1994). *Phys. Rev. Lett.* **73**, 1227.
Wehlitz, R., *et al.* (1991). *Phys. Rev. Lett.* **67**, 3764.
Yang, B., Schafer, K. J., Walker, B., Kulander, K. C., Agostini, P., and DiMauro, L. F. (1993). *Phys. Rev. Lett.* **71**, 3770.
Zangara, R., Fiordilino, E., and Ferrante, G. (1990). *Opt. Acoust. Rev.* **1**, 297.

INFRARED SPECTROSCOPY OF SIZE-SELECTED MOLECULAR CLUSTERS

UDO BUCK

Max-Planck-Institut für Strömungsforschung, 37073 Göttingen, Germany

I. Introduction 121
II. Experimental Methods 124
   A. Size Selection 124
   B. Infrared Photodissociation 127
   C. Double-Resonance Experiments 130
III. Theoretical Methods 132
   A. Structure Calculations 132
   B. Vibrational Spectra 134
   C. Simulation of Temperature Effects 135
IV. Results 136
   A. Spectra and Structures 136
   B. Excitation and Decay Mechanisms 148
   C. Phase Transitions 151
V. Conclusions 155
   References 159

I. Introduction

The investigation of properties and dynamics of molecular clusters has attracted much interest (Benedek *et al.*, 1988; Scoles, 1990). These clusters are usually weakly bound by either van der Waals forces or hydrogen bonds. They are expected to form the link between the behavior of single molecules in the gas phase and that in the bulk condensed phase. Compared with crystalline matter, they are of finite size without the symmetry of the space group, they have a large surface to volume ratio, and they can change their properties when adding or subtracting one constituent. Compared with stable molecules, the clusters form many more isomeric structures which can easily be interchanged. These transitions resemble to a certain extend the phase transitions of the bulk. It is this general transitional behavior which is studied to elucidate our understanding of fundamental problems such as nucleation phenomena and phase transitions as well as solvation problems and cluster specific reactions.

In this chapter we would like to address the behavior of small, mainly homogeneous molecular clusters in the size range from $n = 2$ to $n = 20$. In this range a reliable characterization of molecular clusters is lacking. The smallest complexes, the dimers, are usually well studied in numerous high-resolution spectroscopic experiments (Nesbitt, 1988), and large clusters are very often well described by the behavior of the condensed phase which is explored in electron diffraction experiments (Farges *et al.*, 1988; Bartell *et al.*, 1989, and references cited therein). Aside from the more general questions on the structure of these clusters, also their dynamical behavior is of special interest, since the clusters play a key role in getting the basic information on the microscopic interactions and perturbations between solvents and solutes in solutions. This will include the investigations of the problem of isomeric transitions, the connected question of phase transitions (Berry *et al.*, 1988; Jortner *et al.*, 1988), and the dissociation after the excitation of the molecular vibrations. The flow of energy between an excited intramolecular mode and the motion along the weakly bound intermolecular dissociation coordinate is an extremely interesting mode coupling problem which has attracted, and still attracts, much interest (Janda, 1985; Miller, 1988; Zewail, 1988).

The common tool in all these investigations is infrared spectroscopy which has been and still is one of the classic methods to get structural information about molecules. Therefore it is not surprising that this method has also been widely applied to molecular clusters. In particular, these studies include direct absorption experiments with infrared (Nesbitt, 1988) and far-infrared excitation (Heath *et al.*, 1990; Blake *et al.*, 1991) and the optothermal detection method (Gough *et al.*, 1978; Miller, 1986, 1988; Coker and Watts, 1987), in which the positive (excitation) or negative (dissociation) energy content of the beam is detected by a low-temperature bolometer. All these methods, however, work only for dimers and trimers, since in these cases the spectrum itself can be used for identifying the cluster size based on accompanying calculations.

For larger clusters additional size specific information is necessary. The reason is that the techniques for generating free cluster beams, the supersonic adiabatic expansion or the aggregation in cold gas flows, produce in almost all cases a distribution of cluster sizes (Kappes and Leutwyler, 1988). In principle, the problem could be solved by a size-specific detection method. The most commonly used method for this purpose, however, the ionization and the subsequent mass selection in a mass spectrometer, is hampered by the ubiquitous fragmentation during the ionization process. It is caused by the energy released into the system as the clusters change from their neutral to their ionic equilibrium structure. This excess energy then leads to evaporation of neutral subunits, and thus fragmentation occurs (Haberland, 1985; Saenz *et al.*, 1985; Buck, 1988). For the molecular species considered here, the resulting fragmentation pattern is often modified by fast chemical reactions of the ionized molecular units with neutral partner molecules within the cluster (Buck, 1988, 1990a). In any case, a simple mass spectrum does not characterize at all the neutral cluster distribution.

Therefore selection methods have to be applied to make sure that the experiments are carried out with one neutral cluster size only. One possibility is to use *special ionization techniques* for a certain class of molecules. A well-known example is the two-color resonant two-photon ionization of aromatic molecules (Börnsen et al., 1989; Leutwyler and Bösinger, 1990). In this case the neutral and ionic equilibrium structures are similar, and by carefully adjusting the ionization energy near the threshold region fragmentation is avoided. In addition, the first step, the excitation of a suitable electronic state can be made size specific, so that a very reliable method which is, however, restricted to electronic excitation results. An alternative method is the *momentum transfer* in a scattering experiment with atoms. In this way mass distributions of very large clusters ($n = 10,000$) were obtained by deflection from a crossed jet under multiple collision conditions (Gspann and Vollmer, 1974). In the high-resolution version of this method, small clusters ($n \leqslant 8$) were separated from each other by making use of their different angular and velocity distribution in a single collision experiment (Buck and Meyer, 1986; Buck, 1988; 1990a).

Thus it is the main purpose of this chapter to summarize studies on the structure and dynamics of molecular clusters using the scattering method for the preparation of clusters of one size and combining it with depletion spectroscopy based on infrared photodissociation. The selection method is followed by an appreciable loss of intensity, and thus certain restrictions in the experimental variety occur. But by applying this technique, we have the great advantage that we definitely know that we measure the spectra of one cluster size only.

We start the article in Section II with a description of the experimental techniques used in these experiments, the size selection by momentum transfer, the vibrational predissociation, and double resonance methods. Then the theoretical tools which include calculation of the cluster structures, discussion of the intermolecular potentials involved, and calculation of the corresponding vibrational spectra are presented in Section III.

The simulation of temperature effects which describe the transition between solid-like and liquid-like behavior points to the very attractive field of phase transitions in these finite systems. The rigid–nonrigid transition of clusters is, in general, described as multistate isomerization (Berry et al., 1988; Berry, 1990; Jortner et al., 1988, 1990). The great majority of the investigations are carried out by computer simulation. Although the calculation of IR spectra revealed detailed dependences on the melting phenomena (Eichenauer and LeRoy, 1988; Kmetic and LeRoy, 1991), conclusive experimental information is difficult to get, since the experiments are performed with a distribution of cluster sizes (Gu et al., 1990). In addition, it turned out to be misleading to derive from spectral line shapes alone conclusions on such transitions (Ben-Horin et al., 1992).

Section IV presents the results obtained for size-selected molecular clusters. We start with a detailed presentation of the measured spectra and their relation to the cluster structure. We try to answer the following questions: (1) How can the spectrum, characterized by the line shifts, be related to the structure of the clusters? (2) What is the dependence on cluster size and excited mode? (3) Are there differences for van der Waals or hydrogen-bonded systems and what are the reasons for these? Examples will be presented as case studies for the van der Waals system $(C_2H_4)_n$, the mixed cluster $C_2H_4-(CH_3COCH_3)_m$, the linear hydrogen bonded $(CH_3OH)_n$, the nonlinear hydrogen bonded $(N_2H_4)_n$, and the dipolar interaction bonded $(CH_3CN)_n$. The answers will demonstrate the large differences among these systems. Then we will shortly treat the decay mechanisms and the problems which arise, if more than one photon is necessary for the dissociation of the cluster. Finally calculations for the structural transitions of small methanol clusters which are confirmed by the first measured example for an isomeric transition, that of the methanol hexamer, will be presented.

In Section V conclusions of what is known on the structure and the dynamical behavior of molecular clusters in the investigated size range will be drawn. In addition, we will discuss new experimental and theoretical approaches which include the extension of the wavelength range in IR excitation and new methods like the spectroscopy of molecules and their clusters which are imbedded in large rare-gas clusters.

Several articles and reviews on the spectroscopy with size-selected molecular clusters have already appeared. Some of them are relatively short contributions to conferences (Buck, 1990b, 1992a, 1993) which also address other subjects. Others deal only with a certain aspect of the field or do not contain the latest results which have been obtained both in the interpretation of the measured infrared spectra and in terms of new experiments on isomeric transitions (Huisken, 1991; Buck, 1994a,b).

II. Experimental Methods

A. SIZE SELECTION

The method of size selection by momentum transfer in a scattering experiment with atoms under single collision conditions has been described in detail in the literature (Buck and Meyer, 1986; Buck, 1988, 1990b, 1992b) so that we give here only a short account of the principle. The method is based on the fact that the heavier clusters are scattered into smaller angular ranges

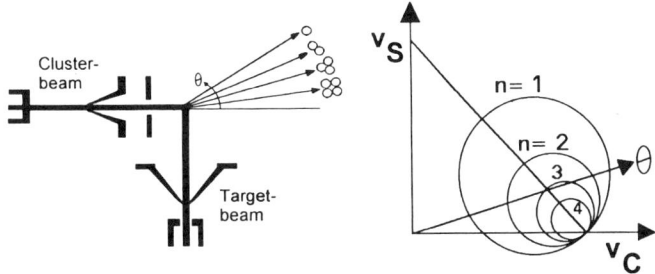

FIG. 1. Schematic arrangement and Newton diagram for the scattering of a cluster beam (velocity v_c) by a target beam (velocity v_s). The circles are the positions of elastically scattered clusters.

with different final velocities compared with the lighter clusters. This is demonstrated pictorially in Fig. 1.

It is best explained by a velocity vector (or Newton) diagram which is constructed on the basis of conservation of momentum and energy. Such a schematic diagram is also shown in Fig. 1, assuming that the cluster beam (velocity, v_c; mass, $n \cdot m_1$) is crossed by the atomic beam (v_s, m_s) at an intersection angle of 90°. The relative velocity is given by the difference vector $\mathbf{g} = \mathbf{v}_c - \mathbf{v}_s$. For elastic scattering all final center-of-mass (cm) velocities u'_c are restricted to end on a sphere around the center-of-mass with the radius

$$u_c^{(n)} = m_s g/(nm_1 + m_s). \tag{1}$$

For the used in-plane geometry of the two beams and the detector, the spheres are reduced to circles. In the case of inelastic scattering, one has to take into account the transferred energy ΔE which changes u'_c according to $u'_c = u_c(1 - \Delta E/E)^{1/2}$, where E is the collision energy.

Since the clusters have nearly the same velocity v_c, also the relative velocity is independent of the cluster size, but the final cm velocities $u_c^{(n)}$ are quite different. With increasing mass of the clusters they decrease and so do the radii of the circles in the diagram (see Eq. (1) and Fig. 1). For the laboratory (lab) angle shown in the diagram only trimers, dimers, and monomers are detected, whereas the maximum scattering angle Θ_4 for tetramers is smaller. It is immediately clear from this picture that larger clusters can be easily excluded by choosing the correct *angle* for detection. For the necessary additional discrimination against smaller clusters, two solutions are used, the measurement of the velocity and the measurement of the mass (Buck and Meyer, 1986; Buck, 1988, 1990a).

In the first method, in addition to the angle, the final *velocity* is specified. For this purpose, either time-of-flight analysis or a mechanical velocity selector is used. This procedure is general and completely independent of the subsequent detection process. In the second method a *mass* spectrometer is used to discriminate against the smaller clusters. This procedure works only if, at least, a small fraction of the cluster M_n is detected at the nominal mass of the ion M_n^+ or at a fragment mass M_m^+ which is larger than that of the next smaller cluster M_{n-1}^+. This method is used for the size selection of the molecular clusters presented here. They are very often detected at the protonated masses, so that, for example, $(CH_3OH)_n$ appears at $(CH_3OH)_{n-1}H^+$, just one mass unit larger than the next smaller nominal cluster mass.

The procedure requires, in general, a high-resolution molecular beam apparatus and intense cluster beams with good expansion conditions to minimize the angular ($\Delta\theta < 0.1°$) and velocity spread ($\Delta v/v < 0.05$) of the colliding beams. The first part is realized by skimmed beams introducing additional apertures. The latter is usually achieved by expanding a mixture of 2 to 10% of the molecule in helium or neon.

The natural limit of this procedure for size selecting clusters is given by the fact that the maximum deflection angles for the different sizes come closer and closer together with increasing n. In the experimental arrangement, it is no problem to select clusters with $n = 6$ for monomer masses around $m_1 = 50$ amu, when He is used as scattering partner ($m_s = 4$ amu). These figures depend, however, very much on the masses and velocities of the particles. Increasing $m_c = nm_1$ and v_c deteriorates the resolution, while increasing m_s and v_s improves it. In the former case the Newton circles shrink, while in the latter case the contrary is valid. For the masses, this result follows directly from Eq. (1). For the velocities, it can be visualized from Fig. 1.

These considerations were demonstrated in experiments with acetonitrile which was seeded in neon instead of helium, which decreases v_c, and scattered from He, Ne, and Ar which increases m_s. At a lab scattering angle of 6° cluster sizes of $n = 5, 12$, and 20 are detected for the scattering from He, Ne, and Ar, respectively. In this way $(CH_3CN)_n$ clusters were selected up to 33 ± 8 (Buck, 1992b; Buck and Ettischer, 1994b). The error range which is calculated from the mass (lower limit) and the deflection angle (upper limit) reflects the largest possible limits and is probably much smaller in reality. In general, the intensity loss which is caused by the scattering process is in the order of 10^{-6} so that size-selected neutral beams of one cluster size with the intensity of 3×10^{12} particles per steradian and second are obtained.

It is noted that, during the scattering process, a certain amount of energy is transferred into the cluster. This is especially valid, if Ne or Ar is used as the scattering partner instead of He. The very fact does not disturb the size selection. As already mentioned earlier, the radii of the Newton circles get smaller and the procedure works as with elastic scattering. The reason is twofold. First, the different cluster sizes are, in general, affected in a similar way by the inelastic scattering, and thus all the circles become smaller, and second, by selecting the deflecting angle for the detection close to the elastic limit, the influence of the internal excitation can be reduced. In any case, this effect can be measured by time-of-flight analysis and taken into account in the data evaluation.

B. INFRARED PHOTODISSOCIATION

The method of size selection is now combined with an adequate spectroscopic procedure. Here we use the well-established depletion technique by vibrational predissociation (Miller, 1988; Gough et al., 1986; Vernon et al., 1982; Hoffbauer et al., 1983). The general process is illustrated in Fig. 2 (LeRoy et al., 1990). The monomer with the vibrational coordinate q and the intramolecular potential $U_m(q)$ is excited within the cluster from the ground state $v = 0$ to the excited state $v' = 1$. The corresponding intermolecular interaction potential $V(q, R)$ of the complex with the stretching coordinate R and the vibrational quantum number n are shown on the right-hand side of Fig. 2. Here the excited potential shifts upward and, in general, is slightly modified. The transition frequency $hv_0 = E'_m - E_m$ shifts,

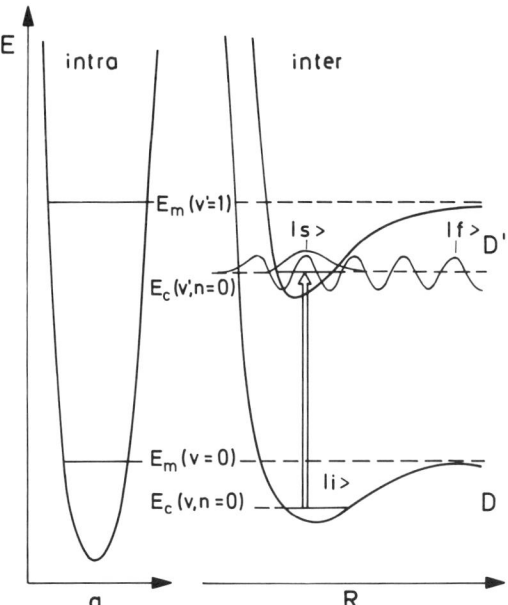

FIG. 2. Schematic diagram of vibrational predissociation of molecular clusters which are first excited from $|i\rangle$ to the bound state $|s\rangle$, and then $|s\rangle$ is coupled to the continuum of the ground state $|f\rangle$. q and R are the intra- and intermolecular coordinates.

if the binding energy of the complex in the excited state $D' = E'_m - E'_c$ is different from that in the ground state $D = E_m - E_c$. According to Fig. 2, the transition of the complex is given by

$$hv = E'_m - D' - (E_m - D) = E'_m - E_m + (D - D') = hv_0 + \Delta hv. \quad (2)$$

The shift Δhv is given by $D - D'$ so that a red (blue) shift occurs, if the binding energy in the excited state is larger (smaller) than that in the ground state.

The dissociation proceeds in two steps. First, the luster is excited from the zeroth-order state $|\Psi_i\rangle$ to $|\Psi_s\rangle$, and then the bound state $|\Psi_s\rangle$ gets coupled to the continuum of the ground state $|\Psi_f\rangle$ with the final translational energy E_t.

In the framework of first-order time-dependent perturbation theory for the electric dipole transition and the golden rule for the decay rate Γ, the dissociation cross-section is expressed by (Beswick, 1987)

$$\sigma(i \to f, v) = \frac{\pi^2}{\epsilon_0} \frac{v}{c} |\langle \Psi_s|\mathbf{me}|\Psi_i\rangle|^2 \frac{2}{\pi} \frac{\Gamma}{(E - E_s)^2 + \Gamma^2} \quad (3)$$

with

$$\Gamma = \pi|\langle \Psi_s|V(q, R)|\Psi_f\rangle|^2. \quad (4)$$

Here, **m** is the transition dipole moment, **e** the polarization vector of the photon, ϵ_0 is the dielectric constant, and c is the velocity of light. This expression reflects the two steps in the dissociation process, the excitation to $v = 1$ and the coupling to the dissociation continuum.

The dissociation spectra contain the following information:

1. The *line shift* Δhv which is caused by the interaction of the excited oscillator with the surrounding molecules gives information on the structure of the clusters.
2. The *linewidth* Γ contains, if homogeneously broadened, information on the lifetime and thus on the dynamical coupling of the intramolecular vibrational mode to the intermolecular modes of the cluster.
3. The *cross-section* σ which describes the excitation and decay process.

The actual experimental arrangement is shown pictorially in Fig. 3. By adjusting the detector to a certain angle and mass, one cluster size is selected. Then in the setup depicted in Fig. 3a the scattered beam is attenuated by the radiation of a laser beam counter-propagating along the beam direction. The difference to the setup depicted in Fig. 3b will be discussed later on. The dissociation is measured by monitoring the decrease in the intensity as a function of laser frequency v and laser fluence F. The

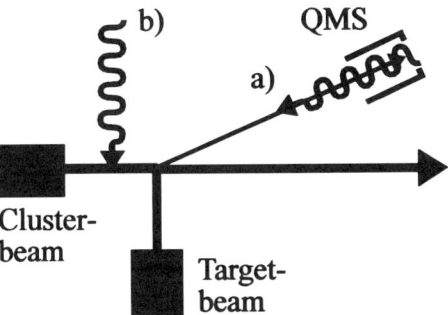

FIG. 3. Experimental arrangements to measure photodissociation depletion spectra with size-selected clusters for (a) collisional excited and (b) cold clusters. The size selection is achieved by deflection and a quadrupole mass spectrometer (QMS).

fraction of dissociating molecules P_{dis} is given by

$$P_{dis}(v) = \frac{N_0 - N(v)}{N_0} = 1 - \sum_i X_i \exp[-\sigma_i(v)F/hv], \qquad (5)$$

where N_0 and $N(v)$ are the cluster signals with laser off and on, X_i is the fraction of isomer i in the beam, v is the laser frequency, and $\sigma_i(v)$ is the dissociation cross-section of the corresponding isomer which is usually expressed by the Lorentzian line shape of Eq. (3). The three quantities Δhv, Γ, and σ are then obtained from the position, the width, and the amplitude of the Lorentzians fitted to the data. In the case of more than one line per cluster, $\sigma_i(v)$ has to be replaced by a sum over these lines. It is noted that the implicit assumption in these derivations, the homogeneity of the linewidth Γ, is not necessarily fulfilled by the measurements which might be inhomogeneously broadened by rotational and librational states. In such a case Γ is simply a measure of the width of the distribution.

Because of the scattering process, the dissociation takes place with internally excited clusters. In order to avoid this excitation, the complementary arrangement (b) also shown in Fig. 3 is used (Huisken, 1991). Here, the laser–molecular beam interaction takes place before the scattering center, where the clusters are still cold. Then the cluster beam is dispersed by the He beam, and cluster-specific detection is obtained as in the first case. Although the laser interacts with all clusters in the beam, only the dissociation of the cluster size to which the detector is adjusted is measured. This arrangement can cause problems, if a larger cluster which is a dissociation product does not leave the beam fast enough and reaches the scattering region, where it is counted as nondissociated and contaminates the detector signal. Thus method of Fig. 3a is better suited for measuring larger clusters, since they are dissociated after the size selection, while method of Fig. 3b is superior for resolving structure in smaller clusters, since the clusters are colder.

This excitation, however, can in turn be used to study the influence of the internal excitation on the measured spectra. The results show that the influence becomes less important the larger the cluster is. Thus we observed no difference in the width of the spectra of $(CH_3CN)_n$ clusters for $n = 8$, $n = 9$, and $n = 24$ which were selected by scattering from He, Ne and Ar, respectively (Buck and Ettischer, 1994b). Apparently, the large number of degrees of freedom of these clusters helps to distribute the energy so that no measurable effect is found in the spectra. This behavior was observed when the spectra of collisionally deflected clusters were compared with those obtained from cold ones. Only for dimers with their restricted possibilities to distribute the energy are large differences observed (Buck et al., 1990a, 1991b; Huisken and Stemmler, 1988, 1992).

The actual experimental setup for measuring depletion spectra of size-selected clusters consists of a high-resolution crossed molecular beam machine coupled with laser excitation. For the excitation in most cases a high-power continuous wave CO_2-laser in the 10-μm range is used in order to achieve a reasonable fraction of dissociating clusters. The wavelength range has been extended to the 3-μm range by employing a Nd:YAG-laser-pumped optical parametric oscillator (Huisken et al., 1991). Thus also the important stretching vibrations in which hydrogen atoms are involved become experimentally accessible by this method.

C. DOUBLE-RESONANCE EXPERIMENTS

An inherent problem of cluster research is the existence of many isomers. In favorable cases their spectral features are well separated from each other so that as in the example of the HCN trimer the linear and the cyclic structures have been identified (Jucks and Miller, 1988). Otherwise the techniques of double resonances or spectral hole burning have to be used (Wittmeyer and Topp, 1991). In such an experimental setup the pump laser reduces or increases the population of a subset of states by a substantial amount, and then this change is probed by a second laser that is scanned through the spectral range of interest. If these spectroscopic techniques are combined with size-selected clusters, direct information on structural isomers is obtained.

Several schemes in which the size-specific detection and the probing are obtained by resonant enhanced two-photon ionization of aromatic molecules have been employed. For the pump process ultraviolet photons (Scherzer et al., 1992), infrared photons (Pribble and Zwier, 1994), or stimulated Raman scattering (Venturo and Felker, 1993) is used. We have combined the size selection by momentum transfer in a scattering process with atoms and the vibrational depletion spectroscopy described earlier with a second infrared laser for the double-resonance condition (Buck and Hobein, 1993).

The experimental arrangement is that of Fig. 3 with both lasers counter-propergating to the scattered beam. The pump laser is fixed at one frequency, v_1, and irradiates the cluster beam constantly. Therefore we have to replace N_0 in Eq. (5) with $N_0^{dr} = N_0 \Sigma_i X_i \exp[-\sigma_i(v_1)F/hv_1]$. Then we get for the dissociated fraction as a function of the probe laser frequency v_p a similar expression,

$$P_{dis}(v_1, v_p) = 1 - \sum_i X_i^{dr} \exp[-\sigma_i(v_p)F/hv] \quad (6)$$

with

$$X_i^{dr} = \frac{X_i \exp[-\sigma_i(v_1)F/hv]}{\Sigma_i X_i \exp[-\sigma_i(v_1)F/hv]}.$$

Assuming that there is only one type of isomer present in the beam ($X_1 = 1$ and $X_i = 0$ for $i > 0$), we get $X_1^{dr} = X_1$, which means that no change in the spectrum occurs. The pump laser increases the absolute number of dissociated clusters but does not change the frequency dependence of the dissociation. In the case of two or more isomers, however, the shape of the spectrum changes, if the different isomers have different line shapes or positions. The result of a simulation is shown in Fig. 4. Here we assume that the cluster beam contains equal amounts ($X_1 = X_2 = 0.5$) of two isomers with overlapping spectral lines centered at 1040 and 1060 cm^{-1} and that both have a width of 15 cm^{-1}. The resulting spectrum is represen-

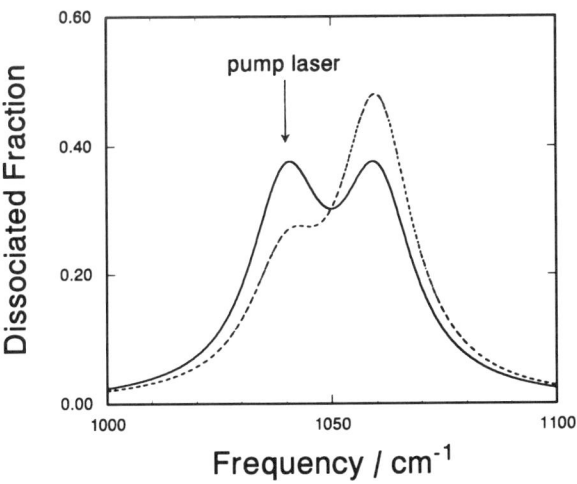

FIG. 4. Simulated photodissociation spectra for demonstrating double resonance experiments. The solid line represents a single laser spectrum using arbitrary Lorentzian parameters and laser fluences (see the text). If a pump laser is added at the fixed frequency marked by the arrow, the dashed line is calculated for the case of two isomers of equal fractions being responsible for the double-peak structure.

ted by the solid line. With a maximum dissociation cross-section of 10^{-18} cm^2 and a fluence of 21 mJ/cm^2 for both lasers, the pump laser is tuned to the maximum of the first peak in order to dissociate isomer 1. This results in a reduced fraction of $X_1^{\text{dr}} = 0.29$ and an increased fraction of $X_2^{\text{dr}} = 0.71$ for isomer 2 which is represented by the dashed line. Here, the advantage of measuring only the *probe* dissociation, relative to the reduced beam signal with a switched-on pump laser, becomes clear: Peak 1 decreases and peak 2 increases in a way that directly reflects the relative changes of the isomer fractions. In this way the measured spectral lines or bands can be unambiguously assigned to several or one isomer.

III. Theoretical Methods

A. STRUCTURE CALCULATIONS

Theoretical calculations of the structure are very important tools for the interpretation of the data, since complete experimental information on the cluster structure comparable to that resulting from X-ray diffraction analysis in solid-state physics is not yet available. For the structure calculations which are based on procedures of finding the minimum of the total energy, a reliable interaction potential has to be used. Complete *ab initio* calculation of the total energy surfaces including correlations are a state-of-the-art task. In many cases model potentials which are based on reliable multicenter pair potentials are used. These potentials are usually written as

$$V = V_{\text{rep}} + V_{\text{dis}} + V_{\text{elec}} + V_{\text{ind}}. \tag{7}$$

V_{rep} is the repulsive first-order contribution to the electronic exchange energy. The distance and orientation dependence can be approximated by a sum of site–site interactions at i and j on different molecules,

$$V_{\text{rep}} = \sum_{i,j} A_{ij} \exp(-B_{ij} R_{ij}). \tag{8}$$

It is usually obtained from *ab initio* calculations at the Hartree–Fock self-consistent field (SCF) level (Greer *et al.*, 1989; Kofranek *et al.*, 1987; Bone *et al.*, 1990) or from other approximative schemes like the test-particle method of Ahlrichs and co-workers (Böhm and Ahlrichs, 1977; Ahlrichs *et al.*, 1990) or the systematic model of Wheatley and Price (1990) which is based on properties of *ab initio* wave functions of the monomer. A further approximative step in which the parameters are fitted to experimental data is done in empirical methods (Nemenoff *et al.*, 1978; Jorgensen, 1986).

The second term represents the attractive long-range dispersion or van der Waals forces which are obtained in second-order perturbation theory and which are often truncated after the first instantaneous dipole–dipole term,

$$V_{\text{dis}} = -\sum_{i,j} C_{ij}/R_{ij}^6 \cdot F_{ij}(R_{ij}) - \cdots. \tag{9}$$

A damping function, $F_{ij}(R_{ij})$, reduces this term at small distances (Ahlrichs et al., 1977; Tang and Toennies, 1984). This contribution is calculated by *ab initio* methods or taken from semiempirical formulas using the known polarizabilities.

The third term is the first-order contribution to the electrostatic energy

$$V_{\text{elec}} = 1/(4\pi\epsilon_0) \sum_{i,j} q_i q_j / R_{ij} \pm \cdots, \tag{10}$$

which is written here as the sum of interactions between point charges q_i. ϵ_0 is the dielectric constant. The site charges can be determined by calculating or measuring the electrostatic multipole potential (Stone, 1981; Wheatley and Price, 1990) which is then reproduced by the point charge model or a truncated series of the multipoles themselves.

The last term is caused by the induction of multipole moments in the polarizable molecule j by the charge distribution of the other molecules at different sites i, i' and vice versa,

$$V_{\text{ind}} = -\frac{1}{2} \sum_{i,i'} \sum_j \alpha_j \frac{q_i q_{i'}}{R_{ij}^2 R_{i'j}^2} - \cdots. \tag{11}$$

This is the only term which is nonadditive and which contributes, in the case of hydrogen-bonded systems, to the well-known cooperative effect, an increase of the incremental bonding energy per molecule with increasing cluster size (Karpfen et al., 1983; Detrich et al., 1984; Curtiss and Blander, 1988). For van der Waals systems it is small, and very often the pairwise additive model is also used for hydrogen-bonded systems. To account for the nonadditive contributions without calculating it, the parameters of the model are sometimes adjusted accordingly to measurements of the condensed phase (Jorgensen, 1986). Once the potential parameters are chosen, the structures are obtained by starting with randomly chosen geometries and minimizing the total energy. We would like to summarize this section by stating that the intermolecular potential is modeled in three different stages of sophistication:

1. *ab initio:* In this case a complete calculation based on the Hartree–Fock approximation with configuration interaction corrections is performed.
2. *systematic model:* Here the formulae just given are used and the parameters are determined from simplified calculations (Böhm and Ahlrichs, 1977; Wheatley and Price, 1990).

3. *empirical:* In this case simple Lennard–Jones or similar potentials together with the electrostatic term which are fitted to calculations and data of the condensed phase are used (Jorgensen, 1986; Nemenoff *et al.*, 1978).

B. Vibrational Spectra

For the calculation of the vibrational line shift a method which is originally derived for solvents in liquids is used (Buckingham, 1960). This procedure can be adapted to also treat molecular clusters: As a starting point both the intra- and the intermolecular potentials are expanded in a Taylor series in dimensionless normal coordinates (Westlund and Lynden-Bell, 1987) of the isolated molecule

$$U_m(q) = \frac{1}{2}\sum_i \omega_i q_i^2 + \frac{1}{6}\sum_{ijk} \phi_{ijk} q_i q_j q_k + \cdots \qquad (12)$$

$$V(q, R) = V_0(R) + \sum_i \frac{\partial V}{\partial q_i} + \frac{1}{2}\sum_{ij} \frac{\partial^2 V}{\partial q_i \partial q_j} q_i q_j + \cdots. \qquad (13)$$

The first equation corresponds to the conventional normal mode approach including cubic anharmonicities. Note that for symmetric molecules some of the cubic force constants ϕ_{ijk} vanish. The second equation describes the intermolecular interaction in terms of the same coordinates. Using nondegenerate perturbation theory one obtains for the shift of the vibrational excitation frequency ($v = 0 \to v = 1$) (Buck and Schmidt, 1990)

$$\Delta \hbar \omega = \frac{1}{2}\frac{\partial^2 V}{\partial q_i^2} - \sum_j \frac{\phi_{iij}}{\omega_j}\frac{\partial V}{\partial q_j}. \qquad (14)$$

The first term, which is obtained in the first order, represents the change of the force constant, while the second term obtained in the second order represents the effect of a force ($-\partial V/\partial q_j$) which shifts the equilibrium value of the normal coordinate multiplied by the corresponding anharmonic force constant. Thus a steeper potential ($\partial^2 V/\partial q_i^2 > 0$) causes a blue shift, while an elongation of the bond ($\partial V/\partial q_i < 0$) leads to a red shift, if the anharmonic force constant is negative (Buck and Schmidt, 1993). In case of homogeneous clusters, the perturbation theory has to be extended to degenerate states and instead of using Eq. (14) the result is obtained by diagonalization of the corresponding perturbation energy matrix (Buck and Schmidt, 1993). In this way the different vibrationally excited modes in the cluster are coupled. This leads to a splitting of the frequencies corresponding to the excitation of different collective modes of vibration. These modes can be viewed as linear combinations of the respective normal modes of the constituent molecules. In the same way also transition dipole moments can be calculated as a vector sum of the moments of the individual molecules. In the second order, the excited modes are coupled to the other vibrational modes of the same molecule. A further improvement in which also the vibrations of the different molecules are coupled has been introduced by Beu (1994). We call this method the *molecular* approach.

The disadvantage of this approach lies in the fact that in the zeroth order the somewhat unrealistic harmonic approximation is used. Therefore we have developed a different procedure that starts from the complete cluster and that includes the anharmonic intramolecular force field of the single molecules and the complete intermolecular potential between the monomer constituents. After the normal mode analysis of the complete cluster, the anharmonic corrections are calculated in the usual perturbation theory up to quartic corrections (Siebers, 1994). The procedure has, in addition, the advantage that the complete set of frequencies for all modes is obtained. This method is called the *cluster* approach.

C. SIMULATION OF TEMPERATURE EFFECTS

For the comparison with real experimental data, the calculations of the structure and the line shifts have to be combined with simulations based on Monte Carlo (MC) sampling or molecular dynamics (MD) calculation for finite temperatures. Such calculations have been performed using the quantum mechanical approach based on perturbation theory and the classical MC method for the ensemble averaging (Buck et al., 1993a).

In order to characterize structural transitions, relative root-mean-square (rms) fluctuations of distances δ_{cm} between the centers of mass of the molecules are calculated,

$$\delta_{cm} = \frac{2}{n(n-1)} \sum_{i<j}^{n} \frac{(\langle r_{ij}^2 \rangle - \langle r_{ij} \rangle^2)^{1/2}}{\langle r_{ij} \rangle}, \tag{15}$$

where r_{ij} is the distance between the center of mass of molecule i and that of molecule j. According to the Lindemann criterion a substance undergoes melting when δ_{cm} reaches about 0.1. In order to get information on the rigid or fluxional behavior of the cluster, the mean square displacements (MSD) of the center-of-mass positions $\vec{r}_i$ of the molecules as a function of time

$$\langle \vec{r}(t)^2 \rangle = \frac{1}{n_{t_0}} \frac{1}{n} \sum_{j=1}^{n_{t_0}} \sum_{i=1}^{n} (\vec{r}_i(t_{0_j} + t) + \vec{r}_i(t_{0_j}))^2, \tag{16}$$

where n_{t_0} is the number of different time origins, are calculated. Note that the slope of the MSD is directly proportional to the diffusion coefficient. Thus a rigid cluster exhibits no dependence on time, while for liquid-like clusters a finite slope results. Because this quantity is time dependent, it cannot be obtained from the MC simulations.

IV. Results

A. SPECTRA AND STRUCTURES

The dissociation spectra of the systems measured in the arrangements described earlier show a large variety of behavior depending on the type of bonding, the excited vibrational mode, and, most importantly, the cluster size. An overview of all the systems measured so far is given in Table I. We will discuss the results for the *line shifts* and their connection to the underlying structure in greater detail for five typical examples which represent different bonding types.

1. A van der Waals System: Ethylene

This system was the first one investigated with size-selected clusters (Buck et al., 1987, 1988a; Huisken and Pertsch, 1987; Ahlrichs et al., 1990) and gave the surprising result that all the spectra in the region of the out-of-plane bending mode (v_7-B_{1u}) at 949 cm^{-1} are shifted by about 3 cm^{-1} to the blue, independent of the cluster size from $n = 2$ to $n = 6$. An attempt to explain this result by the predominance of chainlike structures in the crosswise arrangement of the dimer, which were found in structure calculations of these clusters, was not conclusive (Ahlrichs et al., 1990). No convincing reason could be given as to why the ringlike structures also found in the calculations were not present in the experiment. We performed calculations of the lineshifts of the dimer, trimer, and tetramer using the procedure presented in Section III.B (Buck and Schmidt, 1994). In all cases we found blue shifts between 0.7 and 2.7 cm^{-1}, caused by a slight enhancement of the force constant. The results for the different ring and chain isomers differed by 0.5 cm^{-1} only. Thus the sign and the magnitude of the shifts were within the error limits, in good agreement with the experiment. We conclude that, for this system and the excited mode investigated here, the shifts are small, they do not depend sensitively on the cluster size, and they are similar for the different isomers. Similar results have been found for other van der Waals clusters like those of benzene (de Meijere and Huisken, 1990) and OCS (Krohne et al., 1995), while in the case of SF$_6$ characteristic line splittings are observed as predicted by the dynamically dipole-induced dipole model (Huisken and Stemmler, 1989).

2. A Mixed-Cluster System: Acetone–Ethylene

In contrast to the systems discussed in the last section, the mixed clusters of ethylene molecules with acetone behave differently. The experiments have been carried out near the v_7 mode of ethylene at 949.3 cm^{-1}. After the first

TABLE I

Measured Frequencies of the Investigated Clusters in Size-Selective Experiments, Given in cm^{-1}

Molecule	Mode	$n=1$	$n=2$	$n=3$	$n=4$	$n=5$	$n=6$	Reference
CH$_3$OH	CO str v_8	1033.5	1026.7	1041.0	1044.3	1047.8	1040.4	Buck et al. (1990a); Buck and Hobein (1993).
	CO str v_8	—	1052.0	—	—	—	1052.0	Huisken and Stemmler (1988).
		1033.5	1026.5	1042.2	1044.0	nm	nm	
		—	1051.6	—	—	nm	nm	
	OH str v_1	3681.0	3684.1	3562	nm	nm	nm	Huisken et al. (1991); Huisken and Stemmler (1992).
		—	3574.4	—	nm	nm	nm	
CH$_3$NH$_2$	CH$_3$ roc v_7	1074.5	1071.3	1062.0	1060.6	nm	nm	Huisken et al. (1991).
	CN str v_8	1044.0	1038.3	1046.0	1045.5	1045.4	1046	Buck et al. (1991a).
		—	1047.9	—	—	—	—	
CH$_3$CN	CC str v_4	920.3	—	918.4	918.1	918.0	918.0	Buck and Ettischer (1944b); Buck et al. (1990b); Buck (1992b).
	CH$_3$ roc v_7	1041.8	1046.0	1045.5	1045.6	1046.5	1045.8	
		—	—	—	1036.9	1040.5	1038.7	
		—	—	—	1029.5	nm	nm	
CH$_3$F	CF str v_3	1048.6	1039.6	1023.5	—	nm	nm	Ehrbrecht et al. (1992).
		—	—	1032.7	—	nm	nm	
C$_2$H$_4$	CH$_2$ ben v_7	949.3	951.8	951.2	951.6	951.3	951.9	Buck et al. (1987).
	CH$_2$ ben v_7	949.3	952.9	953.4	953.7	nm	nm	Huisken and Pertsch (1987).
	CH$_2$ ben v_7	949.3	953.2	952.3	952.2	952.2	952.5	Ahlrichs et al. (1990).
C$_6$H$_6$	CH ben v_{18}	1038.3	1037.9	1037.0	1037.5	nm	nm	de Meijere and Huisken (1990).
NH$_3$	umb v_3	950.0	976.9	1006.1	1036.4	nm	nm	Huisken and Pertsch (1988).
		—	1003.4	1016.3	1044.7	nm	nm	
N$_2$H$_4$	NN str v_5	1098.0	1081.9	1088.0	nm	nm	nm	Buck et al. (1989); Beu et al. (1995a).
	NH$_2$ wag v_{12}	937.0	981.9	991.6	1024.6	1025.3	1025.8	
		—	—	1025.2	1045.8	1049.1	1048.6	
OCS	ben 2 v_2	527.0	1045.9	1045.0	1044.7	1044.3	1044.2	Krohne et al. (1995).
SF$_6$	SF ben v_3	948.0	934.4	935.2	936.1	nm	nm	Huisken and Stemmler (1989).
		—	955.6	955.1	959.1	nm	nm	
		—	—	962.0	—	nm	nm	
C$_2$H$_4$–(CH$_3$)$_2$CO	CH$_2$ ben v_7	949.3	950.8	940.5	nm	nm	nm	Buck et al. (1993b).
		—	961.6	958.5	nm	nm	nm	
CH$_3$OH–H$_2$O	CO str v_8	1033.5	1027.8	nm	nm	nm	nm	Huisken and Stemmler (1991).

Note. nm, not measured.

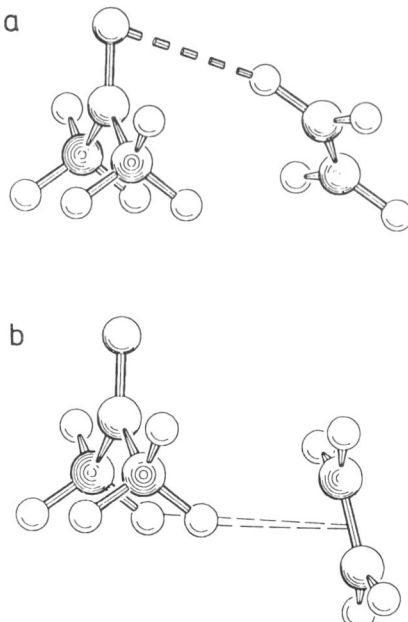

FIG. 5. Calculated minimum energy configurations for the mixed acetone–ethylene dimer. (a) $E = -13.9$ kcal/mol; (b) $E = -12.6$ kcal/mol.

observation of vibrational predissociation spectra, the suggestion was made that the occurrence of two dissociation bands at 950.8 and 961.6 cm^{-1} should be attributed to two isomers of this complex (Buck et al., 1990c). The possibility that the two peaks are caused by the absorption of the ethylene and acetone molecules of the same isomer was ruled out, since pure acetone clusters did not show any absorption in this frequency range. Double-resonance experiment clearly confirmed that two isomers are involved (Buck and Hobein, 1993). The experiments exhibited a behavior similar to that illustrated by the calculated dashed curve in Fig. 4. Calculations of the minimum energy configurations have also been carried out for these complexes using an empirical potential (Buck et al., 1993b). The results are presented in Fig. 5. The minimum at -13.9 kJ/mol (Fig. 5a) belongs to a structure in which one of the H atoms of the ethylene molecule is bound to the O atom of the acetone. In the other isomer (Fig. 5b) with an energy of -12.6 kJ/mol the bond takes place between two hydrogen atoms of acetone and the π-electron cloud of ethylene. The frequency shifts caused by these configurations were calculated by means of the perturbation theory presented in Section III.B in the first order. The theoretical results are in good agreement with the experimental data. For the isomer in Fig. 5a, a frequency shift of $+12.5$ cm^{-1} with respect to the ethylene v_7 gas-phase vibration at 949.3 cm^{-1} is calculated, fitting well to the measured value of 961.6 cm^{-1}.

The isomer in Fig. 5b exhibits a calculated fequency shift of $+2.5\,\text{cm}^{-1}$ which is close to the band centered at $950.8\,\text{cm}^{-1}$. As expected, the larger blue shift occurs for the isomer with the hindered motion of the H atoms of ethylene. The larger clusters which are essentially acetone clusters with one ethylene molecule attached have been analyzed in the same manner, and exhibit similar results.

3. A Nearly Linear Hydrogen Bond: Methanol

Methanol is an important polar solvent with a nearly linear hydrogen bond. The solid phases are known to be made up of parallel hydrogen-bonded chains with coordination number $z = 2$ (Torrie et al., 1989). Similarly, the

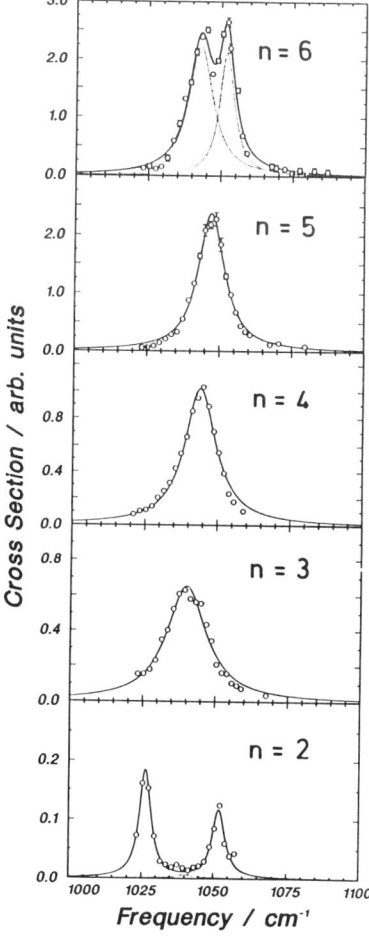

FIG. 6. Measured photodissociation spectra of size-selected methanol clusters near the CO stretch mode at $1033.5\,\text{cm}^{-1}$.

dominating structure of the liquid are long winding chains with occasional branching points. On the other hand, it is well known that properties of the gas phase like the thermal conductivity clearly demonstrate the existence of larger clusters in ethanol vapor (Curtiss and Blander, 1988). Methanol clusters have been thoroughly investigated with the scattering method for both internally excited clusters using CW lasers up to $n = 8$ (Buck et al., 1988b, 1990a) and cold clusters using pulsed lasers up to $n = 4$ (Huisken and Stemmler, 1988, 1992; Huisken et al., 1991). The spectra after the excitation of the CO-stretching mode at $1033.5 \, cm^{-1}$ (which are taken in the experimental arrangement of Fig. 3a) are shown in Fig. 6. Here the laser interacts with slightly excited complexes. In the case of the dimer, the measurements are carried out in the direct beam so that mainly cold ones get probed (Buck and Hobein, 1993). Here the seeding mixture is very dilute to make sure that only dimers are in the beam. The results are in good agreement with the experiment conducted on size-selected, cold species (Huisken and Stemmler, 1988).

The dimer spectrum is characterized by a two-peak structure with one peak shifted by $-6.8 \, cm^{-1}$ to the red and another one shifted by $+18.5 \, cm^{-1}$ to the blue compared with the frequency of the free CO-stretching mode. For the next larger clusters, the trimer, tetramer, and pentamer, only one peak is found shifted to the blue by $+7.5$, $+10.8$, and $+14.3 \, cm^{-1}$, respectively. This behavior changes again with the hexamer. Now a double peak structure appears with one peak shifted by only $+6.9 \, cm^{-1}$ and one peak shifted further to the blue by $+18.5 \, cm^{-1}$. This double structure continues to be observed up to $n = 8$.

The explanation for all these spectra is found in the calculated structures of the clusters, as shown in Fig. 7. These structures are obtained by using the procedure described in Section III. The interaction potentials are empirical site–site model potentials with parameters fitted to data of the complex and the liquid (Jorgensen, 1986). Aside from differences in the total binding energies, several other potential models give nearly the same minimum structures. The dimer exhibits a linear hydrogen bond, the trimer and tetramer are planar or nearly planar rings, and the cyclic structure of the pentamer gets slightly distorted. For the hexamer, finally, several rings are found with the most stable configuration of a symmetric chairlike structure of S_6 symmetry and the energetically second lowest boatlike isomer of C_2 symmetry which are also displayed in Fig. 7. The question arises up to what size a cluster can grow forming ring structures. This was answered in a similar study in which it was found that ring structures continue up to $n = 11$ and then two rings bound by weaker forces appear (Martin et al., 1987).

Now let us try to explain the *line shifts* based on these structure calculations. Here we apply the methods mentioned in Section III.B which are based on the molecular (mol) and cluster (clu) ansatz. In most cases the

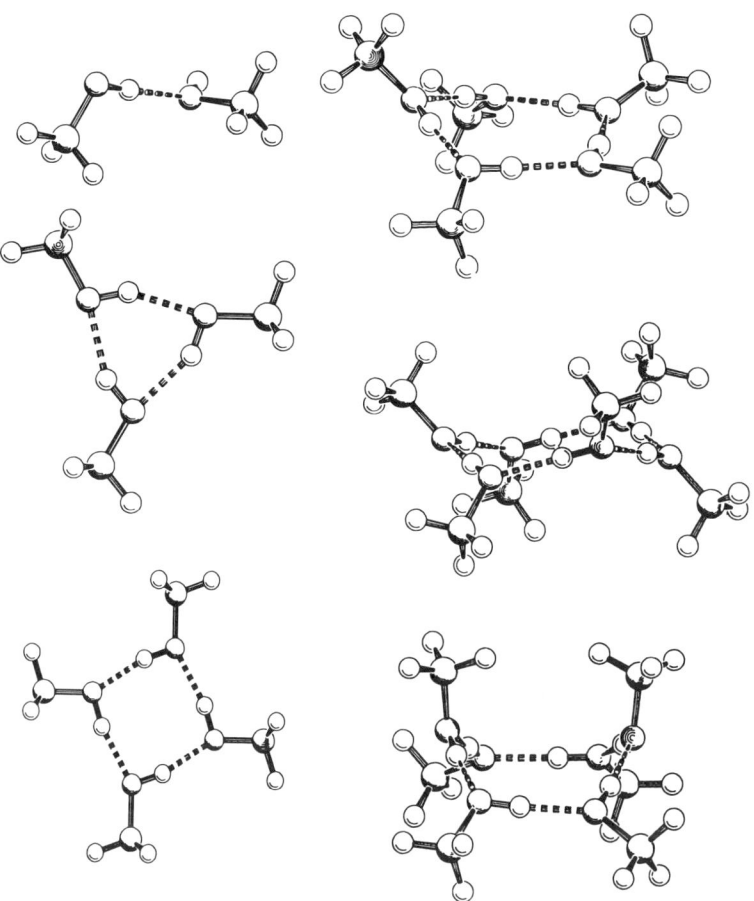

FIG. 7. Calculated minimum energy configurations of methanol clusters using the potential of Jorgensen (1986) from the dimer to the hexamer. For the hexamer two isomers of S_6 (right middle figure) and C_2 symmetry (right lower figure) are shown.

intramolecular force field of Schlegel *et al.* (1977) is used. The results are presented in Table II. The comparison with the experimental results shows, in general, depending somewhat on the chosen method, a good qualitative reproduction of the shifts which allows a unique identification and explanation of the results.

For the dimer, the excited CO-stretching mode is in a nonequivalent position with respect to the hydrogen bond. The O atom of the acceptor participates directly in the bond, while this is not the case for the donor. This explains the line splitting. The calculations give a slight red or blue shift for the acceptor and a large blue shift for the donor. The red shift originates mainly from the elongation of the C–O distance in the attractive hydrogen

TABLE II

Measured (Buck et al., 1990a; Huisken et al., 1991) and Calculated Line Shifts of the CO and OH Stretch Mode of Methanol Clusters in cm^{-1} for Different Methods

Size (n)	Mode	Experimental	mol1-e	mol2-e	clu-e	clu-s	cnm-i
2	OH	3.1	20.6	−42.9	−37.9	−30.1	−3
		−106.6	−239.4	−222.1	−226.5	−108.8	−78
2	CO	+18.5	28.9	17.6	17.6	15.3	12
		−6.8	1.3	2.0	3.1	−9.1	3
3	CO	7.5	3.8	22.6	19.1	9.1	9
						6.1	1
						4.5	−2
4	CO	10.8	24.9	25.9	20.2	4.7	1
						−8.9	−3
5	CO	14.3	27.5	28.0	—	11.4	5
		21.0	24.3	—	2.1	3
		17.2	24.1	—	−10.8	−2
6	CO	18.5	32.8	29.1	18.7	—	4
		6.9	17.0	22.5	12.4	—	−3

Note. mol1 (Buck and Schmidt, 1993), mol2 (Beu, 1994), and clu (Siebers, 1995) use the empirical (e) OPLS potential of Jorgensen (1986) and the systematic (s) model of Wheatley (1995); cnm-i (Bleiber and Sauer, 1995) is based on the normal mode analysis of *ab initio* calculations in the SCF approximation.

bond. The blue shift results from a stronger force constant and the coupling to the OH mode which squeezes the C–O distance.

In the structure calculations, planar rings are found for both the trimer and the tetramer with all the molecules in equivalent positions with respect to the CO excitation. Each molecule acts as a donor as well as an acceptor which explains the line positions between those of the dimer. The fact that only one peak occurs can be rationalized for the planar trimer which has C_{3h} symmetry as follows. The three CO-stretching modes lie in the plane. One mode, the total symmetric combination of the three CO stretches, is not infrared active, and the other two infrared active modes are degenerate. Similar considerations hold for the tetramer. For the pentamer, a nearly planar structure which gives rise to three frequencies which lie closely together such that they are not resolved in the experiment at finite temperatures has been found. For the hexamer, a series of isomers appear. The calculations show that the two peaks are caused by the coupled symmetric and antisymmetric motion of the monomer units of the S_6 isomer only. It is interesting to note that we do not observe in the experiment any indication of the other isomers which have only slightly smaller binding energies but completely different spectra with four to six peaks caused by their reduced symmetry. Double-resonance experiments confirm clearly that the two peaks originate from one isomer only (Buck and Hobein, 1993).

Also, results for the excitation of the OH-stretch mode are available (Huisken et al., 1991). The large red shift of the donor molecule in the dimer is reproduced only in its tendency but not in the absolute value when the empirical potential models are used. To improve this comparison between measurements and calculations, more realistic interaction potential have been constructed. These include *ab initio* calculations (Bleiber and Sauer, 1995) and systematic model potentials (Wheatley, 1995). The results are also shown in Table II. For the dimer, a considerable improvement is achieved. In the case of the trimer, a line splitting which is caused by the nonplanar structure found in these new calculations is predicted. Within the experimental resolution the results also agree with the experiment. For larger clusters the agreement is less satisfactory. In order to clarify whether the remaining discrepancies with respect to the data are caused by the lack of resolution or the probing of other isomers, refined experiments for a more detailed comparison are currently being carried out.

In any case, we find a strong correlation between the line shifts and the structure of the methanol clusters. The comparison with spectra obtained without size selection but with a mass filter clearly shows that the details of the current results are not observed because of the mixing of contributions of fragments from different cluster sizes.

Similar results have been obtained for the excitation of the CN stretch mode of CH_3NH_2 clusters (Buck et al., 1991a). For the dimer a line splitting is observed, while the larger clusters exhibit one line only. Thus the same bonding structure found for methanol is expected. Also, clusters of ethanol and dimethyl ether have been measured by Ehbrecht and Huisken (1994).

4. A Nonlinear Hydrogen Bond: Hydrazine

Hydrazine $[(N_2H_4)_n]$ clusters have been studied (Buck et al., 1989, 1990c), since the two amino groups which are twisted against each other by 90° let us expect an interesting bonding behavior. The photodissociation spectra for $n = 2$ to $n = 6$ are shown in Fig. 8, which also contain new results obtained with isotopically substituted CO_2-lasers (Beu et al., 1995a). In the range of our laser, the antisymmetric NH_2-wagging (v_{12} at 937 cm^{-1}) and the NN-stretching mode (v_5 at 1098 cm^{-1}) are excited. The NN-stretching mode is characterized by one peak which is only slightly shifted to the red. The other mode, however, in which all four hydrogen atoms move in phase with respect to the N–N axis, shows a completely different behavior. Already the dimer exhibits an averaged blue shift of about 45 cm^{-1}. We recognize about three peaks, a result which is also confirmed in experiments with "cold" dimers which are generated in a very dilute mixture in the direct beam (Buck et al., 1989; Beu et al., 1995a). The general trend is continued by the trimer, for which two peaks are observed, with the main peak blue shifted by 88 cm^{-1}. For the tetramer, pentamer, and hexamer the shifts

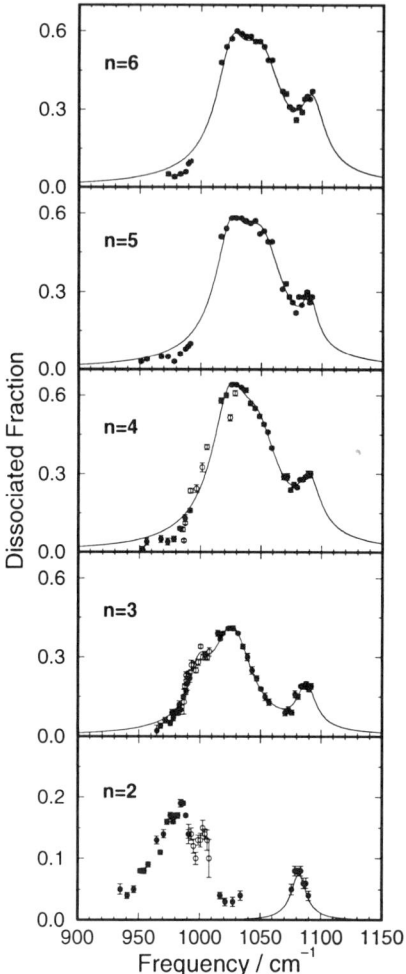

FIG. 8. Measured photodissociation spectra of size-selected hydrazine clusters near the NH_2-wagging (937 cm^{-1}) and the NN-stretching mode (1098 cm^{-1}).

reach 112 cm^{-1}. In general the spectra are quite broad with some indications of underlying structures.

Structure calculations of the lowest energy configurations of the dimer, trimer, and tetramer are shown in Fig. 9 (Beu *et al.*, 1995b). They exhibit indeed different bonding characteristics which might explain the different frequency shifts. The dimer has two hydrogen bonds in a cyclic arrangement. The trimer is also cyclic with one hydrogen bond per molecule, and the tetramer as well as all the larger clusters show three-dimensional structures. There are, in addition many isomeric structures found in the calculations which exhibit bonding structures similar to those displayed in Fig. 9.

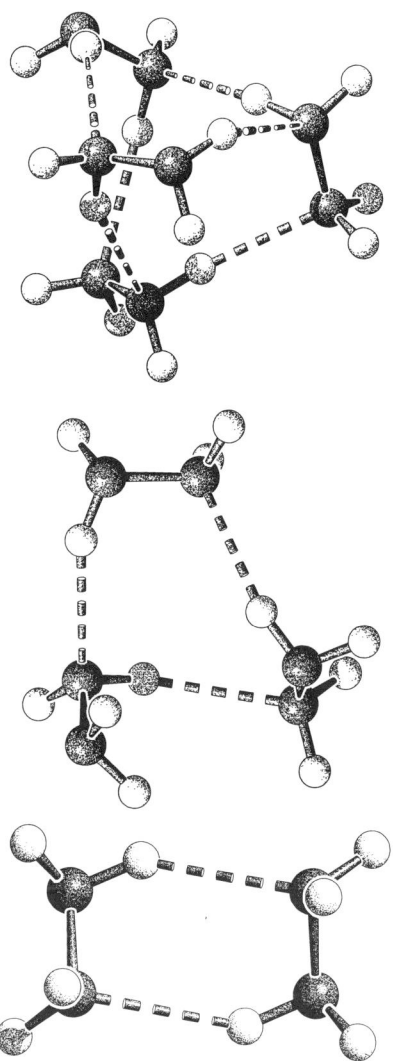

FIG. 9. Calculated minimum energy configurations of hydrazine dimers, trimers, and tetramers.

The lineshift calculations (Beu *et al.*, 1994b) reproduce the frequency shift of the NN stretching quite well, independent of the potential model (Nemenoff *et al.*, 1978; Beu *et al.*, 1994a). For the NH_2-wagging mode, however, the calculations based on the empirical potential (Nemenoff *et al.*, 1978) predict red shifts, while those based on the model potential (Beu *et al.*, 1994a) show the correct blue shifts. It is noted here that this potential takes into account the dependence on the vibrational coordinate, obviously a necessary ingredient to describe the large amplitude motion of the hydrogen atoms correctly.

A closer inspection of the calculated results in comparison with the broad and structured measured distribution reveals that, in the case of the dimer, the contributions of all three isomers are necessary to explain the experimental results. Each isomer leads to a line splitting caused by nonequivalent positions of the two units in the cluster. Accidental coincidences lead finally to three peaks. Similar considerations also hold for the larger clusters. The gross two-peak structure of the trimer is again produced by contributions of several isomers. This trend continues for the larger clusters for which the spectra originate from the three-dimensional structures of several isomers.

In summary, the NH_2 motion is heavily distorted and leads, similar to the results for the umbrella motion of NH_3 (Huisken and Pertsch, 1988), to large blue shifts in the cluster spectra. In the latter case, however, the calculated cyclic structures of the trimer and tetramer (Greer et al., 1989) resemble much more the results obtained for methanol than those obtained for hydrazine, since in both cases only one acceptor molecule is present.

5. A Dipolar Interaction: Acetonitrile

The polar molecule acetonitrile, CH_3CN, plays an important role as a solvent in liquids and it has been concluded on the basis of X-ray and neutron scattering investigations that the pure liquid has a strongly oriented structure with a short range order (Kratochwill et al., 1973). The solid condenses in two different phases, the monoclinic high-temperature phase, starting at 229 K with coordination number $z = 4$ and a nearly antiparallel arrangement of the dipoles (Barrow, 1981), and the orthorhombic low-temperature phase, starting at 216.9 K with $z = 8$ and a paired parallel arrangement of the dipoles (Torrie and Powell, 1992). The dimer and the larger clusters which are held together by multipolar forces (Böhm et al., 1984) have been investigated spectroscopically by far-infrared absorption (Knözinger and Leuthoff, 1981), neutron scattering (Langel et al., 1985), and IR photodissociation (Levandier et al., 1990; Al-Mubarak et al., 1988; Buck et al., 1990b). The dimer showed the expected structure of antiparallel dipoles, while the experiments on the larger clusters were not conclusive, partly because no size selection was applied and partly because the size selection was not complete.

In our first experiment (Buck et al., 1990b), in which the CC stretch ($v_4 = 920 \text{ cm}^{-1}$) and the CH_3 rock mode ($v_7 = 1041 \text{ cm}^{-1}$) were excited, only clusters up to $n = 4$ were separated. In order to look for odd–even effects which might be caused by paired and unpaired dipole moments, a series of new experiments have been carried out. In the new arrangement with much higher resolution photodissociation, spectra were obtained for

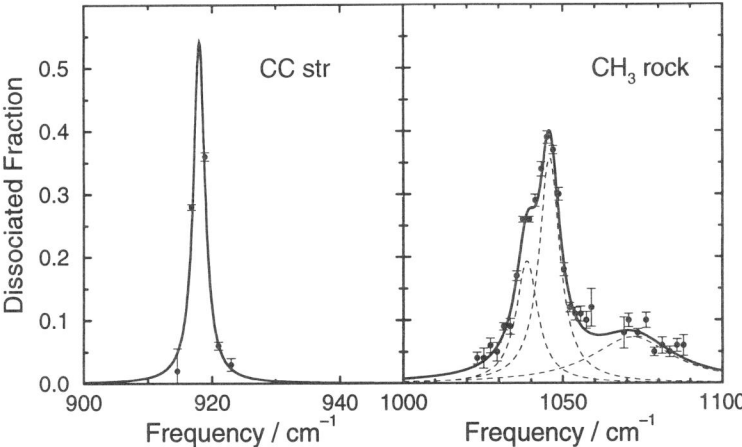

FIG. 10. Measured photodissociation spectra of acetonitrile clusters for $n = 6$. The left figure shows the excitation of the CC-stretching mode, and the right figure, that of the CH_3 rock mode.

single-sized clusters of $n = 4$–9, 12, and 13 as well as for $n = 24 \pm 3$ and $n = 33 \pm 8$ (Buck, 1992b; Buck and Ettischer, 1994b).

For the CC stretch mode a small red shift of $2\,cm^{-1}$ to a frequency of $918\,cm^{-1}$ which does not depend on the cluster size is measured. The narrow line width of less than $2\,cm^{-1}$ is quite remarkable. The CH_3 rock mode exhibits for the even clusters, aside from the main peak at $1046\,cm^{-1}$, a shoulder at smaller wavenumbers of $1038\,cm^{-1}$ and a further low intensity peak at $1075\,cm^{-1}$. This is clearly demonstrated in Fig. 10 which shows the measured spectra of the two modes for $n = 6$ for comparison. For the odd clusters only the main maximum at $1046\,cm^{-1}$ and a shoulder at $1040\,cm^{-1}$ is observed. The complete results are given in Table III. Exceptions from these even–odd relations are the clusters $n = 9$ and $n = 8$ which show, in contrast, the features of even and odd clusters, respectively. The clusters $n = 24 \pm 3$ exhibit again the spectrum of an even cluster as does $n = 33 \pm 8$, but without the third peak at $1078\,cm^{-1}$.

How can we explain, at least in a qualitative way, these results? A comparison of the IR spectra of the even and odd clusters with those of the high-temperature solid shows that the spectra of the odd clusters are in remarkable agreement with them regarding both the frequencies and the amplitudes, if $n = 13$ is considered. Starting with the antiparallel structure of the dimer, it is no surprise that the smaller clusters resemble to a large extent the high-temperature phase with their antiparallel dipole arrangements. Apparently the agreement is better for the more flexible arrangement of odd-numbered clusters with unpaired dipoles. The even clusters represent probably a special stable configuration which is not found in the condensed

TABLE III
MEASURED FREQUENCIES OF THE v_7 MODE OF
ACETONITRILE IN cm^{-1}

n	Peak 1	Peak 2	Peak 2	Peak 3
3	1045.5	—	—	—
4	1045.6	—	1036.9	1078.0
5	1046.4	1040.5	—	—
6	1045.8	—	1038.7	1071.7
7	1046.6	1040.2	—	—
8	1046.6	1040.6	—	1075.0[a]
9	1046.0	—	1038.4	1078.0
12	1045.5	—	1038.1	1072.0
13	1045.5	1040.5	—	—
24 ± 3	1044.5	—	1038.1	—
33 ± 8	1045.5	—	1038.0	—

[a] Very small amplitude.

phase. In this way also the additional small peak may be attributed to a Fermi resonance of v_7 with the $3v_8$ CCN bending mode. The unusual behavior of $n = 8$ and $n = 9$ can be explained by the special stability of $n = 9$ in a paired 3×3 arrangement which was found in molecular dynamics simulations (Del Mistro and Stace, 1993). In any case, these considerations have to be confirmed by calculations of the same type which have been presented for methanol and hydrazine.

Similar results have been obtained for methyl fluoride clusters after the excitation of the CF stretch mode (Ehbrecht et al., 1992). The dimer and tetramer exhibit an antiparallel arrangement of the dipoles with a single line spectrum, while the trimer with one unpaired molecule shows a different spectrum with several lines.

B. Excitation and Decay Mechanisms

Information on the vibrational predissociation rate of the decaying complexes is usually obtained from the measurement of the linewidth using the optothermal method (Miller, 1986, 1988). In a few cases results are also available from direct picosecond measurements by observing the buildup of the nascent products (Breen et al., 1990). It has been demonstrated that also the products of such a dissociation process in the infrared can be detected on a state-to-state basis for dimers (Dayton et al., 1988). The measured lifetimes are found to be extremely mode dependent. A nice example is the decay of the linear (HCN)$_3$ trimer which is found to be larger than 140 ns for the excitation of the free CH stretch and 2.8 ns and 10 ps for the two hydrogen-bonded stretches, respectively (Jucks and Miller, 1988).

In our experiments, there are also great differences in the *linewidths*, especially if the internal excitation varies. A nice example is the methanol dimer for which values between 4.0 and 32 cm^{-1} have been measured (Buck *et al.*, 1991b). Another example is the linewidth of hydrazine and acetonitrile tetramers which, measured under similar experimental conditions, differs by more than a factor of 20, being 34 and 1.6 cm^{-1}, respectively (see Figs. 8 and 10) (Buck *et al.*, 1989; Buck and Ettischer, 1994b). Since, however, the line tunable CO_2-laser does not allow one to determine unambiguously whether the linewidth is homogeneously or inhomogeneously broadened, an explanation without speculation is difficult to get. Possible mechanisms for inhomogeneous line broadening include (1) internal excitation during the cluster formation or the scattering process, (2) unresolved line splitting, (3) contributions from different isomers, and (4) vibrationally hot bands.

In the case of the methanol dimer the excitation energy was varied by using different carrier gases and different experimental arrangements according to Fig. 3. The result was a linear increase of the decay rate Γ with increasing excitation (Buck *et al.*, 1991b). This is a clear indication of an inhomogeneous line broadening. The more initial states are excited, the larger is the final state distribution. We think that most of the measured systems show this effect. On the other hand, all the measured distributions can be fitted by Lorentzian line shapes. Attempts to use Gaussian distributions usually failed. This could be an indication that, with increasing internal energy, also the decay process is faster because the density of states increases too.

The very broad band of hydrazine tetramers after the excitation of the NH_2-wagging mode is certainly partly caused by contributions from several isomers, as was shown in the calculations of the spectra. The reason for the narrow width of the acetonitrile tetramer after the excitation of the CC-stretching mode is still unknown. For the ethylene dimer it was demonstrated using a tunable waveguide laser that aside from the broad band ($\Gamma = 10$ cm^{-1}) also very narrow features of 10^{-4} cm^{-1} were present (Buck *et al.*, 1988a). This result clearly demonstrates that additional measurements are necessary to determine unambiguously the decay rates.

A further interesting experimental result is that for all the systems measured up to now, with one exception only, a linear fluence dependence is measured and thus the validity of applying Eq. (5) is confirmed. Such a result is expected for one-photon absorption processes. It is, however, surprising for a number of clusters, for which the CO_2-laser photon of about 12.5 kJ/mol is not sufficient to dissociate them. Examples are methanol dimers and trimers which were definitely shown in elegant double- and triple-resonance experiments to dissociate only after the absorption of two and three photons, respectively (Bizzari *et al.*, 1990). This is in complete agreement with the latest calculations of the binding energies and the zero-point energies which require 18–20 and 26–30 kJ/mol for the dissocia-

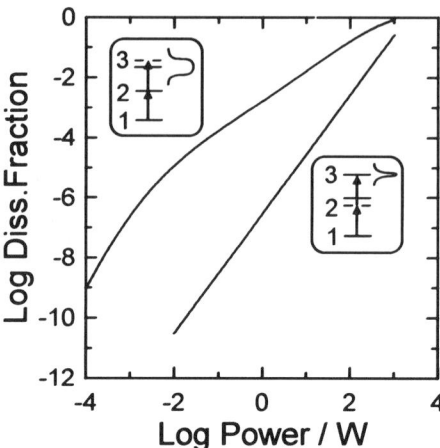

FIG. 11. Simulated dissociation spectra for two photon excitations as a function of the laser power in the double decadic logarithmic scale. The excitation mechanisms which lead to a quadratic and a linear dependence are indicated in the inserts.

tion. Given the laser fluence of about 50 to 100 mJ/cm² of the present experiments, a multiphoton excitation with a nonlinear fluence dependence is expected. This was, however with the exception of the benzene trimer (de Meijere and Huisken, 1990), never observed in all the investigated systems and cluster sizes.

Therefore calculations were carried out by Dam and Reuss (1990) for the coherent two-photon excitation from level 1 over level 2 to level 3 with decay from the upper level 3 to the continuum using the Bloch equation formalism with density matrices (Liedenbaum et al., 1989). The results are presented in Fig. 11. Here the dissociated fraction is plotted against the laser power in a logarithmic scale for two different excitation schemes shown as insets in Fig. 11. The parameters of the calculations are the two-photon detuning energy Δ, the anharmonicity Ω, with the definition $E_3 - E_2 = E_2 - E_1 - 2\Omega$, and the decay rate Γ_3. For $\Omega = 1\,\text{cm}^{-1}$; a small decay rate, $\Gamma_3 = 0.1\,\text{cm}^{-1}$; and no detuning, $\Delta = 0$, we get a perfect two-photon transition and the expected quadratic behavior, as is shown in the lower curve of the figure. For $\Omega = \Delta = 1\,\text{cm}^{-1}$ and a larger decay rate of $\Gamma_3 = 10\,\text{cm}^{-1}$, however, the quadratic power dependence goes over into a linear one for values larger than 0.1 W. Now the first step can be made resonant and the second reaches level 3 within its width inspite of the two-photon energy mismatch. This is demonstrated in the upper insert of the figure. Apparently, most of the investigated systems fulfill the conditions that the anharmonicity Ω is smaller than the decay rate Γ_3. This means that the rate limiting step is the single-photon excitation with a fast decay after the second photon is absorbed. The only exception is apparently the benzene trimer which, aside from the energetic requirement, should have a larger anharmonicity than the decay rate of 10 cm^{-1}.

C. Phase Transitions

The "melting" of clusters is, in general, described as isomerization among a multitude of isomers. The gross features of the size and temperature dependence of this cluster isomerization crucially depend on their interaction and their chemical properties. Weak short-range interactions in rare-gas clusters like Ar_n lead to a large number of isomers, and phase transitions occur preferentially between the lowest energy configurations, usually belonging to an icosahedral growth sequence, and the many other isomers leading to a transition temperature range in which both a solid-like and a liquid-like "phase" are present. In the limit $n \to \infty$ this temperature range converges the melting and condensation temperature of the bulk condensed phase to a single value. Strong long-range interactions in small alkali halide clusters involve isomerization among a very small number of well-characterized isomers, a nice example being the cube → ring transition of $(NaCl)_4$ (Jortner et al., 1990; Martin, 1983; Heidenreich et al., 1992). In both binding types the melting temperature decreases with decreasing cluster size and is lower than the bulk value, a result which had also been observed experimentally for metallic clusters supported on a substrate (Buffat and Borel, 1976) and which can be rationalized by the absence of more and more nearest neighbors with an increasing number of surface atoms.

Similar calculations for weakly bound molecular clusters are rare. Therefore we will first present calculations about the isomeric transitions which occur for the well-investigated methanhol clusters and then demonstrate how they can be measured.

A good indication for such a transition is the relative rms bond length fluctuation δ_{cm} as a function of temperature (see Sec. III.C). According to the well-known Lindemann criterion, a value of more than 0.1 indicates a liquid-like behavior. The results for the small methanol clusters are given in Fig. 12 (Buck et al., 1993a). An almost linear increase of δ_{cm} with increasing T is an indication of the thermal expansion of the system. The molecules of the cluster vibrate about their well-defined minimum positions, but the system cannot pass over any potential barriers. An abrupt rise in δ_{cm}, however, indicates a structural transition. The newly accessed configuration space reveals further local minima on the corresponding potential hypersurface, and the fluctuation between these minima accounts for the sharp rise in δ_{cm}.

For the trimer we observe three steps in the $\delta_{cm}(T)$ curve, a smaller one at 197 K, a more pronounced one at 222 K, and another smaller one at 253 K. They are attributed to a transition from the cyclic lowest energy

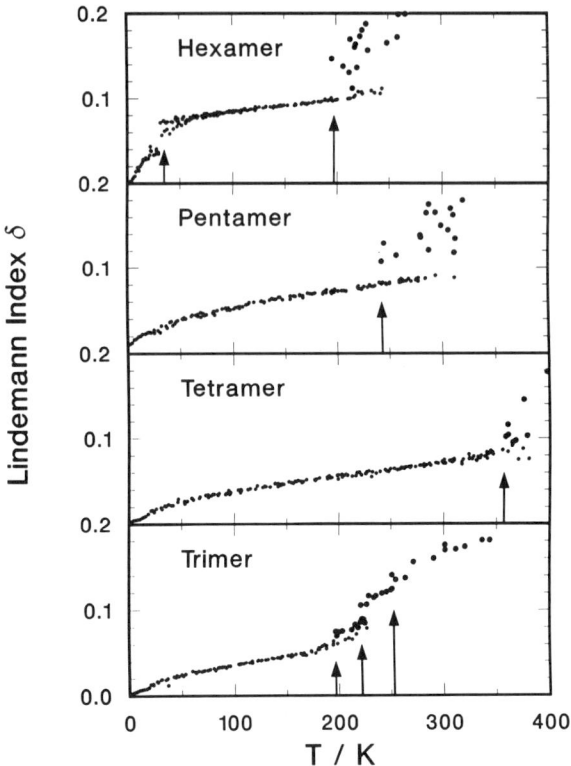

FIG. 12. Calculated relative rms bondlength fluctuations (Lindemann index) which are dimensionless against temperature for small methanol clusters. The arrows indicate single-state or multistate (rigid to fluxional) isomeric transitions.

configuration to different kind of chain structures. This picture of a rigid ring structure and the fluctuating chain structure of the higher lying isomers is further confirmed by looking at the distribution of distances of the center of mass of the molecules to the center of mass of the cluster. The δ_{cm} curves of the tetramer and pentamer show a qualitatively similar step. Therefore we interpret this feature as a structural transition for the tetramer at $T \approx 357$ K and for the pentamer at $T \approx 243$ K to a liquid-like behavior. From the visual inspection of several trajectories we conclude that the tetramer and pentamer spent most of their time in several deformed ringlike structures at high temperatures.

The two lowest lying minima of the hexamer, those with S_6 and C_2 symmetry that are energetically well separated from the other structures (see Fig. 7), characterize the first structural transition of the hexamer. After a sharp increase of δ_{cm} at $T \approx 35$ K, the curve flattens again, signaling that the structural transition has taken place and that there is rapid fluctuation between the two isomeric states. A second "phase transition," showing features similar to those of the tetramer and pentamer, occurs at $T \approx 197$ K.

We note that with n increasing from 4 to 5 and 6 the temperature of the final structural transitions decreases. This is in contrast to what has been found for clusters up to now. In addition, we note that the melting temperature of the bulk at $T = 175.2$ K is well below the above-mentioned transition temperatures. This result is a consequence of the unique binding properties of these clusters which exhibit a maximum of relative binding energies for the tetramer because of the cooperativity effect in the hydrogen bonding (Curtiss and Blander, 1988).

In order to prove whether the clusters are liquid or solid like, we have also calculated the mean-square displacements as a function of time, the slope of which is proportional to the diffusion coefficient. For the hexamer, three curves have been calculated at 20, 97, and 198 K. The first two curves are still in the regime of the solid-like behavior and confirm the interpretation of an isomeric transition between the S_6 and the C_2 structures. Note that the behavior before and after the transition is solid like. The third curve corresponds already to liquid-like behavior and indicates a multistate isomerization similar to the transitions found for the tetramer and the pentamer.

Now we have to answer the question, is there any chance of measuring these isomeric transitions as single-state or multistate behavior. As has already been mentioned in Section IV.B, the infrared spectra are an unambiguous fingerprint of different isomers. A good example is the two energetically lowest isomers of the methanol hexamer of S_6 and C_2 symmetry. While the S_6 isomer shows a double peak structure, four different lines appear for the less symmetric C_2 isomer (Buck et al., 1993a). Thus the only problem left is the change of the temperature. We have simply heated the nozzle, hoping that in competition with evaporative cooling also the cluster temperature in the beam increases. The experimental result confirms this expectation. The measured IR spectra of size-selected methanol hexamers which are produced at two different nozzle temperatures are shown in Fig. 13 (Buck and Ettischer, 1994a). The lower curve taken at 338 K clearly exhibits a two-peak structure which is characteristic for the S_6 isomer. The frequency shifts to 1040 and 1051 cm^{-1} are in very good agreement with results obtained previously (Buck et al., 1990a; Buck and Hobein, 1993) (see also Fig. 6). This means that under the expansion conditions of this experiment the temperature of the clusters is so low that only the energetically lowest isomer, the one of S_6 symmetry, is present in the beam. In contrast, the upper spectrum, taken at the nozzle temperature 493 K, shows a completely different shape. It is a factor of 2 broader, and the onset of absorption in the frequency regions below and above that of the S_6 isomer is stronger than the usual line broadening. We have fitted this spectrum, keeping the two peaks of the S_6 isomer at the values obtained in the

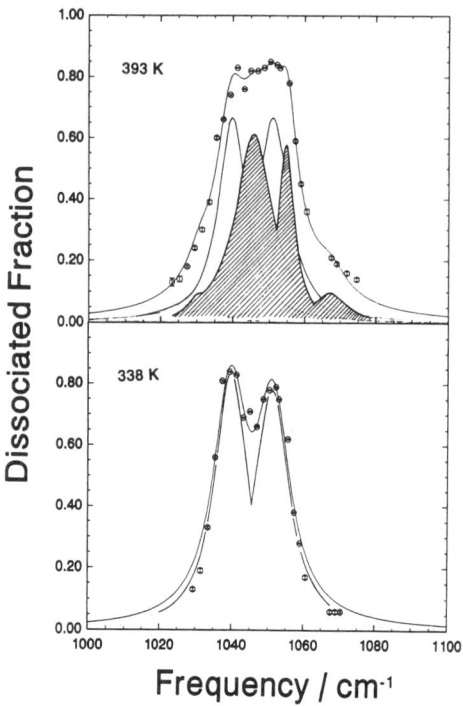

FIG. 13. Measured IR-dissociation spectra of size-selected methanol hexamers at two different nozzle temperatures. The unshaded double-peak structure is caused by the S_6, and the shaded one, by the C_2 isomer.

low-temperature spectrum. Four additional peaks at 1034, 1044, 1054, and 1066 cm^{-1} are necessary to get agreement with the experimental results. Four lines were also predicted in the calculation of the band shifts of the C_2 isomer. For these reasons, the changes are interpreted as a transition from the S_6 symmetry configuration to the second most stable isomer of C_2 symmetry which bears strong IR intensities in these regions.

Further measurements at different temperatures demonstrate that the transition first occurs in the spectrum taken at 368 K. From the fit to the measured data we know the ratio of the population of the C_2 isomer to that of the S_6 isomer. This is plotted against the cluster temperature in Fig. 14. The cluster temperature is estimated by a collisional relaxation model which takes into account the energy released by the formation of the clusters during the expansion (Buck and Ettischer, 1994a).

The transition starts at about 23 K. This result is in agreement with the Monte Carlo simulations of the cluster spectra (Buck, 1994a; Buck and Ettischer, 1994a) for which at 20 K the pure S_6-isomer spectrum is obtained, while at 50 K a broad, unstructured spectrum results. A similar temperature

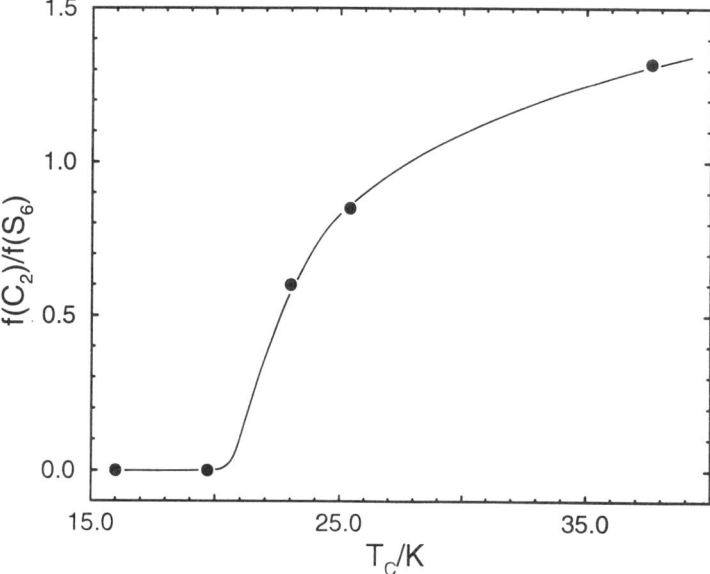

FIG. 14. Ratio of the concentration of the C_2 isomer to the S_6 isomer of the methanol hexamer taken from data similar to those of Fig. 13 as a function of the cluster temperature which is estimated from a relaxation model.

range is also predicted by the Lindemann criterion, which gives a transition temperature of 36 K (Buck et al., 1993a) for the empirical potential of Jorgensen (1986). It is, however, noted here that this model potential exhibits only a small difference in the binding energies of the two isomers: -204.13 and -203.47 kJ/mol. Two SCF calculations for the methanol hexamer gave the same symmetries for the two energetically lowest isomers but much larger differences in the energies: -175.2 and 171.6 kJ/mol (Bleiber and Sauer, 1994) and -189.7 and -185.7 kJ/mol (Wheatley, 1994). For these potentials the transition temperature is larger. In order to discriminate between the different potentials, the model for calculating the temperature of the clusters has to be refined.

V. Conclusions

The field of weakly bound neutral molecular clusters has enjoyed a remarkable growth in the last years. The main experimental tools are combinations of different spectroscopic methods with those size selecting the clusters. Pure mass spectroscopic methods do not give valuable information on the neutral cluster structures and energetics, because strong fragmentation occurs during the ionization process and the measured intensities reflect very often only the stability of the cluster ions after a complicated fragmentation and evaporation process. From the successfully employed experimental methods we have presented here results based on the combination of the size-selective process of the scattering by an atomic beam with infrared photodissociation, resulting in depletion. Although most of the results have been obtained in the size range below $n = 10$, experiments show that $n = 20$ can be reached without problems. Because of the low beam intensities, usually powerful CO_2-lasers in the 10-μm range have been used. First experiments, however, with a pulsed OPO system in the 3-μm range have already been carried out (Huisken et al., 1991) so that now also the very important hydrogen-bonded systems of HF and H_2O clusters are experimentally accessible (Huisken et al., 1995; Kulcke et al., 1994). The same argument is valid for high-resolution experiments with size-selected clusters.

The major effects of the cluster on the spectral properties of the monomer are to shift the peaks and to broaden the width. The former point gives detailed information about the structure, while the latter one is dependent on the decay dynamics. The spectra exhibit a great variety of behaviors. We observe single lines, splitting because of nonequivalent positions, and exciton splitting which is caused by the coupling of the modes from different molecules. The bands undergo only small shifts, and they do not depend on cluster size for several van der Waals systems like C_2H_4, OCS, and C_6H_6 as well as the NN-stretching vibration of N_2H_4 and CC-stretching mode of CH_3CN. They are extremely size dependent for hydrogen-bonded systems like the moderate blue shifts of the CO stretch or the large red shifts of the OH-stretching mode of CH_3OH. They are strongly shifted to the blue for the NH_2-wagging mode of N_2H_4, and the umbrella mode of NH_3. Systems with large dipole moments show a pronounced even–odd effect like CH_3CN and CH_3F. These results already indicate the large sensitivity of the line shifts to the forces involved and to the excited modes. The latter dependence arises as the well-known mode selectivity of the decay rates (Miller, 1988; LeRoy et al., 1990) from the intra- and intermolecular coupling function $V(q, R)$.

All these data are strongly correlated with the structure of the clusters and the underlying interaction potential. Therefore reliable theoretical methods were developed to first calculate the minimum energy configurations which are based on detailed model potentials and to determine from them the spectra which are compared with the data in order to refine the potentials. In this way all the measured spectral features could be explained, and together with data from electron diffraction studies the following general picture on the cluster structures emerges as well. Very often the dimer structure is also found in the solid, since this is the optimum geometry for the interaction of two or an arrangement of many particles which consists of subunits of chains of dimers. The clusters in the size range up to $n = 10$ behave, however, differently. The smaller clusters of hydrogen-bonded systems, the trimers, tetramers, and pentamers, of methanol, water, hydrogen fluoride, ammonia, and methylamine form very stable symmetric and sometimes even planar rings. The reason is simply that the number of bonds which are now available for these structures is larger compared with open chainlike arrangements. Then it does not matter that the stabilization energy per bond is usually smaller, since the ideal bonding configuration of the dimer has to be strained. From the hexamer upward still cyclic structures result, but now they are distorted and strong deformations start to appear. In addition, the number of isomers with configurations nearly as stable as the global cyclic arrangements gets larger with increasing cluster size. For hydrazine clusters with two possible hydrogen bonds, compact three-dimensional structures appear from the tetramer upward and several isomers with comparable bonding energies are present from the beginning on for all clusters sizes. On the other hand, molecules with large dipole moments like acetronitrile and methyl fluoride exhibit a preferred arrangement in paired dipoles in their clusters. In all cases investigated the structures found in the low-temperature solid are not seen for the clusters. The transition to the ordered crystalline structure of the solid usually occurs at much larger cluster sizes in the range of hundred particles per cluster (Bartell et al., 1989). These values are smaller than those found for atomic van der Waals clusters, but much larger than the size range considered here.

It is still an open question how these clusters behave in the size range between $n = 10$ and $n = 20$. It is expected that they consist of combinations of rings and chains with branching points and look like the frozen structure of a liquid or better an amorphous solid (Farges et al., 1988). Based on the promising techniques introduced here, we will probably get the answer soon.

The problem of multiple isomeric configurations has been tackled by applying double-resonance techniques which allow one to distinguish whether a line splitting is caused by nonequivalent positions in one cluster or by two different isomers. In this way we observe the existence of isomers in the mixed acetone–ethylene clusters. In the case of the methanol hexamer, however, only the isomer with S_6 symmetry is found in the beam under normal experimental conditions. The simulation of the structural transitions to the other isomers reveals a pure isomeric one at 35 K to the still solid-like C_2 isomer and as multistate isomerization at 197 K to a liquid-like state. It is noted here that for the tetramer this latter transition takes place at 357 K, a much higher value than the melting temperature of the bulk at 175 K. This is a clear indication of the strong cooperative binding effect predicted for this cluster.

The pure isomeric transition of the methanol hexamer has been observed experimentally using the spectra as fingerprints (Buck and Ettischer, 1994a).

This opens a very attractive field of new experiments, since the calculations demonstrated that these transitions depend sensitively on the interaction potential. In addition very interesting perspectives arise, for example, for acetonitrile, for which even clusters exhibit much higher transition temperatures than odd clusters. We note here also that many of our results indicate that the investigated standard solvent molecules form very stable aggregates so that the interaction between a solvated molecule and the solution very probably takes place with a cluster instead of an ensemble of single molecules.

Finally we would like to point out a new, promising field which is based on nearly the same spectroscopic techniques described here. This is the IR spectroscopy of molecules and molecular clusters imbedded in or absorbed on the surface of large rare gas clusters. It is expected that the medium of the large cluster simplifies the spectra like in the well-known matrix spectroscopy. Under favorable conditions it is even hoped that the size selection by momentum transfer is not necessary anymore, since the ubiquitous fragmentation upon the ionization might be reduced by evaporating the atoms from the rare-gas clusters, and therefore simple mass separation does the job of size selection. The pioneering experiments have been performed by Amar *et al.* (1993) using bolometric detection. Later on also the mass spectrometer was used as a detector by Huisken and Stemmler (1993) which opens the possibility to study the behavior of larger clusters. A further application is the investigation of the solvation dynamics in the time domain with femtosecond resolution (Liu *et al.*, 1993). In the case of the helium clusters which are a quantum liquid with a very fluxional surface and the possibility to undergo a transition to superfluidity, it is expected that the proposed experiments might also be used to learn something about the behavior of the host clusters (Goyal *et al.*, 1993; Fröchtenicht *et al.*, 1944).

Acknowledgments

I acknowledge with gratitude the many contributions of my co-workers on both the experimental and the theoretical side. In particular I thank Dr. C. Lauenstein, Dr. A. Rudolph, Dr. X. J. Gu, M. Hobein, I. Ettischer, and Dr. R. Krohne, as well as Dr. B. Schmidt, J. G. Siebers, and Dr. T. Beu. I am grateful to Dr. C. Lauenstein for his valuable advice in preparing the manuscript and Dr. F. Huisken for helpful discussions. Part of the work was supported by the Deutsche Forschungsgemeinschaft in SFB 357 and the Schwerpunktprogramm "Molecular Clusters," the European Community, and the Max-Planck-Research award.

References

Ahlrichs, R., Penco, R., and Scoles, G. (1977). *Chem. Phys.* **19**, 119.
Ahlrichs, R., Brode, S., Buck, U., DeKieviet, M. Lauenstein, C., Rudolph, A., and Schmidt, B. (1990). *Z. Phys. D* **15**, 341.
Al-Mubarak, A. S., del Mistro, G., Lethbridge, P. G., Abul-Sattar, N. Y., and Stace, A. J. (1988). *Faraday Discuss. Chem. Soc.* **86**, 209.
Amar, F. G., Goyal, S., Levandier, D. J., Perera, L., and Scoles, G. (1993). In *Clusters of Atoms and Molecules* (Haberland, H., ed.), Springer-Verlag, Berlin.
Barrow, M. (1981). *Acta Crystallogr, Sect. B* **B37**, 2239.
Bartell, L. S., Harsanyi, L., and Valente, E. J. (1989). *J. Phys. Chem.* **93**, 6201.
Benedek, G., Martin, T. P., and Pacchioni, G., eds. (1988). *Elemental and Molecular Clusters*. Springer-Verlag, Berlin.
Ben-Horin, N., Even, U., and Jortner, J. (1992). *J. Chem. Phys.* **97**, 5988.
Berry, R. S. (1990). In *The Chemical Physics of Atomic and Molecular Clusters* (Scoles, G., ed.), p. 23, North-Holland Publ., Amsterdam.
Berry, R. S., Beck, T. L., Davies, H. L., and Jellinek, J. (1988). *Adv. Chem. Phys.* **70**, 74.
Beswick, J. A. (1987). In *Structure and Dynamics of Weakly Bound Molecular Complexes* (Weber, A., ed.), p. 563, Reidel, Dordrecht, Netherlands.
Beu, T. (1994). *Z. Phys. D* **31**, 95.
Beu, T., Buck, U., Döderlein de Win, S., Ettischer, I., Hobein, M., Siebers, J. G., and Wheatley, R. J. (1995a). *J. Chem. Phys.* to be published.
Beu, T., Buck, U., Siebers, J. G., and Wheatley, R. (1995b). *J. Chem. Phys.* to be published.
Bizzari, A., Stolte, S., Reuss, J., van Duijnveldt-van de Rijdt, J. G. C. M., and van Duijneveldt, F. B. (1990). *Chem. Phys.* **143**, 423.
Blake, G. A., Laughlin, K. B., Cohen, R. C., Busarow, K. L., Gwo, D.-H., Schmuttenmaer, C. A., Steyert, D. W., and Saycally, R. J. (1991). *Rev. Sci. Instrum.* **62**, 1693.
Bleiber, A., and Sauer, J. (1995). Unpublished results.
Böhm, H. J., and Ahlrichs, R. (1977). *J. Chem. Phys.* **77**, 2028.
Böhm, H. J., Ahlrichs, R., Scharf, P., and Schiffer, H. (1984). *J. Chem. Phys.* **81**, 1389.
Börnsen, K. O., Lin, L. H., Selzle, H. L., and Schlag, E. W. (1989). *J. Chem. Phys.* **90**, 1299.
Bone, R. G. A., Amos, R. D., and Handy, N. C. (1990). *J.C.S. Faraday* **11**, 1931.
Breen, J. J., Willberg, D. M., Gutmann, M., and Zewail, A. H. (1990). *J. Chem. Phys.* **93**, 9180.
Buck, U. (1988). *J. Phys. Chem.* **92**, 1023.
Buck, U. (1990a). In *The Chemical Physics of Atomic and Molecular Clusters* (Scoles, G., ed.), p. 543, North-Holland Publ., Amsterdam.
Buck, U. (1990b). In *Dynamics of Polyatomic van der Waals Complexes* (Halberstadt, N., and Janda, K. C. eds.), NATO ASI Series, p. 42, Plenum, New York.
Buck, U. (1992a). *At. Phys.* **13**, 557.
Buck, U. (1992b). *Ber. Bunsenges. Phys. Chem.* **96**, 1275.
Buck, U. (1993). In *Dynamical Processes in Molecular Physics* (Delgado-Barrio, G., ed.), p. 275, IOP, Bristol.
Buck, U. (1994a). *J. Phys. Chem.* **98**, 5190.
Buck, U. (1994b). In *Clusters of Atoms and Molecules* (Haberland, H., ed.), p. 396, Springer-Verlag, Berlin.
Buck, U., and Ettischer, I. (1994a). *J. Chem. Phys.* **100**, 6974.
Buck, U., and Ettischer, I. (1994b). *Faraday Discuss. Chem. Soc.* **97**, 215.
Buck, U., and Hobein, M. (1993). *Z. Phys. D* **28**, 331.
Buck, U., and Meyer, H. (1986). *J. Chem. Phys.* **84**, 4854.
Buck, U., and Schmidt, B. (1990). *J. Mol. Liq.* **46**, 181.

Buck, U., and Schmidt, B. (1993). *J. Chem. Phys.* **98**, 9410.
Buck, U., and Schmidt, B. (1994). *J. Chem. Phys.* **101**, 6356.
Buck, U., Huisken, F., Lauenstein, C., Meyer, H., and Sroka, R. (1987). *J. Chem. Phys.* **87**, 6276.
Buck, U., Lauenstein, C., Rudolph, A., Heijmen, B., Stolte, S., and Reuss, J. (1988a). *Chem. Phys. Lett.* **144**, 396.
Buck, U., Gu, X. J., Lauenstein, C., and Rudolph, A. (1988b). *J. Phys. Chem.* **92**, 5561.
Buck, U., Gu, X. J., Hobein, M., and Lauenstein, C. (1989). *Chem. Phys. Lett.* **163**, 455.
Buck, U., Gu, X. J., Lauenstein, C., and Rudolph, A. (1990a). *J. Chem. Phys.* **92**, 6017.
Buck, U., Gu, X. J., Krohne, R., and Lauenstein, C. (1990b). *Chem. Phys. Lett.* **174**, 247.
Buck, U., Gu, X. J., Hobein, M., Lauenstein, C., and Rudolph, A. (1990c). *J.C.S. Faraday* **86**, 1923.
Buck, U., Gu, X. J., Krohne, R., Lauenstein, C., Linnartz, H., and Rudolph, A. (1991a). *J. Chem. Phys.* **94**, 23.
Buck, U., Lauenstein, C., and Rudolph, A. (1991b). *Z. Phys. D* **18**, 181.
Buck, U., Schmidt, B., and Siebers, J. G. (1993a). *J. Chem. Phys.* **99**, 9428.
Buck, U., Hobein, M., and Schmidt, B. (1993b). *J. Chem. Phys.* **98**, 9425.
Buckingham, A. D. (1960). *J.C.S. Faraday* **56**, 753.
Buffat, D. A., and Borel, J. P. (1976). *Phys. Rev. A* **13**, 2287.
Coker, D. F., and Watts, R. O. (1987). *J. Phys. Chem.* **91**, 2513.
Curtiss, L. A., and Blander, M. (1988). *Chem. Rev.* **88**, 827.
Dam, N., and Reuss, J. (1990). Unpublished results.
Dayton, D. C., Jucks, K. W., and Miller, R. E. (1988). *J. Chem. Phys.* **90**, 1631.
Del Mistro, G., and Stace, A. J. (1993). *J. Chem. Phys.* **99**, 4656.
de Meijere, A., and Huisken, F. (1990). *J. Chem. Phys.* **92**, 5826.
Detrich, J., Corongiu, G., and Clementi, E. (1984). *Chem. Phys. Lett.* **112**, 426.
Ehbrecht, M., and Huisken, F. (1995). Unpublished results.
Ehbrecht, M., de Meijere, A., Stemmler, M., and Huisken, F. (1992). *J. Chem. Phys.* **97**, 3021.
Eichenauer, D., and LeRoy, R. J. (1988). *J. Chem. Phys.* **88**, 2898.
Farges, J., de Feraudy, M.-F., Raoult, B., and Torchet, G. (1988). *Adv. Chem. Phys.* **70**, 45.
Fröchtenicht, R., Toennies, J. P., and Vilisov, A. (1994). *Chem. Phys. Lett.* **225**, 1.
Gough, T. E., Miller, R. E., and Scoles, G. (1978). *J. Chem. Phys.* **69**, 1588.
Gough, T. E., Knight, D. G., Rowntree, P. A., and Scoles, G. (1986). *J. Chem. Phys.* **90**, 4026.
Goyal, S., Schutt, D. L., and Scoles, G. (1993). *J. Phys. Chem.* **97**, 2236.
Greer, J. C., Ahlrichs, R., and Hertel, I. V. (1989). *Chem. Phys.* **133**, 191.
Gspann, J., and Vollmer, H. (1974). In *Rarified Gas Dynamics* (Karamcheti, K., ed.), p. 261, Academic Press, New York.
Gu, X. J., Levandier, D. J., Zhang, B., Scoles, G., and Zhuang, D. (1990). *J. Chem. Phys.* **93**, 4898.
Haberland, H. (1985). *Surf. Sci.* **156**, 305.
Heath, J. R., Sheekes, R. A., Crosky, A. L., and Saycally, R. J. (1990). *Science* **249**, 855.
Heidenreich, A., Jortner, J., and Oref, I. (1992). *J. Chem. Phys.* **97**, 197.
Hoffbauer, M. A., Liu, K., Giese, C. F., and Gentry, W. R. (1983). *J. Chem. Phys.* **78**, 5567.
Huisken, F. (1991). *Adv. Chem. Phys.* **81**, 63.
Huisken, F., and Pertsch, T. (1987). *J. Chem. Phys.* **86**, 106.
Huisken, F., and Pertsch, T. (1988). *Chem. Phys.* **126**, 213.
Huisken, F., and Stemmler, M. (1988). *Chem. Phys. Lett.* **144**, 391.
Huisken, F., and Stemmler, M. (1989). *Chem. Phys. Lett.* **132**, 351.
Huisken, F., and Stemmler, M. (1991). *Chem. Phys. Lett.* **180**, 332.
Huisken, F., and Stemmler, M. (1992). *Z. Phys. D* **24**, 277.
Huisken, F., and Stemmler, M. (1993). *J. Chem. Phys.* **98**, 7680.
Huisken, F., Kulcke, A., Laush, C., and Lisy, J. M. (1991). *J. Chem. Phys.* **95**, 3924.
Huisken, F., Kaloudis, M., Kulcke, A., and Voelkel, D. (1995). *Infrared Phys. Technol.* **36**, 171.
Janda, K. C. (1985). *Adv. Chem. Phys.* **60**, 201.

Jorgensen, W. L. (1986). *J. Phys. Chem.* **90**, 1276.
Jortner, J., Scharf, D., and Landmann, U. (1988). In *Elemental and Molecular Clusters* (Benedek, G., Martin, T. P., and Pacchioni, G., eds.), p. 148, Springer-Verlag, Berlin.
Jortner, J., Scharf, D., Ben-Horin, N., Even, U., and Landman, U. (1990). In *The Chemical Physics of Atomic and Molecular Clusters* (Scoles, G., ed.), p. 43, North-Holland Publ., Amsterdam.
Jucks, K. W., and Miller, R. E. (1988). *J. Chem. Phys.* **88**, 2196.
Kappes, M., and Leutwyler, S. (1988). In *Atomic and Molecular Beam Methods* (Scoles, G., ed.), p. 380, Oxford Univ. Press, New York.
Karpfen, A., Beyer, A., and Schuster, P. (1983). *Chem. Phys. Lett.* **102**, 289.
Kmetic, M. A., and LeRoy, R. J. (1991). *J. Chem. Phys.* **95**, 6271.
Knözinger, E., and Leuthoff, D. (1981). *J. Chem. Phys.*, **74**, 4812.
Kofranek, M., Karpfen, A., and Lischka, H. (1987). *Chem. Phys.* **113**, 53.
Kratochwill, A., Weidner, J. U., and Zimmermann, H. (1973). *Ber. Bunsenges. Phys. Chem.* **77**, 408.
Krohne, R., Linnartz, H., and Buck, U. (1995). Unpublished results.
Kulcke, A., Kaloudis, M., and Huisken, F. (1994). *Faraday Discuss. Chem. Soc.* **97**, 319.
Langel, W., Kollhoff, H., and Knözinger, E. (1985). *Ber. Bunsenges. Phys. Chem.* **89**, 927.
LeRoy, R. J., Davies, M. R., and Lam, M. E. (1990). *J. Phys. Chem.* **95**, 2167.
Leutwyler, S., and Bösinger, J. (1990). *Chem. Rev.* **90**, 489.
Levandier, D. J., Mengel, M., McCombie, J., and Scoles, G. (1990). In *The Chemical Physics of Atomic and Molecular Clusters* (Scoles, G., ed.), p. 331, North-Holland Publ., Amsterdam.
Liedenbaum, C., Stolte, S., and Reuss, J. (1989). *Phys. Rep.* **178**, 1.
Liu, Q., Wang, J.-K., and Zewail, A. H. (1993). *Nature (London)* **364**, 427.
Martin, T. P. (1983). *Phys. Rep.* **95**, 167.
Martin, T. P., Bergmann, T., and Wassermann, B. (1987). In *Large Finite Systems* (Jortner, J., and Pullmann, B., eds.), Reidel, Dordrecht, Netherlands.
Miller, R. E. (1986). *J. Phys. Chem.* **90**, 3301.
Miller, R. E. (1988). *Science* **240**, 447.
Nemenoff, R. A., Snir, J., and Scheraga, H. A. (1978). *J. Phys. Chem.* **82**, 2504.
Nesbitt, D. J. (1988). *Chem. Rev.* **88**, 843.
Pribble, R. N., and Zwier, T. (1994). *Faraday Discuss.* **97**, 229.
Saenz, J. J., Soler, J. M., and Garcia, N. (1985). *Surf. Sci.* **156**, 121.
Scherzer, W., Selzle, H. L., and Schlag, E. W. (1992). *Chem. Phys. Lett.* **195**, 11.
Schlegel, H. B., Wolfe, S., and Bernardi, F. (1977). *J. Chem. Phys.* **67**, 4181.
Scoles, G., ed. (1990). *The Chemical Physics of Atomic and Molecular Clusters*. North-Holland Publ., Amsterdam.
Siebers, J. G. (1995). Unpublished results.
Stone, A. J. (1981). *Chem. Phys. Lett.* **83**, 233.
Tang, K. T., and Toennies, J. P. (1984). *J. Chem. Phys.* **80**, 3725.
Torrie, B. H., and Powell, B. M. (1992). *Mol. Phys.* **75**, 613.
Torrie, B. H., Weng, S.-X., and Powell, B. M. (1989). *Mol. Phys.* **67**, 575.
Venturo, V. A., and Felker, P. M. (1993). *J. Chem. Phys.* **99**, 748.
Vernon, M. F., Krajnovich, D. J., Kwok, H. S., Lisy, J. M., Shen, J. R., and Lee, J. T. (1982). *J. Chem. Phys.* **77**, 47.
Westlund, O. O., and Lynden-Bell, R. M. (1987). *Mol. Phys.* **60**, 1189.
Wheatley, R. J. (1995). Unpublished results.
Wheatley, R. J., and Price, S. L. (1990). *Mol. Phys.* **71**, 1381.
Wittmeyer, S. A., and Topp, M. (1991). *J. Phys. Chem.* **93**, 4627.
Zewail, A. H. (1988). *Science* **242**, 1645.

FEMTOSECOND SPECTROSCOPY OF MOLECULES AND CLUSTERS[1]

T. BAUMERT and G. GERBER

Physikalisches Institut der Universität, Am Hubland, D-97074 Würzburg, Germany

I. Introduction 163
II. Experimental Setup 165
   A. Molecular/Cluster Beam and TOF Spectrometers 165
   B. The Femtosecond Laser Systems 167
   C. The Pump–Probe Delay Line 170
   D. Data Acquisition 172
III. Results and Discussion of Experiments in Molecular Physics 172
   A. Vibrational Wave-Packet Motion in the $2\,^1\Sigma_u^+$ Double Minimum State of Na_2 172
   B. Dynamics of Multiphoton Ionization of Na_2 178
   C. Control of Na_2^+ versus Na^+ Yield 184
   D. High Laser Field Effects in Multiphoton Ionization of Na_2 . . . 186
IV. Results and Discussion of Experiments in Cluster Physics 188
   A. Dynamics of Na_n^+ Photofragmentation 188
   B. Dynamics of the Neutral Na_4 Resonance at 680 nm 191
   C. Na_n Cluster Resonances and Their Decay Dynamics 195
   D. Experiments with Mercury Clusters and Fullerenes 200
V. Conclusions 205
   References 206

I. Introduction

Many gases in nature are diatomics. The simple way of thinking about a diatomic molecule is the imagination of two spheres connected by a rigid rod, symbolizing the chemical bond. This bond is of course not rigid, since the atoms are vibrating against each other. This dynamical view on molecules is usually not treated in quantum mechanical textbooks because the time-dependent part of the Schrödinger equation is separated and only the time-independent Schrödinger equation is solved, whose solutions are of course time independent and do not describe the dynamical aspects of molecules.

---

[1] Work has been performed at the University of Freiburg, Germany.

By coherent coupling of the vibrational states of a molecule a vibrational wavepacket is formed. The different evolution of the phases of these vibrational eigenfunctions leads to a motion of the wavepacket that resembles the classical oscillatory motion. This was realized by Erwin Schrödinger himself (Schrödinger, 1926) shortly after he had published his Schrödinger equation. In that work he constructed a wavepacket in a harmonic oscillator potential and showed that this wavepacket moves like a point mass according to the laws of classical mechanics.

With the development of ultrafast laser sources with their inherent spectral width it has become a standard technique to couple coherently quantum mechanical eigenstates. The observation of Rydberg wavepackets in atoms (ten Widle et al., 1989; Yeazall et al., 1990), showing the classical Kepler orbits of the electron around the nucleus (Averbukh and Perelman, 1989), and the observation of vibrational wavepackets in molecules (Bowman et al., 1989; Fragnito et al., 1989; Baumert et al., 1991a) are prominent examples.

Besides the beauty of looking at a quantum mechancial system from a classical point of view these time-resolved techniques have influenced many areas in physics, chemistry, biology, and technology (for examples see *Ultrafast Phenomena IX*, 1994). Many formerly unfeasible experiments can now be performed in the time domain, and often the results of time domain experiments are interpreted more directly than results obtained in the frequency domain. Examples from molecular and cluster physics will be discussed in this chapter.

In the molecular physics section we first discuss the evolution of a vibrational wavepacket in a double minimum potential well. Second, we will focus on the dynamics of multiphoton ionization (MPI) in a diatomic, and we will see how the unexpected results give new input into the challenging field of controlling chemical reactions by means of time-resolved laser techniques. As high laser intensities are achieved in focused ultrashort light pulses, their interaction with molecules is of particular interest and will be treated next. For all these experiments we have chosen the Na_2 molecule as a model system, as there is a wealth of spectroscopic and theoretical information available, which facilitates the interpretation of the time domain results considerably.

In the cluster section of this chapter we report the first experiments in cluster physics employing ultrashort laser pulses to time-resolved studies of cluster ionization and fragmentation processes. Clusters and in particular metal clusters have been the fascinating subject of many experimental and theoretical studies. Clusters form the link between solid-state physics and molecular physics. Metal clusters exhibit distinct features ranging from molecular properties seen in small particles to the solid state like behavior of larger aggregates. Studies of cluster properties like geometric structures, the evolution of the electronic states from localized to delocalized in nature, and the real-time dynamics of ionization and fragmentation have not yet been performed in detail as a function of cluster size. Alkali clusters are attractive species to be studied experimentally and theoretically, because there is only one valence electron per atom. We will describe cluster-size-dependent studies of physical properties of sodium clusters such as absorption resonances, lifetimes, and decay channels using tunable femtosecond light pulses in resonance-enhanced multiphoton ionization (REMPI) of the *neutral* cluster size under investigation. Two-photon ionization spectroscopy failed in nanosecond-laser experiments due to the ultrafast decay of the studied neutral clusters. The other metal cluster system we will report on is the mercury clusters, Hg_n. In experiments with single femtosecond laser pulses prompt formation of singly and doubly charged clusters are observed up to $n \approx 60$. In pump-probe experiments the transient multiphoton ionization spectra show a "short"-time wavepacket dynamics, which is identical for all observed singly and doubly charged mercury clusters. The "long"-time fragmentation and recombination dynamics, however, indicating a cage effect behavior, is different for the individual clusters.

II. Experimental Setup

The time-resolved femtosecond laser studies in molecular/cluster physics are performed with a combination of experimental techniques: A supersonic (seeded) molecular/cluster beam provides the molecules/clusters in a collision-free environment and restricts the set of initial states mainly to $v'' = 0, J''$. Time-of-flight (TOF) mass spectrometry is used to determine the mass of the ions, the released kinetic energy of the ionic fragments, and the energy distribution of the ejected electrons. Because the ion and electron detection angles are fixed, the ion and electron angular distributions can be studied by rotating the laser polarization. We are thus able to determine the final continuum states. The collinear femtosecond pump–probe techniques are used to induce and to probe molecular/cluster transitions, to resolve the interactions, and to display the evolution of coherences and populations in real time.

The schematic experimental arrangements of the molecular beam apparatus and the ion and electron TOF spectrometers are shown in Fig. 1. In the following section we will describe the experimental setup in detail.

A. MOLECULAR/CLUSTER BEAM AND TOF SPECTROMETER

The supersonic sodium molecular/cluster beam is produced by a coexpansion of sodium vapor (50–100 mbar) with the inert carrier gas argon (1–8 bar) through a small orifice about 100 μm in diameter. In order to achieve this vapor pressure the oven is usually operated at 1000 K with nozzle temperatures about 50 K higher. This technique provides efficiently

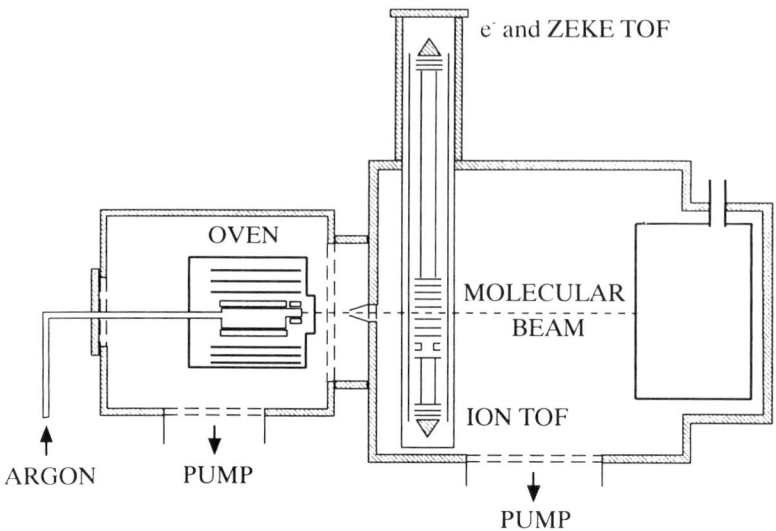

FIG. 1. Sectional drawing of the molecular/cluster beam apparatus and the two different designed TOF spectrometers. The femtosecond lasers (not shown), the molecular/cluster beam axis, and the spectrometer axis are mutually perpendicular to each other.

cooled sodium molecules/clusters. We have measured a vibrational temperature of about 50 K for Na_3 and a rotational temperature of about 15 K for Na_2. With a similar setup and a carrier gas pressure of 10 bar, vibrational temperatures as low as 25 K for Na_3 have been reported (Broyer et al., 1987). To produce mercury clusters, Hg_n, with this technique, the necessary oven temperatures are much lower.

The laser–molecular beam interaction region is placed between parallel ring-shaped plates so that the formed ions and electrons can be detected in opposite directions by two differently designed TOF spectrometers. Both TOF spectrometers are perpendicular to the laser beam axis and the molecular beam axis. Extracting the ions with a low electric field (5–10 V/cm), the ion time of flight depends on the mass and on the projection of the initial fragment velocity onto the spectrometer axis. Thus, by knowing the extraction field, the measured time-of-flight spectrum is transformed into a velocity spectrum, from which upper bounds of the fragment kinetic energy are determined. This experimental technique is widely known (Wiley and McLaren, 1955; Ogorzalek Loo et al., 1988; Baumert et al., 1993a). The mass resolution $m/\Delta m$ of our ion TOF spectrometer is about 100 and is sufficient to study molecules and small metal cluster systems. The energy calibration of the electron TOF spectrometer is based upon several one- and two-photon resonance-enhanced ionizing transitions in atomic sodium, leading to electrons with well-defined kinetic energy. The energy resolution of the electron TOF spectrometer is 25 to 80 meV in the range of 0.1 to 3 eV. With our ZEKE (zero kinetic energy) photoelectron spectrometer, built according to the design of Müller-Dethlefs and Schlag (Müller-Dethlefs et al., 1984), we achieve an electron energy resolution of 0.1 meV (or about 1 cm^{-1}).

B. THE FEMTOSECOND LASER SYSTEMS

Femtosecond light pulses are generated in two different home-built laser systems. Independently tunable femtosecond pulses down to 50 fs time duration and up to 50 μJ energy are generated in the laser system shown in Fig. 2. The tunability of our present system covers the near UV, the complete visible range, and the near IR. The output pulses either of a colliding-pulse mode-locked (CPM) ring dye laser (see below) or of a Ti:sapphire oscillator (see below) are amplified in a bow-tie amplifier, which is pumped by an excimer laser at 308 nm, pulse compressed, and focused into a cell containing methanol to generate a white light continuum. Further amplification schemes for ultrashort dye laser pulses are reviewed in Knox (1988), Simon (1989), and Heist et al. (1990). Pump and probe pulses at specific wavelengths are selected from the white light continuum with a grating, which can also be used to compensate for group velocity dispersion in the subsequent amplification stages. Using adjustable slits for wavelength selection, the bandwidth of the pulses can be chosen depending upon the requirements of the experiment (Noordam et al., 1991). Pump and probe pulses are amplified again in two additional bow-tie amplifiers. Additional wavelength conversion methods like frequency doubling are used to generate tunable ultrashort UV–laser pulses. The pump and probe laser beams are recombined collinearly and focused into the interaction region. A Michelson arrangement is used to delay the probe laser pulse relative to the pump laser pulse.

1. The CPM Oscillator

In our initial setup we decided to use a colliding pulse mode-locked ring dye laser (CPM) (Fork et al., 1981) as a source of the femtosecond laser pulses, because of the ease of getting into the sub-100-fs regime and because of the reliability of this kind of passively mode-locked laser in terms of pulse stability. This choice implied a nanosecond laser for pumping the amplification stages. We decided on an excimer laser (Lambda Physik LPX 120i) with a 200-Hz repetition rate and a maximum of 200 mJ pulse energy in a 17-ns pulse. This laser is a compromise between the low repetition (high pulse energy) Nd:YAG laser systems and the high repetition rate (low energy) copper vapor laser systems. Because of the wavelength of the excimer laser (308 nm, XeCl), pulses even in the blue spectral region — obtained by spectral filtering of the supercontinuum (see above) — can be amplified directly. Building up our CPM laser — indicated at the top of Fig. 2 — we followed the design proposed by Valdmanis and Fork (1986). We

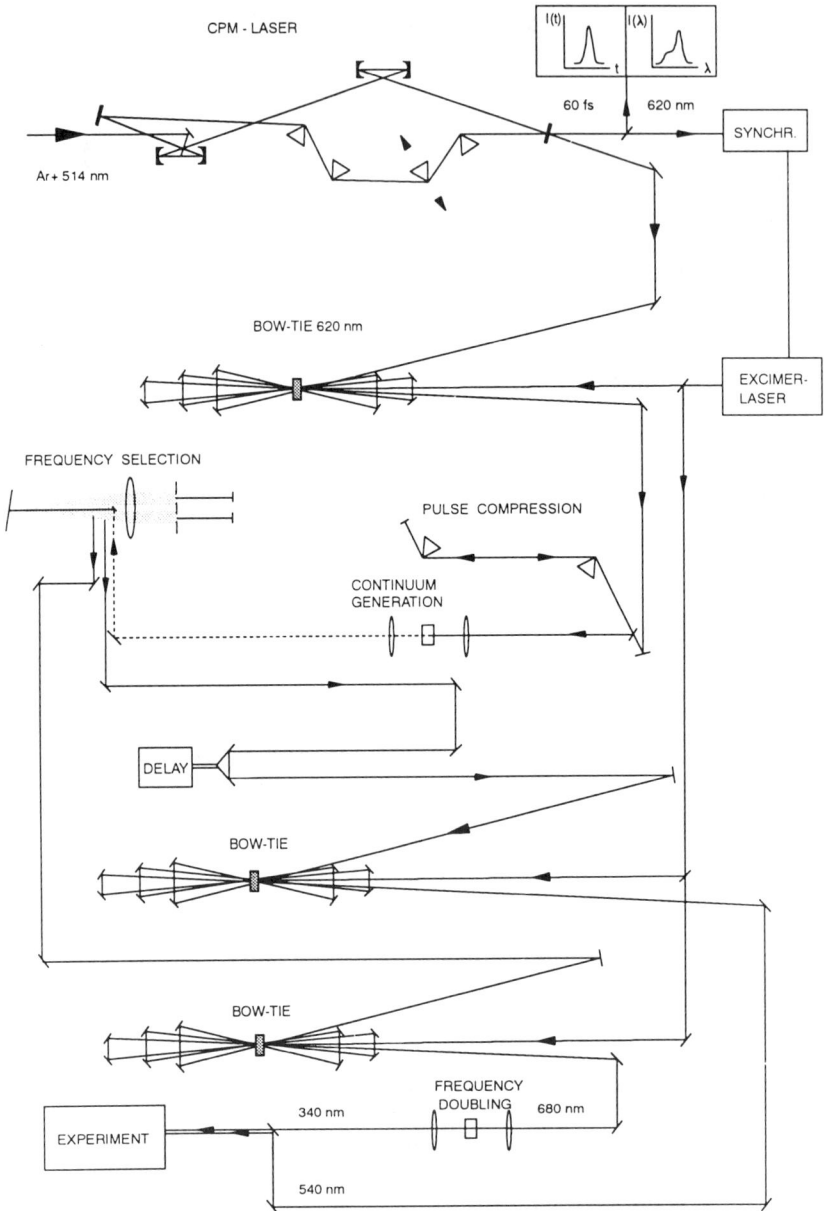

FIG. 2. Femtosecond laser system for independently tunable pump and probe wavelengths.

used broadband dielectric mirrors in the resonator. The folding mirrors have a radius of curvature of 10 cm for the gain and 5 cm for the absorber. The output coupler has a transmission of 3%. Folding angles are 5° (full angle) for the gain folding and 10° for the absorber. They are determined by mirror dimensions. We decided on a 4-m ring resonator, to allow the amplification of only one pulse out of the pulse train of the CPM, during the 17-ns pulse duration of the excimer laser. We use Rhodamine 6G (2.3×10^{-3} molarity dissolved in ethylene glycol) for the gain and DODCI (dissolved in ethylene glycol) for the absorber. The concentration of the DODCI is determined by the desired wavelength and varies in the range of 10^{-4} to 10^{-5} molarity. Standard coherent nozzles are used for the gain and the absorber jets. Compensation of the intracavity group velocity dispersion is accomplished with the standard four-prism design. The fused silica prisms have a distance of 38 cm. The 514.5-nm light of the Ar^+ laser is focused down by a 5-cm folding mirror. With 4 W pump power we obtain pulses of 60 fs $sech^2$ at a center wavelength of 627 nm. The center wavelength of the CPM dye laser can be shifted in the range 613–630 nm by adjusting the saturable absorber concentration, resulting, however, in a slight increase of the output pulse duration if the laser is shifted to shorter wavelengths. The typical output power for operation in the sub-100-fs regime is 10–15 mW. The pump power dependence of our system is similar to that described in Jacobovitz et al. (1986). One output of the ring laser is used to monitor the spectral and temporal behavior of the laser and to derive a synchronization signal for the excimer laser. The other output is used for amplification. The spectral distribution is measured on line with an optical multichannel analyzer (OMA). The pulse duration is determined by the autocorrelation technique using collinear and anticollinear second harmonic generation in a 1-mm KDP crystal and assuming a $sech^2$ pulse shape (Ippen and Shank, 1977). We built the resonator using ABCD resonator calculations (Kogelnik and Li, 1966; Rigrod, 1965; Li, 1982). Having the boundary conditions of 4 m resonator length, spacing of the saturable absorber jet and gain jet of one-quarter of the total round trip, and folding angles of 5° and 10°, we computed the common stability region of tangential and sagittal stability points by varying the absorber folding distance d_1 and the gain folding distance d_2. For these stability points we calculated the location and the spot size of the tangential and sagittal beam waists. Knowing the spot sizes one can perform an s-parameter analysis (New, 1974). This information is not very useful in CPM lasers as the dispersive pulse shaping mechanism is more dominant than pulse shaping by saturable gain and saturable absorption for production of the ultrashort laser pulses. This can be seen because the shortest pulses are often not obtained when the saturable absorber is placed into the focus (largest s-parameter) but a few 100 μm away from the focus. Knowing the location of the beam waists is useful information because straightforward alignment procedures can be derived from this by simply applying Descartes' formula for sagittal and tangential image points. We applied this concept to a 1.9-m CPM resonator and to the 4-m resonator and found in both cases that the calculated alignment leads to stable ring dye laser operation. In the 4-m resonator geometry we took the stability point $d_1 = 10.55$ cm and $d_2 = 5.25$ cm for the alignment. In the gain folding we got for the location l_s and l_t of the beam waist $l_s = 5.45$ cm and

$l_t = 5.43$ cm in the direction toward the saturable absorber folding. In the absorber folding we got $l_s = 2.53$ cm and $l_t = 2.53$ cm in the direction toward the output coupler. To align the gain folding distances we used the fluorescence light from the pump spot of the Ar^+ laser on the gain jet, and to adjust the absorber folding we used a slit to align the distance of the two mirrors. The threshold pump power for cw operation of the ring was then 0.9 W.

2. The Ti:Sapphire Oscillator

We have built an amplified Ti:sapphire laser system based on the chirped pulse amplification (CPA) technique (Strickland and Mourou, 1985). While dye amplifier systems are limited to average power outputs on the order of 10 mW, up-to-date solid-state systems can produce more than 2 W of average power with peak powers of 14 GW (Squier et al., 1993). For the experiments with mercury clusters we used the Ti:sapphire oscillator described below. This Ti:sapphire laser produces 20- to 70-fs light pulses in the wavelength range 700 to 810 nm. Using again the excimer pumped bow-tie amplifiers pulse energies of the order of several 10 μJ are obtained. The setup of this laser system is depicted in Fig. 3. Contemporary Ti:sapphire laser designs are able to produce pulses as short as 11 fs (Asaki et al., 1993; Stingl et al., 1994). Pulse durations as low as 8.2 fs were reported (Stingl, 1994).

Our Ti:sapphire oscillator is shown in the upper part of Fig. 3. The folding mirrors have a radius of curvature of 10 cm. The distance from the left folding mirror to the end mirror is about 80 cm and the distance from the right folding mirror to the output coupler is also 80 cm. The prisms, made of LaKL21, are spaced by 33 cm in order to minimize cubic phase distortions. The 8-mm-long Ti:sapphire rod has Brewster end faces and is doped with 0.1% Titan. Again we performed and ABCD matrix analysis of the resonator: the astigmatism of the laser rod is compensated for a folding angle of 19.25°; stable operation is then achieved in a region from 4.86 to 5.16 cm (distance from folding mirrors to the Brewster end faces of the laser rod). The laser rod is pumped by an all-lines Ar^+ laser focused with a 10-cm focusing lens. Typical pump powers are in the range from 4 to 5 W, resulting in an average output power of 50 to 400 mW of the femtosecond oscillator.

C. THE PUMP–PROBE DELAY LINE

For the one-color pump–probe experiments we used a Michelson arrangement with a 50% beam splitter to delay the probe laser pulse relative to the pump laser pulse. A computer-controlled stepper-motor-driven linear preci-

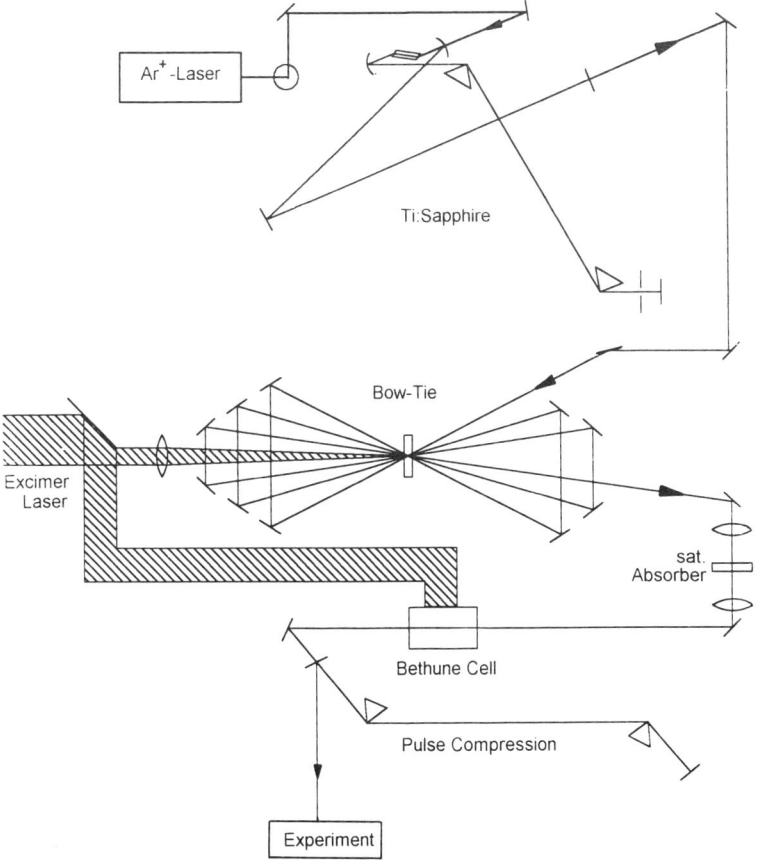

FIG. 3. Schematic setup of the Ti:sapphire femtosecond laser system. In the upper figure the Ti:sapphire oscillator is shown. The slit is used to tune the laser wavelength.

sion actuator with a 0.05-μm feed in one arm of the interferometer sets the time delay. The pulses are recombined collinearly. Thus we obtain identical pulses for the pump and the probe lasers. The pulses are identical in polarization, wavelength, energy, and duration. The advantage of one-color pump–probe experiments is that the dynamics of multiphoton ionization studied with femtosecond laser pulses can directly be compared with one-color nanosecond laser multiphoton experiments. Identical pump and probe laser pulses are crucial for this comparison. A further advantage is that in this case symmetric spectra are always obtained with respect to zero delay time. With an accurate determination of zero delay time, phase information can be derived from the observed wavepacket motion. The disadvantage of a Michelson pump–probe arrangement is the loss of half of the pulse energy at the beam splitter. The same actuator has also been used in two-color pump–probe experiments. A typical two-color pump–probe femtosecond laser system is depicted in Fig. 2.

D. DATA ACQUISITION

The data for the Na^+, Na_2^+, and Na_3^+ transients in the one-color pump–probe experiments were taken in the following way: the MCP (multichannel plates) signal of the TOF spectrometer is fed into a boxcar integrator (SR 255). The boxcar is triggered by a laser reflex monitored by a fast photodiode. The gate of the boxcar is set to the ionic mass to be measured. Note that in a TOF spectrometer the time of flight of an ion is proportional to the square root of the mass of this ion. The integration of the boxcar averager is set to 300 or 1000 laser shots. With a given repetition rate of the amplified pulses the feed of the stepper motor is set such that during an integration over 300 or 1000 laser shots a pump–probe delay time of 20 fs is scanned. The reason for the choice of such an averaging interval (AI) is that AI has to be greater than 1 fs to get rid of interferometric structures from overlapping pump and probe pulses. The same condition is applied to collinear intensity autocorrelation measurements. On the other hand AI has to be much smaller than the pulse duration in order not to lose time resolution. In particular for a fast Fourier transformation (FFT) the sampling interval of the FFT (that corresponds to AI) should be smaller than half of the pulse width to avoid aliasing effects in the FFT spectra. So in our (one-color pump–probe) transient Na^+, Na_2^+, and Na_3^+ spectra each data point corresponds to a pump–probe delay time interval of 20 fs and each of these intervals is averaged over 300 or 1000 laser shots.

The laser intensities in all experiments were monitored with calibrated photodiodes and an additional boxcar averager. The data for the cluster experiments were taken with a CAMAC transient recorder (Transiac 2001 AS, DSP) and averager (4101, DSP).

III. Results and Discussion of Experiments in Molecular Physics

A. VIBRATIONAL WAVE-PACKET MOTION IN THE $2\,^1\Sigma_u^+$ DOUBLE MINIMUM STATE OF Na_2

In this section we will first discuss the propagation of a wavepacket in a double minimum potential well. Then the experimental preparation and detection of a vibrational wavepacket on such an excited electronic surface are considered. Furthermore we will show that frequency spectroscopy can be performed in the time domain as well.

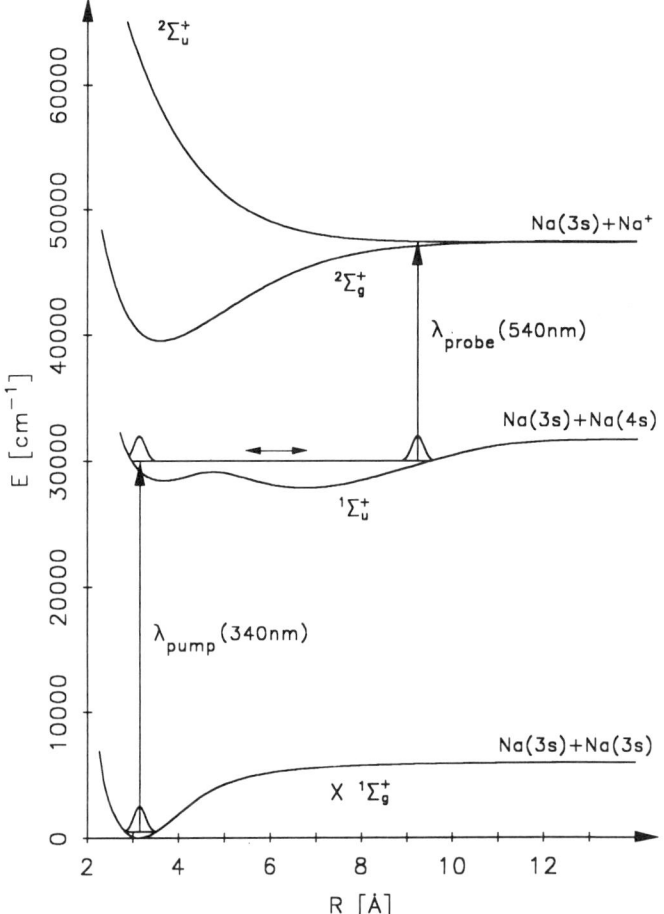

FIG. 4. Potential energy surfaces relevant for the pump–probe experiment on the $2\,^1\Sigma_u^+$ double minimum state of the Na_2. The pump laser prepares a vibrational wavepacket at the inner turning point above the barrier. The probe laser transfers this wavepacket at the outer turning point onto the repulsive $^2\Sigma_u^+$ ionic state of the Na_2. Measuring the Na^+ signal as a function of pump–probe delay therefore resembles the wavepacket dynamics in the double minimum state.

The double minimum structure of the $2\,^1\Sigma_u^+$ state of Na_2 (see Fig. 4) is formed by the avoided crossing of two adiabatic potential curves. One has mainly Rydberg character and the other has a substantial ionic character at large internuclear distances. The inner potential minimum is at an internuclear distance of 3.688 Å and at an energy of 28454.56 cm^{-1}; the outer potential minimum is located at 6.739 Å with an energy of 27879.40 cm^{-1} (Cooper et al., 1984; Vergès et al., 1984). The barrier is at 4.716 Å with an energy of 29132 cm^{-1}. The top of the barrier corresponds to an excitation wavelength of 344.2 nm relative to the $X\,^1\Sigma_g^+$ ($v'' = 0$) state. The $2\,^1\Sigma_u^+$ state was theoretically predicted by Valance and Nguyen Tuan (1982) and confirmed by Jeung (1983). The first experimental observation was reported by Cooper et al. (1984) and Vergès et al. (1984). They employed Fourier transform spectroscopy of the laser-induced infrared fluorescence. Experiments involving the two-photon ionization (TPI) technique were performed by Delacrétaz and Wöste (1985) and by Haugstätter et al. (1988).

By using time-dependent perturbation theory and its implementations into time-dependent molecular physics (Kulander and Heller, 1978; Engel, 1991a) we have calculated the propagation of a wavepacket created by the coherent superposition of vibrational eigenstates at the inner turning point of the $2\,^1\Sigma_u^+$ state with an 80-fs excitation pulse. Note, that no quantum calculations are needed if the only interest is the average classical oscillation period T as T is simply given by $T = 1/\Delta v$, where Δv is the average vibrational spacing in Hertz.

The results of the quantum calculations are displayed in Fig. 5. There is a strong dependence of the wavepacket dynamics with respect to its energy or equivalently to the wavelength of the exciting laser pulse. At $\lambda = 347$ nm the wavepacket is excited below the barrier, and therefore only motion within the inner well is possible. Centering the energy of the wavepacket at the barrier ($\lambda = 344.2$ nm) the wavepacket splits into a transmitted and reflected part as is expected from basic quantum mechanics. At an excitation energy of $\lambda = 341.5$ nm the wavepacket is already above the barrier. At $\lambda = 335$ nm the wavepacket is far above the barrier, and the dynamics represents nicely the classical motion of a point mass in this potential well. Transient electron spectra reflecting wavepacket motion in this $2\,^1\Sigma_u^+$ state were calculated by Meier and Engel (1994a). They used an excitation scheme similar to that described below.

In order to prepare a wavepacket above the barrier experimentally, we excited the molecule with a short pump pulse at $\lambda = 340$ nm. The wavepacket is created at the inner turning point of the $2\,^1\Sigma_u^+$ state, as we start from $v'' = 0$ in the narrow $X\,^1\Sigma_g^+$ state (see Fig. 4). In general, classical transition regions are determined by the Mulliken difference potential analysis (Mulliken, 1971). The application of this analysis to femtosecond pump–probe experiments in molecular physics was discussed by Baumert et al. (1991b). Because of the known average vibrational spacings we expected an oscillation period of about 1 ps, as is also seen in the quantum calculations in Fig. 5. The probe laser wavelength is chosen such that only at the outer turning point a transition onto the repulsive $^2\Sigma_u^+$ state of Na_2^+ is possible by energetic and Franck–Condon arguments. The probe laser will therefore form slow atomic Na^+ fragments. Preparing the wavepacket at the inner turning point of the $2\,^1\Sigma_u^+$ state and transferring it into the ionization continuum at the outer turning point, one observes the first signal

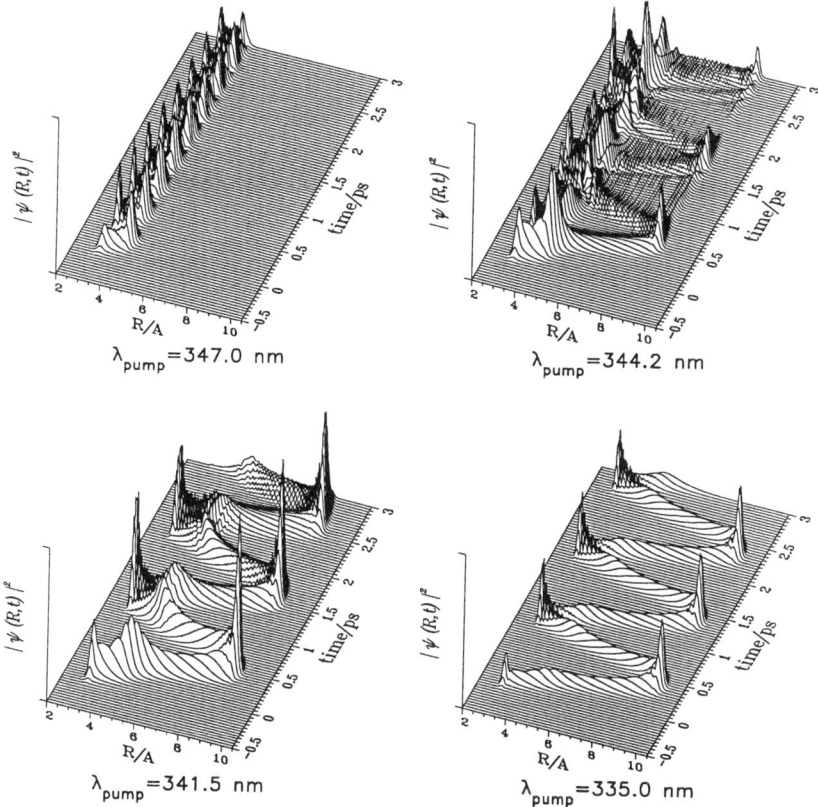

FIG. 5. Dynamical behavior of a wavepacket in the $2\,^1\Sigma_u^+$ double minimum state (Fig. 4) for different excitation energies (see the text): $\lambda_{pump} = 347.0$ nm, below the barrier; $\lambda_{pump} = 344.2$ nm, at the barrier—note that part of the wavepacket is reflected, whereas the other part is transmitted; $\lambda_{pump} = 341.5$ nm, just above the barrier; and $\lambda_{pump} = 335.0$ nm, far above the barrier.

only after half a vibrational period according to the formation of the ionic (Na$^+$) fragments. Depending on the pump–probe delay the ionic (Na$^+$) signal is expected to oscillate with a period of about 1 ps. This indeed is what we observed in the experiment. Figure 6 shows the ionic (Na$^+$) signal versus the pump–probe delay time. The expected 1-ps oscillation is clearly seen. Due to the anharmonicity of the $2\,^1\Sigma_u^+$ potential the vibrational spacings are not constant within the spectral width of the exciting laser pulse. Therefore the amplitude of the observed oscillation decreases with time. This spreading of a vibrational wavepacket takes place in a well-defined way, and the wave packet is restored completely after a certain time, which is known as the recurrence time of a wavepacket. The spreading and recurrence of a vibrational wavepacket motion in the $A\,^1\Sigma_u^+$ state of the sodium dimer was

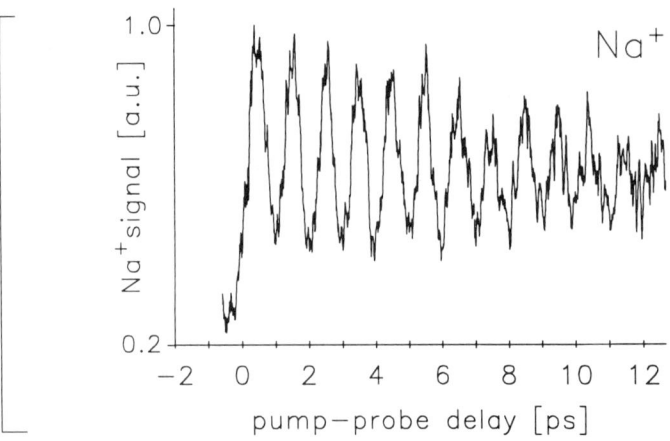

FIG. 6. The measured Na$^+$ signal as a function of pump–probe delay time shows the expected oscillatory behavior, determined by the $2\,^1\Sigma_u^+$-potential curve.

investigated experimentally (Baumert et al., 1992a). A comparison with quantum calculations is also given in that publication.

Now we turn to the topic of frequency spectroscopy in the time domain. At first glance the terms frequency spectroscopy and time domain seem to be contradictory because of the broad spectral distribution of an ultrashort laser pulse. However, the spectroscopic information can be derived by a Fourier transformation from data taken in the time domain. This has been shown for diatomics and diatomic-like molecules by Zewail's group for the system I_2 (Gruebele et al., 1990) and ICN (Janssen et al., 1990). Although the time domain approach cannot compete with the elaborate techniques of high-resolution spectroscopy for bound systems, for predissociating or dissociative systems this approach might sometimes be the only choice to determine spectroscopic data especially in the transition-state region. Another advantage of the time-resolved method is the ease of distinction between vibrational and rotational spectroscopic information, because energy spacings (e.g., oscillation periods) are different in general by two orders of magnitude. For bound systems the achievable resolution is limited only by the scan length. Using a square window in the Fourier transformation, the theoretical resolution limit for the FWHM is $0.1\,\mathrm{cm}^{-1}$ for a scan length of 300 ps. Note that peak positions (frequencies) in such a FFT spectrum can be determined with even higher accuracy.

The correspondence of resolution in the frequency domain and scan length in the time domain is now used in the analysis of our data. The FFT of the transient given in Fig. 6 consists of a frequency distribution centered around $33.4\,\mathrm{cm}^{-1}$ (corresponding to 1 ps), where the individual energy spacings of the vibrational levels the wavepacket is composed of are not

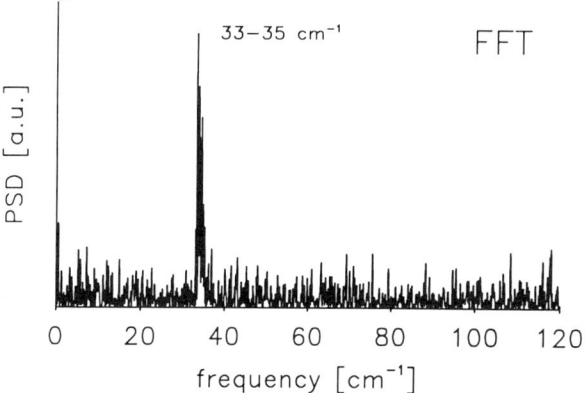

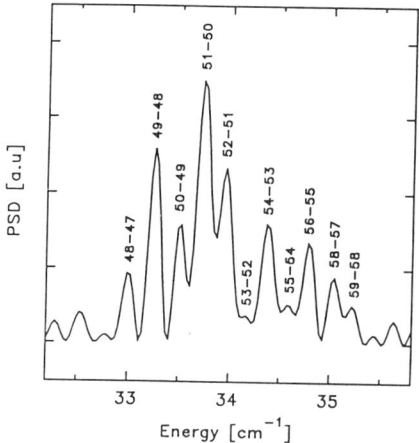

FIG. 7. (upper) Fourier transformation of a 170-ps Na^+ transient measured under the same experimental conditions as the transient displayed in Fig. 6. Only frequencies in the range 33–35 cm^{-1} contribute to the formation of the wavepacket. (lower) Enlargement of the Fourier transformation displayed in the upper figure. The frequency distribution is composed of individual frequencies corresponding to the energy spacings of the vibrational levels forming the wavepacket.

resolved. However, the FFT of a 170-ps-long Na^+ transient, which is not shown, shows frequencies in the range of 33–35 cm^{-1} in the upper part of Fig. 7. In the enlarged lower part of that figure the individual vibrational frequencies are clearly seen. These frequencies are indeed the vibrational energy spacings the wavepacket is composed of, as is proven by the comparison (Table I) with the high-resolution frequency spectroscopy reported by Cooper et al. (1984) and Vergès et al. (1984). The nanosecond

TABLE I
VIBRATIONAL ENERGY SPACINGS $\Delta G(v)$ IN THE $2\,^1\Sigma_u^+$ STATE FROM FEMTOSECOND DATA

v'	REMPI[a] (cm^{-1})	REMPI[b] (cm^{-1})	FTS[c] (cm^{-1})	Femtosecond experiment (cm^{-1})
59–58	—	—	35.19	
58–57	—	—	35.06	35.1
57–56	35.1	35.5	34.91	
56–55	34.1	34.0	34.75	34.78
55–54	34.5	35.4	34.58	34.65(s)
54–53	34.8	34.4	34.39	34.39
53–52	33.8	33.5	34.20	34.16(s)
52–51	34.5	34.7	33.99	33.98
51–50	35.9	35.4	33.76	33.73
50–49	31.5	33.4	33.53	33.51
49–48	33.3	31.6	33.28	33.27
48–47	33.4	33.7	33.00	33.00
47–46	32.5	33.0	32.71	—
46–45	32.5	33.1	32.40	—
⋮	⋮	⋮	⋮	⋮

Note. Vibrational energy spacings $\Delta G(v)$ obtained from the FFT spectrum displayed in Fig. 7. The letter "s" denotes values obtained from shoulders in Fig. 7. The assignment to v' levels in the $2\,^1\Sigma_u^+$ state is based upon high-resolution Fourier transform spectroscopy (FTS) data. Results of nanosecond laser REMPI experiments are also given.
[a]Delacrétaz and Wöste (1985).
[b]Haugstätter et al. (1988).
[c]Cooper et al. (1984).

laser REMPI results of Delacrétaz and Wöste (1985) and of Haugstätter et al. (1988) are also shown for comparison.

The future experimental work will concentrate on studies of the influence of the barrier on the vibrational wavepacket dynamics. The excitation and detection of a vibrational wavepacket in the $4\,^1\Sigma_g^+$ shelf state of the Na$_2$ is another example for performing frequency spectroscopy of high lying electronic states in the time domain (Baumert and Gerber, 1994). Note that from transient spectra of the three-dimensional wavepacket motion in Na$_3$, the normal modes of the excited B-state and of the X-ground state as well have been obtained by an FFT (Baumert et al., 1993b).

B. DYNAMICS OF MULTIPHOTON IONIZATION OF Na$_2$

In this section we will first give a short introduction to the topic of MPI of small molecules before we present and discuss our results on femtosecond pump–probe experiments on the dynamics of MPI and subsequent fragmentation. Again we have chosen the sodium dimer molecule as a prototype.

Multiphoton ionization of small molecules has been studied by a variety of techniques and is generally well understood. The ionization is predominantly due to REMPI processes, whereas nonresonant multiphoton processes play only a minor role. Dynamical aspects of the interaction of laser radiation with molecules and details of the excitation processes and the different decay channels of highly excited states, embedded in the ionization and in the fragmentation continuum, have rarely been studied up to now. We reported on the interaction of a doubly bound excited molecular state with different continua and the competition between the various decay channels (Baumert et al., 1990). In that study we used femtosecond laser pulses as an experimental tool to distinguish between the dissociative ionization of the molecule and the neutral fragmentation with subsequent excited-fragment photoionization. Both processes are difficult to distinguish when using nanosecond or even picosecond laser pulses. This distinction is of particular importance in multiphoton ionization studies of metal cluster systems (Baumert et al., 1992b). The multiphoton ionization and fragmentation of alkali-metal molecules and, in particular, of Na_2 and Na_3 have attracted considerable current interest. In many experiments with Na_2 it has been found that, in conjunction with the formation of Na_2^+ ions, ionic fragments Na^+ are also formed. REMPI processes via the $A\,^1\Sigma_u^+$ or the $B\,^1\Pi_u$ states are responsible for this observation (Keller and Weiner, 1984; Burkhardt et al., 1985).

Dynamical aspects of the interaction of laser radiation with molecules and details of the excitation and ionization processes can be studied directly in the time domain by femtosecond pump–probe techniques. This allows a closer look at the time scales of the absorption of several photons in molecular multiphoton ionization and the photofragmentation of neutral and ionic molecules, as will be demonstrated in the following parts of this section.

We have reported femtosecond time-resolved multiphoton ionization and fragmentation dynamics of the Na_2 molecule (Baumert et al., 1991a, b). In these experiments we applied photons of 2 eV (around 620 nm) to ionize the Na_2 molecule. At that wavelength it is the $X\,^1\Sigma_g^+$ (Kusch and Hessel, 1978), the $A\,^1\Sigma_u^+$ (Gerber and Möller, 1985), and the $2\,^1\Pi_g$ (Taylor et al., 1983) states that participate in MPI of the neutral molecule. Three photons are needed to ionize the molecule into its ionic ground state, $^2\Sigma_g^+$ (Bordas et al., 1989) (see Fig. 9).

From the real-time observation of vibrational wavepacket motions it was concluded that two different physical processes determine the time evolution of multiphoton ionization. The observed femtosecond pump–probe delay spectrum of the molecular ion (Na_2^+) signal is shown in the upper part of Fig. 8. Evident from the beat structure seen in this transient, there are two frequencies involved and therefore there are two contributions to the transient ionization spectrum. The envelope intensity variation reveals them

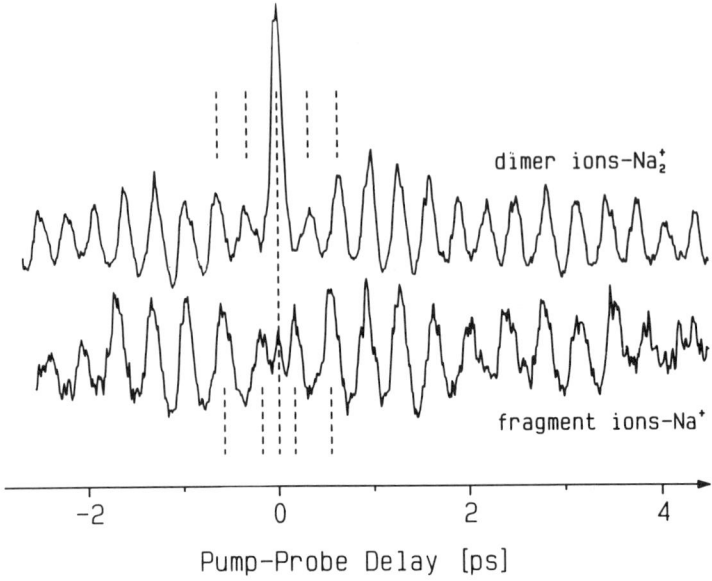

FIG. 8. (upper) Transient Na_2^+ signal obtained as a function of pump–probe delay time between two identical femtosecond laser pulses with 85 fs at $\lambda_{max} = 623$ nm. The envelope intensity variation and the oscillatory structure of this transient Na_2^+ MPI signal reveal two contributions out of phase by 180°. They correspond to independent wavepacket motions in bound molecular potentials with 309 fs ($A\,^1\Sigma_u^+$) and 362 fs ($2\,^1\Pi_g$) oscillation periods. (lower) Transient Na^+ photo fragment signal, obtained under the same experimental conditions as the Na_2^+ transient. The transient shows the dynamics of the 180° phase-shifted $2\,^1\Pi_g$-state wavepacket motion.

to be 180° out of phase. A Fourier analysis of this spectrum yields two groups of frequencies, one centered at 108.1 cm^{-1} and a second one centered at 92.2 cm^{-1}, with an experimental uncertainty of less than 0.5 cm^{-1}. From the observed two oscillation periods, the 180° phase shift and the additionally measured time-resolved Na^+ photofragmentation spectrum (see the lower part of Fig. 8), we concluded that for Na_2 two different MPI processes exist, to require incoherent addition of the intensities to account for the observations.

The direct photoionization of an excited state electron, in which one pump photon creates a vibrational wavepacket in the $A\,^1\Sigma_u^+$ state and two probe photons transfer that motion via the $2\,^1\Pi_g$ state in the ionization continuum, is one (REMPI) process. In this *one-electron* direct photoionization process all three photons are absorbed at the inner turning point. This is therefore an MPI process in which all photons can be absorbed at once or at least within the time duration of the light pulses. The question that has to be addressed now is, why is it that all excitation steps of this MPI process mainly occur at the inner turning point? This can be discussed by classical arguments, with the help of Mulliken's difference potential concept (Baumert et al., 1991a) and by a quantum mechanical treatment as well (Engel et al., 1993). The results of these two approaches are the same. It is in fact the resonance enhancing $2\,^1\Pi_g$ Rydberg state that restricts the two-photon probe transitions to locations near the inner turning point.

The second MPI process involves excitation of *two electrons* and subsequent electronic autoionization. Here two pump photons create a wavepacket at the inner turning point, in the $2\,^1\Pi_g$ Rydberg state, which then propagates to the outer turning point, where the probe laser transfers the motion into the continuum by exciting a second electron, forming a doubly excited bound neutral Na_2^{**} $(nl, n'l')$ molecule. In this case the probe photon is absorbed at the earliest about 180 fs after the pump photons were absorbed.

In Fig. 9 the relevant potential energy curves taking part in these two different MPI processes are displayed. The two-photon-pump and one-photon-probe ionization process which involves excitation and decay of doubly excited states is indicated. The decay of these doubly excited states takes place by electronic autoionization into the $^2\Sigma_g^+$ ground state of Na_2^+, being responsible for the observed 180° phase-shifted $2\,^1\Pi_g$ wavepacket motion in the Na_2^+ transient (upper part of Fig. 8) and by electronic-autoionization-induced fragmentation, leading to slow Na^+ atomic fragments. This interpretation is confirmed by the observed 180° phase-shifted $2\,^1\Pi_g$ state wavepacket dynamics seen in the Na^+ ionic fragment transient displayed in the lower part of Fig. 8.

Preliminary calculations of doubly excited neutral electronic states of Na_2 correlating with the $Na(4s) + Na(3p)$ asymptote performed by Meyer (1992) show a $^1\Pi_u$ state that can be excited from the outer turning point of the $2\,^1\Pi_g$ state by absorption of one probe photon as indicated in Fig. 10. Although this state can decay by electronic autoionization, it cannot decay by electronic-autoionization-induced fragmentation. This is why there has to be a spin–orbit interaction with the nearby $^3\Pi_u$ doubly excited state via which the fragmentation proceeds.

To give a full account of the problem, the direct photoionization out of the $2\,^1\Pi_g$ state has to be included in the discussions. Again classical methods (Baumert et al., 1991a) as well as quantum mechanical calculations (Engel et al., 1993) can be applied. Note that further excitation of the coherently excited v^* levels of the $2\,^1\Pi_g$ state by a photon of 2 eV leads to a total excitation energy above the dissociation limit of the Na_2^+ ($^2\Sigma_g^+$) state, and therefore all possible electron kinetic energies according to Franck–Condon considerations can be produced. Analyzing this situation with Mulliken's difference potential concept leads to the result displayed on the left-hand side of Fig. 11. The table of Franck–Condon factors calculated for this probe transition shows four diagonals with high transition probabilities,

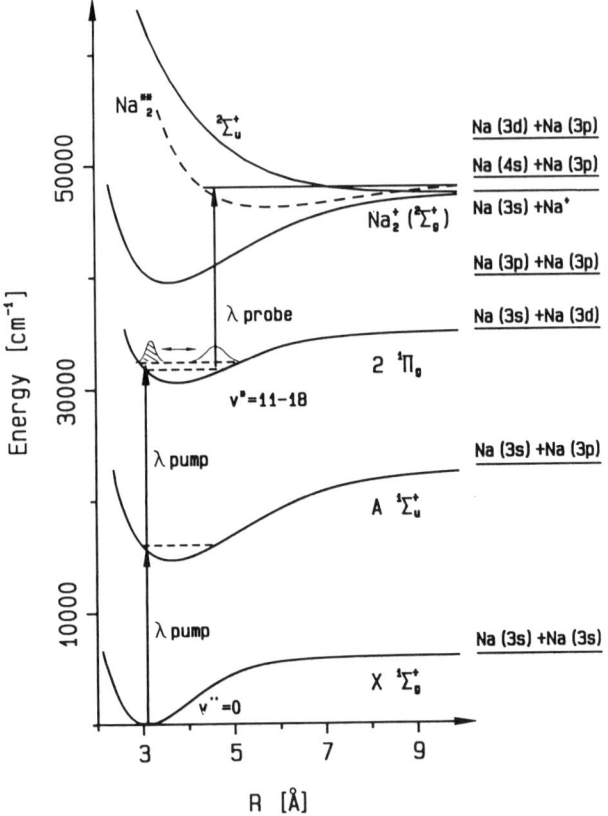

FIG. 9. The relevant potential energy curves of Na_2 taking part in the two different MPI processes are displayed. The two-photon-pump and one-photon-probe ionization processes which involves excitation and decay of doubly excited states is indicated. The dashed line is an estimate for the potential curve of $Na_2^{**}(nl,n'l')$. Calculations of doubly excited states are shown in Fig. 10.

leading essentially to four different electron energies for the considered v^*-vibrational levels. The difference potential $Na_2^+ (^2\Sigma_g^+) - Na_2 (2\,^1\Pi_g)$ for these four groups of electrons, displayed on the left-hand side of Fig. 11, shows that for each group a different range of internuclear distances is involved in the ionizing transition. However, the different ranges of internuclear distances overlap, and the summation leads to an R-independent transition probability, leading to a time-independent Na_2^+ signal for direct ionization of the oscillating $2\,^1\Pi_g$ wavepacket. In a quantum mechanical calculation by Engel (1991b) this direct ionization process is simulated by "tuning" the probe laser in discrete steps up to the energy of the experiment. For low (ionizing) photon energies a pronounced oscillation, with a period of the oscillating $2\,^1\Pi_g$ wavepacket, is seen on the right-hand side of Fig. 11,

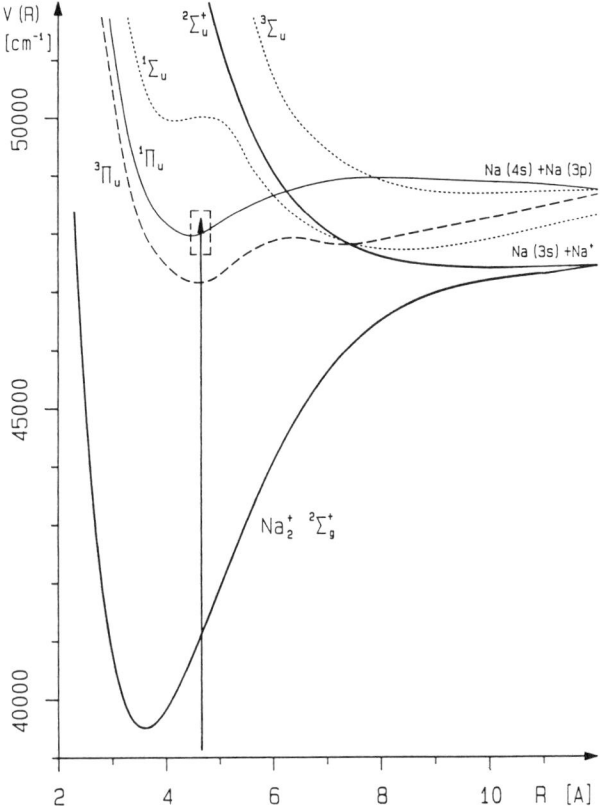

FIG. 10. The potential energy curves of the $^2\Sigma_g^+$ and $^2\Sigma_u^+$ states of Na_2^+ and four doubly excited neutral states of Na_2 are shown. These curves are shown on the basis of unpublished calculations by Meyer (1992). The excitation of the $^1\Pi_u$ doubly excited state by absorption of one probe photon at the outer turning point of the $2\,^1\Pi_g$ state is indicated.

because only low v^+ levels in the Na_2^+ ($^2\Sigma_g^+$) state are accessed. Populating all possible final v^+ states at the experimental laser wavelength, however, leads to a time-independent total Na_2^+ signal.

We performed these experiments with central laser wavelengths of 618 to 627 nm and pulse durations from 70 to 110 fs. For low excitation intensities no change in the global behavior of the measured transients was observed. Only the oscillation periods of the $A\,^1\Sigma_u^+$-state wavepacket and of the $2\,^1\Pi_g$-state wavepacket show slight variations according to the different spectral regions excited.

The Na_2 case is the first example of a femtosecond molecular multiphoton ionization study. It was only through time domain measurements that the existence of a second major ionization process was established. A compre-

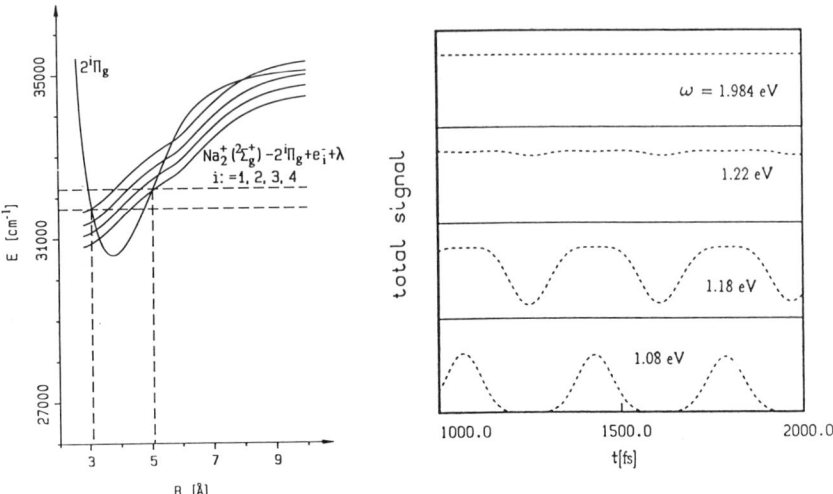

FIG. 11. (left) Difference potential analysis for a direct photoionization process out of the $2\,^1\Pi_g$ state of Na_2. In this bound-continuum transition essentially four different groups of electrons are produced for the coherently excited v^* levels. The electron energies are in the range of 6300–7200 cm^{-1} for laser excitation with 2 eV of photon energy. Although the range of internuclear distances is different for each group, the summation over all electron energies leads to an R-independent ionization signal. (right) Calculated total Na_2^+ signal as a function of pump–probe delay for different laser wavelengths (Engel, 1991b). At 1.984 eV (625 nm) the time dependence for a direct photoionization process out of the $2\,^1\Pi_g$ state of Na_2 has vanished.

hensive discussion based on classical arguments can be found in Baumert *et al.* (1991a), whereas the comparison between experiment and quantum mechanical calculations can be found in Engel *et al.* (1993).

C. Control of Na_2^+ versus Na^+ Yield

In this section we will discuss the Na_2 MPI results from a different point of view. We show that by femtosecond pump–probe techniques basic ideas from the area of control of chemical reactions can be realized.

Controlling a chemical reaction such that a given product is produced at the expense of another, energetically allowed, product is one of the basic issues in physical chemistry. Many publications are devoted to this topic. Some references can be found in a review by Warren *et al.* (1993). As in larger molecules the locally deposited energy redistributes very rapidly — typically on picosecond time scales — throughout the molecule; specially designed pulse shapes and phase-shifted pulses are currently being considered in order to achieve bond selectivity in these systems. For smaller molecular systems, however, the basic ideas of the Tannor–Kosloff–Rice scheme (Tannor et al., 1986) are applicable. These authors have proposed that controlling the duration of propagation of a wavepacket on an excited electronic potential energy surface, by simply controlling the time delay between pump and probe pulses, can be used to generate different chemical products. This idea was realized in an experiment by Zewail and co-workers (Potter et al., 1992a). They used two sequential coherent laser pulses to control the reaction of I_2 molecules with Xe atoms to form XeI. It was shown that the yield of product XeI is modulated as the delay between the pulses is varied, reflecting its dependence on the nuclear motions of the reactants. However, an example of how the propagation of wavepackets can be used to produce one product at the expense of another energetically allowed product is given by our experiments for the first time, according to our knowledge.

In order to better illustrate the topic, let us assume that we focus a nanosecond laser, having a photon energy of about 2 eV, on our molecular beam. After absorption of three photons we will detect Na_2^+ and Na^+ in our TOF spectrometers according to the two ionization processes described before. There are no simple means of producing Na_2^+ at the expense of Na^+ with this nanosecond laser at a fixed intensity and wavelength. Using the time-resolved approach, we know that at the inner turning point of the

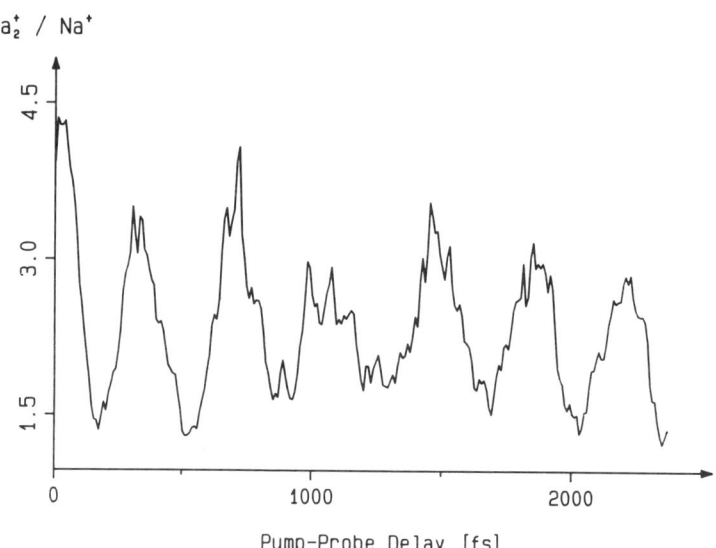

FIG. 12. Controlling a chemical reaction such that a given product is produced at the expense of another, energetically allowed, product is of general interest in physics and chemistry. Here we apply femtosecond pump–probe techniques in order to control the internuclear distance of the molecule Na_2. According to the different MPI processes this leads to a strongly varying Na_2^+/Na^+ ratio with respect to the pump–probe delay time.

$A\,^1\Sigma_u^+$ and $2\,^1\Pi_g$ states in Na_2 the molecule is directly ionized by the probe laser, whereas only at the outer turning point of the $2\,^1\Pi_g$ state fragment ions of Na^+ are produced by exciting the neutral doubly excited state with its subsequent decay channels (Fig. 9). Thus by controlling the duration of propagation of the wavepackets on the $A\,^1\Sigma_u^+$ and $2\,^1\Pi_g$ states in Na_2 we are able to produce Na_2^+ at the expense of Na^+ by adjusting the pump–probe delay time. This is illustrated in Fig. 12, in which we display the ratio of the Na_2^+ signal over the Na^+ signal from Fig. 8. A modulation of this ratio by at least a factor of 2 is seen as a function of pump–probe delay.

D. HIGH LASER FIELD EFFECTS IN MULTIPHOTON IONIZATION OF Na_2

In a further study the dependence of the total Na_2^+ ion signal on the intensity of the femtosecond pulses was investigated in detail (Baumert et al., 1992c). The experimental results, shown in the upper part of Fig. 13, were obtained for three different intensities ($I_0 = 10^{12}\,W/cm^2$, $0.3*I_0$, and $0.1*I_0$). The curves exhibit periodic oscillations with different periods for different laser intensities. The periodic contributions to these transients were analyzed by taking their Fourier transform, displayed in the lower part of Fig. 13. Additionally for higher laser intensities the relative contributions from the $A\,^1\Sigma_u^+$ and the $2\,^1\Pi_g$ states change dramatically, indicating the increasing importance of the two-electron versus the one-electron process. For the strongest fields used in these experiments a vibrational wavepacket motion in the electronic ground state $X\,^1\Sigma_g^+$ is observed. It is created through stimulated emission during the time the ultrashort pump pulse interacts with the molecule. This ground state wavepacket dynamics is monitored by absorption of three photons from the time-delayed probe laser in a direct photoionization process.

Time-dependent quantum calculations were performed to investigate this behavior. They show that for different laser field strengths the electronic states, involved in the MPI and coupled coherently by the laser interaction, are populated differently in a Rabi-type process. For lower intensities the $A\,^1\Sigma_u^+$ state is preferentially populated by the pump pulse and the $A\,^1\Sigma_u^+$ wavepacket motion dominates the ion signal. For the highest intensity used in these experiments, the contribution of the $2\,^1\Pi_g$ state motion dominates. The reason for this is that after the pump pulse is over the $2\,^1\Pi_g$ state is more strongly populated than the $A\,^1\Sigma_u^+$ state. The population in the $A\,^1\Sigma_u^+$ state is initially increasing with the rising part of the pump pulse, but then the Rabi-type process starts to decrease the population. This behavior is nicely illustrated in Fig. 14 for four different laser intensities [for computational details, see Baumert et al. (1992c) and Meier (1992)]. Thus by changing the intensity of the femtosecond pump laser one may selectively control the relative strength of the direct one-electron photoionization versus the two-electron excitation and electronic autoionization process.

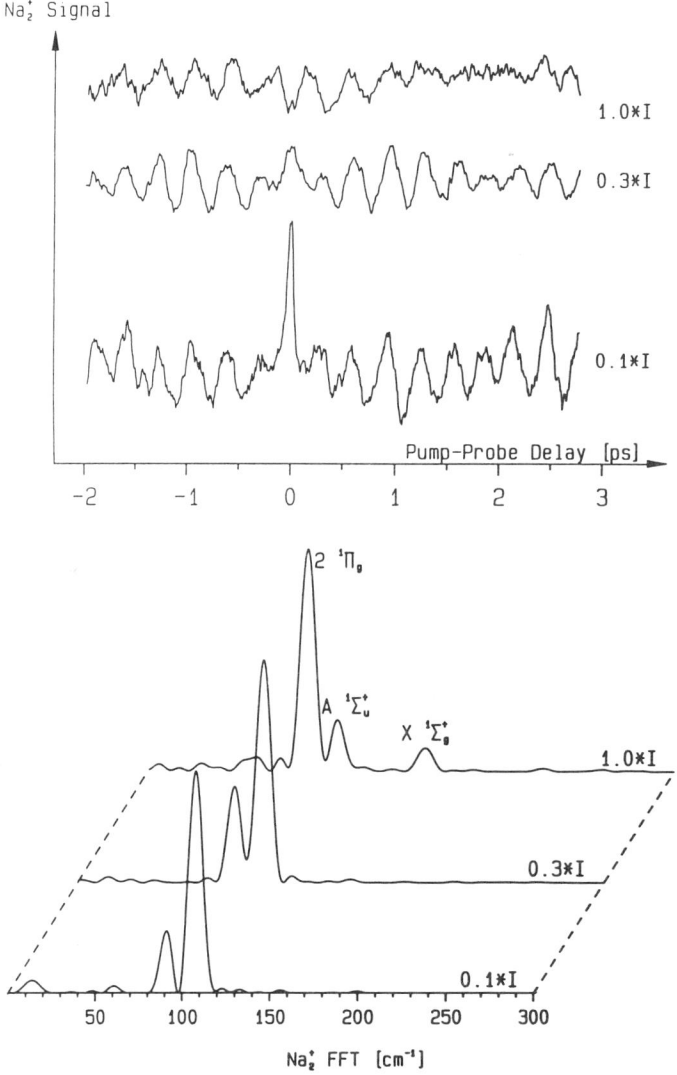

FIG. 13. The upper figure shows transient Na_2^+ spectra as a function of delay time between pump and probe pulses. Different intensities were used as indicated. In the lower figure, the Fourier transforms of the transient spectra are displayed. Note the dramatic change of the Fourier amplitudes as a function of intensity. At $1.0 \cdot I$ a contribution of the $X\,^1\Sigma_g^+$ ground state wavepacket to the transient ionization spectrum is observed in the Fourier spectrum.

This intensity-dependent effect could be used in addition to the delay time variation to optimize the control scheme discussed in the previous section. The coherent coupling of the electronic states as a function of laser intensity is expected to influence the total ionization yield as well as the kinetic energy of the ejected photoelectrons (Meier and Engel, 1994c).

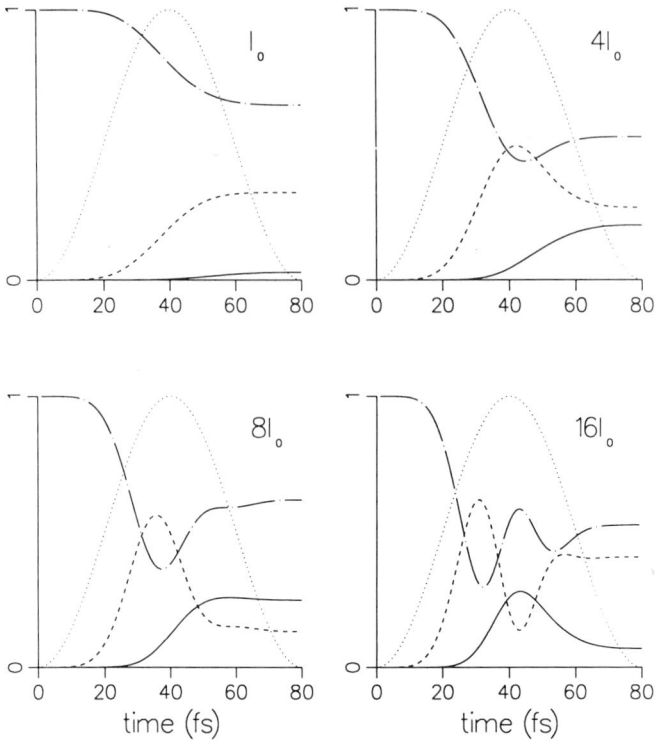

FIG. 14. The electronic states involved in the MPI processes are coupled coherently by the laser transaction and also populated differently in a Rabi-type process: the figure shows the change of population in the $X\,^1\Sigma_g^+$ (dashed–dotted line), $A\,^1\Sigma_u^+$ (dashed line), and $2\,^1\Pi_g$ (solid line) states of Na_2 during the pump pulse (dotted line) interaction for four intensities (Meier and Engel, 1994b). The creation of a ground-state ($X\,^1\Sigma_g^+$) wavepacket by stimulated emission pumping at higher intensities is seen as well as the observed change of the final population of the $A\,^1\Sigma_u^+$ and $2\,^1\Pi_g$ states (I_0, $4I_0$, and $8I_0$). Calculated transients (Baumert et al., 1992c) are in agreement with the measured transients displayed in Fig. 9. For the highest intensity ($16I_0$) the Rabi-type process is clearly seen.

IV. Results and Discussion of Experiments in Cluster Physics

A. Dynamics of Na_n^+ Photofragmentation

The stability and fragmentation dynamics of metal cluster ions formed through laser photoionization are a major issue in cluster physics. In particular for sodium and potassium cluster ions, Bréchignac et al. (1987) have reported that the predominant fragmentation channels, the evaporation of neutral monomers and dimers, respectively, are associated with microsecond fragmentation times due to sequential evaporation of monomers and dimers caused by cluster heating. To improve the understanding of the stability and fragmentation of metal clusters and the dynamics during and immediately after the photoionization event and to determine the size of the ejected particles, experiments with femtosecond time resolution are performed (Baumert et al., 1992b). Ultrashort light pulses of 60 fs duration are used to photoionize the neutral sodium cluster in the beam and to induce the fragmentation. The time-delayed identical probe laser pulses photoionize the neutral photofragments, ejected by the cluster ions. Time-of-flight spectroscopy is used to determine the mass of the ion and the initial kinetic energy of the probe-laser-ionized fragments. Note that for laser pulse durations of 60 fs, which are considerably shorter than the cluster vibrational periods of a few hundred femtoseconds, the clusters are nearly frozen during the photoionization event. So it is questionable whether cluster heating in femtosecond photoionization plays any role and whether sequential evaporative cooling occurs.

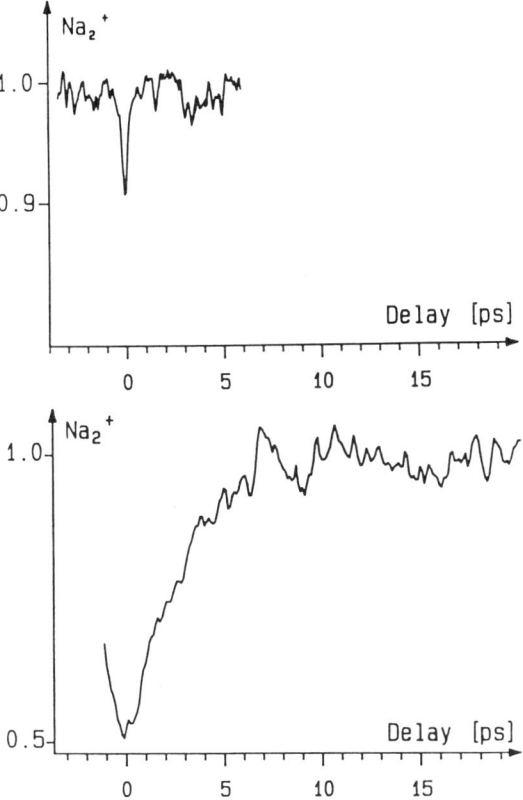

FIG. 15. Transient Na_2^+ signal in a molecular beam (top) and in a cluster beam (bottom). The ejection of Na_2 is due to ionic Na_n^+ cluster fragmentation. The rise time of the Na_2^+ signal is 2.5 ps.

The lower part of Fig. 15 shows the observed transient Na_2 fragmentation spectrum. The signal is symmetric around zero time delay, because we use identical pump and probe laser pulses. The buildup of the Na_2^+ probe signal strongly depends on the pump–probe delay time and shows a rise time of 2.5 ps. This transient was measured with a seed gas pressure of 8 bar for the cluster beam. In order to prove that this rise time is due to ionic cluster fragmentation we reduced the seed gas pressure to 0.1 bar. At that pressure there are virtually no clusters with $n > 3$ in the beam. The corresponding transient Na_2^+ spectrum, displayed in the upper part of Fig. 15, shows only a small narrow dip at time zero. This is due to the fragmentation of Na_2^+ by the enhanced laser intensity from the temporal overlap of pump and probe pulses.

The increase of the Na_2^+ probe signal is due to the fragmentation processes of Na_n^+ clusters ($n \geq 3$). The neutral sodium clusters are ionized by the pump laser, followed by a fragmentation into Na_2 and other channels, as will be described below. The neutral Na_2 fragments are then ionized by the time-delayed probe laser. Contributions to the Na_2^+ transient from neutral fragmentation of the clusters are unlikely, as for 2 eV photon energy there are no strong absorption resonances in sodium clusters, except for the B state of Na_3. However, our time-resolved experiments on the B state of Na_3 (Baumert et al., 1993b) show no sign of a decay on a picosecond time scale. Broyer et al. (1988) determined the lifetime of the B state to be 14 ns. Therefore we believe the following processes are responsible for the observed rise of the Na_2^+ signal.

Pump $Na_n + k \cdot h\nu \rightarrow Na_n^+ \rightarrow Na_{n-2}^+ + Na_2$
Probe $Na_2 + j \cdot h\nu \rightarrow Na_2^+$

In order to ionize Na_n ($n = 2, 4, 6, 8$) at $\lambda = 618$ nm at least three photons are necessary, while for all other masses at least two photons are needed (Kappes et al., 1988). Note that for a probe laser ionization with more than three photons a fragmentation of Na_2^+ into $Na^+ + Na$ is energetically possible.

An increasing question which had not been investigated before was whether neutral fragments such as Na_3 and Na_4 are ejected in dissociative photoionization of larger sodium clusters. As Na_3 can be ionized by two photons of 2 eV photon energy via the B state we performed a pump–probe experiment again, now on the ejected trimers. Figure 16 shows the observed buildup of the Na_3^+ signal with a time constant of 0.4 ps. This experiment clearly proved that neutral Na_3 photofragments are formed and that the ejection time scale is extremely fast. As it has been discussed in Baumert et al. (1992b), we believe the fragmentation of Na_8^+ into Na_5^+ and Na_3 is mainly responsible for the observed neutral trimer signal.

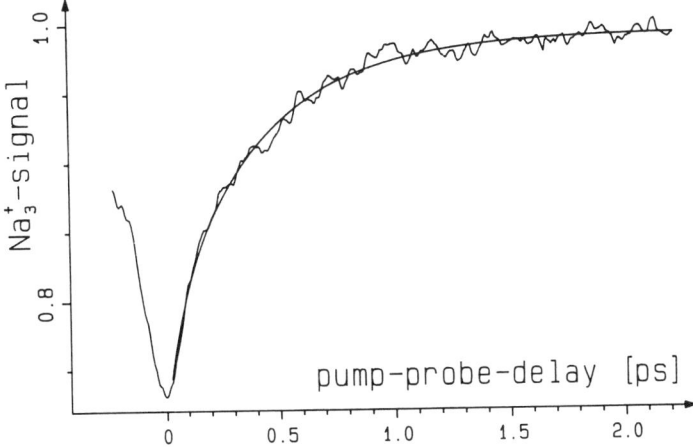

FIG. 16. Pump–probe delay spectrum of the ejected neutral trimer fragments Na_3. The probe signal Na_2^+ buildup time constant of 0.4 ps indicates a direct photoinduced fragmentation of sodium cluster ions rather than a statistical unimolecular decay.

In the discussion of the stability of metal cluster ions against fragmentation, it is often assumed that photoinduced electronic excitation is strongly coupled and relaxed to internal modes. The excess energy from successive absorption of photons in photoionization is thought to be quickly redistributed between the vibrational modes of the cluster ion, leading to statistically dominated fragmentation. However, the observed rise time of 2.5 ps (Na_2^+) and 0.4 ps (Na_3^+) is much too fast for an efficient relaxation of electronic energy in a cluster.

In conclusion, we find that the ejection of dimer Na_2 and trimer Na_3 photofragments occurs on ultrashort time scales of 2.5 and 0.4 ps, respectively. This and the absence of cluster heating reveal that direct photoinduced fragmentation processes are important at short times rather than the statistical unimolecular decay.

B. Dynamics of the Neutral Na_4 Resonance at 680 nm

In this section the cluster Na_4 is taken as an example to demonstrate how we obtain the spectroscopy, the dynamics, and the decay channels of neutral metal clusters by employing femtosecond laser techniques. Figure 17 shows the absorption spectrum of the cluster sizes Na_4 to Na_7, studied by multiphoton ionization with tunable femtosecond laser pulses. The observed spectrum of Na_4 is similar to absorption spectra measured by nanosecond laser depletion spectroscopy (Wang et al., 1990). The Na_4 intermediate

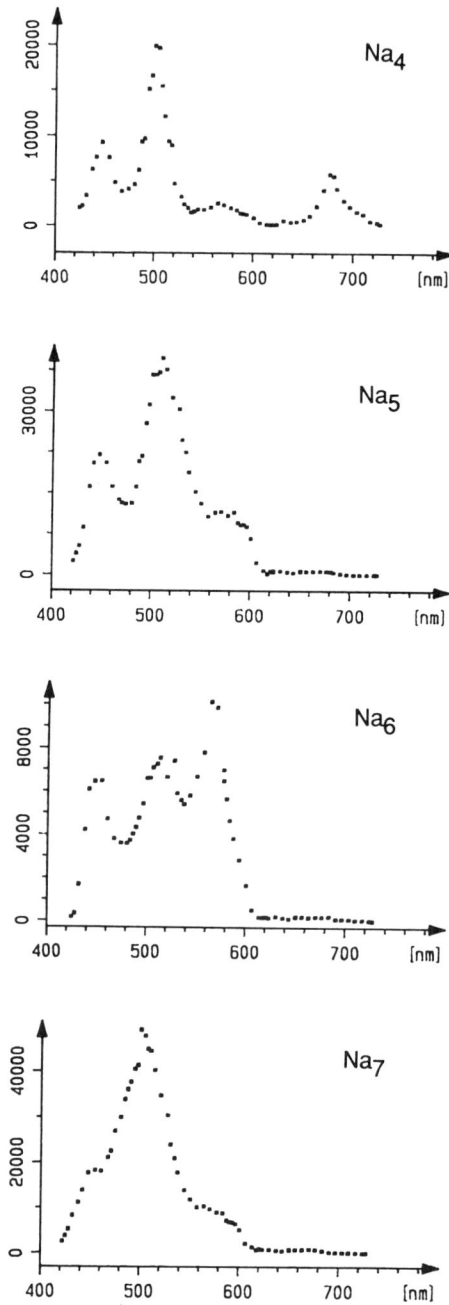

FIG. 17. Absorption spectrum of Na_4 to Na_7 measured by multiphoton ionization with tunable femtosecond laser pulses. Note that only the Na_4 has an absorption band at 680 nm.

resonance at 680 nm is chosen to demonstrate how the femtosecond pump–probe technique is ideally suited to investigate neutral cluster photofragmentation. All larger cluster sizes, Na_n with $5 < n \leqslant 21$, show little photoabsorption strength in this wavelength regime. Therefore contributions from fragmentation of larger cluster masses to the photodissociation signal and to the observed dynamics can be neglected. The actual femtosecond pump–probe experiment to study the decay of the intermediate Na_4^* resonance was carried out at 694 nm and resulted in the transient Na_4^+ signal shown in Fig. 18. Since pump and probe pulses are identical, the spectrum is symmetric with respect to time zero. The decrease of the signal to longer delay times reflects the decay of the intermediate resonance. The decay curve can be represented by a single exponential with a time constant of 0.74 ± 0.05 ps. The question which has to be addressed now is, what is the physical origin of the decay? There are two possibilities (which do not exclude each other): either the energy is redistributed among the other degrees of freedom in the cluster or the decay is attributed to an instantaneous photodissociation. The latter should lead to formation of neutral fragments Na_n ($n < 4$), with a buildup time being comparable to the decay time observed for the Na_4 parent cluster.

The sodium trimer Na_3 is the most studied and best known small metal cluster. Its absorption spectrum was first measured by the Wöste group (Broyer *et al.*, 1986a,b), employing nanosecond two-photon ionization spectroscopy and also depletion spectroscopy and later by the Kappes group (Wang *et al.*, 1990) using depletion spectroscopy.

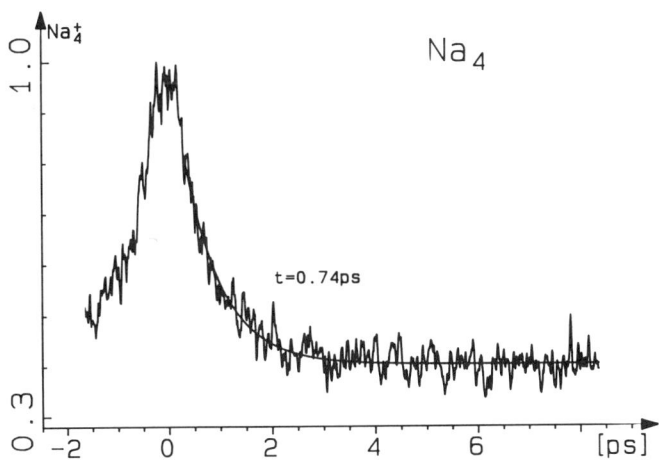

FIG. 18. Decay of the Na_4 absorption resonance (Fig. 17) measured in a pump–probe experiment at 694 nm. The decay time determined by fitting a single exponential is 0.74 ± 0.05 ps.

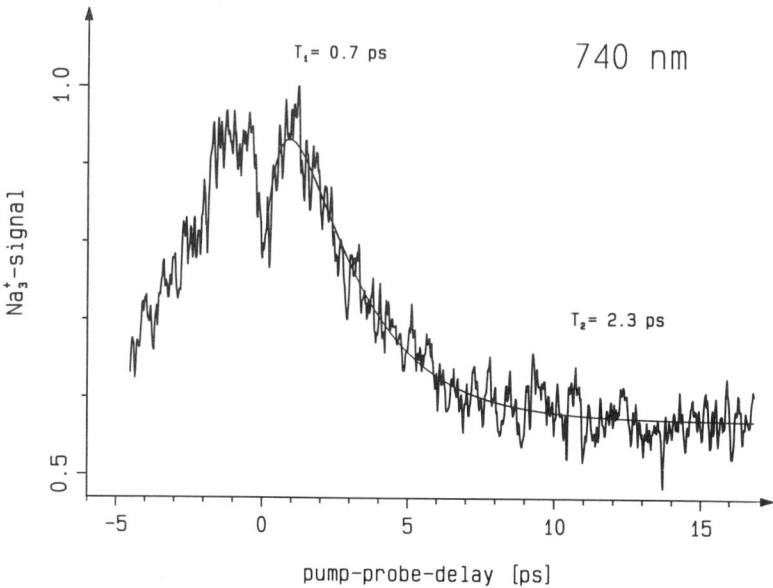

Fig. 19. Transient Na_3 spectrum as a function of pump–probe delay time. The maximum A'-state absorption of Na_3 is at 740 nm. Fitting exponential rise and decay functions to this transient gives a rise time, τ_2, of 0.7 ± 0.1 ps and a decay time, τ_1, of 2.3 ± 0.2 ps. Similar rise times for Na_3 transients were measured at 668 and 731 nm (see the text). The rise time of the Na_3 transients reflects the decay of an Na_4^* resonance (Fig. 18), while the decay time is the lifetime of the $Na_3 A'$ state.

There are two electronically excited states of Na_3, whose energies are close to the Na_4 resonance at 680 nm. To the blue side it is the A state, having a band origin at 675 nm (Broyer et al., 1988), and to the red side, the A' state with a band maximum at 740 nm (Wang et al., 1990; Broyer et al., 1986a,b). For both of these states we have studied the decay dynamics by femtosecond pump–probe spectroscopy. As an example the transient Na_3^+ signal obtained at the A' state maximum ($\lambda = 740$ nm) is shown in Fig. 19. For early pump–probe delay times we first observe a buildup of the transient Na_3^+ signal which is then followed by a decay. An analysis in which exponential rise and decay functions are fitted to this transient curve leads to a rise time, τ_2, of 0.7 ± 0.1 ps and to a decay time, τ_1, of 2.3 ± 0.2 ps. Experiments performed at different wavelengths such as $\lambda = 731$ nm gave $\tau_2 = 0.75 \pm 0.10$ ps and $\tau_1 = 2.35 \pm 0.20$ ps, whereas for $\lambda = 668$ nm we obtained $\tau_2 = 0.74 \pm 0.10$ ps and $\tau_1 = 2.26 \pm 0.20$ ps.

Note that the excitation wavelengths used in these three experiments with Na_3 are close to the 680-nm resonance of Na_4 (Fig. 17). Note also that the derived buildup times, τ_2, of the Na_3^+ transients for the three excitation wavelengths match the decay time of the observed Na_4 transient (Fig. 18).

The conclusion we draw from this is that excitation of Na_4 in this wavelength region leads to an instantaneous fragmentation of the neutral excited Na_4^* into neutral excited Na_3^* and atomic Na.

Now we turn to the question of what geometric structure of Na_4 can be derived from the absorption spectrum measured by femtosecond multiphoton ionization spectroscopy. Configuration interaction (CI) calculations by Bonacic-Koutecky et al. (1990a) lead to two different geometric structures for Na_4. One is a planar rhomboidal structure with D_{2h} symmetry; the other is a slightly tilted three-dimensional tetrahedral structure, being 0.256 eV higher in total energy. The femtosecond multiphoton ionization spectrum of Na_4 (top of Fig. 17) as well as depletion spectroscopy results obtained by Wang et al. (1992) show resonances at 1.80, 2.20, 2.45, and 2.75 eV. Dipole allowed transitions for the planar geometry were calculated for 1.71, 2.07, 2.45/2.46, and 2.76 eV and are in good agreement with the experimental results. The calculated spectrum of the three-dimensional structure consists mainly of only one transition at 2.12 eV and therefore does not agree with the experiments. Results obtained by applying the ZEKE technique to the study of Na_4 are also consistent with a planar rhomboidal structure (Thalweiser, 1992). The same conclusion had already been reached and reported by Wang et al. (1990).

C. Na_n Cluster Resonances and Their Decay Dynamics

One of the topics in metal cluster physics is to pursue the change from molecular to metallic behavior. In that context it is very instructive to investigate the electronic excitations of metal clusters. In a molecular picture the electronic excitations are treated with quantum chemical methods. In the case of sodium clusters, which are convenient for theoretical studies, because there is only one valence electron per atom, full CI calculations were performed on clusters with up to four atoms (Bonacic-Koutecky et al., 1990a). The larger Na_n clusters ($n = 5-20$) were treated by employing *ab initio* effective core potentials (Bonacic-Koutecky et al., 1990b, 1992). In quantum chemical calculations the detailed geometric structure of the cluster is essential. For several sodium cluster sizes Bonacic-Koutecky et al. (1990a, 1992) calculated the electronic excitations belonging to energetically low lying geometric structures.

However, in the past another model, the free electron model, has often been used to discuss the electronic excitations of metal clusters. In that treatment the most simple approach is the jellium model, in which the delocalized electrons are assumed to move in a constant, smeared out positively charged background (Eckardt, 1984a). This approach was reviewed by de Heer (1993) and Brack (1993). In the jellium model the detailed positions of the cluster atomic ionic cores play no role. Eckardt and Penzar (1988), Penzar and Eckardt (1990), and Eckardt (1984b) calculated the electronic structure of Na_n clusters self-consistently within the density functional theory (DFT). The electron correlation was taken into consideration by the local density approximation (LDA) and by time-dependent LDA (TDLDA) (Eckardt and Penzar (1991). Another approach was discussed by Clemenger (1985), who assumed that the cluster electrons are moving in a harmonic oscillator potential and are restricted to spheres and spheroids as the possible cluster geometries. This concept originated from nuclear physics. Selby et al. (1989) extended this model to ellipsoidal clusters. Originally Mie (1908) calculated light absorption and diffraction of small metal spheres within classical electrodynamics. Nowadays a modified Mie model is used to explain the absorption spectra of metal clusters (Wang et al., 1990). For the spherical clusters Na_8 and Na_{20} that model predicts a single homogeneously broadened surface plasmon resonance. Applying more elaborate techniques like time-dependent density–functional theory to the study of the photoabsorption spectrum of closed-shell sodium clusters within the spherical jellium model leads to a splitting of the Na_{20} absorption resonance (Rubio et al., 1992). Configuration interaction calculations of jellium clusters by the nuclear shell model (Koskinen et al., 1994) give a single peak result of the Na_{20} resonance. Only when the calculations are restricted to lower excitations is a two-peak absorption spectrum calculated. In the latter publication no splittings were calculated for the Na_8 resonance.

Experimentally the absorption spectra of metal clusters were measured by means of depletion spectroscopy (Selby et al., 1989; Wang et al., 1990). Due to the very short-lived excited states of Na_n clusters with $n > 3$ the well-known method of REMPI spectroscopy with nanosecond laser pulses fails. In depletion spectroscopy the ion signal decrease (due to cluster fragmentation after the interaction with nanosecond laser pulses) is detected as a function of the laser wavelength.

As discussed in Section IV.B we employed REMPI spectroscopy with tunable femtosecond laser pulses to investigate the absorption resonances in the wavelength region of 420 to 750 nm. Additionally femtosecond pump–probe techniques are introduced to investigate the decay dynamics of intermediate resonances.

In the framework of the jellium model, the Na_8 cluster is predicted to be a spherically symmetric metallic particle. During laser pulses about 100 fs in duration the clusters are excited and ionized by the absorption of photons out of the same laser pulse. Due to the very short pulse duration the clusters do not fragment during the interaction time as would be the case with nanosecond laser pulses. Therefore femtosecond REMPI spectroscopy is, besides depletion spectroscopy, a well-suited technique to measure the absorption spectrum of size-selected neutral metal clusters. It is necessary to notice that, besides the excitation step, the ionization of the excited cluster states may influence the spectral intensity distribution of the cluster spectra.

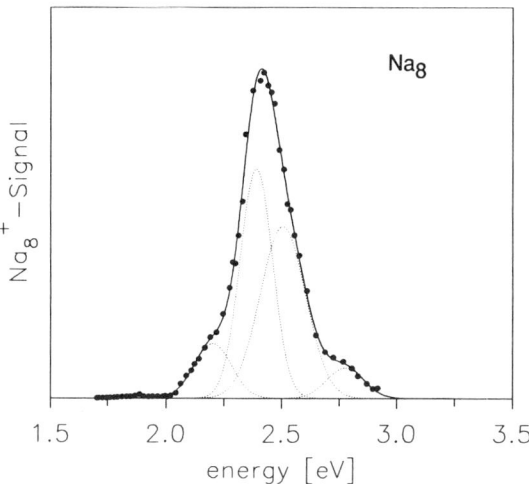

FIG. 20. Absorption spectrum of Na_8 measured by multiphoton ionization with tunable femtosecond laser pulses. Taking Gaussian line shapes into account, four different resonance contributions are derived from the analysis of the measured spectrum.

This may slightly affect the strength of the different spectral components of the cluster resonances. Depletion spectroscopy on the other hand suffers from neutral fragmentation of larger cluster masses, which would affect the observed absorption spectrum of a specific cluster size.

Figure 20 shows the Na_8 femtosecond REMPI spectrum from 1.7 to 2.9 eV. The spectrum is normalized to a constant laser intensity. In addition the absorption spectra for Na_n cluster sizes of $n = 4$ to $n = 21$ were measured. In Fig. 20 the solid line through the data points is based upon a fit of the experimental data with a sum of gaussian functions. The experimental data are best represented by a sum of four gaussian functions. The central energies of the four contributions are 2.2, 2.39, 2.49, and 2.78 eV. Our measurements of the absorption spectra of different cluster sizes agree very well with the corresponding spectral curves obtained by depletion spectroscopy (Selby et al., 1989; Wang et al., 1990). These results from two different experiments clearly disagree with the simple picture of a single homogeneously broadened surface plasmon resonance due to the spherical shape of Na_8. On the other hand the experimentally determined positions of the four Na_8 resonances are in much better agreement with quantum chemical calculations by Bonacic-Koutecky et al. (1990b, 1992). In the case of Na_8 the best agreement between calculation and experimentation is achieved assuming a T_d-cluster geometry. This shows that the detailed geometric structure of this cluster is essential in understanding the multiple resonances.

In Fig. 21 three transient ionization spectra of Na_8 are displayed. Each was obtained with the femtosecond laser tuned to the individual resonances

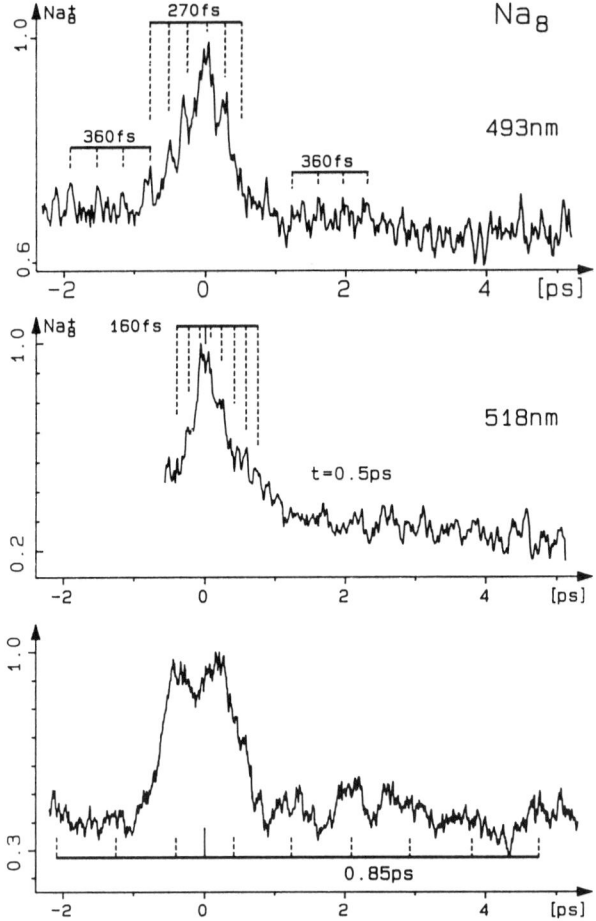

FIG. 21. Femtosecond time-resolved decay of Na_8^* intermediate states. The laser wavelengths are tuned to the individual resonances at 493, 518, and 540 nm as shown in Fig. 20.

in Fig. 20. The pump pulse excites the Na_8 cluster while the time-delayed identical probe pulse probes the coherence and residual population by photoionizing Na_8^*, the intermediate excited electronic states. As already discussed we have applied this technique to Na_2, Na_3, Na_4, and Na_n^+ to investigate wavepacket motion and fragmentation dynamics. The three transient ionization spectra of Na_8^+ in Fig. 21 are symmetric with respect to zero delay time, because pump and probe pulses had the same time duration and intensity. In order to determine the decay time constants we have analyzed the transient spectra with a sum of exponential decay functions. At 518 nm the best agreement between experimentation and simulation is

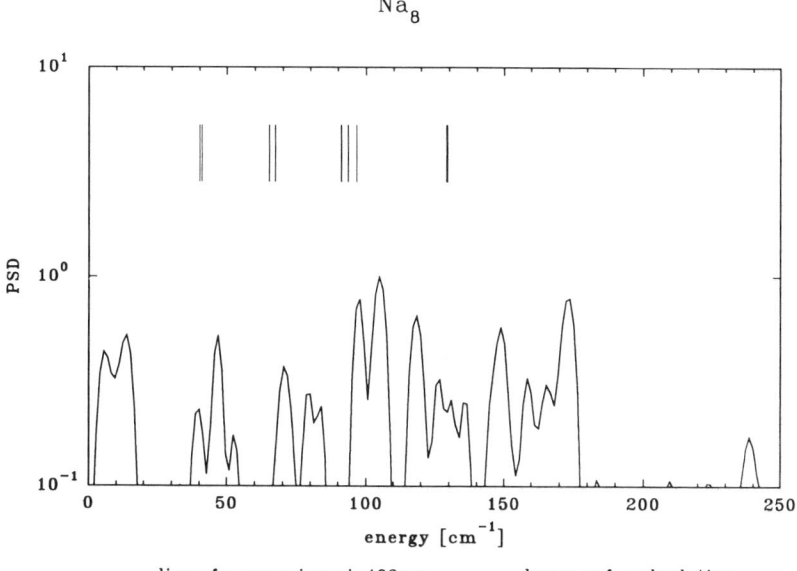

FIG. 22. Fast Fourier transformation of the transient Na_8^+ signal of Fig. 21 (at 493 nm). The obtained frequencies are in the range of the normal modes of small sodium clusters (see the text). The bars represent calculated frequencies of normal modes of the Na_8 ground state (Bonacic-Koutecky, 1994).

achieved by taking into account two contributions with time constants of 0.5 ps and of about 4 ps. For the transient spectrum obtained at 493 nm only a single exponential decay with a time constant of 0.45 ps describes the measured spectrum well. What is, however, clearly demonstrated by the different transient Na_8^+ spectra in Fig. 21 is that each absorption resonance has its own decay dynamics. Besides the observed decay of the signal there is also for each resonance an additional superimposed oscillatory structure. For the 518-nm resonance we find a regular oscillation with a time interval of about 160 fs, while at 493 nm at least two different series with time intervals of 360 and 270 fs are clearly seen. The transient spectrum of the 540-nm resonance is quite different from the two others. Here we observe a much slower dynamics with time constants of 0.8 and 2.8 ps. A fast Fourier transformation (FFT) of the time domain spectra at $\lambda = 493$ nm is displayed in Fig. 22 and shows frequencies corresponding to 360 and 270 fs and some additional frequencies in the range from 30 to 180 cm^{-1}. All these frequencies are in the range of the known vibrational eigenfrequencies of Na_3 and Na_4 as determined by ZEKE-photoelectron spectroscopy (Thalweiser et al., 1993). The bars in Fig. 22 represent calculated normal mode frequencies of the Na_8 ground state (Bonacic-Koutecky, 1994). Electronic ground state vibrational modes can be excited by stimulated emission pumping during the pump laser duration as has been shown in Na_2 (Baumert et al., 1992c; see also Section III.D) and Na_3 (Baumert et al., 1993b). Note that due to symmetry considerations and the ultrashort pulse duration only a small number of the $3N-6$ normal modes could be excited in an electronic transition. Based on this reasoning we believe the observed superimposed fast oscillations in the decay measurements are due to wavepacket motions in the potential surfaces of these metal clusters. The strong wavelength dependence of the decay time constants and the wavepacket motions are much better understood when taking into account molecular structures and excitations rather than considering Na_8 a metal sphere with delocalized electrons and surface plasmon-type excitations.

We have performed time-resolved experiments on all cluster sizes up to $n = 21$. Selected decay curves and absorption resonances obtained for other sodium cluster masses can be found in Baumert et al. (1994a, 1994b). Two-color femtosecond pump–probe experiments on small sodium clusters ($n \leqslant 10$) employing essentially the same experimental technique that we have introduced were reported by Kühling et al. (1994). On the basis of the number of observed cluster absorption resonances, their different energies and bandwidths, and their different decay patterns, we conclude that at least for cluster sizes Na_n with $n \leqslant 21$ molecular excitations and properties prevail over collective excitations and surface plasmon-like properties.

D. EXPERIMENTS WITH MERCURY CLUSTERS AND FULLERENES

In this section we will highlight experiments conducted on mercury clusters and fullerenes in which interesting ionization and fragmentation processes are observed in high-laser-intensity femtosecond single-pulse and pump–probe experiments.

1. Mercury Clusters

Clusters form a new class of materials, which often exhibit unexpected properties. A very interesting situation arises with mercury clusters. The mercury atom has a $5d^{10} 6s^2 np^0$ closed-shell electronic configuration with an ionization potential of 10.4 eV. Diatomic Hg_2 and other small mercury clusters are predominantly van der Waals-bound systems. However, the electronic structure changes strongly with increasing cluster size and finally converges toward the bulk, where the $6s$ and $6p$ bands overlap, giving mercury its metallic properties. This means that for the divalent Hg_n cluster a size-dependent transition from van der Waals to covalent and metallic

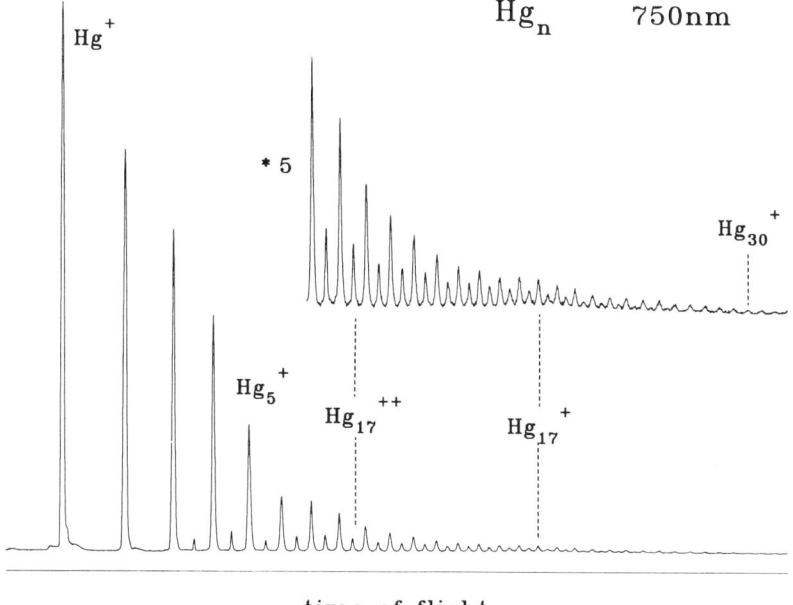

FIG. 23. Time-of-flight mass spectrum of singly and doubly charged mercury cluster ions.

binding exists. Therefore mercury provides the ideal system to study the size-dependent nonmetal–metal transition. For the neutral mercury cluster the ionization potentials are reasonably well known (Rademann et al., 1987), but other optical properties are practically unknown. The situation is much better for singly and doubly ionized Hg_n clusters with the reported ionization potentials and optical absorption spectra (Haberland et al., 1993).

Here we report the first studies of multiphoton ionization and fragmentation of mercury clusters in the femtosecond time domain. We observe the prompt formation of singly and doubly charged cluster ions and measure directly the decay of parent ions due to photofragmentation, together with the subsequent growth of daughter species. Furthermore we observe size-selected ion intensity oscillations in pump–probe measurements, indicating wave packet dynamics in both singly and doubly charged clusters. For these experiments we have used the Ti:sapphire laser, generating a train of light pulses 20–70 fs in duration. The pulses are amplified, compressed, and delayed to form a sequence of pump–probe pairs (see Fig. 3). TOF spectrometry determines the mass of the cluster ion and the initial kinetic energy of the ionic fragments.

Figure 23 shows a TOF mass spectrum of singly and doubly charged Hg_n cluster ions produced by 30-fs pulses at 750 nm. The most striking features of the spectrum are, first, that a multiphoton (6/12 photons) absorption

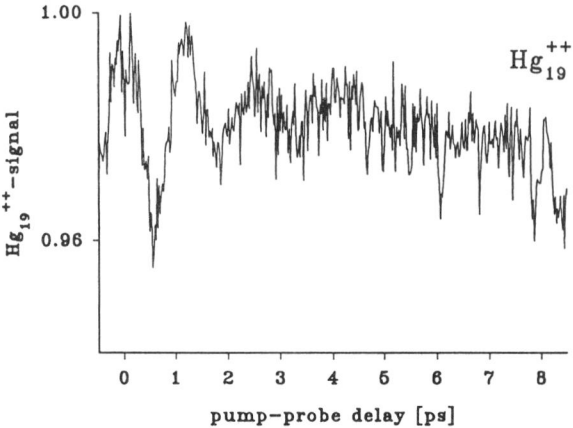

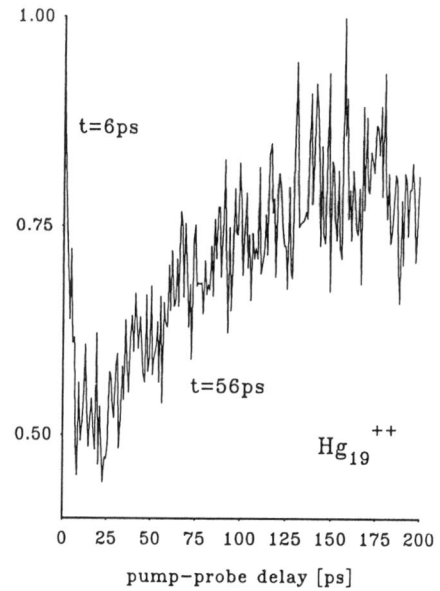

FIG. 24. Short-time dynamics (upper) and long-time dynamics (lower) of doubly charged Hg_{19} clusters.

forms singly and doubly ionized clusters and, second, that the intensity ratio between singly and doubly charged clusters of the same mass is in favor of the doubly charged species. Furthermore this ratio does not change for up to 10 times lower laser intensity, but the ratio does change strongly with wavelength.

A very surprising result of our time-resolved studies of size-selected neutral mercury clusters is shown in Fig. 24. The upper part of the figure details the first 8 ps of the pump–probe measurement of doubly charged Hg_{19}. The oscillatory behavior of this transient—indicating a wavepacket motion in a neutral cluster excited-state manifold—is seen in transient ionization spectra of singly and doubly charged clusters up to $n = 43$, for all $n \geqslant 5$.

The simplicity and similarity of the wavepacket motion over a broad range of cluster masses lead us to propose multiphoton absorption to a core Hg_2^* "chromophore" imbedded within and common to all the Hg_n neutral clusters examined in these experiments. The fact that blocking the probe pulse effectively quenches all ion signals and that the pump–probe spectrum with a weaker pump (or probe) is still symmetric with respect to $t = 0$ means that the pump pulse must excite a manifold of high-lying Rydberg states near but below the individual cluster ionization limit.

The lower part of Fig. 24 shows the long time dynamics of doubly charged Hg_{19} up to a 200-ps pump–probe delay time. Note the decrease in the signal at short delay times and the increase of the signal for long delay times. This type of transient was seen in time-resolved experiments on $I_2^-(CO_2)_n$ clusters in the Lineberger group (Papanikolas et al., 1993) and in I_2Ar_n in the Zewail group (Potter et al., 1992b; Lienau et al., 1993) and is attributed to the cage effect, i.e., recombination induced by a surrounding solvent shell.

From the transient ionization spectra of singly and doubly charged mercury clusters it is clear that a maximum ion signal is observed for both species at precisely zero delay time. Note that this—together with the identical transients (0–8 ps) for singly and doubly charged clusters—suggests that double ionization occurs directly through a transition from the neutral to the doubly charged manifold and not via the singly ionized continuum as it is observed in many high-laser-field experiments with atoms and molecules. The observation that the ratio of observed singly and doubly charged clusters of the same mass is independent with laser intensity for at least a factor of 10 (as noted before) supports this interpretation.

2. Fullerenes

The ultrafast dynamics of isolated fullerenes C_{60} and C_{70} have been investigated in a molecular beam employing ion and electron TOF spectrometry in combination with femtosecond pump–probe laser techniques. Femtosecond pulses are generated in the CPM laser, amplified in bow ties and Bethune cells, and time delayed in a Michelson arrangement as described in Section II.

Multiphoton ionization with 620- and 310-nm femtosecond laser pulses shows no sign of delayed ionization. With nanosecond laser pulses, multiphoton ionization of fullerenes leads to microsecond-delayed ionization,

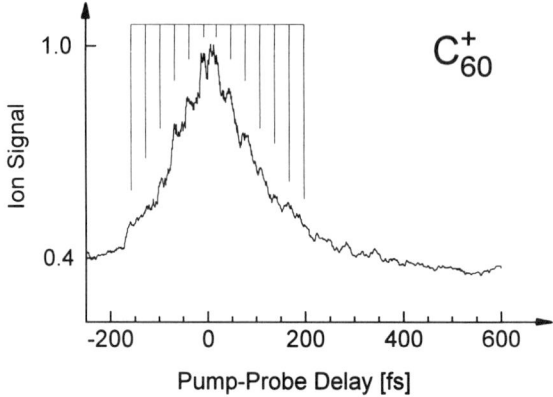

FIG. 25. Transient ionization signal of C_{60} obtained with 620-nm femtosecond pump and probe pulses. The ultrafast decay of the intermediate resonance is modulated by a 30-fs oscillatory structure.

often explained within a model of thermionic electron emission (Campbell et al., 1991; Wurz and Lykke, 1991).

The transient ionization spectra of C_{60} and C_{70} for (2 + 2) photoionization at 620 nm with 60-fs pulses show a very fast decay. In the case of C_{60} — the actual measurement is shown in Fig. 25 — it can be represented by a double exponential decay with about 100-fs and 3-ps time constants, while for C_{70} only the fast decay of about 100 fs is observed. Superimposed on the C_{60} decay curve is a 30-fs oscillatory structure. A Fourier transformation of the C_{60} transient gives frequencies in a range of 500 to 1600 cm^{-1} with a dominant peak at 1128 cm^{-1}. Frequencies in that range have been seen, for instance, in Fourier transform infrared (FTIR) C_{60} thin-film absorption spectra (Meilunas et al., 1991).

The TOF ion spectrum obtained at the threshold of multiphoton ionization of the fullerenes contains only C_{60}^+ and C_{70}^+. For laser intensities of 10^{12} W/cm^2 and higher, doubly and triply charged fullerenes and fragments are observed. The corresponding electron spectra show in both cases only one broad peak. However, it is surprising that the relative signal strengths of the singly, doubly, and triply charged species do not change with increasing laser intensity. This is clearly in contradiction to the conventionally discussed stepwise excitation process. We believe this result indicates a new not yet fully understood ionization mechanism in high laser fields. Moreover we observe only even-numbered neutral and multiple ($n = 1, 2, 3$) charged fragments like C_{58}^{n+}, C_{56}^{n+}, ..., C_{42}^{n+} from C_{60} for both 620- and 310-nm radiation. The measured initial kinetic energy release of the fragments indicates a fission of the parent fullerene rather than a sequential loss, which is reported to be the major channel in nanosecond experiments.

V. Conclusions

The real-time dynamics of multiphoton ionization and fragmentation of sodium and mercury molecules and clusters and of fullerenes have been studied in molecular beam experiments employing femtosecond pump–probe techniques and ion and electron spectroscopy.

Sodium with one valence electron per atom is an experimentally and theoretically very attractive system. Femtosecond time-resolved multiphoton ionization of sodium molecules reveals unexpected features in the dynamics of the absorption of several photons:

In Na_2 a second major REMPI process involving stepwise excitation of *two* electrons and subsequent electronic autoionization is observed in addition to the direct *one*-electron photoionization process. The femtosecond pump–probe technique demonstrates the possibility of controlling reactions by controlling the duration of propagation of a wavepacket on an excited electronic surface, as the ratio of Na_2^+ vs Na^+ varies by more than 100% as a function of pump–probe delay time.

The contribution of the one-electron REMPI process versus the two-electron REMPI process to the total ion yield varies strongly with the applied laser field strength. This can be understood by coherent coupling of the electronic states participating in these REMPI processes, leading to laser-intensity-dependent Rabi-type population switching between these states. At high laser intensities a wavepacket in the electronic ground state of Na_2 is created by stimulated emission pumping during the pump pulse duration. The $2\,^1\Sigma_u^+$ double minimum state of Na_2 is given as an example of how to perform frequency spectroscopy through time domain measurements.

Cluster physics bridge the gap between molecular and solid-state physics. Cluster size-dependent studies of physical properties such as absorption resonances, lifetimes, and decay channels have been performed with tunable, ultrashort light pulses. A major result of our femtosecond experiments is that the conventional view of the optical response of a metal cluster, e.g., the absorption, ionization, and decay processes as well as the corresponding time scales, had to be changed. Our results clearly show that for cluster sizes of Na_n with $n \leqslant 21$, the molecular structure, excitations, and properties prevail over collective excitations and surface plasmon-like properties. It is, however, obvious that for even larger clusters the optical response must finally be dominated by collective interactions.

The preliminary analysis of the time-resolved mercury experiments gives astonishing results. First, the observation of singly and doubly ionized clusters in direct multiphoton ionization transitions and, second, an almost identical vibrational wavepacket motion in both singly and doubly charged clusters up to $n = 43$ are very surprising. Probably an Hg_2^* chromophore imbedded within and common to all Hg_n neutral clusters carries the oscillator strength and determines the short-time wavepacket dynamics.

In multiphoton ionization of isolated fullerenes with high-intensity femtosecond laser pulses singly, doubly, and triply charged fullerenes and fragments are observed. The relative signal strength of the different charged species does not change with increasing laser intensity. This cannot be explained within the conventionally discussed stepwise excitation process. The real-time studies of the dynamics of ionization and fragmentation with femtosecond time resolution open up new and very exciting fields in molecular and cluster physics and yield results which in many cases are not accessible in nanosecond or picosecond laser experiments.

Acknowledgments

We gratefully acknowledge discussions with V. Engel and C. Meier and in particular the contributions of A. Assion, B. Bühler, M. Grosser, B. Lang, D. Schulz, V. Seyfried, R. Thalweiser, V. Weiss, and E. Wiedenmann to the various experiments. This work has been supported by the Deutsche Forschungsgemeinschaft through the Sonderforschungsbereich 276 "Korrelierte Dynamik hochangeregter atomarer und molekularer Systeme" in Freiburg.

References

Asaki, M. T., Huang, C.-P., Garvey, D., Zhou, J., Kapteyn, H. C., and Murnane, M. M. (1993). *Opt. Lett.* **18**, 977.
Averbukh, I. S., and Perelman, N. F. (1989). *Phys. Lett. A* **139**, 449.
Baumert, T., and Gerber, G. (1994). *Is. J. Chem.* **34**, 103.
Baumert, T., Bühler, B., Thalweiser, R., and Gerbert, G. (1990). *Phys. Rev. Lett.* **64**, 733.
Baumert, T., Grosser, M., Thalweiser, R., and Gerber, G. (1991a). *Phys. Rev. Lett.* **67**, 3753.
Baumert, T., Bühler, B., Grosser, M., Thalweiser, R., Weiss, V., Wiedenmann, E., and Gerber, G. (1991b). *J. Phys. Chem.* **95**, 8103.
Baumert, T., Engel, V., Röttgermann, C., Strunz, W. T., and Gerber, G. (1992a). *Chem. Phys. Lett.* **191**, 639.
Baumert, T., Röttgermann, C., Rothenfußer, C., Thalweiser, R., Weiss, V., and Gerber, G. (1992b). *Phys. Rev. Lett.* **69**, 1512.
Baumert, T., Engel, V., Meier, C., and Gerber, G. (1992c). *Chem. Phys. Lett.* **200**, 488.
Baumert, T., Pedersen, S., and Zewail, A. H. (1993a). *J. Phys. Chem.* **97**, 12447.
Baumert, T., Thalweiser, R., and Gerber, G. (1993b). *Chem. Phys. Lett.* **209**, 29.
Baumert, T., Thalweiser, R., Weiss, V., Wiedenmann, E., and Gerber, G. (1994a). In *Femtosecond Reaction Dynamics* (Wiersma, D. A., ed.), p. 29, North-Holland Publ., Amsterdam.
Baumert, T., Thalweiser, R., Weiss, V., and Gerber, G. (1994b). In *Femtosecond Chemistry* (Manz, J., and Wöste, L., eds.), p. 397, Verlag Chemie, Berlin.
Bonacic-Koutecky, V. (1994). Private communication.
Bonacic-Koutecky, V., Fantucci, P., and Koutecky, J. (1990a). *J. Chem. Phys.* **93**, 3802.

Bonacic-Koutecky, V., Kappes, M. M., Fantucci, P., and Koutecky, J. (1990b). *Chem. Phys. Lett.* **170**, 26.
Bonacic-Koutecky, V., Pittner, J., Scheuch, C., Guest, M. F., and Koutecky, J. (1992). *J. Chem. Phys.* **96**, 7938.
Bordas, C., Labastie, P., Chevaleyre, J., and Broyer, M. (1989). *Chem. Phys.* **129**, 21.
Bowman, R. M., Dantus, M., and Zewail, A. H. (1989). *Chem. Phys. Lett.* **161**, 297.
Brack, M. (1993). *Rev. Mod. Phys.* **65**, 677.
Bréchignac, C., Cahuzac, P., Roux, J. P., Pavolini, D., and Spiegelmann, F. (1987). *J. Chem. Phys.* **87**, 5694.
Broyer, M., Delacrétaz, G., Labastie, P., Whetten, R. L., Wolf, J. P., and Wöste, L. (1986a). *Z. Phys.* D **3**, 131.
Broyer, M., Delacrétaz, G., Labastie, P., Wolf, J., and Wöste, L. (1986b). *Phys. Rev. Lett.* **57**, 185.
Broyer, M., Delacrétaz, G., Labastie, P., Wolf, J. P., and Wöste, L. (1987). *J. Phys. Chem.* **91**, 2626.
Broyer, M., Delacrétaz, G., Guoquan, N., Wolf, J. P., and Wöste, L. (1988). *Chem. Phys. Lett.* **145**, 232.
Burkhardt, C. E., Barver, W. P., and Leventhal, J. J. (1985). *Phys. Rev.* A **31**, 505.
Campbell, E. E. B., Ulmer, G., and Hertel, I. V. (1991). *Phys. Rev. Lett.* **67**, 1986.
Clemenger, K. (1985). *Phys. Rev.* B **32**, 1359.
Cooper, D. L., Barrow, R. F., Vergès, J., Effantin, C., and d'Incan, J. (1984). *Can. J. Phys.* **63**, 1543.
de Heer, W. A. (1993). *Rev. Mod. Phys.* **65**, 611.
Delacrétaz, G., and Wöste, L. (1985). *Chem. Phys. Lett.* **120**, 342.
Eckardt, W. (1984a). *Phys. Rev.* B **29**, 1558.
Eckardt, W. (1984b). *Phys. Rev. Lett.* **52**, 1925.
Eckardt, W., and Penzar, Z. (1988). *Phys. Rev.* B **38**, 4273.
Eckardt, W., and Penzar, Z. (1991). *Phys. Rev.* B **43**, 1322.
Engel, V. (1991a). *Comp. Phys. Commun.* **63**, 228.
Engel, V. (1991b). *Chem. Phys. Lett.* **178**, 130.
Engel, V., Baumert, T., Meier, C., and Gerber, G. (1993). *Z. Phys.* D **28**, 37.
Fork, R. L., Greene, B. I., and Shank, C. V. (1981). *Appl. Phys. Lett.* **38**, 671.
Fragnito, H. L., Bigot, J.-Y., Becker, P. C., and Shank, C. V. (1989). *Chem. Phys. Lett.* **160**, 101.
Gerber, G., and Möller, R. (1985). *Chem. Phys. Lett.* **113**, 546.
Gruebele, M., Roberts, G., Dantus, M., Bowman, R. M., and Zewail, A. H. (1990). *Chem. Phys. Lett.* **166**, 459.
Haberland, H., von Issendorf, B., Yufeng, J., Kolar, T., and Thanner, G. (1993). *Z. Phys.* D **26**, 8.
Haugstätter, R., Goerke, A., and Hertel, I. V. (1988). *Z. Phys.* D **9**, 153.
Heist, P., Rudolph, W., and Wilhelmi, B. (1990). *Exp. Tech. Phys.* **38**, 163.
Ippen, E. P., and Shank, C. V. (1977). In *Ultrafast Light Pulses* (Shapiro, S. L., ed.), Topics in Applied Physics, Vol. 18, p. 83, Springer-Verlag, Berlin.
Jacobovitz, G. R., Brito Cruz, C. H., and Scarparo, M. A. (1986). *Opt. Commun.* **57**, 133.
Janssen, M. H. M., Bowman, R. M., and Zewail, A. H. (1990). *Chem. Phys. Lett.* **172**, 99.
Jeung, G. (1983). *J. Phys.* B **16**, 4289.
Kappes, M. M., Schär, M., Röthlisberger, U., Yeretzin, C., and Schumacher, E. (1988). *Chem. Phys. Lett.* **143**, 251.
Keller, J., and Weiner, J. (1984). *Phys. Rev.* A **30**, 213.
Knox, W. H. (1988). *IEEE J. Quantum Electron.* **QE-18**, 101.
Kogelnik, H., and Li, T. (1966). *Appl. Opt.* **5**, 1550.
Koskinen, M., Manninen, M., and Lipas, P. O. (1994). *Phys. Rev.* B **49**, 8418.
Kühling, H., Kobe, K., Rutz, S., Schreiber, E., and Wöste, L. (1994). *J. Phys. Chem.* **98**, 6679.
Kulander, K. C., and Heller, E. J. (1978). *J. Chem. Phys.* **69**, 2439.
Kusch, P., and Hessel, M. M. (1978). *J. Chem. Phys.* **68**, 2591.

Li, K. K. (1982). *Appl. Opt.* **21**, 967.
Lienau, C., Williamson, J. C., and Zewail, A. H. (1993). *Chem. Phys. Lett.* **213**, 289.
Meier, C. (1992). Diplomarbeit, Univ. Freiburg.
Meier, C., and Engel, V. (1994a). *J. Chem. Phys.* **101**, 2673.
Meier, C., and Engel, V. (1994b). Private communication.
Meier, C., and Engel, V. (1994c). *Phys. Rev. Lett.* **73**, 3207.
Meilunas, R., Chang, R. P. H., Liu, S., Jensen, M., and Kappes, M. M. (1991). *J. Appl. Phys.* **70**, 5128.
Meyer, W. (1992). Private communication.
Mie, G. (1908). *Ann. Phys. (Leipzig)* **25**, 377.
Müller-Dethlefs, K., Sander, M., and Schlag, E. W. (1984). *Z. Naturforsch. A* **39A**, 1089.
Mulliken, R. S. (1971). *J. Chem. Phys.* **55**, 309.
New, G. (1974). *IEEE J. Quantum Electron.* **QE-10**, 115.
Noordam, L. D., Joosen, W., Broers, B., ten Wolde, A., Lagendijk, A., Van Linden van den Heuvell, H. B., and Muller, H. G. (1991). *Opt. Commun.* **85**, 331.
Ogorzalek Loo, R., Hall, G. E., Haerri, H.-P., and Houston, P. L. (1988). *J. Phys. Chem.* **92**, 5.
Papanikolas, J. M., Vorsa, V., Nadal, E. M., Campagnola, P. J., Buchenau, H. K., and Lineberger, W. C. (1993). *J. Chem. Phys.* **99**, 8733.
Penzar, Z., and Eckardt, W. (1990). *Z. Phys. D* **17**, 69.
Potter, E. D., Herek, J. L., Pedersen, S., Liu, Q., and Zewail, A. H. (1992a). *Nature (London)* **355**, 66.
Potter, E. D., Liu, Q., and Zewail, A. H. (1992b). *Chem. Phys. Lett.* **200**, 605.
Rademann, K., Kaiser, B., Even, U., and Hensel, F. (1987). *Phys. Rev. Lett.* **59**, 2319.
Rigrod, W. W. (1965). *Bell Syst. Tech. J.* May, p. 907.
Rubio, A., Balbas, L. C., and Alonso, J. A. (1992). *Phys. Rev. B* **46**, 4891.
Schrödinger, E. (1926). *Naturwissenschaften* **14**, 664.
Selby, K., Vollmer, M., Masui, J., Kresin, V., de Heer, W. A., and Knight, W. D. (1989). *Phys. Rev. B* **40**, 5417.
Simon, J. D. (1989). *Rev. Sci. Instrum.* **60**, 3597.
Squier, J., Korn, G., Mourou, G., Vaillancourt, G., and Bouvier, M. (1993). *Opt. Lett.* **18**, 625.
Stingl, A. (1994). *Adriat. Res. Conf. Ultrafast Phenom. Appl. Trieste* (lecture).
Stingl, A., Spielmann, C., and Krausz, F. (1994). *Opt. Lett.* **19**, 204.
Strickland, D., and Mourou, G. (1985). *Opt. Commun.* **56**, 219.
Tannor, D. J., Kosloff, R., and Rice, S. A. (1986). *J. Chem. Phys.* **85**, 5805.
Taylor, A. J., Jones, K. M., and Schawlow, A. L. (1983). *J. Opt. Soc. Am.* **73**, 994.
ten Wilde, A., Noordam, L. D., Muller, H. G., and van Linden van den Heuvell, H. B. (1989). In *Fundamentals of Laser Interaction II* (F. Ehlotzky, ed.), Lecture Notes in Physics, No. 339, p. 194, Springer-Verlag, Berlin.
Thalweiser, R. (1992). Thesis, Univ. Freiburg.
Thalweiser, R., Vogler, S., and Gerber, G. (1993). *SPIE Proc.* **1858**, 196.
Ultrafast Phenomena IX (1994). (Knox, W., Barbara, P., Mourou, G. A., and Zewail, A. H., eds.), Springer Series in Chemical Physics, Vol. 60, Springer-Verlag, Berlin.
Valance, A., and Nguyen, Tuan, Q. (1982). *J. Phys. B* **15**, 17.
Valdmanis, J. A., and Fork, R. L. (1986). *IEEE J. Quantum Electron.* **QE-22**, 112.
Vergès, J., Effantin, C., d'Incan, J., Cooper, D. L., and Barrow, R. F. (1984). *Phys. Rev. Lett.* **53**, 46.
Wang, C. R. C., Pollack, S., Cameron, D., and Kappes, M. M. (1990). *J. Chem. Phys.* **93**, 3787.
Wang, C. R. C., Pollack, S., Dahlseid, T. A., Koretsky, G. M., and Kappes, M. M. (1992). *J. Chem. Phys.* **96**, 7931.
Warren, W. S., Rabitz, H., and Dahleh, M. (1993). *Science* **259**, 1581.
Wiley, W. C., and McLaren, I. H. (1955). *Rev. Sci. Instrum.* **26**, 1150.
Wurz, P., and Lykke, K. R. (1991). *J. Chem. Phys.* **95**, 7008.
Yeazell, J. A., Mallalieu, M., and Stroud, C. R., Jr. (1990). *Phys. Rev. Lett.* **64**, 2007.

CALCULATION OF ELECTRON SCATTERING ON HYDROGENIC TARGETS

IGOR BRAY

Electronic Structure of Materials Centre, School of Physical Sciences, The Flinders University of South Australia, Adelaide 5001, Australia

and

ANDRIS T. STELBOVICS

Centre for Atomic, Molecular, and Surface Physics, School of Mathematical and Physical Sciences, Murdoch University, Perth 6150, Australia

I. Introduction 210
II. Electron Scattering Theories for Hydrogenic Targets 211
   A. Born-Based Approximations 211
   B. Second-Order Born Approximations 213
   C. *R*-Matrix Method 214
   D. Intermediate-Energy *R*-Matrix Method 215
   E. Pseudostate–Close-Coupling Methods 216
   F. Finite-Element Methods 217
   G. *J*-Matrix Method 217
   H. Coupled-Channel Optical Methods 218
   I. Hyperspherical-Coordinate Methods 219
III. Convergent Close-Coupling Method 219
   A. Hydrogen-like Target Approximation 220
   B. Generation of Target States 221
   C. Formulation of the Three-Body Scattering Problem 223
   D. Solving the Coupled Integral Equations 228
IV. Electron–Hydrogen Scattering 234
   A. Temkin–Poet Model 234
   B. Angular Correlation Parameters at 54.4 eV 235
   C. Ionization 236
V. Electron Scattering on the He^+ Ion 241
   A. 2*S* Excitation 241
   B. Ionization 242
VI. Electron–Sodium Scattering 242
   A. Spin Asymmetries and $L_\perp^S$ 244
   B. Differential Cross-Sections 246
   C. Ionization and Total Cross-Sections 247
VII. Concluding Remarks 250
   References 251

I. Introduction

The ability to calculate electron–atom/ion scattering is of both fundamental and practical interest to physicists. For the simplest target of atomic hydrogen, the e–H scattering problem is the fundamental one, involving the interaction of three charged particles, and still attracts considerable attention. The reason for this is that there are considerable difficulties both for theoretical modeling and experimentation. Models have been successful in explaining aspects of the scattering problem for particular experiments, but a full solution to the scattering problem has been slow to emerge as evidenced by several areas in which theory still disagrees with experimentation. By full solution of the scattering problem we mean a solution of as many aspects of the problem as possible and thus demand a formalism that yields accurate results for elastic, excitation, ionization, and total cross-sections simultaneously for each projectile energy. Furthermore, we demand the reliability of our results to be independent of the choice of the projectile energy. In practice it is found that certain projectile energies and target transitions require more computational effort than others. As this chapter will demonstrate, the state of theory development is now such that modern desktop workstations are sufficient for accurately describing most of the current experimental data for electron scattering on hydrogen and hydrogen-like targets.

As it is our aim to develop a practical, general, and reliable scattering theory it is very important that there exists a broad range of appropriate accurate data. While agreement with experimentation does not immediately imply the general validity of a scattering theory, a discrepancy with accurate data invalidates the theory model, leading to further theoretical refinement. Hence a large experimental database for various aspects of electronic collisions is essential when developing general scattering theories, and so it is particularly helpful that such a large base of experimental data for hydrogen and hydrogen-like targets now exists.

We concentrate in this chapter on one of the latest electron scattering methods, the convergent close-coupling (CCC) method, developed by the authors (Bray and Stelbovics, 1992a). It takes the close-coupling (CC) formalism to completeness, and the present indications are that it satisfies the requirements of a general, reliable, and practical scattering theory. The target states are obtained by diagonalizing the target Hamiltonian in a large truncated orthogonal Laguerre basis. This ensures that the obtained target states are all square integrable. The orthogonality of the basis ensures that "completeness" of the expansion is approached as the basis size is increased. We say we have convergence whenever further increase in basis size does not have a significant effect on the results of interest. The CCC method has proved to be one of the most successful in describing electron scattering from light targets. We take this opportunity to combine here the ideas that appear in a number of publications and provide further detail not given elsewhere.

This chapter is structured in the following way. First, we will give an outline of the various electron scattering methods currently in use, then discuss their strengths and weaknesses, and contrast these with the CCC method. This will be followed by a section devoted to the detailed description of the CCC method. Subsequently, various comparisons of experimentation, the CCC method, and other available theories will be presented for a number of targets. We will not attempt to give a comprehensive survey (see Andersen et al., 1995; McCarthy and Weigold, 1990a,b), but rather concentrate on issues of greatest interest to us, namely where treatment of the target continuum is of great importance or where there are unresolved discrepancies with experiment. Last, we will indicate what we consider to be outstanding problems and suggest future directions for our approach to electron scattering problems.

II. Electron Scattering Theories for Hydrogenic Targets

We start with an overview of some of the electron–atom/ion scattering theories currently in common use.

A. BORN-BASED APPROXIMATIONS

All Born-based approximations are conveniently described in terms of the T matrix, which may be defined as the solution of $T = V + VG_0T$. Here V is the matrix of potentials coupling to all channels in the scattering problem at hand, and G_0 is the diagonal matrix of free Green's functions coupling to the channels. The Born approximation is obtained by omitting the kernel term on the right-hand side of the T-matrix equation. Born approximations are easy to calculate and yield excellent results for high incident energies. Formally, the Born approximation does not include exchange or any short-ranged distorting potentials, U. Its simplicity and usefulness have been utilized since the advent of quantum mechanics. For many practical purposes it yields sufficiently accurate results. The reason the Born approximation works so well at high energies is that the Green's function has an inverse energy dependence, so neglecting the kernel is increasingly well justified as the energy increases.

The energy range of validity of the Born approximation is difficult to determine and depends on the transition of interest and the accuracy required. It is probably safe to assume accuracy of the order of a few percent for projectile energies above 1 keV, but often the Born approximation will yield excellent results down to 100 eV for some transitions. It is this uncertainty that makes it difficult to draw conclusions in say the case in which the Born approximation and experimentation differ significantly in the energy range of 100 to 1000 eV.

To extend the energy range, the treatment of exchange and distortion may be added to the Born approximation. This can be done relatively simply and serves a number of useful purposes. For example, if introduction of exchange or variation of distorting potentials yields results very similar to those of the Born approximation, then we can be confident in the accuracy of the approximations. On the other hand, if the results vary significantly, then this is an indication that the Born approximation is likely to be invalid and that the distorted-wave Born approximation (DWBA) should be better if the difference is not very large.

In the DWBA method the potential V is partitioned into two terms, $V = (V - U) + U$, where U is assumed to be dominant and readily calculable. Then the T matrix $\langle \Phi_f | V | \Psi_i^{(+)} \rangle$ is approximated (Gell-Mann and Goldberger, 1953) by $\langle X_f^{(-)} | V - U | X_i^{(+)} \rangle + \langle X_f^{(-)} | U | \Phi_i \rangle$, where the distorted-wave channel functions, $|X_f^{(\pm)}\rangle$, are obtained using the potential U. This approximation treats the potential U to all orders subject to $V - U$, which is assumed to be small, being treated to only first order.

The ideas behind the Born-based approximation play a pivotal role in the CCC method, which requires the solution of the full integral equation for the T matrix. We first require the calculation of the V-matrix elements. These are just the Born $\langle \Phi_f | V | \Phi_i \rangle$ or the distorted-wave Born $\langle X_f^{(-)} | V - U | X_i^{(+)} \rangle$ approximations calculated for a large range of initial and final states. We solve the coupled equations by integrating in momentum-space over the V-matrix elements. These are very smooth as a function of incident or final momentum, allowing for relatively few integration points. This may be contrasted to solving the coupled equations in configuration space, where integrations and differentiation is necessary over oscillatory wave functions. This is a primary reason why the CCC method is often able to couple significantly more states than other methods.

Within the CCC formalism we also use arbitrary short-ranged distorting potentials, U. It is clear that DWBA depends on the choice of U, but the CCC solution for T must be independent of U. In the

CCC formalism distorting potentials are used only as a numerical technique to reduce computational resources required to solve the coupled equations, and the full two-potential formalism of Gell-Mann and Goldberger (1953) is employed without approximation. The BA or the DWBA may be trivially obtained from the "driving" term of the CCC integral equations.

B. SECOND-ORDER BORN APPROXIMATIONS

The second-order distorted-wave Born approximation (DW2BA) improves on the first-order calculation by adding an additional contribution that is obtained by iterating the T-matrix defining equation once and generating $V + VG_0V$ as a second-order approximation; see, for example, Kingston and Walters (1980). The use of a distorting potential has the effect of including higher-order contributions in the iterative series as well, although at least some third-order effects are not included. The method is very much more complicated to implement numerically than DWBA. In fact the same computer code is used in generating the DW2BA results of Madison et al. (1990) as in the CCO method (see below) of Bray et al. (1990). However, there is a good physical reason for going to the trouble of doing DW2BA calculations. One of the fundamental requirements of a general scattering theory is that it should be able to calculate elastic scattering, and, to do this, the effect of dipole polarization must be included. This is not possible within the DWBA framework, as this effect comes in at the second-order level. Other benefits accrue with the inclusion of higher-order terms in the iterative expansion. Since DW2BA contains more of the correct energy dependence through the free Green's function in the iteration the method is more reliable at lower energies than the DWBA approximation.

The DW2BA method has been applied to hydrogen (Madison, 1984; Madison et al., 1990, 1991), for which it yields results comparable to those of more sophisticated theories in the intermediate and high-energy regions. Application to electron–sodium scattering (Madison et al., 1992) has demonstrated that the method is able to obtain good results for the differential cross-sections down to 10 eV, although agreement with the more sensitive spin-resolved measurements is more elusive.

It is important to emphasize that the CCC method always requires the solution of the full T-matrix-coupled equations. For numerical reasons (such as optimizing convergence rates for numerical solutions) we often solve the equations in a distorted-wave formalism, but no approximations are involved as is the case with the DW2BA. In fact DW2BA is a subset of the full solution, but, unlike the full noniterative solution which is independent of the choice of these distorting potentials, this approximation may be significantly affected by their choices (Madison et al., 1992).

C. R-Matrix Method

The R-matrix method was introduced by Wigner and Eisenbud (1947) to study resonance reactions in nuclear physics. Here we concentrate on those R-matrix methods that have dominated the application to electron scattering on hydrogenic targets. General reviews of R-matrix approaches to the electron–atom problems have been presented by Burke and Robb (1975) and Burke (1987). For a discussion of the relation of Wigner–Eisenbud and eigenchannel R-matrix approaches, see Green (1988). In the R-matrix method (see, e.g., Burke and Robb, 1975), the three-body wave function in coordinate space is divided into an internal and external region. The internal region is spherical and is centered on the nucleus of the atom. The radius of the sphere is taken to be large enough so that all initial and final states of interest have negligible probability of being outside the internal region. In the external region, therefore, the overlap between the continuum scattered electron and the bound electrons is neglibible, and so the effects of electron exhange may be safely neglected and the three-body wave function in the external region can be found by solving a coupled channels system of equations without exchange potentials. The external wave function is formed by summing over physical channels only. The internal region wave function is made using an expansion over bound and continuum orbitals formed with the R-matrix boundary conditions. The basis orbitals are nonzero only in the internal region. The diagonalization of the Hamiltonian in this basis gives discrete levels whose lowest members are the physical states and the remaining ones are pseudostates. The expansion of the wave function is restricted to products of orbitals between the bound states and bound with continuum. Continuum-to-continuum products are omitted with the justification that coupling to the continuum is weak, particularly at low energies. The matching produces T matrices that have pseudo-resonances which are removed by a T-matrix-averaging procedure. A numerical averaging procedure was introduced by Burke et al. (1981) and further tested by Scholz (1991). An analytical justification of the procedure has been discussed by Slim and Stelbovics (1987) and Slim and Stelbovics (1988). The latest extended applications of the R-matrix method to atomic hydrogen and singly ionized helium have been performed by Aggarwal et al. (1991a,b). They calculated collision strengths for all transitions among the $n = 1, 2, 3, 4, 5$ states from the scattering threshold to intermediate energies. The method has been applied also by Fon et al. (1992) to determine differential cross-sections and angular correlation parameters just above the ionization threshold at which the continuum correlations of the electrons are thought to be not so important. At intermediate energies the neglect of continuum–continuum couplings for the internal region wave function is more problematical, and for this reason a variant of the R-matrix method has been developed. We turn to a consideration of it now.

D. INTERMEDIATE-ENERGY R-MATRIX METHOD

The intermediate-energy R-matrix method (IERM) has been developed by Burke et al. (1987) to extend the applicability of the method to intermediate energies for which there are a large number of continuum channels open, and neglect of the continuum–continuum orbital couplings is no longer justified when forming a basis over which to expand the internal-region wave function. The method has been applied to intermediate-energy electron scattering from hydrogen (Scott et al., 1989; Scholz et al., 1991) with success. An analysis of the approximations involved in the method have been carried out by Scholz (1991). He noted that approximation of the wave function in the external region by a sum over the pseudostate channels as well as the physical ones (and hence reducing the Q space) dramatically reduces the pseudoresonances in the T-matrix amplitudes; in fact, they are replaced by the minor structures typically observed in extended pseudostate-coupled channel calculations. An average over these structures gives the same smoothed amplitudes as before. Details of the averaging procedures are given in Scholz (1993). The importance of including continuum states in the IERM method has been emphasized by Scott et al. (1993), who compared the IERM integrated cross-sections for electron scattering from hydrogen for the $n = 1$ to $n = 3$ transitions with the standard R-matrix calculations of Aggarwal et al. (1991a) above the ionization threshold. These overestimate the cross-sections significantly even below 1.5Ry incident electron energy, reinforcing the importance of including the transitions to the target continuum for scattering above the ionization threshold.

Essentially, the R-matrix techniques solve the electron–atom/ion scattering problem independent of the projectile energy. The computationally expensive part of the problem for the IERM method is in obtaining the solution in the interior region. Then the work required to match the solutions and to compute the amplitudes is such as to allow one to take the projectile energy at very fine intervals, unlike in the CCC method, which requires a separate calculation for each projectile energy. This makes R-matrix methods ideally suited to the study of resonance phenomena. With the CCC method the effort required to solve scattering problems in the energy region from 0 to 1 keV varies significantly, with the intermediate-energy region requiring the most effort. For example, the number of partial waves, J, that need to be summed in this energy range varies from unity to in excess of a hundred. This is readily implemented in the CCC theory. In the IERM method the full calculation is only implemented for $J \leq 4$. The higher partial waves are treated by means of pseudostate-close-coupling models because of the computational demands of the IERM calculation. A reduced version of the IERM method (Scott and Burke 1993) is able to be applied at higher partial-waves and has also demonstrated the importance of treating the target continuum in the calculation of $n = 2$ to $n = 3$ transitions in e-H scattering (Odgers et al., 1994).

E. Pseudostate–Close-Coupling Methods

The pseudostate-close-coupling methods (PSCC) are those calculations for which, in addition to the treatment of true discrete eigenstates, there are also a number of square-integrable states with positive energies. These are called pseudostates because they are not true eigenstates of the target Hamiltonian, but are usually obtained by diagonalizing the Hamiltonian in a Slater-type orbital (STO) basis. There are very many pseudostate calculations in the literature (see, e.g., Callaway, 1985; van Wyngaarden and Walters, 1986, and references therein). These have proved to be some of the most successful methods in electron–hydrogen scattering. A major difficulty associated with these methods is that as the projectile energy is varied near a pseudothreshold, an unphysical resonance may appear in the cross-sections. Smooth cross-sections can then be obtained by performing various types of energy averaging whose justification relies on the assumption that the physical cross-sections should be smooth in the energy range of interest. Such energy-averaged cross-sections usually lead to very good results when compared with experimentation.

The observation that the effect of pseudoresonances decreased with increasing numbers of STO states was a significant factor in the development of the CCC method. For practical purposes the CCC method can be thought of as a very large PSCC calculation, for which the number of states is so large that the pseudoresonance behavior is greatly diminished. For a study of this near a fixed pseudothreshold, see Bray and Stelbovics (1995). In the CCC method use is made of the fact that the diagonalization of the target Hamiltonian, in a square-integrable basis of size N, results in a set of states, ϕ_{nl}^N, which form a quadrature rule for the sum over the true discrete eigenstates ϕ_{nl} and an integral, $d\mathbf{k}$, over the true continuum states $\chi(\mathbf{k})$. Thus the multichannel expansion of the total wave function using the true discrete and continuum eigenstates may be approximated by a finite expansion ($n = 1, N$) using the set of states ϕ_{nl}^N. As the total wave function depends on the projectile energy so does our quadrature rule. Whereas in the pseudostate methods it is common to choose a particular set of states without ascertaining the convergence of the basis and use them for various projectile energies, in the CCC method we increase the number of states N for a particular energy until satisfactory convergence has been obtained. Using momentum-space techniques we are able to solve the coupled equations for up to 50 states using a desktop workstation. The coordinate-space techniques, as employed in the pseudostate methods, require larger computational resources in dealing with half as many states. In this case energy averaging is a most practical alternative to increasing the size of the calculations. This makes the assumption that the scattering amplitudes must vary smoothly with energy.

One further difference between the CCC and the PSCC methods is that the former uses an orthogonal Laguerre basis for the generation of the target states. This has the advantage over the STO basis in that we do not encounter any linear-dependence problems as the basis size is increased, and so allows for systematic convergence studies. Furthermore, there has been considerable analytical study of the issues surrounding the usage of Laguerre-basis states in electron–atom scattering (Heller and Yamani, 1974b; Yamani and Reinhardt, 1975; Broad, 1978), which provides the formal justification for interpreting the sums over positive-energy pseudostates as a quadrature approximation to the integration over the continuum states of the target.

F. Finite-Element Methods

There are a few other approaches to solving the Schrödinger equation directly which have been discussed in the literature and which may become more popular in the future. One approach is that of Wang and Callaway (1993, 1994), who solve electron–hydrogen scattering by a finite-difference propagation method for the wave function in a triangular domain of configuration space for the two-electron coordinates. Their work extends the ideas of an earlier calculation (Poet, 1980), and yields good agreement with other methods. A second approach is that of finite-element analysis which is popular in engineering applications (see, e.g., Wait and Mitchell, 1985). In this approach the wave function is approximated by simple piecewise polynomial functions, and the scattering boundary conditions are readily implemented. Thus far the method has been applied by Shertzer and Botero (1994) and Botero and Shertzer (1992) below the $n = 2$ threshold for the first few partial waves. There is excellent agreement with the calculations of Wang and Callaway (1993) and with earlier variational and R-matrix results. These approaches warrant further study and application.

G. J-Matrix Method

The J-matrix method relies on the use of a set of square integrable (L^2) functions which span the target space. The functions which are chosen are Laguerre functions, and they are usually chosen in a form in which overlaps of the basis functions form a tridiagonal or Jacobi matrix (Heller and

Yamani, 1974b). The full Hamiltonian is diagonalized in a finite subset of this basis and then matched to the asymptotic forms of the scattering wave functions expanded in the L^2 basis; effectively the method provides a reformulation of the scattering problem from one using plane-wave expansions to one which is an infinite matrix realization in the Laguerre-function basis. The method was studied extensively during the 1970s and early 1980s but only a limited number of calculations were performed. It was found (Heller and Yamani, 1974a) that the method, with the number of basis states limited to 10 or less for the expansion of the Hamiltonian, leads to amplitudes which exhibited pseudoresonance behavior very similar to that observed in pseudostate close-coupling calculations. One fruitful outcome of the analytical studies was the clarification of the role of the pseudostates in representing the true continuum functions of the target. This is very transparent for the Laguerre bases used in the CCC calculations. The method has been reassessed by Konovalov and McCarthy (1994b) in the Temkin–Poet model (Temkin, 1962; Poet, 1978), who found that by increasing the number of basis states in the diagonalization of the target by a factor of up to four times that of the early calculations, accurate amplitudes can indeed be obtained. The effect of the pseudoresonances are then no longer significant. Konovalov and McCarthy (1994a) also applied the method successfully to the calculation of resonances in electron–hydrogen scattering. Extension to more general problems is currently under way.

H. Coupled-Channel Optical Methods

The coupled-channel optical (CCO) method divides the target space into two orthogonal spaces, usually denoted by P and Q ($P + Q = I$). Using the formalism of Feshbach (1962), One may restrict the coupled equations to only P space with the effect of Q space being treated by a complex (in general) nonlocal polarization potential, V_Q. The complexity of the scattering problem due to the target continuum is relegated to this potential. Unfortunately, this potential takes a very complicated form and as yet has not been calculated exactly. As a result there are many versions of the CCO method which have various approximations, too many to list here. In our own group there are three distinctly different versions (McCarthy and Stelbovics, 1983; Bray et al., 1990, 1991b). The latter utilizes P and Q projection operators that are symmetric in the space of the two electrons and has proved most successful when applied to the electron–sodium scattering problem (Bray, 1992; Bray and McCarthy, 1993). Another example of a CCO model applied to $e-\text{He}^+$ scattering, and presented later in this work, is due to Unnikrishnan et al. (1991).

The philolsophy of the CCO model is that the contribution of V_Q should be small, and so an approximation in its calculation should not have a severe effect. Interestingly, Bray (1992) found that in the case of electron–sodium scattering the effect of V_Q can be quite large, but still sufficiently accurate. A detailed examination of these issues is given by Bray (1994b).

In contrast to the CCO model, the CCC method treats both P and Q spaces in exactly the same way, with the proviso that Q space is expanded in a square-integrable basis. In fact the CCC method was developed, in part, after Bray *et al.* (1991a) showed that, for a given CCO model, using square-integrable Q space yielded the same results as using true continuum functions. It was a natural step to drop the CCO model and use the close-coupling formalism directly for both P and Q spaces, without approximation.

I. HYPERSPHERICAL-COORDINATE METHODS

Hyperspherical coordinates for two-electron systems have been used extensively and successfully in classifying and identifying doubly excited resonant states for H$^-$ and He (see, e.g., Sadeghpour, 1992; Tang *et al.*, 1992). In this approach, the hyperspherical coordinates $R = \sqrt{r_1^2 + r_2^2}$ and $\alpha = \arctan(r_2/r_1)$ replace r_1 and r_2 and the three-body scattering wave function is expanded in these coordinates in an interior region and matched to asymptotic forms for the scattered waves. This method is still in its infancy as far as scattering applications above the ionization threshold and intermediate energy are concerned. The initial indications are that the method works well (Watanabe *et al.*, 1993; Kato and Watanabe, 1994) in the Temkin–Poet model up to several hundred electron volts of incident energy. Further, Kato and Watanabe (1994) demonstrated that the total ionization cross-section for the full electron–hydrogen problem can be obtained with excellent agreement with the data of Shah *et al.*, (1987) down to 0.1 atomic unit in energy, and in this respect their hyperspherical model is superior to all other models at these low energies. A different but equally promising approach has been discussed by Rudge (1992), who has included the electron–electron continuum effects using the hyperspherical coordinates in a hybrid close-coupling formalism. The initial calculations have been confined to demonstrating the method works in the elastic-scattering region. These hyperspherical techniques promise to be valuable additions as general electron–atom scattering techniques.

III. Convergent Close-Coupling Method

We give here a detailed description of the convergent close-coupling method for the calculation of electron scattering on light hydrogenic targets. We first present the general framework based on Bray (1994c) and then demonstrate

A. HYDROGEN-LIKE TARGET APPROXIMATION

We begin by defining hydrogen-like targets as those targets for which the frozen-core Hartree–Fock approximation yields sufficiently accurate one-electron energies for the ground and excited states. In other words, our target states are eigenstates of the target Hamiltonian H_2 (the subscript indicating target space),

$$H_2 = K_2 + V_2^{FC}, \tag{1}$$

where

$$V^{FC}\phi_j(\mathbf{r}) = \left(-\frac{Z}{r} + 2\sum_{\psi_{j'}\in C}\int d^3r' \frac{|\psi_{j'}(\mathbf{r}')|^2}{|\mathbf{r}-\mathbf{r}'|}\right)\phi_j(\mathbf{r})$$
$$-\sum_{\psi_{j'}\in C}\int d^3r' \frac{\psi_{j'}^*(\mathbf{r}')\phi_j(\mathbf{r}')}{|\mathbf{r}-\mathbf{r}'|}\psi_{j'}(\mathbf{r}) \tag{2}$$

and where the notation C indicates the set of frozen core states.

In order to obtain the ground and excited target states ψ_j, we first generate the core target states ψ_j by performing a self-consistent-field Hartree–Fock (SCFHF) calculation (Chernysheva et al., 1976) for the ground state of the target T,

$$(K + V^{HF} - \epsilon_j)\psi_j(\mathbf{r}) = 0, \quad \psi_j \in T, \tag{3}$$

where

$$V^{HF}\psi_j(\mathbf{r}) = \left(-\frac{Z}{r} + 2\sum_{\psi_{j'}\in T}\int d^3r' \frac{|\psi_{j'}(\mathbf{r}')|^2}{|\mathbf{r}-\mathbf{r}'|}\right)\psi_j(\mathbf{r})$$
$$-\sum_{\psi_{j'}\in T}\int d^3r' \frac{\psi_{j'}^*(\mathbf{r}')\psi_j(\mathbf{r}')}{|\mathbf{r}-\mathbf{r}'|}\psi_{j'}(\mathbf{r}). \tag{4}$$

It is worth noting that the core states may also be obtained by performing the SCFHF calculation for the ionic core; see McEachran and Cohen (1983), for example. In this way the core states would be optimized for the higher excited states, whereas we have them optimized for the ground state. Variation in the choice of core states provides for a test of the stability of the scattering results.

As the frozen-core model [Eq. (2)] does not allow for core polarization, which is important for heavier targets, we may take an additional phenomenological core polarization potential,

$$V^{\text{pol}}(r) = \frac{-\alpha_d}{2r^4}\{1 - \exp[-(r/\rho)^6]\}. \tag{5}$$

Here α_d is the static dipole polarizability of the core and may be obtained for many of the targets of interest to us from McEachran et al. (1979), for example. The value of ρ is chosen empirically to fit the one-electron ionization energies. In this case the hydrogen-like target one-electron Hamiltonian takes the form

$$H_2 = K_2 + V_2 = K_2 + V_2^{\text{FC}} + V_2^{\text{pol}}, \tag{6}$$

where we use V_2 (and similarly with V_1) for brevity of notation.

It is very important that the results for the scattering problem of interest are not significantly affected by reasonable choices of the core states or the core polarizability constants. In applications thus far, a significant variation in the results has been observed only near excitation thresholds (Bray and McCarthy, 1993). There it is clear that the target state energies must be relatively more accurate than, for example, at energies well above the ionization threshold.

B. Generation of Target States

Having made explicit our approximation for the target structure, we are in a position to generate the target ground and excited states. The true spectrum of the target Hamiltonian [Eq. (6)] contains both discrete and continuum states. In order to treat the continuum states in the close-coupling formalism we proceed by using a basis of orthogonal Laguerre functions to project the full target space onto a Hilbert subspace. The continuum functions can be expressed approximately as a Fourier-series expansion in this basis. Truncating the full basis to the first N functions defines our sequence of approximations to the target space. In the process of forming this finite-dimensional subspace, the target continuum is discretized and hence easily incorporated into the close-coupling formalism. In the limit $N \to \infty$ the discrete and continuous spectrum of the target is recovered. A suitable basis set is given by

$$\xi_{kl}(r) = \left(\frac{\lambda_l(k-1)!}{(2l+1+k)!}\right)^{1/2} \times (\lambda_l r)^{l+1} \exp(-\lambda_l r/2) L_{k-1}^{2l+2}(\lambda_l r), \tag{7}$$

where the $L_{k-1}^{2l+2}(\lambda_l r)$ are the associated Laguerre polynomials and k ranges from 1 to the basis size N_l. Thus, we obtain our target states $|i_n^N\rangle$ by

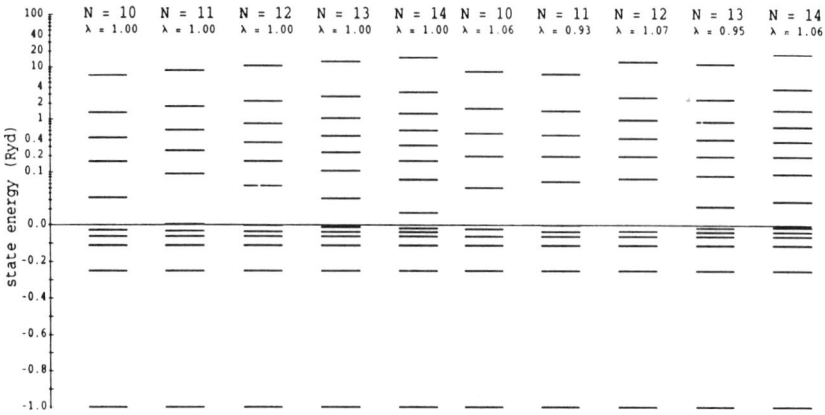

FIG. 1. The spectrum of the hydrogen target s states as obtained by diagonalizing the target Hamiltonian on the Laguerre basis of the indicated parameters [see Eq. (7)]. On the left side we keep $\lambda = 1$. On the right side λ is varied so as to keep one of the positive energy levels constant at 0.2 Rydberg. The latter is useful when performing $(e, 2e)$ calculations. A logarithmic scale is used for the positive energies.

diagonalizing the target Hamiltonian [Eq. (6)] in the ξ_{kl} basis

$$\langle i_m^N | H_2 | i_n^N \rangle = \epsilon_n^N \delta_{mn}, \tag{8}$$

where, expanding the notation $|i_n^N\rangle$,

$$\langle \mathbf{r} | i_n^N \rangle \equiv \langle \mathbf{r} | i_{nlm}^{N_l} \rangle = r^{-1} \phi_{nl}^{N_l}(r) Y_{lm}(\hat{\mathbf{r}}), \tag{9}$$

we have

$$\phi_{nl}^{N_l}(r) = \sum_{k=1}^{N_l} C_{nk}^l \xi_{kl}(r). \tag{10}$$

The coefficients C_{nk}^l are obtained after the diagonalization.

We keep the superscript N_l on $\phi_{nl}^{N_l}(r)$ to indicate that as the basis size is altered so is the nl target state. However, the states with negative energies approach the true target discrete states with increasing N_l. In all our calculations that involve transitions between true discrete eigenstates, say $\phi_{nl}(r)$ and $\phi_{ml'}(r)$, the basis sizes N_l and $N_{l'}$ must be sufficiently large that $\langle \phi_{nl} | \phi_{nl}^{N_l} \rangle \approx \langle \phi_{ml'} | \phi_{ml'}^{N_{l'}} \rangle \approx 1$. An example of what happens to the energy levels as the N_l and λ_l are varied for $l = 0$ states of hydrogen is shown in Fig. 1.

The Laguerre basis is ideal for generating states of atomic hydrogen in the sense that for a relatively small N_l a number of exact discrete eigenstates are readily generated. However, for the heavier hydrogenic targets such as sodium or potassium, much larger bases are required. See Bray (1994c) for an example of the generation of sodium s states.

C. Formulation of the Three-Body Scattering Problem

Having defined the one-electron target Hamiltonian [Eq. (6)] we may write the full three-body Hamiltonian H as

$$H = H_1 + H_2 + V_{12}, \qquad (11)$$

where the indices 1 and 2 denote the projectile and target spaces, respectively, and V_{12} is the electron–electron potential. We make the assumption that the center of mass is the center of the nucleus of the target and that any relativistic effects can be neglected. The total two-electron wave function has the total spin S as a good quantum number and satisfies

$$\Psi^S(\mathbf{r}_1, \mathbf{r}_2) = (-1)^S \Psi^S(\mathbf{r}_2, \mathbf{r}_1) \qquad (12)$$

by the Pauli principle.

Let us define the identity operator I in terms of a complete set (excluding core states) of target eigenstates $|i_n\rangle$ and the identity-like operator I^N in terms of our L^2 states [Eq. (9)] by

$$\begin{aligned} I &= \sum_n^\uparrow |i_n\rangle\langle i_n| \\ &= \sum_{n=1}^N |i_n^N\rangle\langle i_n^N| + \sum_n^\uparrow |\bar{i}_n^N\rangle\langle \bar{i}_n^N| \\ &= I^N + \bar{I}^N, \end{aligned} \qquad (13)$$

where the states $|\bar{i}_n^N\rangle$ are orthogonal to $|i_m^N\rangle$ and are defined by $\bar{I}^N = I - I^N$. By the completeness of the Laguerre basis we have

$$\lim_{N\to\infty} I^N = I. \qquad (14)$$

Without loss of generality we can write Eq. (12) as

$$|\Psi^S\rangle = (1 + (-1)^S P_r)|\psi^S\rangle, \qquad (15)$$

where P_r denotes the space exchange operator. With such an expansion no symmetry requirements on $|\psi^S\rangle$ are necessary. However, it is clear that such an expansion is too general since one can add to $|\psi^S\rangle$ any function satisfying $(1 + (-1)^S P_r)|\phi^S\rangle = 0$ without altering Eq. (15).

At this point it is convenient to introduce the multichannel expansion

$$\begin{aligned} |\psi^S\rangle &\approx \sum_{n=1}^N |i_n^N f_n^{SN}\rangle \\ &= |\psi^{SN}\rangle, \end{aligned} \qquad (16)$$

where $|f_n^{SN}\rangle = \langle i_n^N|\psi^S\rangle$. Having made this expansion we are in a position to address the problem of nonuniqueness of $|\psi^S\rangle$ in Eq. (15). The functions

$|f_n^{SN}\rangle$ may be expanded as

$$|f_n^{SN}\rangle = I^N|f_n^{SN}\rangle + \bar{I}^N|f_n^{SN}\rangle$$
$$= |F_n^{SN}\rangle + |\bar{F}_n^{SN}\rangle. \tag{17}$$

Then we may write

$$|\psi^{SN}\rangle = \sum_{n=1}^{N} |i_n^N F_n^{SN}\rangle + \sum_{n=1}^{N} |i_n^N \bar{F}_n^{SN}\rangle. \tag{18}$$

In the first term both electrons are expanded using states $|i_n^N\rangle$, whereas in the second term the $|\bar{F}_n^{SN}\rangle$ are spanned by a set of states orthogonal to the $|i_n^N\rangle$. Therefore, symmetric or antisymmetric functions, leading to nonuniqueness for $|\Psi^{SN}\rangle$ when added to Eq. (18), can take only the form of the first term. As we are going to be solving for $|\psi^{SN}\rangle$ numerically, it is important to eliminate any contribution to $|\psi^{SN}\rangle$ by say $|\phi^{SN}\rangle$ for which $(1 + (-1)^S)P_r|\phi^{SN}\rangle = 0$. We do this by imposing the condition

$$\sum_{n=1}^{N} |i_n^N F_n^{SN}\rangle = (-1)^S P_r \sum_{n=1}^{N} |i_n^N F_n^{SN}\rangle$$
$$= (-1)^S \sum_{n=1}^{N} |F_n^{SN} i_n^N\rangle. \tag{19}$$

In other words, we specify that $|\psi^{SN}\rangle$ has the same symmetry as Eq. (12) in the two-electron space spanned by the $|i_n^N\rangle$. This may be written, using $\langle i_m^N|\bar{F}_n^{SN}\rangle = 0$ for $m, n = 1, \ldots N$, as

$$\langle i_n^N i_m^N|\psi^{SN}\rangle = (-1)^S \langle i_m^N i_n^N|\psi^{SN}\rangle$$

or

$$\langle i_n^N|f_m^{SN}\rangle = (-1)^S \langle i_m^N|f_n^{SN}\rangle. \tag{20}$$

Thus, we have established some restrictions on the multichannel expansion functions so that the expansion [Eq. (15)] using Eq. (18) is unique. Taking finite N, $|\psi^{SN}\rangle$ does not have the symmetry [Eq. (12)] in all of the two-electron space, but

$$|\Psi^{SN}\rangle = (1 + (-1)^S P_r)|\psi^{SN}\rangle \tag{21}$$

does.

Using Eq. (21) the Schrödinger equation for the three-body problem is

$$(E^{(+)} - H)(1 + (-1)^S P_r)|\psi^{SN}\rangle = 0, \tag{22}$$

where $(+)$ denotes incoming plane/Coulomb wave and outgoing spherical wave boundary conditions and where the limit of large N is implicit. We rearrange this to give

$$(E^{(+)} - K_1 - U_1 - H_2)|\psi^{SN}\rangle = V^S|\psi^{SN}\rangle, \tag{23}$$

where

$$V^S|\psi^{SN}\rangle = [V_1 - U_1 + V_{12} + (-1)^S(H - E)P_r]|\psi^{SN}\rangle, \quad (24)$$

and $U_1 = U(r_1)$ is such that $K_1 + U_1 + H_2$ is the asymptotic (large r_1) free Hamiltonian with

$$(K_1 + U_1 - \epsilon_k^{(\pm)})|\mathbf{k}^{(\pm)}\rangle = 0 \quad (25)$$

defining the complete set of bound and continuous distorted/distorted-Coulomb waves $|\mathbf{k}^{(\pm)}\rangle$. For $r_1 \to \infty$, $U(r_1) \to U_a(r_1) = -Z_a/r_1$, where Z_a is zero for a neutral target and is the asymptotic charge for an ion. By premultiplying Eq. (23) by $\langle \mathbf{k}_n^{(\pm)} i_n^N |$ and setting $V^S = 0$ we define a set of N^O on-shell momenta $\epsilon_{k_n} = k_n^2/2$ such that

$$E = \epsilon_n^N + k_n^2/2. \quad (26)$$

The projectile continuum states are expanded as

$$\langle \mathbf{r}|\mathbf{k}^{(\pm)}\rangle = (2/\pi)^{1/2}(kr)^{-1} \sum_{L,M} i^L e^{\pm i(\sigma_L + \delta_L)} u_L(k, r) Y_{LM}(\hat{\mathbf{r}}) Y_{LM}^*(\hat{\mathbf{k}}), \quad (27)$$

where σ_L is the Coulomb phase shift, $u_L(k, r)$ is real and has the asymptotic form

$$u_L(k, r) \to F_L(kr) \cos \delta_L + G_L(kr) \sin \delta_L, \quad (28)$$

and for asymptotic charge Z_a (Coulomb parameter $\eta = -Z_a/k$) the $F_L(kr)$ and $G_L(kr)$ are the regular and irregular Coulomb functions, respectively. The $u_L(k, r)$ are calculated using Numerov's method, starting at the origin and matching to the correct boundary conditions as soon as $U_1(r) \approx -Z_a/r$ to a sufficient accuracy.

Supposing that our electron–hydrogenic target scattering system has initially the target in the state i_0 and the projectile momentum $\mathbf{k}_0$, we rewrite Eq. (23) as an integral Lippmann–Schwinger equation,

$$|\psi_0^{SN}\rangle = |i_0 \mathbf{k}_0^{(+)}\rangle + \sum_{n=1}^{N} \oint_k \frac{|i_n^N \mathbf{k}^{(-)}\rangle \langle \mathbf{k}^{(-)} i_n^N | V^S | \psi_0^{SN}\rangle}{E^{(+)} - \epsilon_n^N - \epsilon_k}, \quad (29)$$

where the total energy $E = \epsilon_0 + k_0^2/2$ in atomic units. Thus far we have not implemented the condition [Eq. (20)]. This is now done by considering the V-matrix element proportional to the energy term. Furthermore, on account of the symmetry condition [Eq. (20)] one may write more generally

$$\langle i_n^N | f_m^{SN} \rangle = (1 - \theta)\langle i_n^N | f_m^{SN}\rangle + (-1)^S \theta \langle i_m^N | f_n^{SN}\rangle, \quad (30)$$

which implements Eq. (20) for any nonzero constant θ. Following this, some simple algebra (Bray, 1994c) shows that we may replace the operator V^S above with

$$V^{SN}(\theta) = V_1 - U_1 + V_{12} - \theta I_1^N E + (-1)^S(H - E(1 - \theta))P_r. \quad (31)$$

This introduces explicit θ and N dependence to the V-matrix elements. An important check of our calculations is provided by the fact that given a particular N our results must be independent of the choice of any nonzero θ (see Bray and Stelbovics, 1995, for some examples). For a more detailed analysis of these issues, see Stelbovics (1990).

Rather than solving Eq. (29) for $|\psi_0^{SN}\rangle$ we use it to form a set of coupled integral Lippmann–Schwinger equations for the T^{SN} matrix:

$$\langle \mathbf{k}_f^{(-)} i_f^N | T^{SN} | i_0 \mathbf{k}_0^{(+)} \rangle = \langle \mathbf{k}_f^{(-)} i_f^N | V^{SN}(\theta) | \psi_0^{SN} \rangle \tag{32}$$

$$= \langle \mathbf{k}_f^{(-)} i_f^N | V^{SN}(\theta) | i_0 \mathbf{k}_0^{(+)} \rangle$$
$$+ \sum_{n=1}^{N} \fint_k \frac{\langle \mathbf{k}_f^{(-)} i_f^N | V^{SN}(\theta) | i_n^N \mathbf{k}^{(-)} \rangle \langle \mathbf{k}^{(-)} i_n^N | T^{SN} | i_0 \mathbf{k}_0^{(+)} \rangle}{E^{(+)} - \epsilon_n^N - \epsilon_k}. \tag{33}$$

It is prudent to confirm that using Eq. (32) in defining our T matrix is appropriate for target eigenstates $|i_f^N\rangle = |i_f\rangle$. Without loss of generality we can take $\theta = 0$ (nonzero θ in Eq. (31) leads to terms which simply cancel below) and write

$$\langle \mathbf{k}_f^{(-)} i_f | V^{SN}(\theta) | \psi_0^{SN} \rangle$$
$$= \langle \mathbf{k}_f^{(-)} i_f | V_1 - U_1 + V_{12} + (-1)^S (H - E) P_r | \psi_0^{SN} \rangle$$
$$= \langle \mathbf{k}_f^{(-)} i_f | (V_1 - U_1 + V_{12})(1 + (-1)^S P_r) + (-1)^S (K_1 + U_1 + H_2 - E) P_r | \psi_0^{SN} \rangle$$
$$= \langle \mathbf{k}_f^{(-)} i_f | (V_1 - U_1 + V_{12})(1 + (-1)^S P_r) + (-1)^S (\epsilon_{k_f} + \epsilon_f - E) P_r | \psi_0^{SN} \rangle$$
$$= \langle \mathbf{k}_f^{(-)} i_f | (V_1 - U_1 + V_{12})(1 + (-1)^S P_r) | \psi_0^{SN} \rangle, \tag{34}$$

since here we are on the energy shell. This is just the required form for the case of $U_1(r_1) = U_a(r_1) = -Z_a/r_1$ as then the $|\mathbf{k}^{(\pm)}\rangle$ are plane waves or pure Coulomb waves for nonzero Z_a. However, in the event that there is an additional short-ranged distorting potential part in U_1, we refer to the T-matrix elements in Eq. (33) as the distorted-wave T-matrix elements, and these must then be related to the required physical T-matrix elements. Before doing this it is worth noting that for $\epsilon_{k_f} \neq E - \epsilon_f$, i.e., off the energy shell, the energy term leads to nonconvergence for the half-off-shell K matrix (Bray and Stelbovics, 1995).

To relate the distorted-wave T matrix to the physical T matrix we follow Gell-Mann and Goldberger (1953) and write $U_1 = U_s + U_a$, i.e., a sum of the short-ranged and asymptotic potentials, and use $V_a^{SN}(\theta)$ for the case $U_s = 0$. The required T matrix is then written as

$$\langle \mathbf{q}_f^{(-)} i_f^N | V_a^{SN}(\theta) | \psi_0^{SN} \rangle = \langle \mathbf{q}_f^{(-)} i_f^N | T^{SN} | i_0 \mathbf{q}_0^{(+)} \rangle, \tag{35}$$

where the plane/Coulomb waves $|\mathbf{q}^{(\pm)}\rangle$ satisfy

$$(K_1 + U_a - \epsilon_q^{(\pm)}) |\mathbf{q}^{(\pm)}\rangle = 0. \tag{36}$$

To relate the distorted-wave T matrix in Eq. (33) with the physical T matrix in Eq. (35) we require two relations. The first is obtained from Eq. (29) by setting $U_s = 0$, premultiplying by $\langle i_f^N |$, and not using the expansion over the complete set of the distorted waves

$$\langle i_f^N | \psi_0^{SN} \rangle = |q_0^{(+)}\rangle \delta_{f0} + \frac{1}{E^{(+)} - \epsilon_f^N - K_1 - U_a} \langle i_f^N | V_a^{SN}(\theta) | \psi_0^{SN} \rangle. \quad (37)$$

The second relates the distorted waves $|\mathbf{k}^{(\pm)}\rangle$ to plane/Coulomb waves $|\mathbf{q}^{(\pm)}\rangle$ by setting $V^S = 0$ in Eq. (23) and writing

$$\langle \mathbf{k}^{(\mp)} i_f^N | = \langle \mathbf{q}^{(\mp)} i_f^N | + \langle \mathbf{k}^{(\mp)} i_f^N | U_s \frac{1}{E^{(\pm)} - \epsilon_f^N - K_1 - U_a}. \quad (38)$$

From these two relations direct substitution for $\langle \mathbf{q}_f^{(-)} i_f^N |$ below results in

$$\langle \mathbf{q}_f^{(-)} i_f^N | V_a^{SN}(\theta) | \psi_0^{SN} \rangle = \langle \mathbf{k}_f^{(-)} i_f^N | T^{SN} | i_0 \mathbf{k}_0^{(+)} \rangle + \delta_{f0} \langle \mathbf{k}_f^{(-)} | U_s | \mathbf{q}_0^{(+)} \rangle, \quad (39)$$

where we used $V^{SN} = V_a^{SN} - U_s$. Thus, in order to obtain the physical T matrix only the elastic channel of the distorted-wave T matrix needs an additional term.

So far we have specified that $U_1 \to U_a$ for large r_1, but have not discussed the choice of U_s. In distorted-wave Born approximations short-ranged distorting potentials play a very important role, and the results vary with the choice of these potentials. In the close-coupling formalism the results must be independent of the choice of such potentials. So what then is the purpose of having nonzero U_s? They are used as a purely numerical technique to make it easier to solve the coupled integral equations [Eq. (33)]. With a suitable choice of U_s one can reduce the number of k-quadrature points in the integration, thereby considerably reducing the computational effort. A suitable choice for U_1 as a whole may be extracted from Eq. (6) by taking

$$U_1(r) = -\frac{Z}{r} + V^{\text{pol}}(r) + 2 \sum_{\psi_j \in C} \int d^3 r' \frac{|\psi_j(\mathbf{r}')|^2}{|\mathbf{r} - \mathbf{r}'|} + \int d^3 r' \frac{|\phi_j(\mathbf{r}')|^2}{|\mathbf{r} - \mathbf{r}'|}, \quad (40)$$

where we typically take ϕ_j to be the ground state, or we may in fact have j dependence $U_1 \equiv U_j$. If the target is asymptotically neutral (i.e., there are $N_c = Z - 1$ core electrons), then $U_1 = U_s$ and $U_a = 0$. For charged targets $Z_a = Z - 1 - N_c$ and $U_a(r_1) = -Z_a/r_1$.

The form of the U_1 potential above removes a considerable number of terms occurring in V_1 of Eq. (31). In particular, it removes the $-Z/r_1$ behavior for small r_1, which has the effect of allowing us to concentrate our k-quadrature points around smaller values of k in Eq. (33).

The complete set of distorted waves $|\mathbf{k}^{(\pm)}\rangle$ typically has not only continuum states, but also a few discrete states in the case of an atomic target and an infinite set in the case of ionic targets. We include as many of these as necessary for the required accuracy in Eq. (33).

D. Solving the Coupled Integral Equations

The coupled integral equations [Eq. (33)] are expanded in partial waves of the total orbital angular momentum J. The V- and T-matrix elements in Eq. (33) are related to their reduced counterparts by (e.g., for V)

$$\langle \mathbf{k}^{(-)} i_n^N | V^{SN} | i_{n'}^N \mathbf{k}^{(+)} \rangle$$
$$\equiv \langle \mathbf{k}_f^{(-)} i_{nlm}^N | V^{SN} | i_{n'l'm'}^N \mathbf{k}^{(+)} \rangle$$
$$= \sum_{\substack{L,L',J \\ M,M',M_J}} C_{LlJ}^{MmM_J} C_{L'l'J}^{M'm'M_J} Y_{LM}(\hat{\mathbf{k}}) Y_{L'M'}^*(\hat{\mathbf{k}}') \langle Lk^{(-)} ln \| V_{J\Pi}^{SN} \| n'l'k'^{(+)}L' \rangle, \quad (41)$$

where C denotes the Clebsch-Gordan coefficient and $\Pi = (-1)^{l+L} = (-1)^{l'+L'}$ is the parity. The partial-wave expansion [Eq. (41)] may be inverted to give the reduced matrix elements via

$$\langle Lk^{(-)} ln \| V_{J\Pi}^{SN} \| n'l'k'^{(+)}L' \rangle$$
$$= \sum_{\substack{M,m \\ M',m'}} C_{LlJ}^{MmM_J} C_{L'l'J}^{M'm'M_J}$$
$$\times \int d\hat{\mathbf{k}} \int d\hat{\mathbf{k}}' Y_{LM}^*(\hat{\mathbf{k}}') Y_{L'M'}(\hat{\mathbf{k}}') \langle \mathbf{k}^{(-)} i_{nlm}^{N_l} | V^{SN} | i_{n'l'm'}^{N_{l'}} \mathbf{k}'^{(+)} \rangle. \quad (42)$$

Substituting these definitions into Eq. (33) we obtain one-dimensional coupled integral equations for the reduced T matrix

$$\langle Lk_{nl}^{(-)} ln \| T_{J\Pi}^{SN} \| n_0 l_0 k_0^{(+)} L_0 \rangle$$
$$= \langle Lk_{nl}^{(-)} ln \| V_{J\Pi}^{SN} \| n_0 l_0 k_0^{(+)} L_0 \rangle \quad (43)$$
$$+ \sum_{l',L'} \sum_{n'=1}^{N_{l'}} \oint_{k'} \frac{\langle Lk_{nl}^{(-)} ln \| V_{J\Pi}^{SN} \| n'l'k'^{(-)}L' \rangle}{E^{(+)} - \epsilon_{n'l'} - \epsilon_{k'}} \langle L'k'^{(-)}l'n' \| T_{J\Pi}^{SN} \| n_0 l_0 k_0^{(+)} L_0 \rangle.$$

For scattering on a nS state we require matrix elements corresponding only to the "natural" parity $\Pi = (-1)^J$. If scattering on target states with nonzero orbital angular momentum the separate set of equations corresponding to the "unnatural" parity $\Pi = (-1)^{J+1}$ must also be solved.

Application of the partial-wave expansion reduces the problem to solving the coupled equations in only the single scalar linear momentum space, although for as many J as may be required. For a particular J there are two major steps in solving the coupled integral equations. They are the calculation of the reduced V-matrix elements [Eq. (31)] followed by the solution of the coupled equations.

1. Calculation of the Reduced V-Matrix Elements

Before we begin, it is helpful to subdivide $U_s = U_1 - U_a$ into two parts,

$$U_s(r) = U_c(r) + U_v(r), \tag{44}$$

corresponding to short-ranged distorting potentials due to the core electrons, and the single valence electron. We also define V_p to be the potential on an electron due to a single proton ($V_1 = V_p$, $U_c = 0$ in the case of atomic hydrogen target). The three-dimensional V-matrix elements [Eq. (31)] may be written as

$$\langle \mathbf{k}'^{(-)} i_{n'}^N | V^{SN}(\theta) | i_n^N \mathbf{k}^{(\pm)} \rangle = \langle \mathbf{k}'^{(-)} i_{n'}^N | V_1 - V_p - U_a - U_c | i_n^N \mathbf{k}^{(\pm)} \rangle \tag{45}$$

$$+ \langle \mathbf{k}'^{(-)} i_{n'}^N | V_p + V_{12} - U_v - \theta E I_1^N | i_n^N \mathbf{k}^{(\pm)} \rangle \tag{46}$$

$$+ (-1)^S \langle \mathbf{k}'^{(-)} i_{n'}^N | V_{12} | \mathbf{k}^{(\pm)} i_n^N \rangle \tag{47}$$

$$+ (-1)^S \langle \mathbf{k}'^{(-)} i_{n'}^N | H_1 + H_2 - E(1-\theta) | \mathbf{k}^{(\pm)} i_n^N \rangle. \tag{48}$$

We now define the corresponding reduced matrix elements.

In the case of the existence of core electrons we define, using Eq. (40), the core distorting potential $U_c(r)$ to be

$$U_c(r) = -\frac{Z - Z_a - 1}{r} + 2 \sum_{n_c, l_c \in C} (2l_c + 1) R^0_{n_c l_c n_c l_c}(r) + V^{\text{pol}}(r), \tag{49}$$

where we use the notation

$$R^\lambda_{\alpha' l' \alpha l}(r) = \int_0^\infty dr' \psi_{\alpha' l'}(r') \psi_{\alpha l}(r') \frac{r_<^\lambda}{r_>^{\lambda+1}} \tag{50}$$

and where for $\alpha = n$ we take $\psi_{\alpha l}(r) = \phi^{Nl}_{nl}(r)$ of Eq. (9) and for $\alpha = k$ we take $\psi_{\alpha l} = u_l(k, r)$ of Eq. (27). Note that $U_c(r)$ is asymptotically neutral due to the number of electrons in the core, i.e., $N_c = Z - Z_a - 1$.

The reduced matrix elements corresponding to the core terms [Eq. (45)] are

$$\langle L' k'^{(-)} l' n' \| V_c \| nl k^{(\pm)} L \rangle$$

$$= \delta_{nn'} \delta_{ll'} \delta_{LL'} \frac{2}{\pi} \frac{i^{L-L'}}{kk'} e^{i(\sigma_{L'} + \delta_{L'} \pm \sigma_L \pm \delta_L)} \left[\int_0^\infty dr u_{L'}(k', r) u_L(k, r) \right.$$

$$\times \left(-\frac{Z - Z_a - 1}{r} + V^{\text{pol}}(r) - U_c(r) + 2 \sum_{n_c, l_c \in C} (2l_c + 1) R^0_{n_c l_c n_c l_c}(r) \right)$$

$$\left. - \int_0^\infty dr u_{L'}(k', r) \sum_{n_c, l_c \in C} \frac{2l_c + 1}{2L + 1} \sum_\lambda C^{000}_{l_c \lambda L} C^{000}_{l_c \lambda L} R^\lambda_{n_c l_c k L}(r) \phi^{N_{l_c}}_{n_c l_c}(r) \right]. \tag{51}$$

The special choice for $U_c(r)$ in Eq. (49) removes all of the "direct" core potential.

The second term in Eq. (46) is a direct potential matrix element whose reduced elements are

$$\langle L'k'^{(-)}l'n' \| V_d \| nlk^{(\pm)}L \rangle$$
$$= \frac{2}{\pi} \frac{i^{L-L'}}{kk'} e^{i(\sigma_{L'} + \delta_{L'} \pm \sigma_L \pm \delta_L)} \left[\int_0^\infty dr\, u_{L'}(k', r) u_L(k, r) (-1)^{J+l+L'} \hat{L}\hat{l} \right.$$
$$\times \sum_\lambda C^{000}_{l\lambda l'} C^{000}_{L\lambda L} W(LL'll'\lambda J)(R^\lambda_{n'l'nl}(r) - \delta_{\lambda 0}(U_v(r) + 1/r))$$
$$\left. - \delta_{nn'}\delta_{ll'}\delta_{LL'}\theta E \sum_{n_i,l_i} \int_0^\infty dr'\, u_{L'}(k', r') \phi^{N_{l_i}}_{n_i l_i}(r') \int_0^\infty dr\, \phi^{N_{l_i}}_{n_i l_i}(r) u_L(k, r) \right]. \quad (52)$$

The sum over the n_i, l_i is a sum over all of the target states used in the multichannel expansion. The notation $\hat{l}$ is used to mean $\sqrt{2l+1}$, and W denotes a Racah coefficient. The function $R^\lambda_{n'l'nl}(r)$ is readily evaluated and, after subtraction of the nuclear term for $\lambda = 0$, has at worst $1/r^2$ behavior for large r. This necessitates allowance of our radial integrations to go out to distances of order $r_{max} = 200$ arbitrary units, i.e., at which $1/r^2_{max}$ is sufficiently small. We use variation in r_{max} and the size of the radial step as a check on the accuracy of the calculation of the matrix elements. Subtraction of the nuclear term introduces $-1/r$ behavior at the origin. The distorting potential $U_v(r)$ is used to remove this. We may vary the definition of $U_v(r)$ from something as simple as

$$U_v(r) = -e^{-cr}/r, \quad (53)$$

for some constant c, to

$$U_v(r) = (-1)^{J+l+L} \hat{L}\hat{l} \sum_\lambda C^{000}_{l\lambda l} C^{000}_{L\lambda L} W(LLll\lambda J)(R^\lambda_{nlnl}(r) - \delta_{\lambda 0}/r). \quad (54)$$

The latter form removes all of the (summed over λ) diagonal radial part in Eq. (52). A difficulty associated with the latter form is that it leads to $U_v(r)$ and therefore $U_1(r)$ that has $1/r^2$ behavior for large r, which may make the calculation of the distorted waves less accurate.

The reduced matrix elements corresponding to the "exchange" potential matrix elements [Eq. (47)] are given by

$$\langle L'k'^{(-)}l'n' \| V_e \| nlk^{(\pm)}L \rangle = \frac{2}{\pi} \frac{i^{L-L'}}{kk'} e^{i(\sigma_{L'} + \delta_{L'} \pm \sigma_L \pm \delta_L)} \int_0^\infty dr\, u_{L'}(k', r) \phi^{N_l}_{nl}(r)$$
$$\times (-1)^{L'+l} \hat{L}\hat{l} \sum_\lambda C^{000}_{L\lambda l'} C^{000}_{l\lambda L'} W(lL'Ll'\lambda J) R^\lambda_{n'l'kL}(r) \quad (55)$$

There are no significant difficulties with evaluating these matrix elements due to the existence of exponentially decreasing functions in the integrands.

CALCULATION OF ELECTRON SCATTERING ON HYDROGENIC TARGETS 231

To obtain the reduced matrix elements corresponding to Eq. (48) we first expand H_1 and H_2 and using Eq. (25) obtain

$$\langle \mathbf{k}'^{(-)} i_{n'}^N | H_1 + H_2 - E(1 - \theta) | \mathbf{k}^{(\pm)} i_n^N \rangle$$
$$= \langle \mathbf{k}'^{(-)} | V_1 - U_1 | i_n^N \rangle \langle i_{n'}^N | \mathbf{k}^{(\pm)} \rangle$$
$$+ \langle \mathbf{k}'^{(-)} | i_n^N \rangle \langle i_{n'}^N | V_2 - U_2 | \mathbf{k}^{(\pm)} \rangle$$
$$+ (\varepsilon_{k'} + \varepsilon_k - E(1 - \theta)) \langle \mathbf{k}'^{(-)} | i_n^N \rangle \langle i_{n'}^N | \mathbf{k}^{(\pm)} \rangle, \quad (56)$$
$$= (\epsilon_n + \epsilon_{n'} - E(1 - \theta)) \langle \mathbf{k}'^{(-)} | i_n \rangle \langle i_{n'} | \mathbf{k}^{(\pm)} \rangle \quad (57)$$

in the case of true eigenstates $i_{n'}^N = i_{n'}$ and $i_n^N = i_n$. The form [Eq. (56)] is valid in the general case, whereas we use Eq. (57) in the case of true discrete eigenstates to test the computational coding of the general expression. The reduced one-electron exchange matrix elements corresponding to the first term of Eq. (56) are

$$\langle L' k'^{(-)} l' n' \| V_e^1 \| nlk^{(\pm)} L \rangle$$
$$= \delta_{l'L} \delta_{lL'} (-1)^{L+l+J} \frac{2}{\pi} \frac{i^{L-L'}}{kk'} e^{i(\sigma_{L'} + \delta_{L'} \pm \sigma_L \pm \delta_L)} \left[\int_0^\infty dr u_{L'}(k', r) \phi_{nl}^{N_l}(r) \right.$$
$$\times \left(-\frac{Z - Z_a}{r} + V^{\text{pol}}(r) - U_c(r) - U_v(r) + 2 \sum_{n_c, l_c \in C} (2l_c + 1) R_{n_c l_c n_c l_c}^0(r) \right)$$
$$- \int_0^\infty dr u_{L'}(k', r) \sum_{n_c, l_c \in C} \frac{2l_c + 1}{2L + 1} \sum_\lambda C_{l_c \lambda L}^{000} C_{l_c \lambda L}^{000} R_{n_c l_c nl}^\lambda(r) \phi_{n_c l_c}^{N_{l_c}}(r) \right]$$
$$\times \int_0^\infty dr' u_L(k, r') \phi_{n'l'}^{N_{l'}}(r'). \quad (58)$$

Similarly, we obtain $\langle L' k'^{(-)} l' n' \| V_e^2 \| nlk^{(\pm)} L \rangle$ from the second term in Eq. (56). This leaves the simple third term which yields reduced matrix elements

$$\langle L' k'^{(-)} l' n' \| V_e^3 \| nlk^{(\pm)} L \rangle$$
$$= \delta_{l'L} \delta_{lL'} (-1)^{L+l+J} \frac{2}{\pi} \frac{i^{L-L'}}{kk'} e^{i(\sigma_{L'} + \delta_{L'} \pm \sigma_L \pm \delta_L)} (\varepsilon_{k'} + \varepsilon_k - E(1 - \theta))$$
$$\times \int_0^\infty dr u_{L'}(k', r) \phi_{nl}^{N_l}(r) \int_0^\infty dr' u_L(k, r') \phi_{n'l'}^{N_{l'}}(r'). \quad (59)$$

In summary, the reduced matrix elements that are needed in the solution of Eq. (43) are the combination of Eqs. (51), (52), (55), (58), and (59) given by

$$\langle L' k'^{(-)} l' n' \| V_{JII}^{SN} \| nlk^{(-)} L \rangle$$
$$= \langle L' k'^{(-)} l' n' \| V_c + V_d \| nlk^{(\pm)} L \rangle$$
$$+ (-1)^S \langle L' k'^{(-)} l' n' \| V_e + V_e^1 + V_e^2 + V_e^3 \| nlk^{(\pm)} L \rangle. \quad (60)$$

The remaining reduced matrix element that we require is due to the contribution to the elastic T matrix element (see Eq. (39)) from the short-ranged distorting potential $U_s(r)$. This is

$$\langle Lk^{(-)}ln \| T_s \| nlk^{(+)}L \rangle = \frac{2}{\pi} \frac{e^{i(2\sigma_L+\delta_L)}}{k^2} \int_0^\infty dr u_L(k,r) U_s(r) u'_L(k,r)$$

$$= -\frac{e^{i(2\sigma_L+\delta_L)}}{\pi k} \sin \delta_L, \quad (61)$$

where $u'_L(k,r)$ is a pure plane or Coulomb wave and is obtained by setting $U_1 = U_a$ in Eq. (25). A derivation of this relation is given on page 15 of Bransden (1983). If we take $U_1 = U_a$ in the full calculation, then $\delta_L = 0$, and we have no contribution from the above term.

All of these matrix elements may be readily calculated to a desired precision. Although they depend on the choice of the short-ranged distorting potentials $U_c(r)$ and $U_v(r)$ and the arbitrary constant θ, nevertheless the T matrix resultant from solution of Eq. (43), with the elastic term modified by Eq. (61), is independent of these choices.

2. Solving Coupled Lippman–Schwinger Equations

Having reduced the coupled three-dimensional Lippmann–Schwinger equations into partial-wave form we note that any complex phases in the potential matrix elements may be trivially factored out. The remaining complex arithmetic is contained in the Green's function with the outgoing spherical-wave boundary conditions. However it is easy to apply standing-wave boundary conditions by introducing the K matrix. We drop the $(\pm)$ notation below and convert our intermediate integrations to the principal-value type. Thus the treatment of charged targets (nonzero σ_L) and the introduction of the distorted-wave formalism (nonzero δ_L) do not remove our ability to solve the coupled equations using real arithmetic. We define the K matrix by means of the identity

$$\langle Lkln \| K^{SN}_{J\Pi} \| n_0 l_0 k_0 L_0 \rangle = \sum_{l',L'} \sum_{n'=1}^{N^{on}_{l'}} \langle Lkln \| T^{SN}_{J\Pi} \| n'l'k_{n'l'}L' \rangle \quad (62)$$

$$\times (\delta_{l'l_0}\delta_{L'L_0}\delta_{n'n_0} + i\pi k_{n'l'} \langle L'k_{n'l'}l'n' \| K^{SN}_{J\Pi} \| n_0 l_0 k_0 L_0 \rangle),$$

where k_{nl} is defined for $1 \leq n \leq N^{on}_l \leq N_l$, for which

$$k_{nl} = \sqrt{2(E - \epsilon_{nl})} \quad (63)$$

is real. In this case we say that the channel nlL is open, and, if $E < \epsilon_{nl}$, we say that this channel is closed. For a particular l the number of states which lead to open channels is N^{on}_l.

Substitution of Eq. (62) into Eq. (43) results in a set of real coupled integral equations of the principal-value type for the K-matrix amplitudes

$$\langle Lk_{nl}ln \| K_{J\Pi}^{SN} \| n_0 l_0 k_0 L_0 \rangle$$
$$= \langle Lk_{nl}ln \| V_{J\Pi}^{SN} \| n_0 l_0 k_0 L_0 \rangle$$
$$+ \sum_{l',L'} \sum_{n'=1}^{N_{l'}} \mathcal{P} \int_{k'} \frac{\langle Lk_{nl}ln \| V_{J\Pi}^{SN} \| n'l'k'L' \rangle \langle L'k'l'n' \| K_{J\Pi}^{SN} \| n_0 l_0 k_0 L_0 \rangle}{E - \epsilon_{n'l'} - \epsilon_{k'}}. \quad (64)$$

The intermediate sum over target space l' has $0 \leq l' \leq l_{\max}$. We use N to denote the total number of states used in the close-coupling expansion, which is the sum of all $N_{l'}$ for $0 \leq l' \leq l_{\max}$. The sum and integral in Eq. (64) represent a sum over bound states of the U_1 potential and a principal-value integral over the continuum projectile momenta.

It is important to be able to solve Eq. (64) as efficiently as possible as this allows us to take the largest possible N within finite computational resources. Given N_l for $0 \leq l \leq l_{\max}$, we need to perform the principal-value integrals with as few quadrature points as possible. We replace the sum and integral in Eq. (64) by introducing a quadrature rule k_j, w_j such that

$$\sum_{l',L'} \sum_{n'=1}^{N_{l'}} \mathcal{P} \int_{k'} \frac{\| n'l'k'L' \rangle \langle L'k'l'n' \|}{E - \epsilon_{n'l'} - \epsilon_{k'}} \approx \sum_{l',L'} \sum_{n'=1}^{N_{l'}} \sum_{\epsilon_{k'}<0}^{N_L^b} \frac{\| n'l'k'L' \rangle \langle L'k'l'n' \|}{E - \epsilon_{n'l'} - \epsilon_{k'}} \quad (65)$$
$$+ \sum_{l',L'} \sum_{n'=1}^{N_{l'}} \sum_{j=1}^{N_L^c} w_j k_j'^2 \frac{\| n'l'k_j'L' \rangle \langle L'k_j'l'n' \|}{E - \epsilon_{n'l'} - k_j'^2/2}.$$

Note that we allow both the number of bound N_L^b and the continuum N_L^c states to depend on the orbital angular momentum of the projectile L'. The quadrature rule must be such that for an arbitrary function, $f(k)$, we may replace

$$\mathcal{P} \int_0^\infty dk \frac{k^2 f(k)}{E - \epsilon_{n'l'} - k^2/2} \approx \sum_{j=1}^{N_L^c} w_j k_j^2 f(k_j). \quad (66)$$

There are a number of suitable ways to choose the weights w_j and the corresponding knots k_j. A discussion of these issues has been presented by Bransden et al. (1993), and references therein. Unfortunately, no single method appears to be superior to all others. There are a number of difficulties that need to be addressed. First, the quadrature rule must be able to handle the singularity, which varies in position with $n'l'$. We do this by taking an even number of Gaussian points in an interval which is symmetric about the singularity. This requires that we allow the w_j and k_j to be different in each channel $n'l'L'$.

After the discretization of the continuum integral, the coupled integral equations [Eq. (64)] take the form of a large set of linear equations $AX = B$. These are solved using standard linear algebra techniques (Anderson et al., 1992). For details of the method of solution and minimization of computer memory usage, see Bray (1994c) and Bray and Stelbovics (1995). Typically we find that the amount of computational time for the two processes to calculate the V-matrix elements and to solve the resultant linear equations is similar.

The calculation of the lower partial waves J tends to dominate the total calculation time. This is because for the higher J only the direct matrix elements survive, and furthermore these are highly peaked near $k = k'$ (Allen et al., 1988). In fact, with increasing J accurate solutions to Eq. (64) are obtained by simply taking the first or second term in the iterative expansion. The CCC computer code has been used with $J \leqslant 100$.

IV. Electron–Hydrogen Scattering

We will not attempt to provide a comprehensive review of applications of the CCC or other methods to various electron–atom/ion scattering processes. Instead, we will concentrate on areas in which theory still has questions to answer. These include collision processes for which the effects of the target continuum coupling are important and for which there are unresolved discrepancies with experiment.

The atomic–hydrogen target is fundamental for the purpose of testing electron scattering theories. None of the complexities of charged targets, or approximate treatments of core electrons as described above, are necessary. Assuming nonrelativistic dynamics and that the center of mass is at the center of the proton, we present an extensive range of results that strongly suggest the CCC method solves the electron–hydrogen scattering problem without further approximation.

A. Temkin–Poet Model

The Temkin–Poet model (Temkin, 1962; Kyle and Temkin, 1964; Poet, 1978, 1980, 1981) is a simplification of the full electron–hydrogen scattering problem that treats only states with zero orbital angular momentum and has been solved to a high accuracy. From our point of view it serves as a classical test of any general scattering theory. It contains most of the difficulties associated with treating exchange and the target continuum and has been used by a number of groups (Callaway and Oza, 1984; Bhatia et al., 1993; Konovalov and McCarthy, 1994b; Robicheaux et al., 1994) to test their general scattering theories.

One of our earliest applications of the CCC method was to this problem (Bray and Stelbovics, 1992b). This demonstrated two very important points. The first was that the CCC method yielded convergent results as the basis size was increased. The pseudoresonances, typically associated with square-integrable expansions of the target continuum, diminished and disappeared with increasing basis size. This has been studied in greater detail in the vicinity of a pseudothreshold (Bray and Stelbovics, 1995). The second point was that convergence was to the correct result, independent of the projectile energy or the transition studied. These facts give us considerable confidence in applying the CCC method to the full electron–hydrogen scattering problem (Bray and Stelbovics, 1992a).

B. Angular Correlation Parameters at 54.4 eV

One of the reasons for developing the CCC method in the first place was an attempt to resolve the long-standing discrepancies between theory and experimentation for the 2p angular correlation parameters at 54.4 eV projectile energy. For the λ and R parameters there are two experiments by Weigold et al. (1979) and Williams (1981). Both use similar experimental techniques, and their measurements are in broad agreement. None of the very many attempts to calculate these parameters agreed with the measurements at large angles. It was to our great disappointment to find that the application of the CCC method (Bray and Stelbovics, 1992a) did not resolve these discrepancies. The CCC results proved to be very similar to the results obtained with other sophisticated theories. Importantly though, the CCC method allowed us to place an estimate on the convergence of our correlation parameters; the discrepancy can no longer be attributed to a deficiency in the representation of the coupling to the target continuum states. Thus at present we have the unsatisfactory situation of broad consistency among different theories and disagreement with the data. It appears timely for further experimental work to attempt to clarify this situation.

In Fig. 2 we present our study of convergence for the λ, R, and I parameters of the 2p excitation of atomic hydrogen by 54.4-eV electrons. The CCC results are also compared with experimental results and results obtained with some of the other more sophisticated theories. In the figure, the left column has calculations which include s, p, and d states, demonstrating convergence as the basis size N_l is increased for each of these l. The center column fixes the basis size for each l and demonstrates convergence as l is increased from s and p states to include d and f states. The right column compares our results with those of other theories and experiments. Examination of the left and middle columns shows satisfactory convergence in both the basis sizes N_l and l_{max}. In other words, we believe that any larger CCC calculations will not yield significantly different results. The right column shows that the theories are in quite good agreement with each other, but not with the experiment.

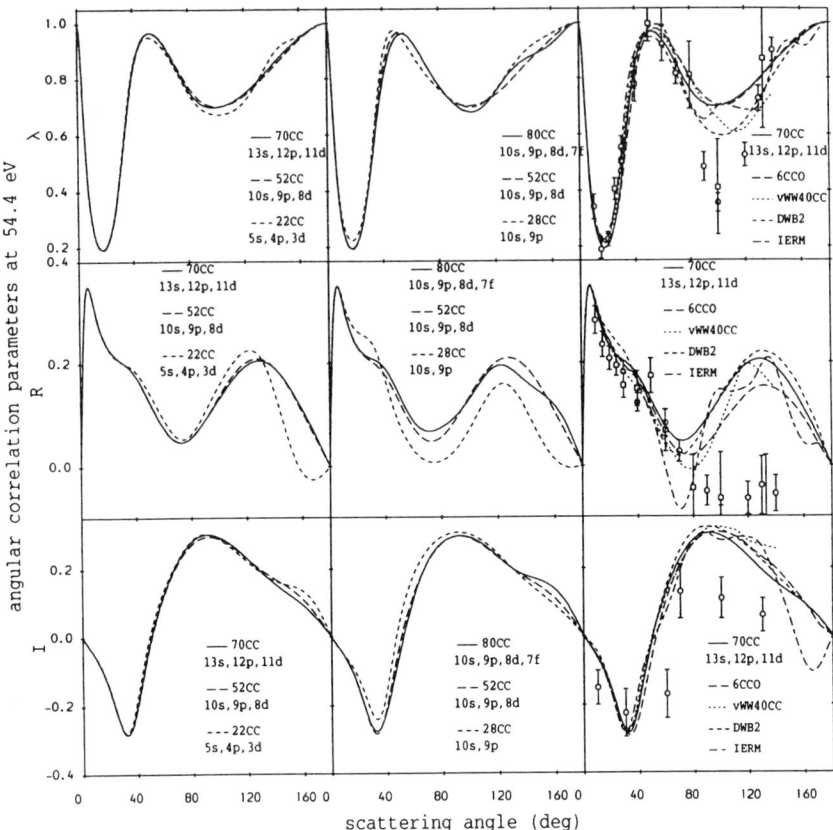

FIG. 2. Angular correlation parameters for the 2p excitation of hydrogen at 54.4 eV calculated with the indicated number of Laguerre basis states (Bray and Stelbovics, 1992a). The coupled-channel optical calculation of Bray et al. (1991c), the intermediate energy R-matrix method of Scholz et al. (1991), the distorted wave second Born calculation of Madison et al. (1991), and the 40-channel pseudostate method of van Wyngaarden and Walters (1986) are denoted by 6CCO, IERM, DWB2 and vWW40CC, respectively. The measurements denoted by ○ are due to Williams (1981, 1986); those denoted by □ are due to Weigold et al. (1979).

Even though our results have not agreed with experimental results in this case, we are still confident of the validity of the CCC approach. We shall see that in the considerably more complicated electron–sodium scattering system these same parameters, which have caused us so much trouble here, are quantitatively well described.

C. IONIZATION

One of the stronger tests of any method that claims to be able to treat the target continuum is application of the method to ionization phenomena.

Apart from giving T-matrix elements for transitions between states with negative energies, the CCC method also yields T-matrix elements to positive-energy states. Of course, since these states are L^2 it is only after some renormalization procedure that they can be interpreted in terms of transitions to the true continuum states. There are many ways a renormalization can be defined. Indeed it is possible to interpret the sum over the positive-energy states as an approximation to the subspace of the target continuum, for which the weights of the quadrature sum are directly related to the renormalization of the L^2 states. For total ionization cross-sections one therefore does not have to calculate renormalizations but merely has to sum the cross-sections for excitation to the positive-energy states. Thus one expects the CCC method to be able to provide convergent total-ionization cross-sections as the Laguerre basis is increased to completeness.

1. Total Ionization Cross-Section

We consider first the calculation of the total ionization cross-section for the electron impact of atomic hydrogen. As discussed above, it is obtained from our calculations by summing the cross-sections of positive-energy pseudo-states. This test is very stringent due to the very high accuracy of the experimental data of Shah *et al.* (1987). For this reason it is particularly encouraging to find excellent agreement with this experiment (Bray and Stelbovics, 1993). Furthermore, the measurements of the total ionization spin asymmetries are also in excellent agreement with the CCC predictions. These results are presented in Fig. 3 and indicate that the CCC method

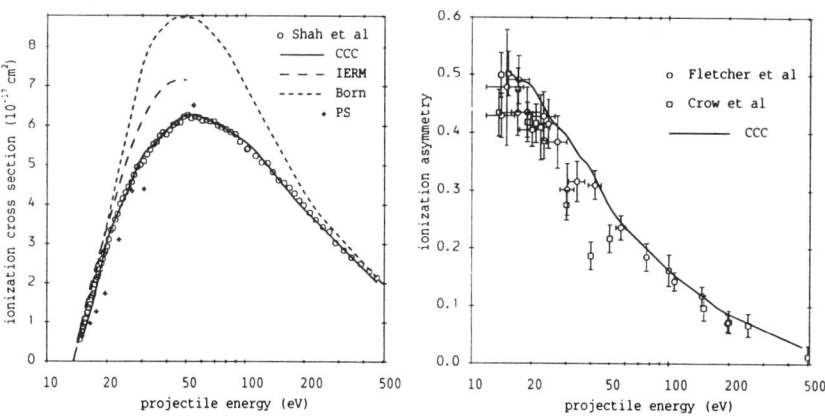

FIG. 3. The total ionization cross-section and spin asymmetries for the electron impact of atomic hydrogen. The total ionization measurements of Shah *et al.* (1987) are denoted by ○. The spin asymmetry measurements denoted by ○ and □ are due to Fletcher *et al.* (1985) and Crowe *et al.* (1990), respectively. The IERM results are due to Scholz *et al.* (1990), and the pseudostate (PS) results are due to Callaway and Oza (1979). The convergent close-coupling results (Bray and Stelbovics, 1993) are denoted by CCC.

apportions the correct amount of electron flux in both the singlet and the triplet ionization channels, in contrast to the IERM method of Scholz et al. (1990) and the pseudostate method of Callaway and Oza (1979). The CCC method also permits calculation of ionization from excited states We have lately obtained quantitative agreement with the total ionization cross-section measurements (Defrance et al., 1981b) for electron impact ionization of the metastable 2s state of hydrogen. These results will be presented elsewhere.

2. Differential Ionization Cross-Sections

Konovalov et al. (1994) have extended the CCC method to calculate the singly differential ionization cross-section. This is obtained by dividing the positive-energy pseudostate cross-sections into terms that we will denote σ_{nl}. The corresponding energies lie in the range $0 < \epsilon_{nl} < \epsilon_{n+1\,l} < E$, where E is the total energy. A procedure suggested by Bransden and Stelbovics (1984) is used to obtain properly renormalized differential cross-sections from the pseudostates. Symmetry requires that the singly differential cross-section $d\sigma_l/dE$ should be symmetric about $E/2$ for each l; in other words, in the experiment the two electrons are indistinguishable. This is not the case with the CCC method, in which one of the electrons is treated by a square-integrable expansion and other, by a plane wave. The resultant $d\sigma_l/dE$ is highly peaked for small ϵ_{nl} and falls off monotonically to zero at E. However, upon explicitly symmetrizing $d\sigma_l/dE$ about $E/2$ and summing over all l used in the calculation, impressive agreement with experimentation is achieved. This is particularly interesting since at the $E/2$ point, at which both continuum electrons have the same energy, symmetrization does not alter the original result. See Konovalov et al. (1994) for more detail. These ideas require further consideration since the method relies on the ansatz of explicit symmetrization.

A stronger test of the CCC method is provided by calculating triply differential ionization cross-sections. The definition of the $(e, 2e)$ T matrix is unambiguous in the CCC formalism. Having made the approximation of expanding one electron in a set of square-integrable states, the other must asymptotically be a plane wave. There is no room for incorporation of three-body boundary conditions (Brauner et al., 1989). We use the T matrix [Eq. (33)] that comes directly from the close-coupling formalism. To get the normalization we multiply it by the overlap $\langle \chi(\mathbf{k}_f)|i_f^N \rangle$, where $\langle \chi(\mathbf{k}_f)|$ is a pure Coulomb wave with $k_f^2/2 = \epsilon_f$. This procedure is equivalent to replacing each positive-energy L^2 state by a Coulomb wave of the same energy projected onto the N-dimensional target space.

Now, unlike the case of scattering to a discrete excited state, in which the target final state has fixed orbital angular momentum l, in $(e, 2e)$, describing the two continuum electrons can require many l depending on the magnitude of their energies, with each partial wave having the same energy. Let us suppose that we have an incident electron of energy E_0 and the two outgoing electrons have energies $E_a \geqslant E_b$. In the CCC formalism l_{max} denotes the largest l used in the square-integrable expansions. Thus for each $0 \leqslant l \leqslant l_{max}$ we generate square-integrable states such that there is one state with $\epsilon_{nl} = E_b$. This is achieved by minor variations of the exponential fall-off parameter λ_l (see Fig. 1). If the slowest of the two outgoing electrons is sufficiently well described with such l, then no further approximation is necessary. However, there are cases in which for computational limitations we are unable to include as many l as we would wish in the CCC calculations. In this case we obtain the T matrix for the larger l used in the expansion of $\langle \chi(\mathbf{k}_f)|$ by replacing $\langle i_f^N|$ in Eq. (33) with $\langle \chi(\mathbf{k}_f)|$ and dropping the exchange terms. The latter are dropped because they lead to divergent integrals in the V-matrix elements. In our view, inclusion of exchange can be consistently implemented only in the CC formalism using L^2 states. It is for this same reason that close coupling using true continuum states has not yet been implemented. It may seem that this approximation for obtaining the T-matrix elements for the larger l is rather severe. Fortunately, we are able to test it by simply varying l_{max} in the CCC calculations.

These ideas have been applied (Bray et al., 1994b) to the calculation of triply differential cross-sections for 54.4- and 150-eV electron-impact ionization of atomic hydrogen. In the latter case, coplanar measurements by Ehrhardt et al. (1986) were performed for three choices of E_b, 3, 5, and 10 eV, with three choices for the fast electron scattering angle, $\theta_a = 4°$, 10°, and 16°. These measurements were put on an absolute scale using experimental techniques and so may be regarded as absolute. Agreement between the CCC and other theories with these measurements is satisfactory. Given the large incident projectile energy, effects of exchange are fairly small, and this makes it a less stringent test of theory.

The coplanar measurements (Brauner et al., 1991) at the incident projectile energy of 54.4 eV were performed for a single choice of the slow outgoing electron energy, $E_b = 5$ eV, and our angles of the fast electron, $\theta_a = 4°$, 10°, 16°, and 23°. These data have proved much more challenging for theorists. In Fig. 4 we present this relative data normalized to the CCC theory together with a number of calculations. There are two distorted-wave calculations that incorporate the three-body boundary condition. The calculations of Brauner et al. (1991) use the same functions in both the external and the internal regions, whereas those of Jones et al. (1989) also incorporate a further correlation factor. The pseudostate calculations of Curran and Walters (1987) used just nine states, which we find to be insufficient at this energy.

It can be seen that only the CCC theory can claim to have good qualitative agreement with experimentation. In fact the quality of the agreement at this energy is similar to that for the 150-eV results. The figure

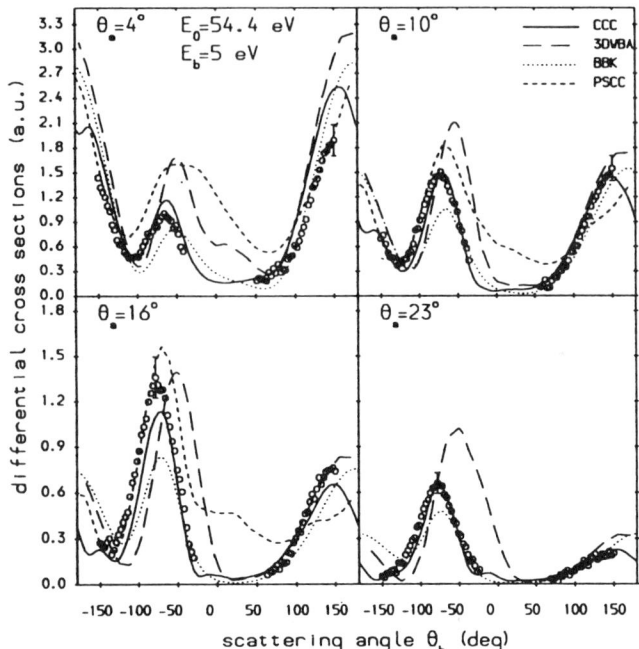

FIG. 4. Triply differential cross-sections of e–H ionization at incident electron energy $E_0 = 54.4$ eV. The relative measurements of Brauner et al. (1991) have been normalized using the best visual fit to the CCC (Bray et al., 1994b) theory using a single multiplicative constant. The calculations denoted by 3DWBA, BBK, and PSCC are due to Jones et al. (1989), Brauner et al. (1991), and Curran and Walters (1987), respectively.

indicates the importance of having a variety of measurements for different angles of the fast electron, while keeping the same normalization constant. This results in a considerably more stringent test of theory. The three other calculations, although working well at some angles, generally show significant discrepancy with experimentation, suggesting that they are primarily reliable at higher energies. A similar conclusion was also found by Konovalov (1994), who has developed a unified approach to calculations of this type using direct numerical integration, which allows any trial correlation function to be incorporated.

Brauner et al. (1991) have also performed coplanar measurements at incident energies of 27.2, 17.6, and 15.6 eV, at which both outgoing electrons have the same energy of 6.8, 2, and 1 eV, respectively. It is of particular importance to test the CCC method with these results since, in the formalism, the two equal-energy electrons are not treated in a symmetric manner. Initial application of the CCC method at these energies has demonstrated the need for incorporation of large l_{max}. In particular, the smaller the energy the more l we require. This suggests that the electron–

electron correlation becomes more important with diminishing energies, as one would expect. Thus far, initial indications are that we will be able to get convergence in our calculations with $l_{max} = 5$ on our desktop workstations at 27.2 eV and expect to obtain good agreement with experimentation. It is not yet clear as to what will happen for the two smaller energies.

V. Electron Scattering on the He$^+$ Ion

Another excellent target for testing general electron–atom/ion scattering theory is provided by the He$^+$ ion. Like the case of atomic hydrogen the single electron is very accurately described by the nonrelativistic one-electron Hamiltonian. The extra complexity over hydrogen is primarily numerical. We have to use Coulomb-wave rather than plane-wave boundary conditions. In addition, solution of the coupled equations [Eq. (64)] involves a sum over an infinite number of bound states of the U_1 potential. This sum must be truncated, and the bound states, found numerically. These problems simply require a little more careful numerical analysis.

There is not a great deal of experimental data for electron scattering on the He$^+$ ion. However, one particular subject is of considerable interest, namely, the long-standing discrepancy between theory and experimentation for the near-threshold 2S excitation.

A. 2S Excitation

There are two sets of independent measurements for this cross-section made by Dance et al. (1966) and Dolder and Peart (1973). They are in a satisfactory agreement with each other, and in the vicinity of the 40.8-eV threshold they are a factor of 2 lower compared with those obtained with existing theories. In Fig. 5 we present the measurements and a number of calculations for the 2S excitation from threshold to 700 eV. The CCC calculations (Bray et al., 1993) were generated with 36 states, but they provide only minor improvement over the 6-state calculations in the near-threshold region. However, at the higher energies the treatment of the target continuum has a substantial effect on the resultant 2S cross-section, bringing about good agreement with experimentation. This can be seen by comparison of the 15CC results, which treat only exact discrete eigenstates, with the CCC or the UCO calculations. The latter calculations, due to Unnikrishnan et al. (1991), use a CCO model for an approximate way of treating the target continuum which gives a substantial improvement over the CC calculations.

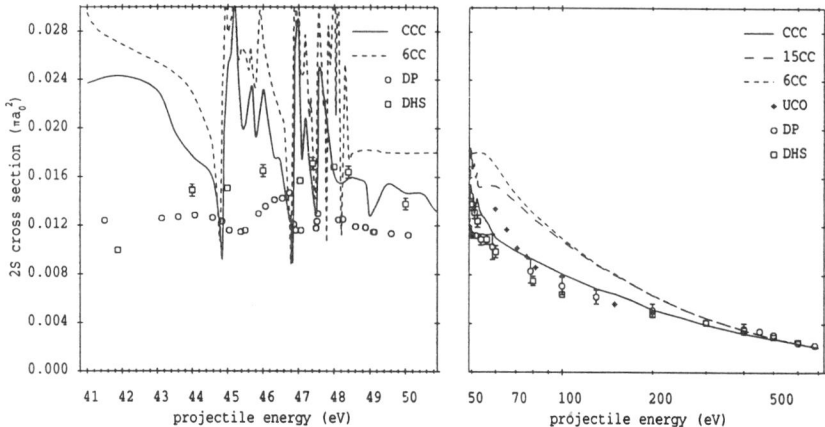

FIG. 5. Direct 2S excitation cross-section from the ground state of He$^+$ by electron impact. The CCC, 15CC, and 6CC results are due to Bray et al. (1993). The calculations of Unnikrishnan et al. (1991) are denoted by UCO. The measurements of Dolder and Peart (1973) and those of Dance et al. (1966) are denoted by DP and DHS, respectively. See the text for the description of the calculations.

We are unable to suggest any reason for the discrepancy with experimentation in the near-threshold region. The problem is a very long-standing one. For a detailed discussion surrounding these issues, see Seaton (1975).

B. IONIZATION

The calculations that were used to generate the 2S cross-section also yield the total ionization cross-section. This is presented in Fig. 6 and is compared with the measurements of Peart et al. (1969) and Defrance et al. (1981a). We find excellent quantitative agreement between the CCC theory and both experiments. On the other hand, the UCO calculation, although quite good for the 2S cross-section, is a factor of 2 larger. This shows that often even an approximate treatment of the continuum is sufficient when calculating discrete excitations. The quality of the agreement of the CCC theory with experimentation here shows, once more, that the target continuum may be adequately treated by the use of L^2 states.

VI. Electron–Sodium Scattering

We now look at applications of the CCC method to electron scattering on targets for which the frozen-core Hartree–Fock approximation is made for the treatment of the core electrons. As soon as one makes this approxi-

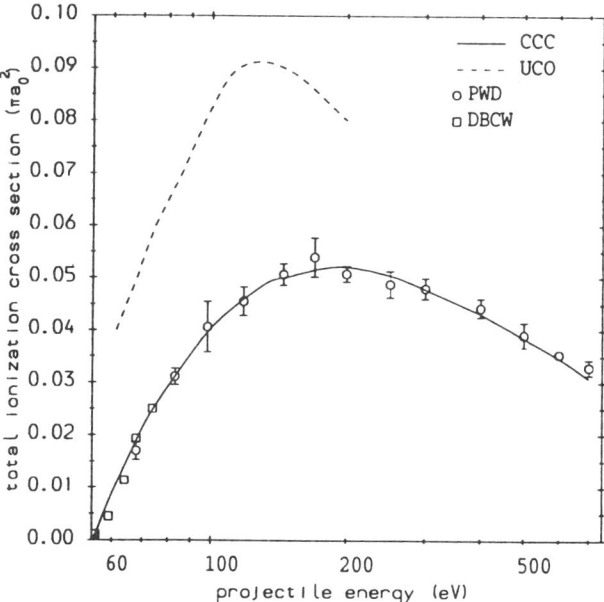

FIG. 6. Total ionization cross-section of the ground state of He$^+$ by electron impact. The calculations of Unnikrishnan et al. (1991) are denoted by UCO. The measurements of Peart et al. (1969) and those of Defrance et al. (1981a) are denoted by PWD and DBCW, respectively. The convergent close-coupling results (Bray et al., 1993) are denoted by CCC.

mation it should be understood that quantitative agreement with experimentation is not necessarily guaranteed. The quality of this approximation depends on the nature of the application. For example, with this approximation the cross-section for double ionization by electron impact is identically zero. This is because no allowance for core excitation, other than virtual, is made. If there is poor agreement with accurate experimental data we may be confident that this is due to the poor treatment of the core electrons rather than that of the valence electron. Having made this point, we find that one of the more spectacular successes of the CCC method has been for the calculation of electron–sodium scattering, where as many as 10 electrons are approximated by a single nonlocal potential.

There is an extensive body of data for electron–sodium scattering on which to test the frozen-core CCC model. The electron–atom scattering community is particularly well served by the works of McClelland et al. (1989), Scholten et al. (1991), Kelley et al. (1991), Lorentz et al. (1991), and McClelland et al. (1992). In the energy range from 1 to 54.4 eV they measured the ratio of triplet-to-singlet differential cross-sections for both the elastic and the 3P excitation channels. For the excitation channel, the angular momentum transferred to the atom perpendicular to the scattering plane $L_\perp^S$ has been measured for both singlet ($S = 0$) and triplet ($S = 1$) spin states. Apart from an overall normalization factor, the absolute differential cross-section, the measurements provide the magnitudes of the spin- and magnetic-sublevel-dependent scattering amplitudes for the two channels under consideration. The resolution of spin provides a more stringent test of scattering theories. For electron scattering on hydrogenic targets the singlet and triplet amplitudes result from independent calculations, and it is helpful to have data to test them individually. The design of the experiment is of such quality that the "ratio" data, with only marginal statistical error, were gathered up to very large scattering angles. In light of the previously mentioned difficulties in reconciling theory and experimentation for the angular correlation parameters at 54.4 eV with the hydrogen target, the highly accurate correlation measurements have proved to be very helpful in confirming the accuracy of the CCC method.

A comprehensive application of the CCC method to electron–sodium scattering may be found in Bray (1994c). Here we summarize the more significant results from this work.

A. Spin Asymmetries and $L_\perp^S$

The spin asymmetry A is defined in terms of the ratio r of triplet-to-singlet cross-sections by $A = (1 - r)/(1 + 3r)$. It always stays finite, taking the value $A = 1$ when the triplet cross-section is zero and $A = -\frac{1}{3}$ when the singlet cross-section is zero. For the elastic channel, A or r is sufficient in testing the magnitudes of the singlet and triplet amplitudes because normalization of theory is a relatively trivial aspect, and most sophisticated calculations give similar spin-averaged cross-sections. Agreement with the ratio of the magnitudes of the two amplitudes essentially implies correct theoretical magnitudes for both and thus the differential cross-section. Similarly for the 3P channel, for which there are four independent amplitudes if there is excellent agreement with A, $L_\perp^0$, and $L_\perp^1$, it can be safely assumed that the theoretical differential cross-section will be accurate as well. For various relations among these quantities, the scattering amplitudes, and angular correlation parameters, see Bray (1994c) and references therein.

In Fig. 7 are shown results for a projectile energy of 10 eV, which is about 5 eV above the ionization threshold. The CCC theory is compared with the previous best model which was the CCO method of Bray and McCarthy (1993), discussed in Section II.H. Also given is a 15-state CC theory that is convergent in the subspace of discrete eigenstates. Comparison of CC with CCC or CCO indicates the importance of including the continuum. From the figure it is clear that the CCC and CCO theories give an excellent description of the spin asymmetries and $L_\perp^1$. The CC theory is not too bad for the asymmetries and is also in excellent agreement with $L_\perp^1$. However, for the $L_\perp^0$ it is only the CCC theory that agrees with experimentation at all

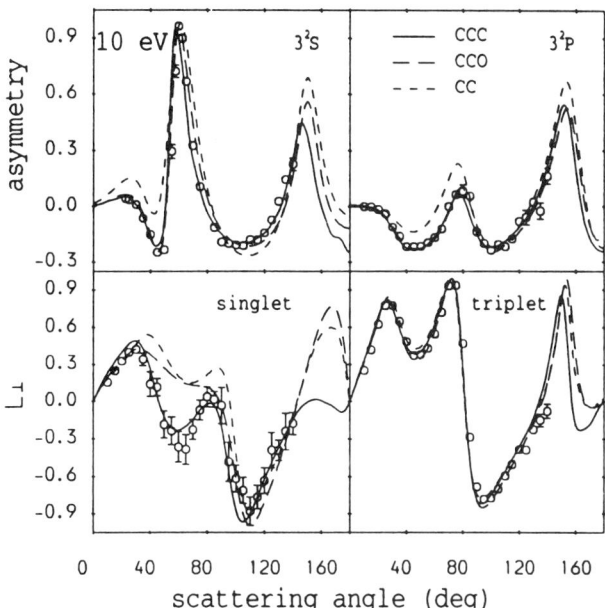

FIG. 7. Elastic and inelastic spin asymmetry and singlet and triplet $L_\perp$ for electron scattering on sodium at 10 eV. The theory is from Bray (1994c), and the measurements are summarized in Kelley et al. (1991).

scattering angles. This confirms the importance of treating continuum effects accurately. These conclusions are possible only due to the existence of such high-precision measurements and indicate just how difficult the e–Na scattering problem of 3P excitation can be. The CCC results suggest that any theory that does not allow for correct coupling within the continuum will not be able to obtain correct $L_\perp^0$ for this particular system.

The effect of the treatment of the continuum is even more apparent at the projectile energy of 20 eV, shown in Fig. 8. Here both the CCC and the CCO theories agree with the measurements, the former being a little better at backward angles for the spin asymmetries. Most noteworthy is how poor the CC theory is in describing the asymmetries. Bray (1994b) demonstrated using the CCO theory that this was more likely to be due to the lack of allowance for continuum-electron flux in the close-coupling formalism, rather than significant intermediate continuum excitation in the experiment.

The earlier work of Bray (1994c) also shows fine quantitative agreement with the measurements of Scholten et al. (1993) for the reduced Stokes parameters P_1, P_2, and P_3, which are trivially related to the λ, R, and I parameters. As all three CC, CCO, and CCC theories yield similar results they are not presented here. Furthermore, the CCC theory is in good agreement with all of the spin-resolved measurements at the other energies

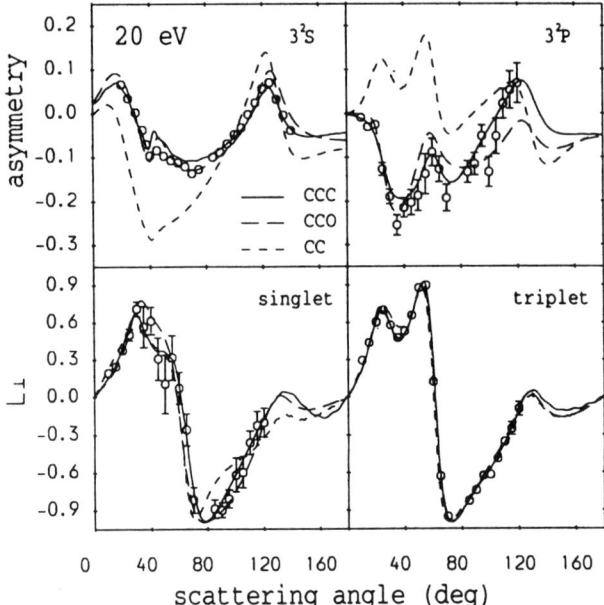

FIG. 8. Elastic and inelastic spin asymmetry and singlet and triplet $L_\perp$ for electron scattering on sodium at 20 eV. The theory is from Bray (1994c), and the measurements are summarized in Kelley et al. (1991).

(Bray, 1994c). The significance of this fine agreement with experimentation for angular correlations in e–Na scattering should not be overlooked. The statistical errors are much smaller than for the corresponding hydrogen correlations. We can see no theoretical reason why the quality of agreement with hydrogen should be different. In fact we take the view that the success of the CCC method for e–Na scattering is due to the correct treatment of the e–H scattering system.

B. DIFFERENTIAL CROSS-SECTIONS

The differential cross-sections are a less strict test of theory models than are angular correlations or spin asymmetries. There is substantial agreement with all close-coupling models of reasonable quality. Unfortunately, this is in direct contrast to experimentation, in which absolute measurements are very difficult, particularly when the cross-section falls many orders of magnitude from the forward to the backward angles. There have been many attempts to measure electron–sodium scattering differential cross-sections with results varying by as much as an order of magnitude at the backward angles (Buckman and Teubner, 1979; Srivastava and Vušković, 1980;

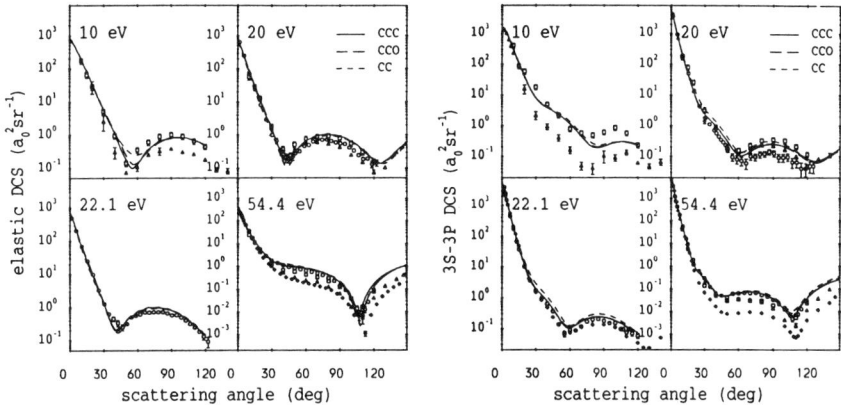

FIG. 9. The electron–sodium scattering differential cross-sections at a range of projectile energies. The theory is due to Bray (1994c) and Bray and McCarthy (1993). The measurements of Lorentz and Miller (1989) are denoted by ○. Those of Srivastava and Vušković (1980) are denoted by □. The measurements of Allen et al. (1987), Teubner et al. (1986), and Buckman and Teubner (1979) are denoted by ◇. The results of Marinković et al. (1992) are denoted by △. All measurements have been normalized to the theory. Error bars are plotted only if they are larger than the size of the symbol denoting the experiment.

Teubner et al., 1986; Allen et al., 1987; Lorentz and Miller, 1989; Marinković et al., 1992). At the forward angles the shapes and magnitudes are satisfactory (see also Jiang et al., 1992). Thus for the purpose of detailed testing of scatering theories, the measurements of the differential cross-sections are less useful than those discussed in the previous section.

In Fig. 9 some of the available measurements for the differential cross-sections for the elastic and $3P$ excitation are shown at a range of energies. The theories all yield similar results. The measurements plotted in the figure have been normalized to the theoretical integrated cross-section, in order to make the variation in shape more apparent at the large scattering angles. A set of integrated cross-sections has been given in Bray (1994c).

C. Ionization and Total Cross-Sections

We now consider the total ionization σ_i and total σ_t cross-sections. The total cross-section may be obtained either by summing the individual σ^{nl} cross-sections for states included in the calculation or by using the unitarity of the close-coupling formalism via the optical theorem. The total ionization cross-section may be obtained by subtracting the cross-sections corresponding to states with negative energies from the total cross-section. In other words we have $\sigma_t = \sigma_i + \sigma_{nb}$, where σ_{nb} is the nonbreakup cross-section.

Each of these may be subdivided into contributions for each l, i.e., $\sigma_t^l = \sigma_i^l + \sigma_{nb}^l$.

In the study of convergence, as in Section IV.B, our computational requirements for testing convergence with increasing N_l given a particular l_{max} rapidly escalate with increasing l_{max}. This is because the maximum number of channels in the CCC calculation generated by a particular N_l is $(l + 1)N_l$. Concentrating on convergence with increasing l_{max}, we would expect that, having obtained convergence in $\sigma_{nb,i,t}$, $\sigma_{nb,i,t}^{l_{max}} \ll \sigma_{nb,i,t}$. That is, if convergence with increasing l_{max} has been demonstrated, we would expect the sequence of partial contributions $\sigma_{nb,i,t}^l$ for $0 \leq l \leq l_{max}$ to be the smallest for $l = l_{max}$. We found (Bray, 1994a) that this is the case for the $\sigma_{nb,t}$, but not σ_i. In Fig. 10 we give the results for electron scattering from the ground and the $3P$ states of sodium. Thick and thin lines indicate that the result comes from a calculation having $l_{max} = 3$ and $l_{max} = 2$, respectively. Partial contributions for each l are denoted by broken lines, with the shorter dashes corresponding to larger l. For the $\sigma_{nb,t}$, convergence is as one might expect, with the shorter-dashed lines being of the smallest magnitude; but for the ionization case, in which convergence is within the difference between the two solid lines, $\sigma_i^{l_{max}}$ is often by far the largest contribution to σ_i.

It is the unitarity of the close-coupling formalism that allows us to obtain relatively convergent total ionization cross-sections without requiring that convergence be obtained for the individual σ_i^l. Via the optical theorem one is guaranteed convergence in the total cross-section once convergence in the elastic amplitudes has been established. This may be achieved with relatively small l_{max}. The nonbreakup cross-section is dominated by elastic and excitation scattering to the first few low-l states and hence may also be obtained accurately with small l_{max}. Therefore converged ionization cross-sections can be obtained with relatively small l_{max}, even though this does not mean convergence within each σ_i^l. It should be mentioned that for the case of ionization from hydrogen, the convergence of each partial ionization cross-section as l_{max} is increased is more evident. Future ionization calculations, when we have even better computers, will be extended to larger l_{max} to confirm that this will also be the case for sodium.

Since publication of Bray (1994a), we have been provided (Johnston, 1983; Johnston and Burrow, 1994) with a set of measurements of the total ionization cross-section for scattering from the ground state. These are in better agreement with our results, compared with the older measurements which are a factor of 2 larger. As these measurements have not yet been published* we describe here briefly the experimental setup. The apparatus (Johnston and Burrow, 1983) incorporates a magnetically collimated electron beam crossed with a sodium beam. Two modes of operation were used, permitting the observation of either the production of positive ions or the excitation function of the first excited state of sodium near threshold, using

---

*Now published: Johnston, A. R. and Burrow, P. D., (1995). *Phys. Rev. A* **51**, R1735.

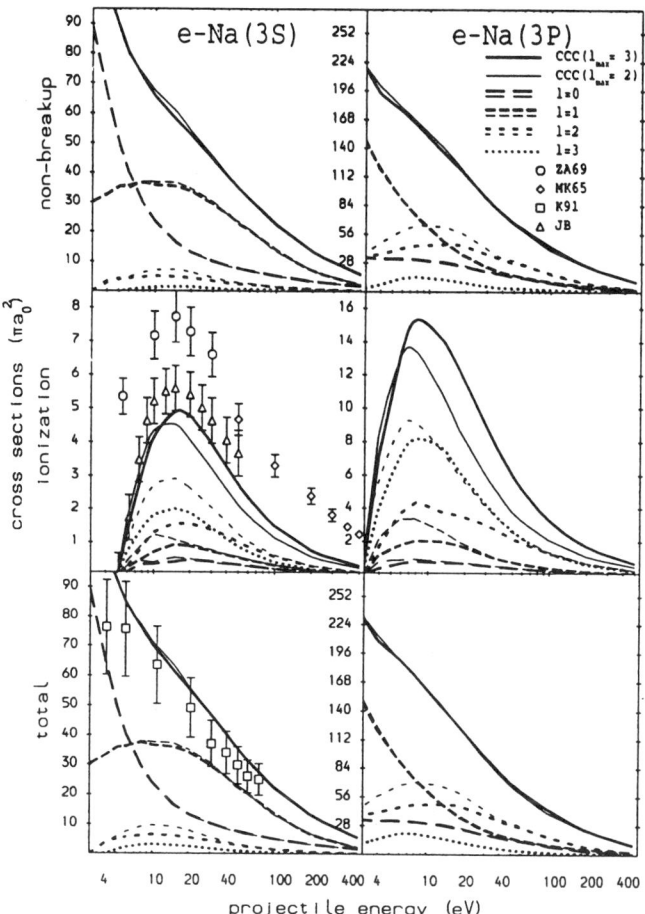

FIG. 10. The total σ_t, nonbreakup σ_{nb}, and total ionization σ_i cross-sections for the electron impact of the sodium 3S or 3P states calculated using the CCC method (Bray, 1994a). The thick and thin lines correspond to calculations having target states with maximum orbital angular momentum $l_{max} = 3$ and $l_{max} = 2$, respectively. The solid lines are the results of adding the partial contributions $\sigma_{nb,i,t}^l$, which are denoted by the broken lines. The shorter dashes denote the larger l contributions. The measurements denoted by ZA69, MK65, and K91 are from McFarland and Kinney (1965), Zapeschonyi and Aleksakhin (1969), and Kwan et al. (1991), respectively. The measurements denoted by JB are as yet unpublished (Johnston, 1983; Johnston and Burrow, 1994). See the text for a short description of the experimental technique.

the trapped-electron method (Schulz, 1958a, b). The relative values of these two cross-sections have been measured, and, by normalizing the excitation function to the theoretical results of Moores and Norcross (1972), an absolute cross-section for positive ionization was determined.

These latest measurements indicate that the CCC method is also able to provide reasonably accurate total-ionization cross-sections for the sodium target. Finally it is worthwhile reiterating that due to the treatment of the core by the frozen-core Hartree–Fock model we have no way of dealing with either autoexcitation or double (or more) ionization. In particular, direct ionization of the core electrons is not possible within the current CCC framework. These effects become significant at larger energies and have been estimated in the Born approximation (McGuire, 1971).

VII. Concluding Remarks

We have concentrated on providing the details of the CCC method for electron scattering on hydrogenic targets and have demonstrated its general applicability to describe elastic, inelastic, and ionization collisions in a unified formalism. Where applicable we have compared its results with the most sophisticated comparable calculations. Because of the limitations of space we have not carried out an exhaustive appraisal of the other methods. For a review concentrating on spin-resolved phenomena, see Andersen et al. (1995).

We are confident in the basic premises of the CCC approach to electron–atom/ion scattering. We believe that we are able to solve the full nonrelativistic electron–hydrogen scattering problem accurately. An exciting aspect of the applications has been the excellent agreement the method gives with experimentation, in which treatment of the target continuum is essential, and it is a clear indication that perhaps theory is now able to master the detailed inclusion of target continuum channels. This has been one area in which other sophisticated methods have not done as well to date. It is our opinion that the CCC method is able to yield accurate results for electron scattering on hydrogenic targets from 0 to, say, 1 keV, for those transitions which are dominated by valence-electron excitation.

There still remain a few significant discrepancies between theory and experimentation. One is the angular correlations of the $2p$ state of atomic hydrogen at backward angles, and another is the excitation of the He^+ ion from the ground state to the $2s$ state, for which experimentation is a factor of 2 lower at the threshold for excitation. We do not consider that these discrepancies invalidate our formalism. Perhaps it is timely to consider further experiments in these two problem areas. We base this suggestion on the fact that the new body of much more detailed data for correlations in the sodium target reveals no discrepancies. There appears to be no theoretical argument that can be advanced to explain why the heavier target can be described well and the lighter one, poorly. The existing experiments are at least a decade old, and new data would be welcomed by the atomic-physics community.

A number of issues relating to the use of the CCC method for calculating $(e, 2e)$ cross-sections are still under investigation. For example, it is not clear what is the formally correct way of symmetrizing the singly differential CCC ionization cross-sections. Furthermore, the applications to the triply differential cross-sections involved no symmetrization and yielded correct absolute results. Application of the CCC method to the low energy $(e, 2e)$ measurements should be very informative.

The CCC method has been extended to the helium target with great success (Bray et al., 1994a, 1995). Excellent agreement with measured differential cross-sections and angular correlation parameters has been obtained. This gives us further confidence that the CCC results are just as accurate for e–H scattering at 54.4 eV. It is our aim to continue the extension of the method to helium-like targets, followed by application to inert gases.

Acknowledgments

We thank Paul Burrow for communicating data prior to publication and for the explanation of the experimental technique. We also thank Chris Greene, Don Madison, Tim Scholz, Yudong Wang, and Shinichi Watanabe for helpful communication. The sudden passing of Joe Callaway, whose work has in many ways been of benefit to us, is deeply regretted. The support of the Australian Research Council is gratefully acknowledged.

References

Aggarwal, K. M., Berrington, K. A., Burke, P. G., Kingston, A. E., and Pathak, A. (1991a). *J. Phys. B* **24**, 1385.
Aggarwal, K. M., Berrington, K. A., Kingston, A. E., and Pathak, A. (1991b). *J. Phys. B* **24**, 1757.
Allen, L. J., Brunger, M. J., McCarthy, I. E., and Teubner, P. J. O. (1987). *J. Phys. B* **20**, 4861.
Allen, L. J., Bray, I., and McCarthy, I. E. (1988). *Phys. Rev. A* **37**, 49.
Andersen, N., Bartschat, K., and Broad, J. T. (1995). *Phys. Rep.* to be published.
Anderson, E., Bai, Z., Bischof, C., Demmel, J., Dongarra, J., Croz, J. D., Greenbaum, A., Hammarling, S., McKenney, A., Ostrouchov, S., and Sorensen, D. (1992). *LAPACK User's Guide*. Soc. Ind. Appl. Math., Philadelphia.
Bhatia, A. K., Schneider, B. I., and Temkin, A. (1993) *Phys. Rev. Lett.* **70**, 1936.
Botero, J., and Shertzer, J. (1992). *Phys. Rev. A* **46**, R1155.
Bransden, B. H. (1983). *Atomic Collision Theory*, Lecture Notes and Supplements in Physics, 2nd Ed., Benjamin/Cummings, New York.
Bransden, B. H., and Stelbovics, A. T. (1984). *J. Phys. B* **17**, 1877.
Bransden, B. H., Noble, C. J., and Hewitt, R. N. (1993). *J. Phys. B* **26**, 2487.
Brauner, M., Briggs, J. S., and Klar, H. (1989). *J. Phys. B* **22**, 2265.
Brauner, M., Briggs, J. S., Klar, H., Broad, J. T., Rösel, T., Jung, K., and Ehrhardt, H. (1991). *J. Phys. B* **24**, 657.
Bray, I. (1992). *Phys. Rev. Lett.* **69**, 1908.

Bray, I. (1994a). *Phys. Rev. Lett.* **73**, 1088.
Bray, I. (1994b). *Z. Phys. D* **30**, 99.
Bray, I. (1994c). *Phys. Rev. A* **49**, 1066.
Bray, I., and McCarthy, I. E. (1993). *Phys. Rev. A* **47**, 317.
Bray, I., and Stelbovics, A. T. (1992a). *Phys. Rev. A* **46**, 6995.
Bray, I., and Stelbovics, A. T. (1992b). *Phys. Rev. Lett.* **69**, 53.
Bray, I., and Stelbovics, A. T. (1993). *Phys. Rev. Lett.* **70**, 746
Bray, I., and Stelbovics, A. T. (1995). *Comp. Phys. Commun.* **85**, 1.
Bray, I., Madison, D. H., and McCarthy, I. E. (1990). *Phys. Rev. A* **41**, 5916.
Bray, I., Konovalov, D. A., and McCarthy, I. E. (1991a). *Phys. Rev. A* **43**, 1301.
Bray, I., Konovalov, D. A., and McCarthy, I. E. (1991b). *Phys. Rev. A* **43**, 5878.
Bray, I., Konovalov, D. A., and McCarthy, I. E. (1991c). *Phys. Rev. A* **44**, 5586.
Bray, I., McCarthy, I. E., Wigley, J., and Stelbovics, A. T. (1993). *J. Phys. B* **26**, L831.
Bray, I., Fursa, D. V., and McCarthy, I. E. (1994a). *J. Phys. B* **27**, L421.
Bray, I., Konovalov, D. A., McCarthy, I. E., and Stelbovics, A. T. (1994b). *Phys. Rev. A* **50**, R2818.
Bray, I., Fursa, D. V., and McCarthy, I. E. (1995). *Phys. Rev. A* **51**, 500.
Broad, J. T. (1978). *Phys. Rev. A* **18**, 1012.
Buckman, S. J., and Teubner, P. J. O. (1979). *J. Phys. B* **12**, 1741.
Burke, P. G. (1987). *Atomic Physics*, Vol. 10, p. 243, Elsevier, Amsterdam.
Burke, P. G., and Robb, W. D. (1975). *Adv. At. Mol. Phys.* **11**, 143.
Burke, P. G., Berrington, K. A., and Sukumar, C. V. (1981). *J. Phys. B* **14**, 289.
Burke, P. G., Noble, C. J. and Scott, M. P. (1987). *Proc. R. Soc. London, Ser. A.* **410**, 289.
Callaway, J. (1985). *Phys. Rev. A* **32**, 775.
Callaway, J., and Oza, D. H. (1979). *Phys. Lett.* **72A**, 207.
Callaway, J., and Oza, D. H. (1984). *Phys. Rev. A* **29**, 2416.
Chernysheva, L. V., Cherepkov, N. A., and Radojevic, V. (1976). *Comp. Phys. Commun.* **11**, 57.
Crowe, D. M., Guo, X. Q., Lubell, M. S., Slevin, J., and Eminyan, M. (1990). *J. Phys. B* **23**, L325.
Curran, E. P., and Walters, H. R. J. (1987). *J. Phys. B* **20**, 337.
Dance, D. F., Harrison, M. F. A., and Smith, A. C. H. (1966). *Proc. R. Soc. London, Ser. A* **290**, 74.
Defrance, P., Brouillard, F., Claeys, W., and Van Wassenhove, G. (1981a). *J. Phys. B* **14**, 103.
Defrance, P., Claeys, W., Cornet, A., and Poulaert, G. (1981b). *J. Phys. B* **14**, 111.
Dolder, K. T., and Peart, B. (1973). *J. Phys. B* **6**, 2415.
Ehrhardt, H., Jung, K., Knoth, G., and Schlemmer, P. (1986). *Z. Phys. D* **1**, 3.
Feshbach, H. (1962). *Ann. Phys. (N.Y.)* **19**, 287.
Fletcher, G. D., Alguard, M. J., Gay, T. J., Wainwright, P. F., Lubell, M. S., Raith, W., and Hughes, V. W. (1985). *Phys. Rev. A* **31**, 2854.
Fon, W. C., Aggarwal, K. M., and Ratnavelu, K. (1992). *J. Phys. B.* **25**, 2625.
Gell-Mann, M., and Goldberger, M. L. (1953). *Phys. Rev.* **91**, 398.
Green, C. H. (1988). In Briggs, J. S., Kleinpoppen, H., and Lutz, H. O., eds. *Fundamental Processes of Atomic Dynamics* p. 105, Plenum, New York.
Heller, E. J., and Yamani, H. A. (1974a). *Phys. Rev. A* **9**, 1209.
Heller, E. J., and Yamani, H. A. (1974b). *Phys. Rev. A* **9**, 1201.
Jiang, T. Y., Ying, C. H., Vušković, L., and Bederson, B. (1992). *Phys. Rev. A* **42**, 3852.
Johnston, A. R. (1983). Ph. D. Thesis, Univ. of Nebraska, Lincoln.
Johnston, A. R., and Burrow, P. D. (1983). *J. Phys. B* **16**, 613.
Johnston, A. R., and Burrow, P. D. (1994). Private communication.
Jones, S., Madison, D. H., Franz, A., and Altick, P. L. (1989). *J. Phys. B* **22**, 2265.
Kato, D., and Watanabe, S. (1995). *Phys. Rev. Lett.* **74**, 2443.
Kelley, M. H., McClelland, J. J., Lorentz, S. R., Scholten, R. E., and Celotta, R. J. (1991). *Correlation and Polarization in Electronic and Atomic Collisions and (e, 2e) Reactions*, p. 23, Inst. Phys., Adelaide.

Kingston, A. E., and Walters, H. R. J. (1980). *J. Phys. B* **13**, 4633.
Konovalov, D. A. (1994). *J. Phys. B* **27**, 5551.
Konovalov, D. A., and McCarthy, I. E. (1994a). *J. Phys. B* **27**, 2407.
Konovalov, D. A., and McCarthy, I. E. (1994b). *J. Phys. B* **27**, L407.
Konovalov, D. A., Bray, I., and McCarthy, I. E. (1994). *J. Phys. B* **27**, L413.
Kwan, C. K., Kauppila, W. E., Lukaszew, R. A., Parikh, S. P., Stein, T. S., Wan, Y. J., and Dababneh, M. S. (1991). *Phys. Rev. A* **44**, 1620.
Kyle, H. L., and Temkin, A. (1964). *Phys. Rev.* **129**, A600.
Lorentz, S. R., and Miller, T. M. (1989). *Abst. Int. Conf. Phys. Electron. At. Collisions*, 16th, Amsterdam, p. 198, North-Holland Publ., New York.
Lorentz, S. R., Scholten, R. E., McClelland, J. J., Kelley, M. H., and Celotta, R. J. (1991). *Phys. Rev. Lett.* **67**, 3761.
Madison, D. H. (1984). *Phys. Rev. Lett.* **53**, 42.
Madison, D. H., Bray, I., and McCarthy, I. E. (1990). *Phys. Rev. Lett.* **64**, 2265.
Madison, D. H., Bray, I., and McCarthy, I. E. (1991). *J. Phys. B* **24**, 3861.
Madison, D. H., Bartshat, K., and McEachran, R. P. (1992). *J. Phys. B* **25**, 5199.
Marinković, B., Pejčev, V., Filipović, D., Čadež, I., and Vušković, L. (1992). *J. Phys. B* **25**, 5179.
McCarthy, I. E., and Stelbovics, A. T. (1983). *Phys. Rev. A* **28**, 2693.
McCarthy, I. E., and Weigold, E. (1990a). *Adv. At. Mol. Phys.* **27**, 201.
McCarthy, I. E., and Weigold, E. (1990b). *Adv. At. Mol. Phys.* **27**, 165.
McClelland, J. J., Kelley, M. H., and Celotta, R. J. (1989). *Phys. Rev. A* **40**, 2321.
McClelland, J. J., Lorentz, S. R., Scholten, R. E., Kelley, M. H., and Celotta, R. J. (1992). *Phys. Rev. A* **46**, 6079.
McEachran, R. P., and Cohen, M. (1983). *J. Phys. B* **16**, 3125.
McEachran, R. P., Stauffer, A. D., and Greita, S. (1979). *J. Phys. B* **12**, 3119.
McFarland, R. H., and Kinney, J. D. (1965). *Phys. Rev. A* **137**, 1058.
McGuire, E. J. (1971). *Phys. Rev. A* **3**, 267.
Moores, D. L., and Norcross, D. W. (1972). *J. Phys. B* **5**, 1482.
Odgers, B. R., Scott, M. P., and Burke, P. G. (1994). *J. Phys. B* **27**, 2577.
Peart, B., Walton, D. S., and Dolder, K. T. (1969). *J. Phys. B* **2**, 1347.
Poet, R. (1978). *J. Phys. B* **11**, 3081.
Poet, R. (1980). *J. Phys. B* **13**, 2995.
Poet, R. (1981). *J. Phys. B* **14**, 91.
Robicheaux, F., Wood, R. P., and Greene, C. H. (1994). *Phys. Rev. A* **49**, 1866.
Rudge, M. H. R. (1992). *J. Phys. B* **25**, L117.
Sadeghpour, H. R. (1992). *J. Phys. B* **25**, L29.
Scholten, R. E., Lorentz, S. R., McClelland, J. J., Kelley, M. H., and Celotta, R. J. (1991). *J. Phys. B* **24**, L653.
Scholten, R. E., Shen, G. F., and Teubner, P. J. O. (1993). *J. Phys. B* **26**, 987.
Scholz, T. T. (1991). *J. Phys. B* **24**, 2127.
Scholz, T. T. (1993). *Comp. Phys. Commun.* **74**, 256.
Scholz, T. T. Walters, H. R. J., and Burke, P. G. (1990). *J. Phys. B* **23**, L467.
Scholz, T. T., Walters, H. R. J., Burke, P. G., and Scott, M. P. (1991). *J. Phys. B* **24**, 2097.
Schulz, G. J. (1958a). *Phys. Rev.* **112**, 150.
Schulz, G. J. (1958b). *Phys. Rev.* **116**, 1141.
Scott, M. P., Scholz, T. T., Walters, H. R. J., and Burke, P. G. (1989). *J. Phys. B* **22**, 3055.
Scott, M. P., and Burke, P. G. (1993). *J. Phys. B* **26**, L191.
Scott, M. P., Odgers, B. R., and Burke, P. G. (1993). *J. Phys. B* **26**, L827.
Seaton, M. J. (1975). *Adv. At. Mol. Phys.* **11**, 83.
Shah, M. B., Elliot, D. S., and Gilbody, H. B. (1987). *J. Phys. B* **20**, 3501.
Shertzer, J., and Botero, J. (1994). *Phys. Rev. A* **49**, 3673.
Slim, H. A., and Stelbovics, A. T. (1987). *J. Phys. B* **20**, L211.
Slim, H. A., and Stelbovics, A. T. (1988). *J. Phys. B* **21**, 1519.

Srivastava, S. K., and Vušković, L. (1980). *J. Phys. B* **13**, 2633.
Stelbovics, A. T. (1990). *Phys. Rev. A* **41**, 2536.
Tang, J., Watanabe, S., and Matsuzawa, M. (1992). *Phys. Rev. A* **46**, 2437.
Temkin, A. (1962). *Phys. Rev.* **126**, 130.
Teubner, P. J. O., Riley, J. L., Brunger, M. J., and Buckman, S. J. (1986). *J. Phys. B* **19**, 3313.
Unnikrishnan, K., Callaway, J., and Oza, D. H. (1991). *Phys. Rev. A* **43**, 5966.
van Wyngaarden, W. L., and Walters, H. R. J. (1986). *J. Phys. B* **19**, 929.
Wait, R., and Mitchell, A. R. (1985). *Finite Element Analysis and Applications*, 1st Ed. Wiley, New York.
Wang, Y. D., and Callaway, J. (1993). *Phys. Rev. A* **48**, 2058.
Wang, Y. D., and Callaway, J. (1994). *Phys. Rev. A* **50**, 2327.
Watanabe, S., Hosada, Y., and Kato, D. (1993). *J. Phys. B* **26**, L495.
Weigold, E., Frost, L., and Nygaard, K. J. (1979). *Phys. Rev. A* **21**, 1950.
Wigner, E. P., and Eisenbud, L. (1947). *Phys. Rev.* **72**, 606.
Williams, J. F. (1981). *J. Phys. B* **14**, 1197.
Williams, J. F. (1986). *Aust. J. Phys.* **39**, 621.
Yamani, H. A., and Reinhardt, W. P. (1975). *Phys. Rev. A* **11**, 1144.
Zapesochnyi, I. P., and Aleksakhin, I. S. (1969). *Sov. Phys.*—JETP (Engl. Transl.) **28**, 41.

RELATIVISTIC CALCULATIONS OF TRANSITION AMPLITUDES IN THE HELIUM ISOELECTRONIC SEQUENCE

W. R. JOHNSON, D. R. PLANTE, and J. SAPIRSTEIN

Department of Physics, University of Notre Dame, Notre Dame, Indiana

I. Introduction	255
II. No-Pair Transition Amplitudes	258
A. No-Pair Hamiltonian	258
B. Electromagnetic Interaction Hamiltonian	261
C. Multipole Potentials and Transition Amplitudes	262
D. A Commutator Identity	265
E. Numerical Calculations and Tests of Transition Amplitudes	267
III. S-Matrix Theory for Decay Rates	270
IV. Application of Perturbation Theory to Helium-like Ions	276
A. Basic Equations	276
B. Angular Reduction	279
C. Negative-Energy Components of Second-Order Amplitudes	282
V. Results and Comparisons	286
A. Comparison with Other Theories	286
B. Comparison with Experiment	291
C. Compilation of Transition Rates	294
D. Future Prospects	295
Appendix: Useful Identities	326
Acknowledgments	327
References	327

I. Introduction

The helium atom and helium-like ions occupy a special position in atomic physics as the simplest multielectron systems. Unlike atoms and ions with many electrons, the fact that only two electrons are present allows calculations of extremely high accuracy to be carried out in the nonrelativistic case. At low Z, the leading relativistic and QED corrections can be calculated as perturbations. However, at higher Z it is desirable to start from a completely relativistic formalism. A well-known problem of any such formalism is the correct treatment of negative-energy states, which, if included improperly, lead to the *continuum dissolution* problem discussed by Sucher [1]. This problem can be avoided by using the *no-pair* Hamiltonian [1,2], which excludes negative-energy states. In Ref. [3], configuration interaction (CI) techniques were used to carry out accurate calculations of energies of $n = 1$ and $n = 2$ states along the helium isoelectronic sequence, starting from the no-pair Hamiltonian. Similar calculations were carried out in Ref. [4] using iterative techniques. It is the purpose of this chapter to extend these calculations to evaluate transition amplitudes along the helium isoelectronic sequence. Specifically, we present highly accurate theoretical calculations of single-photon transition amplitudes and transition rates for 2–2 and 2–1 transitions in helium-like ions with charges in the range $Z = 2$–100. Amplitudes for the electric-dipole (E_1) transitions $2^1P_1 \to 1^1S_0$, $2^1P_1 \to 2^1S_0$, $2^3P_{0,1,2} \to 2^3S_1$, $2^1P_1 \to 2^3S_1$, and $2^3P_1 \to 1^1S_0$ are calculated in length form and in velocity form. Amplitudes for the magnetic-dipole (M_1) transitions $2^3S_1 \to 1^1S_0$ and the magnetic-quadrupole (M_2) transitions $2^3P_2 \to 1^1S_0$ are evaluated in velocity form.

Although the amplitudes are calculated using wave functions determined from previous work, there are several subtle points about the calculation that require particular care. First, the amplitudes, although calculated using exact eigenfunctions of the no-pair Hamiltonian, depend on the gauge of the electromagnetic field, as first noted in Ref. [5]. This gauge dependence is reflected in slight differences between length-form and velocity-form transition amplitudes. The gauge dependence is a direct consequence of the fact that negative-energy contributions are omitted in the no-pair Hamiltonian. We will derive a commutator identity expressing the difference between length- and velocity-form amplitudes in terms of the commutator of the many-body potential with the dipole operator. This commutator is nonvanishing in the no-pair theory because positive-energy projection operators surround the many-body potential. The numerically calculated amplitudes are found to satisfy the commutator identity precisely. In Section II, we describe the calculation of the transition amplitudes and give examples emphasizing the length-velocity differences.

Since transition amplitudes determined using wave functions from the no-pair Hamiltonian are inherently gauge dependent, it becomes important to evaluate the missing negative-energy contributions and to make appropriate corrections to the amplitudes. For this purpose, we turn to the perturbation expansion of the S-matrix in QED. The S-matrix is gauge invariant order by order. The gauge invariance of the first-order S-matrix $S^{(1)}$ follows from an elementary identity involving the one-electron Dirac Hamiltonian. There are, of course, no negative-energy contributions to $S^{(1)}$. Negative-energy contributions first appear in electron propagators in the third-order S-matrix $S^{(3)}$. A second subtle point is that one must add *derivative terms* [6] to the usual third-order diagrams to establish the gauge-invariance of $S^{(3)}$. The origin of these derivative terms is illustrated in Section III by a third-order QED calculation of transitions in hydrogen-like ions carried out using the Gell–Mann–Low formalism [7]. These terms play a dual role in the calculation. In addition to establishing gauge invariance, they also act to effectively change the phase space available for the decay. A standard problem faced when calculating decay rates with approximate energies and wave functions is whether to use the theoretical or experimental photon energy. We will show that the derivative terms act to modify the lowest-order theoretical energies to include higher-order perturbation theory corrections, so that a fundamental justification for the use of corrected theoretical energies (which agree quite precisely with experimental energies) is provided.

The structure of $S^{(3)}$ inferred from the example in Section III is used in Section IV to determine the first two terms in a gauge-invariant perturbation expansion of the transition amplitudes for helium-like systems. We set up the perturbation theory calculation in Section IV and give specific examples, concentrating particularly on the sensitive $2^3P_{0,2} \rightarrow 2^3S_1$ transitions. In these examples, contributions to the amplitudes from the positive-energy and negative-energy components of the propagator are determined separately. The positive- and negative-energy components of the amplitudes depend on gauge, but the sum of the positive- and negative-energy components is gauge independent. We find that, for $2^3P_0 \rightarrow 2^3S_1$ transitions, contributions from negative-energy terms to the velocity-form amplitude range up to 10% of the total amplitude for highly charged ions. Negative-energy contributions to the length-form amplitudes remain small (10^{-7}–10^{-8} atomic units (a.u.)) for all of the electric-dipole transitions studied, in agreement with the conclusions reached in Ref. [5]. The length-form amplitudes from the no-pair wave functions agree to within a few percent with the positive-energy components of the amplitudes from perturbation theory, the difference being primarily from omitted higher-order corrections in the perturbation expansion. Moreover, adding the negative-frequency components determined from perturbation theory to the no-pair amplitudes brings the velocity-form no-pair amplitudes into close agreement with the length-form amplitudes. Since there are negligible contributions to the length-form amplitudes from negative-energy states, we use the length-form no-pair amplitudes in our final tabulations for E_1 transitions.

For magnetic-multipole transitions, the no-pair amplitudes again agree at the 1% level of accuracy with the positive-energy contributions from perturbation theory. We account for the missing negative-energy contributions by adding the negative-energy contributions from perturbation theory to the no-pair amplitudes. Corrections to the wave functions from the Breit interaction significantly modify the negative-energy contributions, as found in an earlier calculation of the M_1 decay of helium-like argon [8]. We note that the relatively large negative-energy state effects were correctly included in earlier work on M_1 decays based on perturbation theory [9] and the RRPA [10]. However, because differential equation techniques were used in those works, which automatically include both positive- and negative-energy states, this feature of the calculation was not noted at that time.

In Section V, we present a summary of results, giving tables of amplitudes, line strengths, and rates for the 2–2 and 2–1 transitions. Where possible, our calculations are compared with previous calculations and with measurements. We believe that the current values for transition amplitudes and rates are as accurate as, or more accurate than, all previous values.

II. No-Pair Transition Amplitudes

A. No-Pair Hamiltonian

For a many-electron system, the no-pair Hamiltonian [1, 2] may be written $H = H_0 + V_C + V_B$, where H_0 is the unperturbed Hamiltonian given in second-quantized form by

$$H_0 = \sum_i \epsilon_i a_i^\dagger a_i. \qquad (1)$$

The quantity ϵ_i is an eigenvalue of the one-electron Dirac equation

$$h\phi_i = \epsilon_i \phi_i, \qquad (2)$$

with

$$h = c\boldsymbol{\alpha}\cdot\mathbf{p} + (\beta - 1)c^2 + V_{\text{nuc}}(r) + U(r). \qquad (3)$$

In Eq. (3), $V_{\text{nuc}}(r)$ is the Coulomb potential of the nucleus incorporating nuclear finite-size corrections and $U(r)$ is a local model potential that approximates the electron–electron interaction. We make two choices of $U(r)$ here, namely,

$$U(r) = \begin{cases} 0, & \text{Coulomb-field case} \\ v_0(1s, r), & \text{Hartree case,} \end{cases} \qquad (4)$$

where $v_0(1s, r)$ is the self-consistent potential of the ground state of the helium-like ion. The potential $U(r)$ is fixed by a ground-state HF calculation and treated as a local potential when solving the one-electron Dirac equation [Eq. (2)]. The index i in Eq. (2) represents the set of one-electron quantum numbers: the principal quantum number n_i, the angular momentum quantum number $\kappa_i[\kappa_i = \mp(j_i + \frac{1}{2})$ for $j_i = (l_i \pm \frac{1}{2})]$, and the magnetic quantum number m_i. The Coulomb interaction V_C is given by

$$V_C = \tfrac{1}{2} \sum_{ijkl} g_{ijkl} a_i^\dagger a_j^\dagger a_l a_k - \sum_{ij} U_{ij} a_i^\dagger a_j, \tag{5}$$

where the sum is over positive-energy states only. This restriction implements the positive-energy projection operators in Refs. [1,2]. The quantities g_{ijkl} in Eq. (5) are two-electron Coulomb matrix elements

$$g_{ijkl} = \left\langle ij \left| \frac{1}{r_{12}} \right| kl \right\rangle, \tag{6}$$

and

$$U_{ij} = \langle i | U(r) | j \rangle. \tag{7}$$

The Breit interaction V_B is given by

$$V_B = \tfrac{1}{2} \sum_{ijkl} b_{ijkl} a_i^\dagger a_j^\dagger a_l a_k, \tag{8}$$

where again the summation is restricted to positive-energy states. Here b_{ijkl} are two-electron matrix elements of the instantaneous Breit operator,

$$b_{ijkl} = -\left\langle ij \left| \frac{\boldsymbol{\alpha}_1 \cdot \boldsymbol{\alpha}_2 + \boldsymbol{\alpha}_1 \cdot \hat{\mathbf{r}}_{12} \boldsymbol{\alpha}_2 \cdot \hat{\mathbf{r}}_{12}}{2r_{12}} \right| kl \right\rangle. \tag{9}$$

Later, in our discussion of perturbation theory, we use the notation

$$v_{ijkl} = g_{ijkl} + b_{ijkl} \tag{10}$$

$$\tilde{v}_{ijkl} = v_{ijkl} - v_{ijlk}. \tag{11}$$

The no-pair Hamiltonian includes all corrections of order $\alpha^2 Z^2$ a.u., but does not include QED corrections of order $\alpha^3 Z^4$ a.u., such as the Lamb shift. Such corrections must be calculated separately as discussed, for example, in Ref. [4].

A two-electron state vector describing an atomic state with angular momentum J, M may be written

$$\Psi_{JM} = \sum_{i<j} c_{ij} \Phi_{ij}, \tag{12}$$

where the quantities c_{ij} are expansion coefficients and where the configuration state vectors Φ_{ij} are defined by

$$\Phi_{ij} = \eta_{ij} \sum_{m_i m_j} \langle j_i m_i, j_j m_j | JM \rangle a_i^\dagger a_j^\dagger | 0 \rangle, \tag{13}$$

with

$$\eta_{ij} = \begin{cases} 1, & i \neq j \\ 1/\sqrt{2}, & i = j. \end{cases} \tag{14}$$

The quantities c_{ij}, Φ_{ij}, and η_{ij} are independent of magnetic quantum numbers. Therefore, a particular state i is uniquely determined by n_i and κ_i. To construct a state of even or odd parity, one must require the sum $l_i + l_j$ to be either even or odd, respectively. From the symmetry properties of the Clebsch–Gordan coefficients, it can be shown that

$$\Phi_{ij} = (-1)^{j_i + j_j + J + 1} \Phi_{ji}. \tag{15}$$

This relation, in turn, implies that Φ_{ii} vanishes unless J is even. The wave-function normalization condition has the form

$$\langle \Psi_{JM} | \Psi_{JM} \rangle = \sum_{i<j} c_{ij}^2 = 1. \tag{16}$$

Substituting Ψ_{JM} into the Schrödinger equation $(H_0 + V)\Psi_{JM} = E\Psi_{JM}$, one obtains the following set of linear equations for the expansion coefficients c_{ij}

$$(\epsilon_i + \epsilon_j) c_{ij} + \sum_{k \leq l} \eta_{ij} V(ij; kl) \eta_{kl} c_{kl} = E c_{ij}. \tag{17}$$

The potential matrix in Eq. (17) is

$$V(ij; kl) = \sum_L (-1)^{j_i + j_k + L + J} \begin{Bmatrix} j_i & j_j & J \\ j_l & j_k & L \end{Bmatrix} X_L(ijkl)$$

$$+ \sum_L (-1)^{j_j + j_k + L} \begin{Bmatrix} j_i & j_j & J \\ j_k & j_l & L \end{Bmatrix} X_L(ijlk) - \delta_{\kappa_i \kappa_k} \delta_{jl} U_{ik}$$

$$- \delta_{\kappa_j \kappa_l} \delta_{ik} U_{jl} + (-1)^{j_i + j_k + J} [\delta_{\kappa_j \kappa_k} \delta_{il} U_{jk} + \delta_{\kappa_i \kappa_l} \delta_{jk} U_{il}], \tag{18}$$

where the quantities $X_L(ijkl)$ are given by

$$X_L(ijkl) = (-1)^L \langle \kappa_i \| C_L \| \kappa_k \rangle \langle \kappa_j \| C_L \| \kappa_l \rangle R_L(ijkl). \tag{19}$$

The quantities $\langle \kappa_i \| C_L \| \kappa_j \rangle$ are reduced matrix elements of normalized spherical harmonics, and the quantities $R_L(ijkl)$ are relativistic Slater integrals defined in Refs. [3,4]. For the case in which both the Coulomb and Breit interactions are included in the Hamiltonian,

$$X_L(ijkl) \rightarrow X_L(ijkl) + M_L(ijkl) + N_L(ijkl) + O_L(ijkl), \tag{20}$$

where $M_L(ijkl)$, $N_L(ijkl)$, and $O_L(ijkl)$ are the magnetic integrals defined in Ref. [11].

We can solve the eigenvalue problem in Eq. (17) to high accuracy using iterative techniques, as described in Ref. [4], or as a large-scale CI problem, using the methods given in Ref. [3].

B. Electromagnetic Interaction Hamiltonian

The interaction Hamiltonian for a single electron with an electromagnetic field described by the vector potential $\mathbf{A}(\mathbf{r}, t)$ and the scalar potential $\phi(\mathbf{r}, t)$ is

$$h_1(\mathbf{r}, t) = e\{-c\boldsymbol{\alpha} \cdot \mathbf{A}(\mathbf{r}, t) + \phi(\mathbf{r}, t)\}, \tag{21}$$

where $e = -|e|$ is the electron charge. If we assume that h_1 has time dependence $e^{-i\omega t}$ associated with an incoming photon field, we can make the replacement

$$h_1(\mathbf{r}, t) \to h_1(\mathbf{r}, \omega)e^{-i\omega t},$$

where

$$h_1(\mathbf{r}, \omega) = e\{-c\boldsymbol{\alpha} \cdot \mathbf{A}(\mathbf{r}, \omega) + \phi(\mathbf{r}, \omega)\}. \tag{22}$$

We will be considering gauge invariance of transition amplitudes in this chapter. We distinguish this kind of gauge invariance from the usual statement of the gauge invariance of QED. In that way of defining gauge invariance, the field theoretic Lagrangian remains invariant under the simultaneous transformation $\psi(x) \to e^{i\chi(x)}\psi(x)$, $A_\mu(x) \to A_\mu(x) - (1/e)\partial_\mu\chi(x)$, where the change in A_μ in the interaction part of the Lagrangian is compensated by the $i\partial_\mu$ part of the free part of the Lagrangian acting on the $e^{i\chi(x)}$ factor. Here, however, when we refer to gauge invariance, we mean that matrix elements of the interaction Hamiltonian remain unchanged when the above gauge transformation is made. The phase change of the electron fields in this case has no effect, but we will nevertheless show that the change of the Hamiltonian has a vanishing matrix element when taken between wave functions. We illustrate the present notion of gauge invariance for one-electron matrix elements in the following paragraph.

A gauge transformation,

$$\mathbf{A}(\mathbf{r}, \omega) \to \mathbf{A}'(\mathbf{r}, \omega) = \mathbf{A}(\mathbf{r}, \omega) + \nabla\chi(\mathbf{r}, \omega) \tag{23}$$

$$\phi(\mathbf{r}, \omega) \to \phi'(\mathbf{r}, \omega) = \phi(\mathbf{r}, \omega) + i\omega\chi(\mathbf{r}, \omega), \tag{24}$$

induces a change in the interaction Hamiltonian,

$$h_1(\mathbf{r}, \omega) \to h_1'(\mathbf{r}, \omega) = h_1(\mathbf{r}, \omega) + \Delta h_1(\mathbf{r}, \omega), \tag{25}$$

where

$$\Delta h_1(\mathbf{r}, \omega) = e\{-c\boldsymbol{\alpha} \cdot \nabla\chi(\mathbf{r}, \omega) + i\omega\chi(\mathbf{r}, \omega)\}. \tag{26}$$

This equation can be rewritten in the more convenient form

$$\Delta h_1 = e\left\{-i\frac{c}{\hbar}\boldsymbol{\alpha} \cdot \mathbf{p}\chi + i\omega\chi\right\} \tag{27}$$

$$= -i\frac{e}{\hbar}\{[h, \chi] - \hbar\omega\chi\}, \tag{28}$$

where h is the one-electron Dirac Hamiltonian given in Eq. (3). The requirement that matrix elements be independent of the gauge of the external field can be written

$$\langle b|\Delta h_1|a\rangle = -i\frac{e}{\hbar}\langle b|[h, \chi] - \hbar\omega\chi|a\rangle = -i\frac{e}{\hbar}(\epsilon_b - \epsilon_a - \hbar\omega)\langle b|\chi|a\rangle = 0. \tag{29}$$

This condition is obviously satisfied for states $|a\rangle$ and $|b\rangle$ that satisfy $\hbar\omega = \epsilon_b - \epsilon_a$. Therefore, transition amplitudes calculated using single-particle orbitals in a local potential are gauge invariant, provided the energy of the incoming photon equals the difference in energies of the final and initial orbitals. Later, we will require matrix elements of derivatives of Δh_1 with respect to ω. These matrix elements can be simplified with the aid of the above identity to give

$$\left\langle b\left|\frac{d\Delta h_1}{d\omega}\right|a\right\rangle = -i\frac{e}{\hbar}\left\langle b\left|\left[h, \frac{d\chi}{d\omega}\right] - \hbar\omega\frac{d\chi}{d\omega} - \hbar\chi\right|a\right\rangle = ie\langle b|\chi|a\rangle, \tag{30}$$

where the equality on the right-hand side is valid only for states satisfying $\hbar\omega = \epsilon_b - \epsilon_a$.

The electromagnetic interaction of a many-electron system is just the sum of the interactions of the individual electrons given above. In second quantization, the electromagnetic interaction Hamiltonian is given by

$$H_1(\omega) = \sum_{ij} (h_1)_{ij} a_i^\dagger a_j, \tag{31}$$

where $(h_1)_{ij} = \langle i|h_1(\mathbf{r}, \omega)|j\rangle$.

C. Multipole Potentials and Transition Amplitudes

The vector and scalar potential for a photon in the transverse gauge are

$$\mathbf{A}(\mathbf{r}, \omega) = \hat{\epsilon} e^{i\mathbf{k}\cdot\mathbf{r}}$$

$$\phi(\mathbf{r}, \omega) = 0,$$

where $\hat{\epsilon}$ is the photon's polarization vector and $\mathbf{k}$ is its wave vector. To simplify the calculation of transition amplitudes, we expand $\mathbf{A}(\mathbf{r}, \omega)$ in a multipole series,

$$\mathbf{A}(\mathbf{r}, \omega) = 4\pi \sum_{JM\lambda} i^{J-\lambda} \mathbf{Y}_{JM}^{(\lambda)}(\hat{\mathbf{k}}) \cdot \hat{\epsilon} \mathbf{a}_{JM}^{(\lambda)}(\mathbf{r}), \tag{32}$$

where the vector spherical harmonics, $\mathbf{Y}_{JM}^{(\lambda)}(\hat{k})$, are those defined in Akhiezer and Berestetskii [12]. The multipole components of the vector potential are given in the transverse gauge by

$$\mathbf{a}_{JM}^{(0)}(\mathbf{r}) = j_J(kr)\mathbf{Y}_{JM}^{(0)}(\hat{r})$$

$$\mathbf{a}_{JM}^{(1)}(\mathbf{r}) = \left[j'_J(kr) + \frac{j_J(kr)}{kr}\right]\mathbf{Y}_{JM}^{(1)}(\hat{r}) + \sqrt{J(J+1)}\,\frac{j_J(kr)}{kr}\,\mathbf{Y}_{JM}^{(-1)}(\hat{r}). \quad (33)$$

In these equations, $\lambda = 0$ designates magnetic multipoles and $\lambda = 1$ designates electric multipoles. The parity of the multipole potential $\mathbf{a}_{JM}^{(\lambda)}(\mathbf{r})$ is $(-1)^{J+1-\lambda}$. Matrix elements of the electric-dipole interaction in the transverse gauge reduce to velocity-form dipole amplitudes in the nonrelativistic limit. Later we refer to the transverse-gauge amplitudes as velocity-form amplitudes.

Gauge transformations [Eqs. (23) and (24)] have no effect on the magnetic-multipole potentials, but they do modify electric-multipole potentials. In particular, a gauge transformation of the transverse-gauge potentials using the gauge function,

$$\chi_{JM}(\mathbf{r}, \omega) = -\frac{1}{k}\sqrt{\frac{J+1}{J}}\,j_J(kr)Y_{JM}(\hat{r}), \quad (34)$$

gives transition amplitudes that reduce to the length form in the nonrelativistic limit. We refer to the resulting potentials as *length-gauge* potentials in the sequel. The electric-multipole potentials in the length gauge are

$$\mathbf{a}_{JM}^{(1)}(\mathbf{r}) = -j_{J+1}(kr)\left[\mathbf{Y}_{JM}^{(1)}(\hat{r}) - \sqrt{\frac{J+1}{J}}\,\mathbf{Y}_{JM}^{(-1)}(\hat{r})\right]$$

$$\phi_{JM}^{(1)}(kr) = -ic\sqrt{\frac{J+1}{J}}\,j_J(kr)Y_{JM}(\hat{r}). \quad (35)$$

In either gauge, the multipole-interaction Hamiltonian can be written in terms of dimensionless multipole-transition operators $t_{JM}^{(\lambda)}(\mathbf{r}, \omega)$,

$$[h_1(\mathbf{r}, \omega)]_{JM} = -iec\sqrt{\frac{(2J+1)(J+1)}{4\pi J}}\,t_{JM}^{(\lambda)}(\mathbf{r}, \omega), \quad (36)$$

where the one-electron reduced matrix elements $\langle i\|t_J^{(\lambda)}\|j\rangle$ are, for the *transverse gauge*,

$$\langle \kappa_i\|t_J^{(0)}\|\kappa_j\rangle = \langle -\kappa_i\|C_J\|\kappa_j\rangle \int_0^\infty dr\,\frac{\kappa_i + \kappa_j}{J+1}\,j_J(kr)[G_i(r)F_j(r) + F_i(r)G_j(r)]$$

$$(37)$$

$$\langle \kappa_i\|t_J^{(1)}\|\kappa_j\rangle = \langle \kappa_i\|C_J\|\kappa_j\rangle \int_0^\infty dr\,\left\{-\frac{\kappa_i - \kappa_j}{J+1}\left[j'_J(kr) + \frac{j_J(kr)}{kr}\right]\right.$$

$$\left.\times [G_i(r)F_j(r) + F_i(r)G_j(r)] + J\,\frac{j_J(kr)}{kr}\,[G_i(r)F_j(r) - F_i(r)G_j(r)]\right\}, \quad (38)$$

and, for the *length gauge*,

$$\langle\kappa_i\|t_J^{(1)}\|\kappa_j\rangle = \langle\kappa_i\|C_J\|\kappa_j\rangle \int_0^\infty dr \left\{ j_J(kr)[G_i(r)G_j(r) + F_i(r)F_j(r)] \right.$$
$$\left. + j_{J+1}(kr) \left[\frac{\kappa_i - \kappa_j}{J+1} [G_i(r)F_j(r) + F_i(r)G_j(r)] + [G_i(r)F_j(r) - F_i(r)G_j(r)] \right] \right\}.$$
(39)

In Eqs. (38) and (39), the functions $G_i(r)$ and $F_i(r)$ are the large and small components, respectively, of the radial Dirac wave functions for the orbital with quantum numbers (n_i, κ_i).

The multipole-transition operators $t_J^{(\lambda)}(\mathbf{r}, \omega)$ are related to the frequency-dependent multipole-moment operators $q_J^{(\lambda)}(\mathbf{r}, \omega)$ by

$$q_J^{(\lambda)}(\mathbf{r}, \omega) = \frac{(2J+1)!!}{k^J} t_J^{(\lambda)}(\mathbf{r}, \omega). \tag{40}$$

Both the transition operators and the multipole-moment operators are irreducible tensor operators. For a many-body system, the multipole-transition operators are given by

$$T_{JM}^{(\lambda)} = \sum_{ij} (t_{JM}^{(\lambda)})_{ij} a_i^\dagger a_j, \tag{41}$$

where $(t_{JM}^{(\lambda)})_{ij} = \langle i|t_{JM}^{(\lambda)}(\mathbf{r}, \omega)|j\rangle$. The Einstein B coefficient, given the absorption probability per unit time of a photon of multipolarity $(J\lambda)$ leading from a state I with angular momentum J_I to a state F with angular momentum J_F, is

$$B_J^{(\lambda)} = 2\alpha\omega \frac{[J]}{[J_I]} \frac{J+1}{J} |\langle F\|T_J^{(\lambda)}\|I\rangle|^2 = 8\pi k \frac{[J]}{[J_I]} \frac{J+1}{J} |\langle F\|T_J^{(\lambda)}\|I\rangle|^2 cR_\infty,$$
(42)

where we use the notation $[J] = 2J + 1$. The corresponding Einstein A coefficient, giving the spontaneous emission probability for transitions from state I to state F with $E_I - E_F = \hbar\omega$ is, of course, given by

$$A_J^{(\lambda)} = B_J^{(\lambda)}. \tag{43}$$

Using wave functions given in Eq. (12) for both the initial and the final states and carrying out the sums over magnetic substates, one obtains

$$\langle F\|T_J^{(\lambda)}\|I\rangle = \sqrt{[J_I][J_F]}(-1)^J \sum_{\substack{m\leqslant n \\ r\leqslant s}} \eta_{rs}\eta_{mn} c_{rs}^{(F)} c_{mn}^{(I)}$$

$$\times \left\{ (-1)^{j_r + j_s + J_I} \begin{Bmatrix} J & J_I & J_F \\ j_s & j_r & j_m \end{Bmatrix} \langle r\|t^{(\lambda)}\|m\rangle \delta_{ns} \right.$$

$$+ (-1)^{j_r+j_n} \begin{Bmatrix} J & J_I & J_F \\ j_s & j_r & j_n \end{Bmatrix} \langle r||t^{(\lambda)}||n\rangle \delta_{ms}$$

$$+ (-1)^{J_F+J_I+1} \begin{Bmatrix} J & J_I & J_F \\ j_r & j_s & j_m \end{Bmatrix} \langle s||t^{(\lambda)}||m\rangle \delta_{nr}$$

$$+ (-1)^{j_r+j_n+J_F} \begin{Bmatrix} J & J_I & J_F \\ j_r & j_s & j_n \end{Bmatrix} \langle s||t^{(\lambda)}||n\rangle \delta_{mr} \Bigg\}. \tag{44}$$

Equations (42)–(44) are used in the following sections to evaluate multipole-transition amplitudes and rates.

D. A Commutator Identity

For electric multipole transitions ($\lambda = 1$), we consider the difference between the transition amplitude in length gauge $(T_{JM}^{(1)})_l$ and that in velocity gauge $(T_{JM}^{(1)})_v$. Let us introduce the operator representing the difference between the multipole-transition operators in the two gauges,

$$\Delta T_{JM}^{(1)} = (T_{JM}^{(1)})_l - (T_{JM}^{(1)})_v. \tag{45}$$

This difference can be written in terms of the multipole-interaction Hamiltonian using Eq. (36),

$$\Delta T_{JM}^{(1)} = \frac{i}{ec} \sqrt{\frac{4\pi J}{(J+1)(2J+1)}} \Delta H_{JM}, \tag{46}$$

where

$$\Delta H_{JM} = \sum_{ij} \langle i|(\Delta h_I)_{JM}|j\rangle a_i^\dagger a_j.$$

Equation (46) can be rewritten with the aid of Eq. (28) to give

$$\Delta H_{JM} = -i\frac{e}{\hbar}\{[H_0, X_{JM}] - \hbar\omega X_{JM}\}$$

$$= -i\frac{e}{\hbar}\{[H, X_{JM}] - \hbar\omega X_{JM} - [V, X_{JM}]\}, \tag{47}$$

where $V = V_C + V_B$ is the sum of the Coulomb and Breit interactions and where

$$X_{JM} = \sum_{ij} (\chi_{JM})_{ij} a_i^\dagger a_j$$

is the many-electron gauge operator. For eigenstates of H that satisfy $E_F - E_I = \hbar\omega$, it follows from Eq. (47) that

$$\langle \Psi_F | \Delta H_{JM} | \Psi_I \rangle = i\frac{e}{\hbar} \langle \Psi_F | [V, X_{JM}] | \Psi_I \rangle. \tag{48}$$

Using Eqs. (34) and (46), we can formulate Eq. (48) as the commutator identity

$$\langle \Psi_F | \Delta T_{JM} | \Psi_I \rangle = \frac{1}{\hbar\omega} \langle \Psi_F | [V, \Xi_{JM}] | \Psi_I \rangle, \tag{49}$$

where

$$\Xi_{JM} = \sum_{ij} (\xi_{JM})_{ij} a_i^\dagger a_j,$$

with

$$\xi_{JM}(\mathbf{r}, \omega) = \sqrt{\frac{4\pi}{2J+1}} j_J(kr) Y_{JM}(\hat{r}).$$

Since the no-pair potential V in Eq. (49) contains only positive-energy components, the commutator of V and Ξ_{JM} does not vanish. Using the wave functions given in Eq. (12) for the case $U(r) = 0$, the reduced matrix element of $[V, \Xi_{JM}]$ is found to be

$$\langle F \| [V, \Xi_J] \| I \rangle = \sqrt{[J_F][J_I]} \sum_{\substack{m \leq n \\ r \leq s}} \eta_{rs} \eta_{mn} c_{rs}^{(F)} c_{mn}^{(I)}$$

$$\times \sum_{i,L} \Bigg[(-1)^{j_n+j_s+J_F+J_I+L} \begin{Bmatrix} j_i & j_n & J_F \\ j_s & j_r & L \end{Bmatrix} \begin{Bmatrix} j_i & j_n & J_F \\ J_I & J & j_m \end{Bmatrix} X_L(srni) \langle i \| \xi_J \| m \rangle$$

$$+ (-1)^{j_n+j_s+J_I+L} \begin{Bmatrix} j_i & j_n & J_F \\ j_r & j_s & L \end{Bmatrix} \begin{Bmatrix} j_i & j_n & J_F \\ J_I & J & j_m \end{Bmatrix} X_L(rsni) \langle i \| \xi_J \| m \rangle$$

$$+ (-1)^{j_n+j_s+J_F+L} \begin{Bmatrix} j_i & j_m & J_F \\ j_s & j_r & L \end{Bmatrix} \begin{Bmatrix} j_i & j_m & J_F \\ J_I & J & j_n \end{Bmatrix} X_L(srmi) \langle i \| \xi_J \| n \rangle$$

$$+ (-1)^{j_n+j_s+L} \begin{Bmatrix} j_i & j_m & J_F \\ j_r & j_s & L \end{Bmatrix} \begin{Bmatrix} j_i & j_m & J_F \\ J_I & J & j_n \end{Bmatrix} X_L(rsmi) \langle i \| \xi_J \| n \rangle \tag{50}$$

$$+ (-1)^{j_m-j_r+L} \begin{Bmatrix} j_i & j_s & J_I \\ j_n & j_m & L \end{Bmatrix} \begin{Bmatrix} j_i & j_s & J_I \\ J_F & J & j_r \end{Bmatrix} \langle r \| \xi_J \| i \rangle X_L(ismn)$$

$$+ (-1)^{j_m-j_r+J_I+L} \begin{Bmatrix} j_i & j_s & J_I \\ j_m & j_n & L \end{Bmatrix} \begin{Bmatrix} j_i & j_s & J_I \\ J_F & J & j_r \end{Bmatrix} \langle r \| \xi_J \| i \rangle X_L(isnm)$$

$$+ (-1)^{j_m-j_r+J_F+L} \begin{Bmatrix} j_i & j_r & J_I \\ j_n & j_m & L \end{Bmatrix} \begin{Bmatrix} j_i & j_r & J_I \\ J_F & J & j_s \end{Bmatrix} \langle s \| \xi_J \| i \rangle X_L(irmn)$$

$$+ (-1)^{j_m-j_r+J_F+J_I+L} \begin{Bmatrix} j_i & j_r & J_I \\ j_m & j_n & L \end{Bmatrix} \begin{Bmatrix} j_i & j_r & J_I \\ J_F & J & j_s \end{Bmatrix} \langle s \| \xi_J \| i \rangle X_L(irnm) \Bigg].$$

From the definition of the no-pair potential, it follows that the sum over i in Eq. (50) is restricted to positive-energy single-particle orbitals. Thus, *electromagnetic transition amplitudes calculated using the no-pair Hamiltonian are gauge dependent.* If we artificially extend the sum over i to include the complete set of positive- and negative-energy states, then, as is shown in the Appendix, the commutator vanishes identically.

Nonvanishing values are a result of the restriction to positive-energy states in the no-pair Hamiltonian. In Section II.E, we calculate reduced matrix elements of the E_1 transition operator for helium-like ions in length and velocity gauges as well as the reduced matrix element of the commutator from Eq. (50). The length–velocity difference is found to be equal to the matrix element of the commutator to the numerical accuracy of the calculation.

E. Numerical Calculations and Tests of Transition Amplitudes

To determine the transition amplitudes numerically, we first solve the CI equations [Eq. (17)] either iteratively [4] or directly as a large-scale eigenvalue problem [3]. After obtaining wave functions and energies, we evaluate the reduced matrix elements of the transition amplitudes using Eq. (44). For concreteness, we consider only calculations carried out starting from the potential $U(r) = 0$ in this section.

In our numerical calculations, we set the goal of obtaining amplitudes accurate to five significant figures. Since it is relatively expensive and time consuming to carry out the multiple sums in Eq. (44), we limit our basis sets as much as possible, within this accuracy constraint. For the allowed E_1 transitions $2^1P_1 \to 1^1S_0$, $2^1P_1 \to 2^1S_0$ and $2^3P_{0,1,2} \to 2^3S_1$, as well as the $2^3S_1 \to 1^1S_0$ (M_1) and $2^3P_2 \to 1^1S_0$ (M_2) transitions, we found that the following restrictions on basis sets gave amplitudes of five-figure accuracy. For nuclear charges in the range $Z = 2$–10, the basis orbitals (*nmrs*) in Eq. (44) were limited to those having orbital angular momentum $l \leqslant 4$. For $Z = 11$–30, it was sufficient to include orbitals with $l \leqslant 3$; for $Z > 30$ only those orbitals with $l \leqslant 2$ were required. For $Z \leqslant 30$, we summed only the first 20 out of a set of 40 basis orbitals for each l value considered. For $Z > 30$, it was necessary to sum the first 25 orbitals from a set of 40 for each l.

In Table I, we compare amplitudes evaluated with the limited basis sets described above with those obtained using complete basis sets [3] for nuclear charges $Z = 10, 20, 30,$ and 40. For this comparison, the Breit interaction was omitted entirely from the calculation. As seen from this table, the difference between the truncated and the exact amplitudes is at worse a few parts in the fifth significant digit. For the forbidden E_1 transitions $2^3P_1 \to 1^1S_0$ and $2^1P_1 \to 2^3S_1$, the basis sets described above were too small to give five-figure accuracy for $Z < 50$. For $Z \leqslant 30$, we

TABLE I

COMPARISON OF CALCULATIONS OF REDUCED MATRIX ELEMENTS USING THE LIMITED NUMBER OF CONFIGURATIONS DESCRIBED IN THE TEXT WITH COMPLETE CI CALCULATIONS

Z	Transition	Limited configurations	Complete configurations	Percentage difference
10	$2^1P_1 - 1^1S_0$	0.178458	0.178461	1.8(−3)
	$2^1P_1 - 2^1S_0$	0.577485	0.577483	4.1(−4)
	$2^1P_1 - 2^3S_1$	0.012193	0.012195	1.6(−2)
	$2^3P_1 - 1^1S_0$	0.003913	0.003913	1.6(−2)
	$2^3P_0 - 2^3S_1$	0.324405	0.324405	1.3(−4)
	$2^3P_1 - 2^3S_1$	0.561963	0.561962	1.3(−4)
	$2^3P_2 - 2^3S_1$	0.726198	0.726198	1.3(−4)
	$2^3S_1 - 1^1S_0$	0.001767	0.001767	1.8(−3)
	$2^3P_2 - 1^1S_0$	0.881752	0.881737	1.6(−3)
20	$2^1P_1 - 1^1S_0$	0.088423	0.088422	1.4(−3)
	$2^1P_1 - 2^1S_0$	0.269181	0.269179	8.7(−4)
	$2^1P_1 - 2^3S_1$	0.041923	0.041932	2.1(−2)
	$2^3P_1 - 1^1S_0$	0.014052	0.014055	2.1(−2)
	$2^3P_0 - 2^3S_1$	0.154662	0.154662	8.7(−5)
	$2^3P_1 - 2^3S_1$	0.265030	0.265025	1.7(−3)
	$2^3P_2 - 2^3S_1$	0.347534	0.347534	6.9(−5)
	$2^3S_1 - 1^1S_0$	0.007208	0.007208	6.5(−4)
	$2^3P_2 - 1^1S_0$	0.432032	0.432027	1.0(−3)
30	$2^1P_1 - 1^1S_0$	0.055224	0.055221	4.8(−3)
	$2^1P_1 - 2^1S_0$	0.166329	0.166325	2.2(−3)
	$2^1P_1 - 2^3S_1$	0.062073	0.062082	1.4(−2)
	$2^3P_1 - 1^1S_0$	0.021048	0.021051	1.4(−2)
	$2^3P_0 - 2^3S_1$	0.100630	0.100630	6.0(−5)
	$2^3P_1 - 2^3S_1$	0.163590	0.163584	3.4(−3)
	$2^3P_2 - 2^3S_1$	0.227606	0.227606	4.7(−5)
	$2^3S_1 - 1^1S_0$	0.016392	0.016392	5.7(−4)
	$2^3P_2 - 1^1S_0$	0.282034	0.282032	6.5(−4)
40	$2^1P_1 - 1^1S_0$	0.038251	0.038248	6.4(−3)
	$2^1P_1 - 2^1S_0$	0.115910	0.115908	2.1(−3)
	$2^1P_1 - 2^3S_1$	0.061313	0.061317	6.9(−3)
	$2^3P_1 - 1^1S_0$	0.020851	0.020852	7.1(−3)
	$2^3P_0 - 2^3S_1$	0.073786	0.073786	1.1(−4)
	$2^3P_1 - 2^3S_1$	0.113168	0.113161	6.8(−3)
	$2^3P_2 - 2^3S_1$	0.168470	0.168469	6.9(−5)
	$2^3S_1 - 1^1S_0$	0.029440	0.029440	3.2(−4)
	$2^3P_2 - 1^1S_0$	0.205929	0.205929	0

Note. Results do not include Breit corrections to the wave function.

needed orbitals with $l \leq 6$, while for $31 \leq Z \leq 40$, we needed orbitals with $l \leq 4$. For $41 \leq Z \leq 49$, orbitals with $l \leq 3$ were required. For all Z, we summed over 25 of 40 basis functions.

To facilitate comparisons of values in Table I and in the other tables in this chapter with previous calculations, we factor $k^J/(2J + 1)!!$ from the transition amplitudes to obtain the multipole-moment operators $Q_{JM}^{(\lambda)}$,

$$Q_{JM}^{(\lambda)} = \frac{(2J+1)!!}{k^J} T_{JM}^{(\lambda)}.$$

For E_1 transitions, we tabulate reduced matrix elements of $Q_{1M}^{(1)}$, while for M_1 and M_2 transitions, we tabulate reduced matrix elements of the magnetic-multipole operator

$$M_{JM} = 2cQ_{JM}^{(0)}.$$

It should be noted that the magnetic-quadrupole amplitude defined in this way is the conventional moment given in many textbooks [13]. It differs from that used in some compilations of transition amplitudes [14] by a factor of $\frac{3}{2}$.

Tests of the commutator identity are shown in Table II, in which values of the reduced matrix elements of $Q_{1M}^{(1)}$ are given for ions with $Z = 20$, 50, and 100. As can be seen by comparing the length-form and velocity-form amplitudes in the third and fourth columns of the table, the amplitudes differ in the two gauges. The differences, which are tabulated in the fifth column

TABLE II

Reduced Matrix Elements of the Electric-Dipole Operator in Both the Length and the Velocity Gauges Obtained Using CI Wave Functions from the Breit–Coulomb, No-Pair Hamiltonian

Z	Transition	$\langle F\|Q_1\|I\rangle_l$	$\langle F\|Q_1\|I\rangle_v$	ΔQ	$[V,\Xi]$
20	$2^1P_1 - 1^1S_0$	0.088563	0.088594	−0.000031	−0.000033
	$2^3P_1 - 1^1S_0$	0.014967	0.014972	−0.000005	−0.000005
	$2^1P_1 - 2^3S_1$	0.045270	0.045221	0.000049	0.000048
	$2^3P_0 - 2^3S_1$	0.154748	0.154398	0.000350	0.000360
	$2^3P_1 - 2^3S_1$	0.264555	0.264030	0.000525	0.000537
	$2^3P_2 - 2^3S_1$	0.347554	0.347051	0.000503	0.000514
50	$2^1P_1 - 1^1S_0$	0.029263	0.029291	−0.000028	−0.000028
	$2^3P_1 - 1^1S_0$	0.018526	0.018542	−0.000016	−0.000017
	$2^1P_1 - 2^3S_1$	0.055209	0.055072	0.000137	0.000140
	$2^3P_0 - 2^3S_1$	0.057633	0.056537	0.001097	0.001106
	$2^3P_1 - 2^3S_1$	0.084451	0.083239	0.001212	0.001228
	$2^3P_2 - 2^3S_1$	0.133095	0.132905	0.000190	0.000192
100	$2^1P_1 - 1^1S_0$	0.011478	0.011499	−0.000021	−0.000021
	$2^3P_1 - 1^1S_0$	0.008181	0.008193	−0.000012	−0.000012
	$2^1P_1 - 2^3S_1$	0.026731	0.026689	0.000042	0.000042
	$2^3P_0 - 2^3S_1$	0.022804	0.020437	0.002367	0.002394
	$2^3P_1 - 2^3S_1$	0.032173	0.028439	0.003734	0.003757
	$2^3P_2 - 2^3S_1$	0.059723	0.059725	−0.000002	−0.000001

Note. ΔQ is the difference between the matrix elements in the two gauges; the final column $[V,\Xi]$ gives values (in a.u.) of the difference predicted by the commutator identity [Eq. (49)].

of the table, are typically in the fourth significant figure, although differences occur in the second figure for the $2^3P_{0,1} \to 2^3S_1$ transitions at high Z. To convince ourselves that these differences are of a physical origin, we compare them with the values predicted by the commutator identity [Eq. (49)], which are listed in the last column of the table. The equality of the entries in the last two columns confirms that the length–velocity differences in the numerical amplitudes are just those expected for calculations carried out using the no-pair theory.

It should be mentioned that the multiple sums in the expression for the commutator are also evaluated using a reduced basis set. We restrict the orbitals (*nmrs*) in Eq. (50) to include those having angular momentum quantum number $l \leqslant 2$. We further restrict the orbitals by summing only the first eight states in a basis set of 40 orbitals. The sums over L are also restricted to values of $L \leqslant 2$, while the sum over i includes all 40 positive-energy basis functions and all values of angular momentum permitted by the selection rules. Because of the large number of indices, these sums are time consuming. Nevertheless, in selected cases, we verify that, on increasing the limits above sufficiently, the differences in the transition amplitudes agree to five figures with the values predicted by the commutator identity. As a further check on our numerical methods, we extend the sum over i to include 40 positive-energy orbitals and 40 negative-energy orbitals and find that the reduced matrix element of the commutator vanishes to more than six figures for each of the cases considered in Table II.

From the values in Table II, it is clear that transition amplitudes determined from the no-pair Hamiltonian are gauge dependent and that the gauge dependence is a result of excluding negative-energy states in the no-pair Hamiltonian. In the following section, we examine electromagnetic transitions in hydrogen-like ions in a gauge-invariant perturbation theory calculation based on QED. This analysis will be used in Section III to construct gauge-invariant amplitudes for helium-like ions in perturbation theory.

III. S-Matrix Theory for Decay Rates

Although we are interested in helium-like ions, in this section we consider the simpler problem of obtaining gauge-independent transition rates for one-electron ions in an external potential, which is chosen to be a Coulomb potential of charge Z_0 together with a perturbing potential $V(r)$. We use the resulting theoretical analysis as a paradigm for constructing gauge-invariant amplitudes for transitions in helium-like ions. To investigate the transition amplitudes in a gauge-invariant framework, we turn to the QED S-matrix.

Analysis of transitions from the point of view of QED shows that transition amplitudes are given by the usual Feynman diagrams from perturbation theory, together with *derivative* terms that have no diagrammatic representation. We use this example primarily to elucidate the structure of such derivative terms.

Specifically, we will work with an interaction Hamiltonian, $H_I = H_I^A + H_I^B$, where

$$H_I^A = e \int d^3 r \bar{\psi}(r) \gamma_\mu A^\mu(r) \psi(r), \tag{51}$$

and

$$H_I^B = \int d^3 r \bar{\psi}(r) V(r) \psi(r). \tag{52}$$

Here $V(r)$ represents an external potential that is in principle arbitrary. One choice that illustrates the important features of the theory is

$$V(r) = -\frac{\delta Z}{r} \gamma_0. \tag{53}$$

Here $\delta Z \equiv Z - Z_0$, where Z is the actual charge of the nucleus of a hydrogen-like ion and Z_0 is a nearby charge that defines the potential in which the calculation is carried out. For example, to calculate decay rates of hydrogen-like thallium ($Z = 81$) as perturbations of the decay rate of hydrogen-like mercury ($Z = 80$), we would have $\delta Z = 1$. Since both decay rates are known, the difference must be accounted for dominantly by first-order perturbation theory. Another choice of $V(r)$, which allows energy levels of $n = 2$ states of lithium-like uranium to be calculated as perturbations from hydrogen-like uranium with $V(r)$ coming from the potential of the two 1s electrons, has been described by Indelicato and Mohr [15].

In order to calculate decay rates, we recall the well-known fact that these rates are related to energy level shifts ΔE by $\Gamma = -2 \text{Im} \Delta E$. This approach was used by Barbieri and Sucher [6] in a paper closely related to the present work to calculate radiative corrections to M_1 decays in hydrogen-like ions. In order to calculate ΔE, we use S-matrix techniques. The basic equation used is the symmetrized version of the Gell-Mann–Low formula [7] given by Sucher [16],

$$\Delta E_n = \frac{i\epsilon}{2} \lim_{\epsilon \to 0} \frac{\partial}{\partial \lambda} \ln \langle n | S_{\epsilon,\lambda} | n \rangle, \tag{54}$$

where

$$S_{\epsilon,\lambda} = e^{-i\lambda \int d^4 x e^{-\epsilon |x_0|} H_I(x)}. \tag{55}$$

We apply these formulas to a hydrogenic ion with an electron in state v. The lowest-order contributions to the S matrix come from a single action of H_1^B,

$$S_{\epsilon,\lambda}^{(1)} = -\frac{2i\lambda}{\epsilon} E^{(1)}(v), \tag{56}$$

and two actions of H_1^A,

$$S_{\epsilon,\lambda}^{(2)} = -\frac{i\lambda^2}{\epsilon} \Sigma_{vv}(\epsilon_v). \tag{57}$$

Here we have introduced

$$E^{(1)}(v) = \int d^3r \bar{\psi}_v(r) V(r) \psi_v(r) \tag{58}$$

and defined a generalization of the usual self-energy

$$\Sigma_{mn}(\epsilon_v) \equiv -i\alpha \int d^3x d^3y \int \frac{d^4k}{(2\pi)^4} \frac{e^{i\vec{k}\cdot(\vec{x}-\vec{y})}}{k^2+i\delta} \bar{\psi}_m(\vec{x}) \gamma_\mu S_F(\vec{x},\vec{y};\epsilon_v-k_0) \gamma^\mu \psi_n(\vec{y}). \tag{59}$$

We ignore the other radiative correction associated with two actions of H_1^A describing vacuum polarization because it is real.

We can now determine the lowest-order decay rate from the imaginary part of the Lamb shift, $\Sigma_{vv}(\epsilon_v)$, illustrated in Fig. 1. That imaginary part arises solely from the "pole terms" [17], which arise when a Wick rotation, $k_0 \to ik_4$, is carried out, and result from the contour passing through poles associated with electrons of energy lower than that of the valence electron v, which we will refer to as m. Specifically, the pole terms are, in Feynman gauge,

$$\Delta E_{\text{pole}} = \alpha \sum_{m<v} \int \frac{d^3x d^3y}{|\vec{x}-\vec{y}|} e^{i(\epsilon_v-\epsilon_m)|\vec{x}-\vec{y}|} \bar{\psi}_v(\vec{x}) \gamma_\mu \psi_m(\vec{x}) \bar{\psi}_m(\vec{y}) \gamma^\mu \psi_v(\vec{y}). \tag{60}$$

The real part of this contributes to the Lamb shift, and the imaginary part reproduces the lowest-order decay rate. Note that there is also a half-pole term when $m = v$, which enters with a factor of $\frac{1}{2}$, but is purely real. It is a simple matter to evaluate the imaginary part of this expression, and agreement with the standard result is found. Because the self-energy is gauge invariant, the same rates must be obtained if the Coulomb gauge is used

FIG. 1. Feynman diagram for $\Sigma_{vv}(E)$. the imaginary part of this diagram gives the leading contribution to the decay rate.

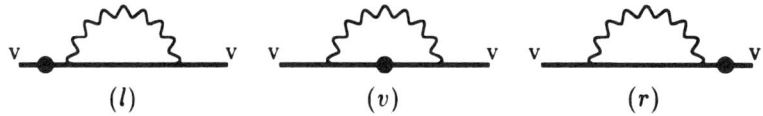

FIG. 2. Feynman diagrams for the three contributions to the third-order S-matrix, leading to the dominant correction to the decay rate. The ● represents the interaction with the external field V. The labels (l), (r), and (v) represent left, right, and the vertex, respectively.

instead of the Feynman gauge. For decays of multipolarity $(J\lambda)$, we obtain

$$\text{Im}\,\Delta E(\omega) = -\alpha\omega \frac{2J+1}{2j_v+1} R^2(\omega), \tag{61}$$

where $\hbar\omega = \epsilon_v - \epsilon_m$ and where

$$R(\omega) = \sqrt{\frac{J+1}{J}} \langle v\|t_J^{(\lambda)}\|m\rangle. \tag{62}$$

The reduced matrix elements $\langle v\|t_J^{(\lambda)}\|m\rangle$ here are precisely those defined in Eqs. (38) and (39). Equation (61) leads to a decay rate in agreement with the general expression given in Eq. (43).

We now consider corrections of higher order. We will be concerned with two actions of H_1^A and one of H_1^B. A first contribution of this order comes from the lowest-order formulas for $S^{(1)}$ and $S^{(2)}$ given above, which give an energy shift that diverges in the limit $\epsilon \to 0$,

$$\Delta E^{\text{div}} = \frac{i}{\epsilon}(E^{(1)}(v) + \Sigma_{vv}(\epsilon_v))(2E^{(1)}(v) + \Sigma_{vv}(\epsilon_v)). \tag{63}$$

We will be concerned here only with the contribution $3i/\epsilon E^{(1)}(v)\Sigma_{vv}(\epsilon_v)$. It clearly must be canceled by a term from the third-order S matrix, which we will show is connected with the derivative terms we are interested in. There are three contributions to the S matrix represented by the Feynman diagrams of Figs. 2 [(l), (v), and (r)]. (Here (v) stands for vertex, (l) for side left, and (r) for side right.) We begin with

$$S_{(r)} = -32\pi\alpha\epsilon^3\lambda^3 \int d^3x d^3y d^3z \int \frac{dE_1}{2\pi} \int \frac{dE_2}{2\pi} \int \frac{d^4k}{(2\pi)^3} \frac{e^{i\vec{k}\cdot(\vec{x}-\vec{y})}}{k^2} \frac{1}{\epsilon^2+(k_0+E_1-\epsilon_v)^2}$$
$$\frac{1}{\epsilon^2+(E_1-E_2+k_0)^2} \frac{1}{\epsilon^2+(E_2-\epsilon_v)^2}$$
$$\times \bar{\psi}_v(\vec{x})\gamma_\mu S_F(\vec{x},\vec{y};E_1)\gamma^\mu S_F(\vec{y},\vec{z},E_2)V(z)\psi_v(\vec{z}). \tag{64}$$

In the above, the terms with ϵ^2 in the denominator act to emphasize the region of integration in which $E_1 = \epsilon_v - k_0$ and $E_2 = \epsilon_v$, and, when it is possible to use these values in the Feynman propagators S_F, the integrations over E_1 and E_2 can be immediately carried out. However, if we now make a spectral decomposition of the second Feynman propagator, reexpressing $S_{(r)}$ as

$$S_{(r)} = -8\pi i \epsilon^3 \lambda^3 \sum_m \int d^3z \int \frac{dE_1}{2\pi} \int \frac{dE_2}{2\pi} \frac{1}{\epsilon^2 + (k_0 + E_1 - \epsilon_v)^2} \frac{1}{\epsilon^2 + (E_1 - E_2 + k_0)^2}$$

$$\times \frac{1}{\epsilon^2 + (E_2 - \epsilon_v)^2} \Sigma_{vm}(E_1) \bar{\psi}_m(z) V(z) \psi_v(\bar{z}) \frac{1}{E_2 - \epsilon_m(1 - i\delta)}, \quad (65)$$

we see that more care is needed when $m = v$. We now concentrate on this case. If we make the approximation $\Sigma_{vm}(E_1) = \Sigma_{vm}(\epsilon_v - k_0)$, it is possible to directly carry out the E_1 and E_2 integrations, which results in the divergent energy

$$\Delta E_{(r)}^{\text{div}} = -\frac{3}{2\epsilon} E^{(1)}(v) \Sigma_{vv}(\epsilon_v). \quad (66)$$

When taken together with an identical term from the (l) calculation, it is seen that the divergent contributions to the energy cancel, as they must. However, because we have found divergences, terms normally safely dropped can contribute a finite amount to the energy. The only approximation we made in the above was to replace E_1 with $\epsilon_v - k_0$ in the self-energy. In order to pick up the derivative term, we now expand

$$\Sigma_{vv}(E_1) = \Sigma_{vv}(\epsilon_v - k_0) + (E_1 - \epsilon_v + k_0) \left. \frac{\partial \Sigma_{vv}(E)}{\partial E} \right|_{E = \epsilon_v - k_0} + \cdots. \quad (67)$$

The second term can now be easily evaluated, as can an identical term from the (l) case, and we find

$$\Delta E_{\text{deriv}}^{(l)+(r)} = E^{(1)}(v) \Sigma'_{vv}(\omega), \quad (68)$$

where we have used the fact that the energy depends on ϵ_v through $\omega = \epsilon_v - \epsilon_m$.

The remaining part of the (l) and (r) terms are easily evaluated and can be shown to coincide with the results of ordinary perturbation theory. Turning to the vertex term, we first note that, while it has no $1/\epsilon$ divergences, it also contains a derivative term. This can be seen most clearly when a spectral decomposition of both electron propagators in Fig. 2 [(v)] is made, in which case we find

$$\Delta E_{(v)} = -4\pi i \alpha \sum_{ij} \int d^3 z \bar{\psi}_i(\bar{z}) V(z) \psi_j(\bar{z}) \int \frac{d^4 k}{(2\pi)^4} \int d^3 x \int d^3 y \frac{e^{i\vec{k}\cdot(\vec{x}-\vec{y})}}{k^2}$$

$$\times \frac{\bar{\psi}_v(\vec{x}) \gamma_\mu \psi_i(\vec{x}) \bar{\psi}_j(\vec{y}) \gamma^\mu \psi_v(\vec{y})}{(\epsilon_v - k_0 - \epsilon_m(1 - i\delta))(\epsilon_v - k_0 - \epsilon_n(1 - i\delta))}. \quad (69)$$

As with the self-energy, the only source of an imaginary part of $\Delta E_{(v)}$ is pole terms, in which one or the other of the states i and j is more deeply bound than the state v. The derivative term comes from the term in which both i and j are the state m, defined above. But in that case, we see that the integral over d^3z in the above equation gives $E^{(1)}(m)$. It is then straightforward to show that, defining $\delta\omega \equiv E^{(1)}(v) - E^{(1)}(m)$, the combined derivative terms taken together with the lowest-order energy give

$$\Sigma_{vv}(\omega) + \Delta E_{\text{deriv}} = \left(1 + \delta\omega \frac{\partial}{\partial E}\right) \Sigma_{vv}(E)\Big|_{E=\omega}, \quad (70)$$

which is clearly the beginning of the Taylor expansion of $\Sigma_{vv}(\omega + \delta\omega)$. In this way, we see that the derivative terms are indeed acting to correct the lowest-order energy difference ω to its actual value. We emphasize here that the standard procedure of using experimental energies in decay rate calculations already accounts for the effect of the derivative term, but the above discussion provides a rigorous theoretical justification for the procedure. In addition to the derivative terms, there are, of course, the standard perturbation theory contributions. They can be accounted for by adding to $R(\omega)$ the quantity

$$\delta R(\omega) = \sqrt{\frac{J+1}{J}} \left(\sum_{i \neq m} \frac{\langle v \| t_J^\lambda \| i \rangle \langle i | U | m \rangle}{\epsilon_m - \epsilon_i} + \sum_{i \neq v} \frac{\langle v | U | i \rangle \langle i \| t_J^\lambda \| m \rangle}{\epsilon_v - \epsilon_i} \right), \quad (71)$$

with the understanding that only cross-terms in $(R(\omega) + \delta R(\omega))^2$ are to be kept.

To illustrate how this calculation works, we consider the calculation of the M_1 decay of the $2s$ state of hydrogen-like thallium as a perturbation of hydrogen-like mercury. In atomic units, the imaginary part of the energy for mercury is -4.927×10^{-4}, and for thallium, -5.648×10^{-4}, so perturbation theory must pick up -0.721×10^{-4} a.u. The nonderivative parts of the calculation, carried out in Feynman gauge, give -6.410×10^{-4} a.u. The derivative terms contribute 5.736×10^{-4} a.u., for a combined -0.674×10^{-4}. Note that, while the bulk of the difference is picked up in this order of perturbation theory, a smaller difference is left over that is consistent with the expected size of the next order of perturbation theory.

While we have employed a fully field theoretic approach to derive the derivative terms, we now note that, if one treats energy-dependent potentials or amplitudes of energy-dependent operators in ordinary Rayleigh–Schrödinger perturbation theory, such terms automatically appear. This is simply because, when we expand $E = E^{(0)} + E^{(1)} + \cdots$, an energy-dependent potential or operator must also be expanded as $T(E) = T(E^{(0)}) + E^{(1)} T'(E^{(0)}) + \cdots$. Thus, if an amplitude has energy dependence, as is the case for electromagnetic transitions, derivative terms will arise. Returning to our field-theoretic discussion, we note that $\text{Im}\,\Delta E \propto \omega R^2(\omega)$, where $R(\omega)$ is essentially the reduced matrix element of the transition operator $t_J^{(\lambda)}(\omega)$. The derivative acting on Im ΔE gives a contribution $2\omega R(\omega)R'(\omega)$ that could be obtained from the arguments just given about Rayleigh–Schrödinger perturbation theory. However, the derivative also acts on the linear factor ω and can clearly be accounted for by replacing ω with $\omega + \delta\omega$ in the expression for Im ΔE. We have been unable to find an alternative derivation for this effect and, therefore, rely on the more elaborate field theory approach given above to explain it.

In carrying out perturbation theory for the decays of helium in the next section, we adopt the following procedure. In lowest order, we will use the above equation with the matrix elements $R(\omega)$ suitably generalized to helium. We will show that in second-order perturbation theory the derivative terms not only are present, but also play a crucial role in preserving gauge invariance. Finally, although we have not carried out a complete field theoretic analysis for helium (which involves treating the imaginary part of all two-photon diagrams), we will change the outside factor ω. However, instead of including only first-order perturbation theory, as done in this section, we also include all the higher-order perturbation theory, recoil, and QED effects calculated in Refs. [3, 4], which amounts to using the experimental energies.

IV. Application of Perturbation Theory to Helium-like Ions

In this section, we carry out a gauge-invariant second-order perturbation theory calculation of amplitudes for transitions in helium-like ions fashioned after the QED calculation of Section III.

A. Basic Equations

We consider single-configuration states such as 2^3P_0 or 1^1S_0 and carry out our analysis in the uncoupled representation. Let us consider two unperturbed wave functions, $\Psi_{va}^{(0)}$ and $\Psi_{wb}^{(0)}$,

$$\Psi_{va}^{(0)} = a_v^\dagger a_a^\dagger |0\rangle \tag{72}$$

$$\Psi_{wb}^{(0)} = a_w^\dagger a_b^\dagger |0\rangle. \tag{73}$$

Here, the orbital labels a and b refer to 1s electrons and the labels v and w refer to excited electrons. We allow for the possibility that v is also a 1s state, in which case the wave function $\Psi_{va}^{(0)}$ represents the $(1s)^2$ ground state. Later, these states will be coupled to give a good value of angular momentum. The energies of the states are $E_{va}^{(0)} = \epsilon_v + \epsilon_a$ and $E_{wb}^{(0)} = \epsilon_w + \epsilon_b$. The energy of the absorbed photon is then, in lowest order, $\hbar\omega = E_{wb}^{(0)} - E_{va}^{(0)} = \epsilon_w - \epsilon_v$.

The lowest-order amplitude for the transition induced by the interaction Hamiltonian H_I of Eq. (31) is

$$T^{(1)} = \langle \Psi_{wb}^{(0)} | H_I | \Psi_{va}^{(0)} \rangle = (h_I)_{wv} \delta_{ba} - (h_I)_{wa} \delta_{bv}. \tag{74}$$

The amplitude $T^{(1)}$ is obviously gauge invariant, since according to Eq. (29), the two single-particle matrix elements are themselves independent of gauge for $\hbar\omega = \epsilon_w - \epsilon_v$.

We write the first-order correction to $\Psi_{va}^{(0)}$ as

$$\Psi_{va}^{(1)} = \sum_{ij \neq va} \rho_{ijva}^{(1)} a_i^\dagger a_j^\dagger |0\rangle. \tag{75}$$

This perturbed wave function satisfies

$$(H_0 - E_{va}^{(0)}) \Psi_{va}^{(1)} = (E_{va}^{(1)} - V_I) \Psi_{va}^{(0)}. \tag{76}$$

We use the local potential $U(r)$ given in Eq. (4) to approximate the electron–electron interaction in the single-particle Dirac equation. This is expected to improve the quality of the approximate wave functions as well as the convergence of the perturbation expansion. We find for the first-order energy

$$E_{va}^{(1)} = \tilde{v}_{vava} - U_{vv} - U_{aa}, \tag{77}$$

where $\tilde{v}_{ijkl}$ is defined in Eq. (11). The first-order correction to the photon energy is

$$\delta\omega = \frac{E_{wb}^{(1)} - E_{va}^{(1)}}{\hbar}.$$

For $ij \neq va$, we find that the expansion coefficients in the first-order wave function are

$$\rho_{ijva}^{(1)} = -\frac{v_{ijva}}{\epsilon_i + \epsilon_j - \epsilon_v - \epsilon_a} + \frac{U_{iv} \delta_{ja}}{\epsilon_i - \epsilon_v} + \frac{U_{ja} \delta_{iv}}{\epsilon_j - \epsilon_a}. \tag{78}$$

The corresponding second-order transition amplitude is given by

$$T^{(2)} = \langle \Psi_{wb}^{(1)} | H_I | \Psi_{va}^{(0)} \rangle + \langle \Psi_{wb}^{(0)} | H_I | \Psi_{va}^{(1)} \rangle + \delta\omega \frac{dT^{(1)}}{d\omega}$$

$$= \langle \Psi_{wb}^{(1)} | H_I | \Psi_{va}^{(0)} \rangle + \langle \Psi_{wb}^{(0)} | H_I | \Psi_{va}^{(1)} \rangle + \delta\omega \left\langle \Psi_{wb}^{(0)} \left| \frac{dH_I}{d\omega} \right| \Psi_{va}^{(0)} \right\rangle, \tag{79}$$

where the third term on the right-hand side is the derivative term discussed in Section III. With the first-order wave function [Eq. (75)], we obtain

$$T^{(2)} = \sum_{ij \neq wb} \rho_{ijwb}^{(1)*} \langle 0 | a_j a_i H_I a_v^\dagger a_a^\dagger | 0 \rangle + \sum_{ij \neq va} \rho_{ijva}^{(1)} \langle 0 | a_b a_w H_I a_i^\dagger a_j^\dagger | 0 \rangle$$

$$+ \delta\omega \left\langle 0 \left| a_b a_w \frac{dH_I}{d\omega} a_v^\dagger a_a^\dagger \right| 0 \right\rangle. \tag{80}$$

This expression can, in turn, be rewritten as

$$T^{(2)} = \sum_{ij \neq wb} \rho_{ijwb}^{(1)*}[(h_1)_{iv}\delta_{ja} - (h_1)_{ia}\delta_{jv} + (h_1)_{ja}\delta_{iv} - (h_1)_{jv}\delta_{ia}]$$
$$+ \sum_{ij \neq va} \rho_{ijva}^{(1)}[(h_1)_{wi}\delta_{bj} - (h_1)_{bi}\delta_{wj} + (h_1)_{bj}\delta_{wi} - (h_1)_{wj}\delta_{bi}]$$
$$+ \delta\omega\left[\left(\frac{dh_1}{d\omega}\right)_{wv}\delta_{ba} - \left(\frac{dh_1}{d\omega}\right)_{wa}\delta_{bv}\right] \quad (81)$$

$$= \sum_{ia \neq wb} \frac{v_{wbia}(h_1)_{iv} - v_{wbai}(h_1)_{iv}}{\epsilon_w - \epsilon_i} - \sum_{iv \neq wb} \frac{v_{wbvi}(h_1)_{ia} - v_{wbiv}(h_1)_{ia}}{\epsilon_i + \epsilon_v - \epsilon_w - \epsilon_b}$$
$$+ \sum_{ib \neq va} \frac{(h_1)_{wi}v_{ibva} - (h_1)_{wi}v_{biva}}{\epsilon_v - \epsilon_i} - \sum_{i} \frac{(h_1)_{bi}v_{wiva} - (h_1)_{bi}v_{iwva}}{\epsilon_i + \epsilon_w - \epsilon_v - \epsilon_a}$$
$$+ \sum_{i \neq w} \frac{U_{wi}(h_1)_{iv}\delta_{ba} - U_{wi}(h_1)_{ia}\delta_{bv}}{\epsilon_i - \epsilon_w}$$
$$+ \sum_{i \neq v} \frac{(h_1)_{wi}U_{iv}}{\epsilon_i - \epsilon_v}\delta_{ba} - \sum_{i \neq a} \frac{(h_1)_{wi}U_{ia}}{\epsilon_i - \epsilon_a}\delta_{bv}$$
$$+ \delta\omega\left[\left(\frac{dh_1}{d\omega}\right)_{wv}\delta_{ba} - \left(\frac{dh_1}{d\omega}\right)_{wa}\delta_{bv}\right]. \quad (82)$$

Equation (82) is the basic equation for the second-order transition amplitude. In contrast to the no-pair approach, we sum over both positive- and negative-energy states as we would if we were evaluating the transition amplitude from the QED point of view.

We are now in a position to prove the gauge invariance of $T^{(2)}$. For this purpose, we replace h_1 with $h_1 + \Delta h_1$ everywhere in the expression above and examine the resulting change, $\Delta T^{(2)}$. With the aid of Eq. (29), we obtain

$$\Delta T^{(2)} = -\frac{ie}{\hbar}\left\{-\sum_{ia \neq wb} \tilde{v}_{wbia}(\chi)_{iv} - \sum_{iv \neq wb} \tilde{v}_{wbvi}(\chi)_{ia}\right.$$
$$+ \sum_{ib \neq va} (\chi)_{wi}\tilde{v}_{ibva} + \sum_{i} (\chi)_{bi}\tilde{v}_{wiva}$$
$$+ \sum_{i \neq w} [U_{wi}(\chi)_{iv}\delta_{ba} - U_{wi}(\chi)_{ia}\delta_{bv}]$$
$$- \sum_{i \neq v} (\chi)_{wi}U_{iv}\delta_{ba} + \sum_{i \neq a} (\chi)_{wi}U_{ia}\delta_{bv}$$
$$\left. - (E_{wb}^{(1)} - E_{va}^{(1)})[(\chi)_{wv}\delta_{ba} - (\chi)_{wa}\delta_{bv}]\right\}. \quad (83)$$

This expression can be rewritten as

$$\Delta T^{(2)} = -\frac{ie}{\hbar} \left\{ \sum_i [-\tilde{v}_{wbia}(\chi)_{iv} - \tilde{v}_{wbvi}(\chi)_{ia} + (\chi)_{wi}\tilde{v}_{ibva} + (\chi)_{bi}\tilde{v}_{wiva}] \right.$$

$$+ \sum_i [(U_{wi}(\chi)_{iv} - (\chi)_{wi}U_{iv})\delta_{ba} - (U_{wi}(\chi)_{ia} - (\chi)_{wi}U_{ia})\delta_{bv}]$$

$$+ [\tilde{v}_{wbwb} - U_{ww} - U_{bb} - \tilde{v}_{vava} + U_{vv} + U_{aa}](\chi)_{wv}\delta_{ba}$$

$$- [\tilde{v}_{wbwb} - U_{ww} - U_{bb} - \tilde{v}_{vava} + U_{vv} + U_{aa}](\chi)_{wa}\delta_{bv}$$

$$\left. - (E^{(1)}_{wb} - E^{(1)}_{va})[(\chi)_{wv}\delta_{ba} - (\chi)_{wa}\delta_{bv}] \right\}. \tag{84}$$

The sums on the first two lines vanish identically if completeness of the intermediate states i is invoked. This fact is demonstrated in the Appendix. Furthermore, from Eq. (77), the terms on the third and fourth lines are seen to cancel the term on the last line. We, therefore, find that

$$\Delta T^{(2)} = 0,$$

provided one sums over both positive- and negative-energy intermediate states.

If we were carrying out an MBPT calculation of the amplitude using the no-pair interaction, the negative-energy states in Eq. (82) would not appear, and we would lose the gauge invariance. We expect the dominant contributions to the all-order no-pair amplitudes from negative-energy states to be accurately predicted by the negative-energy contributions to Eq. (82). Moreover, if the negative-energy contributions from perturbation theory are added to the exact no-pair amplitudes in length gauge and in velocity gauge, then we expect the "corrected" no-pair amplitudes in these two gauges to agree with one another to a high degree of accuracy. We verify the correctness of these conjectures in Section IV.C.

B. Angular Reduction

To compare the analysis given in the previous subsection with that for the no-pair amplitudes given in Section II, we must first replace the interaction Hamiltonian h_I used in the equations of the previous subsection with the transition operator $t_J^{(\lambda)}$ defined in Section II.C. We couple the orbitals v and a in the initial-state wave function to angular momentum J_I, M_I,

$$|J_I M_I\rangle = \eta_{va} \sum_{m_v m_a} \langle j_v m_v j_a m_a | J_I M_I \rangle \Psi_{va}, \tag{85}$$

where $\eta_{va} = 1/\sqrt{2}$ for the identical particle case ($n_a = n_v$ and $\kappa_a = \kappa_v$) and $\eta_{va} = 1$, otherwise. Similarly, we couple the final-state orbitals w and b to angular momentum J_F, M_F,

$$|J_F M_F\rangle = \sum_{m_w m_b} \langle j_w m_w j_b m_b | J_F M_F \rangle \Psi_{wb}. \tag{86}$$

Carrying out the sums over magnetic quantum numbers, we obtain the following expression for the reduced matrix element of the first-order transition amplitude,

$$\langle F\|T_J^{(\lambda)}\|I\rangle^{(1)} = \eta_{va}\sqrt{[J_F][J_I]}\left\{(-1)^{J_I+J+w+a}\begin{Bmatrix} J_F & J_I & J \\ v & w & b \end{Bmatrix}\langle w\|t_J^{(\lambda)}\|v\rangle\delta_{ba}\right.$$
$$\left.+(-1)^{J+w+a}\begin{Bmatrix} J_F & J_I & J \\ a & w & b \end{Bmatrix}\langle w\|t_J^{(\lambda)}\|a\rangle\delta_{bv}\right\}. \tag{87}$$

Similarly, the reduced-matrix element of the transition amplitude in second-order perturbation theory is found to be

$$\langle F\|T_J^{(\lambda)}\|I\rangle^{(2)} = \eta_{va}\sqrt{[J_F][J_I]}$$

$$\times\left\{\sum_{ia \neq wb, L}(-1)^{J_I+J+J_F+L}\begin{Bmatrix} w & b & J_F \\ a & i & L \end{Bmatrix}\begin{Bmatrix} J_F & J_I & J \\ v & i & a \end{Bmatrix}\frac{X_L(wbia)\langle i\|t_J^{(\lambda)}\|v\rangle}{\epsilon_w - \epsilon_i}\right.$$

$$+\sum_{ia \neq wb, L}(-1)^{J_I+J+L}\begin{Bmatrix} w & b & J_F \\ i & a & L \end{Bmatrix}\begin{Bmatrix} J_F & J_I & J \\ v & i & a \end{Bmatrix}\frac{X_L(wbai)\langle i\|t_J^{(\lambda)}\|v\rangle}{\epsilon_w - \epsilon_i}$$

$$+\sum_{iv \neq wb, L}(-1)^{J+L+b+a}\begin{Bmatrix} w & b & J_F \\ i & v & L \end{Bmatrix}\begin{Bmatrix} J_F & J_I & J \\ a & i & v \end{Bmatrix}\frac{X_L(wbvi)\langle i\|t_J^{(\lambda)}\|a\rangle}{\epsilon_i + \epsilon_v - \epsilon_w - \epsilon_b}$$

$$+\sum_{iv \neq wb, L}(-1)^{J_F+J+L+b+a}\begin{Bmatrix} w & b & J_F \\ v & i & L \end{Bmatrix}\begin{Bmatrix} J_F & J_I & J \\ a & i & v \end{Bmatrix}\frac{X_L(wbiv)\langle i\|t_J^{(\lambda)}\|a\rangle}{\epsilon_i + \epsilon_v - \epsilon_w - \epsilon_b}$$

$$+\sum_{ib \neq va, L}(-1)^{J+L+w-v}\begin{Bmatrix} v & a & J_I \\ b & i & L \end{Bmatrix}\begin{Bmatrix} J_F & J_I & J \\ i & w & b \end{Bmatrix}\frac{\langle w\|t_J^{(\lambda)}\|i\rangle X_L(ibva)}{\epsilon_v - \epsilon_i}$$

$$+\sum_{ib \neq va, L}(-1)^{J_I+J+L+w-v}\begin{Bmatrix} v & a & J_I \\ i & b & L \end{Bmatrix}\begin{Bmatrix} J_F & J_I & J \\ i & w & b \end{Bmatrix}\frac{\langle w\|t_J^{(\lambda)}\|i\rangle X_L(biva)}{\epsilon_v - \epsilon_i}$$

$$+\sum_{i,L}(-1)^{J_I+J+J_F+L+w+v}\begin{Bmatrix} v & a & J_I \\ i & w & L \end{Bmatrix}\begin{Bmatrix} J_F & J_I & J \\ i & b & w \end{Bmatrix}\frac{\langle b\|t_J^{(\lambda)}\|i\rangle X_L(wiva)}{\epsilon_i + \epsilon_w - \epsilon_v - \epsilon_a}$$

$$+\sum_{i,L}(-1)^{J+J_F+L+w+v}\begin{Bmatrix} v & a & J_I \\ w & i & L \end{Bmatrix}\begin{Bmatrix} J_F & J_I & J \\ i & b & w \end{Bmatrix}\frac{\langle b\|t_J^{(\lambda)}\|i\rangle X_L(iwva)}{\epsilon_i + \epsilon_w - \epsilon_v - \epsilon_a}$$

$$+(-1)^{J_I+J+w+a}\begin{Bmatrix} J_F & J_I & J \\ v & w & a \end{Bmatrix}\delta_{ba}\sum_{i \neq w}\delta_{\kappa_i \kappa_w}\frac{U_{wi}\langle i\|t_J^{(\lambda)}\|v\rangle}{\epsilon_i - \epsilon_w}$$

$$\left.+(-1)^{J+w+a}\begin{Bmatrix} J_F & J_I & J \\ a & w & v \end{Bmatrix}\delta_{bv}\sum_{i \neq w}\delta_{\kappa_i \kappa_w}\frac{U_{wi}\langle i\|t_J^{(\lambda)}\|a\rangle}{\epsilon_i - \epsilon_w}\right.$$

$$+(-1)^{J_t+J+w+a}\begin{Bmatrix}J_F & J_I & J\\ v & w & a\end{Bmatrix}\delta_{ba}\sum_{i\neq v}\delta_{\kappa_i\kappa_v}\frac{\langle w\|t_J^{(\lambda)}\|i\rangle U_{iv}}{\epsilon_i-\epsilon_v}$$

$$+(-1)^{J+w+a}\begin{Bmatrix}J_F & J_I & J\\ a & w & v\end{Bmatrix}\delta_{bv}\sum_{i\neq a}\delta_{\kappa_i\kappa_a}\frac{\langle w\|t_J^{(\lambda)}\|i\rangle U_{ia}}{\epsilon_i-\epsilon_a}$$

$$+\frac{E_{wb}^{(1)}-E_{va}^{(1)}}{\hbar}\left[(-1)^{J_t+J+w+a}\begin{Bmatrix}J_F & J_I & J\\ v & w & a\end{Bmatrix}\left\langle w\left\|\left(\frac{dt_J^{(\lambda)}}{d\omega}\right)\right\|v\right\rangle\delta_{ba}\right.$$

$$\left.+(-1)^{J+w+a}\begin{Bmatrix}J_F & J_I & J\\ a & w & v\end{Bmatrix}\left\langle w\left\|\left(\frac{dt_J^{(\lambda)}}{d\omega}\right)\right\|a\right\rangle\delta_{bv}\right]\Bigg\}. \qquad (88)$$

With the notation

$$\langle F\|T_J^{(\lambda)}\|I\rangle^{(1+2)} = \langle F\|T_J^{(\lambda)}\|I\rangle^{(1)} + \langle F\|T_J^{(\lambda)}\|I\rangle^{(2)},$$

the Einstein A-coefficients are given by

$$A_J^{(\lambda)} = 2\alpha(\omega+\delta\omega)\frac{[J]}{[J_I]}\frac{J+1}{J}|\langle F\|T_J^{(\lambda)}\|I\rangle^{(1+2)}|^2. \qquad (89)$$

For comparison with no-pair theory, we need expressions for the reduced matrix elements of the multipole-moment operators. These are related to the reduced matrix elements of the transition operators by the expressions

$$\langle F\|T_J^{(\lambda)}\|I\rangle^{(1+2)} = \frac{\alpha^J(\omega+\delta\omega)^J}{(2J+1)!!}\begin{cases}\langle F\|Q_J^{(1)}\|I\rangle^{(1+2)} & \text{for }\lambda=1\\ (\alpha/2)\langle F\|M_J\|I\rangle^{(1+2)} & \text{for }\lambda=0.\end{cases} \qquad (90)$$

As a specific example, let us consider the $2^3P_0 \to 2^3S_1$ transition. We use Eqs. (87) and (88) to evaluate the transition amplitudes and Eq. (90) to obtain the reduced matrix element of the electric-dipole moment operator. In Table III, we list the contributions to length-form and velocity-form electric-dipole moment matrix elements for $Z = 20$, 40, and 80. We see that the first-order matrix elements are identical in length and velocity forms. The sum over intermediate states in Eq. (88) is seen to contribute very little to the length-form moments, but substantially to the velocity-form matrix elements. By contrast, the derivative terms in Eq. (88) are seen to contribute substantially to the length form, but give only tiny contributions to the velocity form. When the contributions from both parts of the second-order calculation are added, we obtain gauge-invariant second-order matrix elements. Summing the first-order and second-order contributions gives values that are not only gauge invariant, but are also in good quantitative agreement with the length-form no-pair results presented, for example, in Table II.

TABLE III
PERTURBATION THEORY APPLIED TO THE REDUCED MATRIX ELEMENT OF THE DIPOLE-MOMENT OPERATOR IN LENGTH FORM AND IN VELOCITY FORM FOR THE $2^3P_0 \to 2^3S_1$ TRANSITION

Z	Contribution	$\langle F\|Q_1\|I\rangle_l$	$\langle F\|Q_1\|I\rangle_v$
20	1st	0.132096	0.132096
	Σ_i	−0.001058	0.022721
	$dT^{(1)}/d\omega$	0.023780	0.000001
	2nd	0.022722	0.022722
	1st + 2nd	0.154818	0.154818
40	1st	0.061919	0.061919
	Σ_i	−0.000246	0.011911
	$dT^{(1)}/d\omega$	0.012157	0.000001
	2nd	0.011911	0.011911
	1st + 2nd	0.073830	0.073830
80	1st	0.024229	0.024229
	Σ_i	−0.000049	0.007958
	$dT^{(1)}/d\omega$	0.008008	0.000001
	2nd	0.007959	0.007959
	1st + 2nd	0.032188	0.032188

Note. The contribution "1st" is from first-order perturbation theory. The contributions "Σ_i" and "$dT^{(1)}/d\omega$" are from the sum over states and the derivative term in second-order perturbation theory. The rows labeled "2nd" and "1st + 2nd" are the second-order contributions and the sum of the first- and second-order contributions, respectively. Results are calculated for the case in which $U(r) = v_0(1s, r)$ and do not include the Breit interaction.

C. NEGATIVE-ENERGY COMPONENTS OF SECOND-ORDER AMPLITUDES

In this subsection, we examine the decomposition of the transition amplitudes calculated in second-order perturbation theory into positive-energy and negative-energy components. The negative-energy contributions to the transition amplitude arise from those terms in the sum over states i in the second-order amplitude given in Eq. (88) for which $\epsilon_i < -2mc^2$.

The value of the sum over i and the corresponding contributions from negative-energy states depend on the potential used in defining the lowest-order Hamiltonian. Thus, the negative-energy contributions to the amplitudes obtained in a calculation starting from $U(r) = 0$ (the Coulomb-field case) will be quite different from the contributions in a calculation starting from $U(r) = v_0(1s, r)$ (the Hartree case). The reason for this difference is obvious on inspecting Eq. (88). In the Hartree case, there are substantial cancellations among the sums on the first eight lines and the counter-terms with U_{ik} on the following four lines; such cancellations cannot occur in the Coulomb case. Since we wish to use the negative-energy contributions from perturbation theory to estimate the negative-energy terms omitted in the no-pair calculations, it is important to carry out both the perturbation-theory calculation and the no-pair calculation starting from the same

TABLE IV
Comparison of Length-Form $\langle F\|Q_1\|I\rangle_l$ and Velocity-Form $\langle F\|Q_1\|I\rangle_v$ Dipole Matrix Elements from No-Pair Calculations and from Second-Order Perturbation Theory, "pert-th"

Z	Type	$\langle F\|Q_1\|I\rangle_l^{(+)}$	$\langle F\|Q_1\|I\rangle_l^{(-)}$	$\langle F\|Q_1\|I\rangle_v^{(+)}$	$\langle F\|Q_1\|I\rangle_v^{(-)}$	$\Delta Q^{(+)}$
			$\langle 2^3P_0 - 2^3S_1\rangle$			
20	no-pair	0.154750		0.154400		0.000350
20	pert-th	0.148663	1(−12)	0.148281	0.000382	0.000382
50	no-pair	0.057636		0.056543		0.001093
50	pert-th	0.056582	5(−10)	0.055457	0.001125	0.001125
100	no-pair	0.022809		0.020442		0.002367
100	pert-th	0.022378	2(−07)	0.020006	0.002374	0.002372
			$\langle 2^3P_2 - 2^3S_1\rangle$			
20	no-pair	0.347546		0.347046		0.000500
20	pert-th	0.337611	4(−09)	0.337055	0.000557	0.000556
50	no-pair	0.133056		0.132865		0.000192
50	pert-th	0.132769	6(−08)	0.132569	0.000203	0.000200
100	no-pair	0.059372		0.059365		0.000007
100	pert-th	0.059368	2(−07)	0.059360	−9(−08)	0.000008

Note. Here, superscripts (+) and (−) refer to contributions from positive- and negative-energy states, respectively. $\Delta Q^{(+)}$ is the difference in the positive-energy parts of the length-form and velocity-form amplitudes. All calculations are carried out on a Coulomb basis, $U(r) = 0$. The notation $a(-b)$ designates $a \times 10^{-b}$.

potential. The *energies* and *length-form matrix elements* obtained from the no-pair Hamiltonian are found to be independent of $U(r)$ to better than six significant figures. However, the *velocity-form* matrix elements obtained from a no-pair calculation starting from the Hartree potential are quite different from those obtained from a corresponding calculation starting from a Coulomb potential.

These points are illustrated in Tables IV and V, in which no-pair and perturbation theory amplitudes for the sensitive $2^3P_0 \to 2^3S_1$ and $2^3P_2 \to 2^3S_1$ transitions at $Z = 20$, 50, and 100 are decomposed into positive- and negative-energy components and compared. Table IV gives the amplitudes obtained starting from a Coulomb potential and Table V gives those obtained starting from a Hartree potential. The positive-energy part of the length-form and velocity-form amplitudes from perturbation theory are seen to differ from the corresponding no-pair amplitudes by less than 4% in the Coulomb case and less than 0.2% in the Hartree case. For both cases, it is seen that there are no contributions to the length-form amplitudes at the level of accuracy of interest here (five significant figures). There are, however, nonnegligible negative-energy contributions to the velocity-form amplitudes in perturbation theory. If we use the negative-energy contributions to the perturbation theory velocity-form amplitudes to estimate the

TABLE V
Comparison of Length-Form $\langle F\|Q_1\|I\rangle_l$ and Velocity-Form $\langle F\|Q_1\|I\rangle_v$ Dipole Matrix Elements from No-Pair Calculations and from Second-Order Perturbation Theory, "pert-th"

Z	Type	$\langle F\|Q_1\|I\rangle_l^{(+)}$	$\langle F\|Q_1\|I\rangle_l^{(-)}$	$\langle F\|Q_1\|I\rangle_v^{(+)}$	$\langle F\|Q_1\|I\rangle_v^{(-)}$	$\Delta Q^{(+)}$
			$\langle 2^3P_0 - 2^3S_1 \rangle$			
20	no-pair	0.154750		0.154831		−0.000081
20	pert-th	0.154936	−8(−09)	0.155018	−0.000082	−0.000082
50	no-pair	0.057636		0.057345		0.000291
50	pert-th	0.057668	−7(−08)	0.057373	0.000295	0.000295
100	no-pair	0.022809		0.021401		0.001408
100	pert-th	0.022772	−2(−07)	0.021349	0.001421	0.001423
			$\langle 2^3P_2 - 2^3S_1 \rangle$			
20	no-pair	0.347546		0.346656		0.000890
20	pert-th	0.347808	2(−08)	0.346932	0.000877	0.000876
50	no-pair	0.133056		0.132655		0.000401
50	pert-th	0.133068	1(−07)	0.132670	0.000399	0.000398
100	no-pair	0.059372		0.059311		0.000061
100	pert-th	0.059378	4(−07)	0.059321	0.000052	0.000057

Note. Here, superscripts (+) and (−) refer to contributions from positive- and negative-energy states, respectively. $\Delta Q^{(+)}$ is the difference in the positive-energy parts of the length-form and velocity-form amplitudes. All calculations are carried out in a Hartree basis, $U(r) = v_0(1s, r)$. The notation $a(-b)$ designates $a \times 10^{-b}$.

missing negative-energy corrections to the velocity-form no-pair amplitudes, then the corrected velocity-form no-pair amplitudes go into agreement with the length-form no-pair amplitudes to four significant figures in the Coulomb-field case and to five significant figures in the Hartree case. Since for either starting potential, the length-form no-pair potental is insensitive to either negative-energy corrections or to the starting potential, this form for the amplitude will be adopted for our final tabulations.

For magnetic-multipole transitions, we have only one form (velocity form) of the transition operator to consider. Again, we carry out no-pair calculations and perturbation-theory calculations in two potentials, the nuclear Coulomb potential and the Hartree potential. The no-pair amplitudes are found to differ somewhat in the two potentials, because the missing negative-energy contributions to the amplitudes depend on the starting potential. It is found that the positive-energy amplitudes from perturbation theory agree with the no-pair amplitudes at the 1% level of accuracy for the Coulomb case and at the 0.1% level for the Hartree case. We use the negative-frequency contributions from perturbation theory to estimate the missing negative-energy contributions from the no-pair amplitudes. The results are shown in Table VI, where we list the no-pair amplitudes, the

TABLE VI

Reduced Matrix Elements of the Magnetic Dipole Moment for $2^3S_1 \to 1^1S_0$ Transitions and the Magnetic Quadrupole Moment for $2^3P_2 \to 1^1S_0$ Transitions

Z	Hartree no-pair (+)	Hartree pert-th (−)	Hartree $\langle F \| M \| I \rangle$	Coulomb no-pair (+)	Coulomb pert-th (−)	Coulomb $\langle F \| M \| I \rangle$
			$\langle 2^3S_1 \to 1^1S_0 \rangle$			
5	4.26756(−4)	−9.99661(−7)	4.25756(−4)	4.26020(−4)	−2.32841(−8)	4.25997(−4)
10	1.76852(−3)	−1.44150(−6)	1.76708(−3)	1.76749(−3)	−1.62401(−7)	1.76733(−3)
20	7.21056(−3)	−8.04420(−7)	7.20976(−3)	7.21122(−3)	−1.24108(−6)	7.20997(−3)
30	1.63948(−2)	2.70616(−6)	1.63975(−2)	1.64020(−2)	−4.23978(−6)	1.63978(−2)
40	2.94451(−2)	9.29513(−6)	2.94544(−2)	2.94647(−2)	−1.04395(−5)	2.94543(−2)
50	4.65451(−2)	1.82441(−5)	4.65633(−2)	4.65844(−2)	−2.15904(−5)	4.65628(−2)
60	6.79538(−2)	2.79272(−5)	6.79817(−2)	6.80207(−2)	−4.01290(−5)	6.79806(−2)
70	9.40213(−2)	3.57720(−5)	9.40571(−2)	9.41243(−2)	−6.94242(−5)	9.40549(−2)
80	1.25217(−1)	3.81507(−5)	1.25255(−1)	1.25366(−1)	−1.14086(−4)	1.25252(−1)
90	1.62145(−1)	3.02091(−5)	1.62175(−1)	1.62351(−1)	−1.80357(−4)	1.62171(−1)
100	2.05548(−1)	5.64640(−6)	2.05554(−1)	2.05824(−1)	−2.76660(−4)	2.05547(−1)
			$\langle 2^3P_2 \to 1^1S_0 \rangle$			
5	1.19578	1.08648(−5)	1.19578	1.19579	−9.37185(−6)	1.19578
10	5.88007(−1)	1.15662(−5)	5.88019(−1)	5.88025(−1)	−7.24570(−6)	5.88018(−1)
20	2.88201(−1)	1.18155(−5)	2.88213(−1)	2.88219(−1)	−6.36802(−6)	2.88213(−1)
30	1.88203(−1)	1.17979(−5)	1.88215(−1)	1.88221(−1)	−6.03153(−6)	1.88215(−1)
40	1.37464(−1)	1.16429(−5)	1.37476(−1)	1.37481(−1)	−5.84629(−6)	1.37475(−1)
50	1.06295(−1)	1.13411(−5)	1.06306(−1)	1.06311(−1)	−5.79091(−6)	1.06305(−1)
60	8.48580(−2)	1.08573(−5)	8.48689(−2)	8.48745(−2)	−5.90008(−6)	8.48686(−2)
70	6.89535(−2)	1.01491(−5)	6.89636(−2)	6.89700(−2)	−6.22291(−6)	6.89638(−2)
80	5.64791(−2)	9.17127(−6)	5.64883(−2)	5.64948(−2)	−6.81199(−6)	5.64880(−2)
90	4.62644(−2)	7.88189(−6)	4.62723(−2)	4.62797(−2)	−7.71841(−6)	4.62720(−2)
100	3.76016(−2)	6.24541(−6)	3.76078(−2)	3.76165(−2)	−8.98877(−6)	3.76075(−2)

Note. Values in columns labeled "Hartree" are evaluated using basis functions created in the Hartree potential $U(r) = v_0(1s, r)$, and those listed in columns "Coulomb" are calculated in a Coulomb-field basis. The no-pair moments are modified by adding negative-energy corrections from perturbation theory to give the corrected matrix elements listed in the columns labeled $\langle F \| M \| I \rangle$. The notation $a(-b)$ designates $a \times 10^{-b}$.

negative-energy contributions from perturbation theory, and the resulting modified amplitudes for M_1 transitions $2^3S_1 \to 1^1S_0$ and M_2 transitions $2^3P_2 \to 1^1S_0$, calculated in two potentials. The resulting modified amplitudes are seen to be independent of the starting potential to better than five figures for all M_2 transitions and for those M_1 transitions with $Z \geqslant 30$. For $Z < 30$, we expect the modified no-pair amplitudes calculated using the Hartree potential to be more accurate than those using the Coulomb potential. We estimate the error in the low-Z M_1 amplitudes to be less than 0.1% for $Z < 30$.

In summary, we adopt length-form no-pair amplitudes in our final tabulation of E_1 transitions, and we adopt no-pair amplitudes modified by adding negative-energy contributions from perturbation theory for magnetic-multipole transitions. For M_1 transitions with $Z < 30$, our amplitudes are calculated starting from the Hartree potential. In the following section, we present our final results and compare them with other calculations and with experimentation.

V. Results and Comparisons

In our final tabulations, we use length-form E_1 transition amplitudes calculated with exact no-pair wave functions, including both the Coulomb and the Breit interactions. By using the length gauge, we obtain amplitudes that are independent of the starting potential and that have negligible corrections from negative-energy states. In our calculations, we use theoretical energies obtained from the no-pair Hamiltonian that include QED corrections. As discussed in Refs. [3, 4], these energies are in very close agreement with experimental energies. Amplitudes for the M_1 and M_2 transitions are calculated in the velocity gauge. Negative-energy corrections, which are important in the velocity gauge, are obtained from second-order perturbation theory.

A. Comparison with Other Theories

Over the years, numerous nonrelativistic calculations of E_1 amplitudes and rates have been made for the 2–1 and 2–2 transitions along the helium isoelectronic sequence. Among the most precise of these calculations are those of Schiff et al. [21], who evaluated oscillator strengths of the $2^1P \to 1^1S$ and $2^3P \to 2^3S$ transitions for ions with nuclear charges in the range $Z = 2$–10 using variational wave functions. These benchmark calculations were confirmed and extended by Kono and Hattori [22] and by Sanders and Knight [23]. Cann and Thakkar [24] have carried out high-precision variational calculations of oscillator strengths for S–P and

TABLE VII
Comparison of the Present Relativistic Calculations of Oscillator Strengths with the High-Precision Nonrelativistic Calculations of Ref [24]

	$2^1P \to 1^1S$		$2^3P \to 2^3S$	
Z	Nonrelativistic	Present	Nonrelativistic	Present
2	0.27617	0.2761	0.5391	0.5391
3	0.45663	0.4565	0.30794	0.3081
4	0.55156	0.5513	0.21314	0.2134
5	0.60892	0.6085	0.16263	0.1632
6	0.64707	0.6463	0.13138	0.1321
7	0.67420	0.6731	0.11018	0.1112
8	0.69445	0.6929	0.09486	0.09619
9	0.71013	0.7080	0.08327	0.08595
10	0.72262	0.7198	0.07420	0.07625

P–D transitions for ions in the range $Z = 2$–10. The nonrelativistic oscillator strengths from Ref. [24] are rounded to five digits and compared with the present calculations in Table VII. The current results for the triplet transitions are obtained by summing the oscillator strengths of the three fine-structure components to give a composite relativistic oscillator strength. As can be seen from the table, the differences between the relativistic and the nonrelativistic values range from one part in the fourth figure at $Z = 2$ to three parts in the third figure at $Z = 10$. This results in an error of about 0.4% in the nonrelativistic oscillator strength for the singlet transition and an error of 3% for the corresponding triplet transition at $Z = 10$. The source of these errors is the use of nonrelativistic energies and wave functions in the calculations of Ref. [24].

Accurate nonrelativistic multiconfiguration Hartree–Fock (MCHF) calculations of the oscillator strength of the $2^3P \to 2^3S$ transition in Li^+ ($Z = 3$) were reported in Ref. [25]. These calculations were in close agreement with those of Refs. [21–24] when nonrelativistic energies were used but came into agreement with the present calculation when relativistic corrections to the energy were made. Such energy corrections are ambiguous for triplet transitions and not at all possible for intercombination transitions.

Drake, in his unified model [26], devised a scheme for smoothly joining precise nonrelativistic calculations of the $2^1P_1 \to 1^1S_0$ oscillator strengths at low Z to uncorrelated, but relativistic, oscillator strengths at high Z. Using this method, accurate oscillator strengths for both the allowed transitions $2^1P_1 \to 1^1S_0$ and the intercombination transitions $2^3P_1 \to 1^1S_0$ were obtained throughout the isoelectronic sequence. In Table VIII, we compare the oscillator strengths from the unified model with values from the present calculation. For low values of Z, the unified model removes the discrepan-

TABLE VIII

Comparison of the Present Relativistic Calculations of Oscillator Strengths for $2^1P_1 \to 1^1S_0$ and $2^3P_1 \to 1^1S_0$ with the Values from the Unified Model of Ref [26]

	$2^1P_1 \to 1^1S_0$		$2^3P_1 \to 1^1S_0$	
Z	Present	Unified	Present	Unified
2	0.2761	0.2762	2.810(−8)	2.774(−8)
5	0.6085	0.6084	7.093(−6)	7.082(−6)
10	0.7198	0.7196	4.425(−4)	4.424(−4)
20	0.7454	0.7452	2.192(−2)	2.193(−2)
30	0.6629	0.6628	1.080(−1)	1.079(−1)
40	0.5743	0.5743	1.849(−1)	1.847(−1)
50	0.5161	0.5162	2.216(−1)	2.213(−1)
60	0.4727	0.4728	2.356(−1)	2.352(−1)
70	0.4333	0.4332	2.382(−1)	2.376(−1)
80	0.3932	0.3928	2.343(−1)	2.335(−1)
90	0.3505	0.3493	2.260(−1)	2.248(−1)
100	0.3042	0.3016	2.137(−1)	2.119(−1)

cies with the nonrelativistic variational calculations, and for high Z the unified model remains in agreement with the present calculations at the 1% level of accuracy. Similarly, the unified model gives oscillator strengths for the intercombination transitions that agree with the present calculations at the 1% level throughout the isoelectronic sequence, with the exception of very low values of Z. For the intercombination transition, our values disagree with those of Drake by more than 1% for $Z = 2$, and agreement is not obtained until $Z = 7$. We attribute this discrepancy to Drake's first-order treatment of the Breit interaction. While the Breit correction to the matrix element of the $2^1P_1 - 1^1S_0$ transition is less than a percent of the total matrix element for the entire isoelectronic sequence, this is not so for the spin-forbidden transition. In fact, for $Z = 2$, the Breit contribution to the matrix element is actually 52% of the total matrix element! This contribution drops to 12% for $Z = 10$. Clearly for very low values of Z, the treatment of the Breit as a first-order perturbation is inadequate. Since we include the Breit contribution to the matrix element exactly, within the numerical limits previously stated, our values of the rates for the $2^3P_1 - 1^1S_0$ transition are expected to be precise.

In addition to the unified model, a variety of other theoretical methods have been used to carry out relativistic calculations of transition probabilities along the helium isoelectronic sequence. Among these are the relativistic random-phase approximation (RRPA) [10], the multiconfiguration Dirac–Fock method (MCDF) [27], relativistic CI expansions [28], the relativistic many-body theory [8], and the QED perturbation theory [9, 29, 30]. In Table IX we compare the decay rates predicted by the present calculation

TABLE IX

COMPARISON OF THE PRESENT E_1 DECAY RATES (s^{-1}) FOR THE $2^1P_1 \to 1^1S_0$ AND $2^3P_1 \to 1^1S_0$ TRANSITIONS WITH THOSE OF OTHER RELATIVISTIC THEORIES

	$2^1P_1 \to 1^1S_0$				$2^3P_1 \to 1^1S_0$			
Z	Present	Unified [26]	RRPA [10]	Krause [28]	Present	Unified [26]	RRPA [10]	Krause [28]
2	1.798(9)	1.799(9)	1.71(9)	1.86(9)	1.757(2)	1.764(2)	2.33(2)	1.57(2)
5	3.719(11)	3.719(11)		4.05(11)	4.210(6)	4.220(6)	4.36(6)	4.15(6)
10	8.850(12)	8.851(12)	8.85(12)	9.28(12)	5.347(9)	5.356(9)	5.40(9)	5.36(9)
20	1.642(14)	1.643(14)	1.65(14)	1.68(14)	4.777(12)	4.786(12)	4.85(12)	4.93(12)
30	7.764(14)	7.773(14)	7.81(14)	7.88(14)	1.251(14)	1.252(14)	1.22(14)	1.31(14)
40	2.212(15)	2.216(15)		2.23(15)	7.008(14)	7.017(14)		8.30(14)
50	5.057(15)	5.071(15)	5.08(15)	5.10(15)	2.120(15)	2.123(15)	2.12(15)	2.19(15)
60	1.006(16)	1.010(16)		1.02(16)	4.845(15)	4.853(15)		4.99(15)
70	1.805(16)	1.813(16)		1.82(16)	9.460(15)	9.480(15)		9.72(15)
80	2.986(16)	3.000(16)	3.01(16)	3.02(16)	1.668(16)	1.672(16)	1.68(16)	1.71(16)
90	4.616(16)	4.638(16)		4.68(16)	2.732(16)	2.741(16)		2.80(16)
100	6.733(16)	6.760(16)	6.79(16)	6.85(16)	4.231(16)	4.250(16)	4.27(16)	4.34(16)

TABLE X

COMPARISON OF THE PRESENT $2^3S_1 \to 1^1S_0$ M_1 DECAY RATES (s^{-1}) WITH THOSE OF OTHER THEORIES

Z	Present	pert.-th. [9]	RRPA [10]	Drake [32]	Krause [28]
2	1.266(−4)	1.253(−4)	1.73(−4)	1.272(−4)	8.737(−5)
5	6.696	6.731	6.90	6.695	6.407
10	1.092(4)	1.098(4)	1.10(4)	1.087(4)	1.081(4)
20	1.412(7)	1.418(7)	1.42(7)	1.383(7)	1.410(7)
30	8.981(8)	9.023(8)	9.01(8)		9.000(8)
40	1.719(10)	1.729(10)	1.73(10)		1.727(10)
50	1.726(11)	1.737(11)	1.74(11)		1.737(11)
60	1.161(12)	1.171(12)	1.17(12)		1.171(12)
70	5.968(12)	6.032(12)	6.03(12)		6.040(12)
80	2.540(13)	2.577(13)	2.58(13)		2.582(13)
90	9.439(13)	9.639(13)	9.63(13)		9.661(13)
100	3.181(14)		3.29(14)		3.299(14)

of the $2^{1,3}P_1 \to 1^1S_0$ transitions with the relativistic calculations of Refs. [10, 26, 28]. All of the calculations agree at the few percent level for high values of Z, for which correlation is unimportant. The $2^3P_1 \to 1^1S_0$ rates at low values of Z are very sensitive to correlation effects, leading to large errors in the approximate calculations of Refs. [10, 28]. The situation is similar for other E_1 transitions; the allowed transition rates are determined to within a few percent in the approximate calculations, but the more sensitive intercombination transitions are poorly described by the approximate calculations, especially at low Z.

A great deal of attention has been given to the M_1 transitions $2^3S_1 \to 1^1S_0$ during the past 25 years. The interest stemmed from the identification of intense M_1 lines in the spectrum of the sun by Gabriel and Jordan [31]. Calculations of the M_1 decay rate based on an expansion of the M_1 matrix element in powers of $Z\alpha$ were carried out in Refs. [29, 32, 33], and other relativistic calculations were reported in Refs. [8–10, 27, 28, 30]. In Table X, we compare the present calculations with some of these previous calculations. For low Z ions, the approximate RRPA [10] and Krause [28] calculations are seriously in error because of inadequate treatment of correlation corrections. At intermediate Z, all M_1 calculations agree to within 1%. At high Z, the M_1 amplitudes are sensitive to the atomic wave functions inside the nucleus; this has the consequence that the present M_1 calculations, which include nuclear finite size corrections, deviate from the previous calculations, which were carried out for point nuclei, by 2 to 3% for ions with $Z > 80$.

The calculation closest in spirit to the present one is that of Lindroth and Salomonson [8], who evaluated the M_1 transition in Ar^{+16} using "exact" no-pair wavefunctions. The Breit interaction and negative-energy correc-

TRANSITION AMPLITUDES IN THE HELIUM ISOELECTRIC SEQUENCES   291

TABLE XI

COMPARISON OF THE PRESENT $2^3P_2 \to 1^1S_0$ M_2 DECAY RATES (s^{-1}) WITH THOSE OF OTHER THEORIES

Z	Present	Drake [34]	RRPA [10]	Kundu et al. [35]	Krause [28]
2	3.271(−1)	3.27(−1)	3.94(−1)	3.62(−1)	1.183(−1)
5	5.014(3)	5.01(3)	5.09(3)	4.96(3)	4.082(3)
10	2.257(6)	2.26(6)	2.27(6)		2.078(6)
20	7.510(8)		7.56(8)		7.254(8)
30	2.104(10)		2.12(10)		2.064(10)
40	2.208(11)		2.24(11)		2.187(11)
50	1.365(12)		1.39(12)		1.360(12)
60	6.060(12)		6.17(12)		6.070(12)
70	2.146(13)		2.19(13)		2.160(13)
80	6.457(13)		6.63(13)		6.529(13)
90	1.718(14)		1.77(14)		1.746(14)
100	4.156(14)		4.32(14)		4.254(14)

tions were treated perturbatively. The contributions to the M_1 matrix element from positive- and negative-energy states in the two calculations are given in the following table.

$10^5 \times M$	Present	Ref. [8]
(+)	2.1255	2.1226
(−)	−0.0004	−0.0001
Sum	2.1251	2.1225

The photon energy used in Ref. [8] was $\omega = 114.072$ a.u., in good agreement with the value $\omega = 114.077$ a.u. used in the present calculation. The lifetime of the 2^3S_1 state from Ref. [8] is $\tau = 209.4 \pm 0.4$ ns, compared to the present value of $\tau = 208.9 \pm 0.2$ ns. The principal difference between the two calculations appears to be the treatment of the Breit interaction.

There have been various calculations of the $2^3P_2 \to 1^1S_0$ M_2 rate as well [10, 27, 28, 34, 35]. In Table XI, we compare our calculations of the M_2 rate with some of these earlier calculations. We find excellent agreement for low Z with the results of Drake [34] and for high Z with the results of Krause [28].

B. COMPARISON WITH EXPERIMENT

The precision of laboratory measurements of the $2^3S_1 \to 1^1S_0$ (M_1) transition rate has increased dramatically over the past few decades. Currently, the most accurately measured M_1 rate is for C^{4+} [55]. The measured

TABLE XII
COMPARISON OF EXPERIMENTAL M_1 DECAY RATES OF THE 2^3S_1 STATE A^{exp} WITH DECAY RATES A^{th} FROM THE PRESENT CALCULATION

Z	A^{exp}	A^{th}	Unit (s^{-1})	Reference
2	1.11 ± 0.33	1.266	10^{-4}	47
3	1.71 ± 0.38	2.035	10^{-2}	49
6	4.857 ± 0.011	4.860	10^{1}	55
7	2.561 ± 0.031	2.537	10^{2}	55
10	1.105 ± 0.018	1.092	10^{4}	53
16	1.41 ± 0.17	1.426	10^{6}	48
17	2.82 ± 0.19	2.661	10^{6}	48
18	4.93 ± 0.29	4.787	10^{6}	50
22	3.88 ± 0.20	3.750	10^{6}	45
23	5.92 ± 0.24	5.913	10^{6}	46
26	2.08 ± 0.26	2.075	10^{8}	46
35	4.46 ± 0.14	4.360	10^{9}	52
36	5.848 ± 0.075	5.822	10^{9}	56
41	2.200 ± 0.008	2.217	10^{10}	57
47	9.01 ± 0.16	9.083	10^{10}	54
54	3.92 ± 0.12	3.846	10^{11}	51

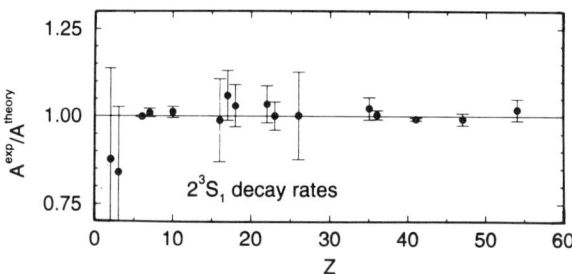

FIG. 3. Ratio of experimental M_1 decay rates of the 2^3S_1 state A^{exp} to those from the present calculation A^{theory}.

rate $A_{exp} = 48.57 \pm 0.11 \text{ s}^{-1}$ differs with the present theoretical value $A_{th} = 48.60 \text{ s}^{-1}$ by less than one standard deviation. The only other high-precision measurement of the M_1 rate is for Nb^{39+} [57], for which the measured rate differs from the present theoretical value by two standard deviations. Comparisons of the present calculations with other measured M_1 rates are given in Table XII. The measured rates, which vary by 15 orders of magnitude for Z in the range 2–54, are also compared with the present calculations in Fig. 3.

The decay of the 2^3P_2 state consists of two branches; $2^3P_2 \to 2^3S_1$ (E_1) and $2^3P_2 \to 1^1S_0$ (M_2). For those ions having nuclei with nonvanishing angular momenta, there is a third branch caused by hyperfine mixing of the

TABLE XIII
Comparison of Experimental Decay Rates A^{exp} of the 2^3P_2 State with Decay Rates A^{th} from the Present Calculation

Z	A^{exp}	A^{th}	Unit (s^{-1})	Reference
6	5.99 ± 0.22	5.705	10^7	71
7	6.71 ± 0.32	6.901	10^7	71
8	8.26 ± 0.14	8.182	10^7	66
9	10.0 ± 1.0	9.598	10^7	62
	9.58 ± 0.14	9.598	10^7	66
13	1.852 ± 0.069	1.877	10^8	65
15	2.778 ± 0.077	2.904	10^8	67
	2.94 ± 0.26	2.904	10^8	68
16	4.00 ± 0.32	3.743	10^8	59
17	5.38 ± 0.29	4.933	10^8	60
18	5.9 ± 1.0	6.618	10^8	58
	6.17 ± 0.30	6.618	10^8	63
22	2.27 ± 0.15	2.326	10^9	63
	2.48 ± 0.25	2.326	10^9	69
24	4.65 ± 0.76	4.347	10^9	69
26	9.1 ± 1.6	7.929	10^9	61
	8.9 ± 1.0	7.929	10^9	64
	8.3 ± 1.0	7.929	10^9	70
28	1.429 ± 0.061	1.404	10^{10}	73
29	2.13 ± 0.23	1.849	10^{10}	72
47	8.06 ± 0.72	9.140	10^{11}	74

Note. Contributions to A^{th} from hyperfine mixing have been omitted.

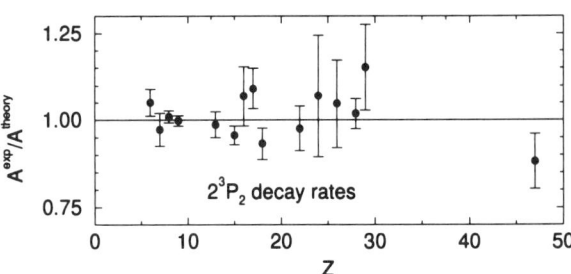

FIG. 4. Ratio of experimental $M_2 + E_1$ decay rates of the 2^3P_2 state A^{exp} to those from the present calculation A^{theory}.

2^3P_2 state with the 2^3P_1 state. This branch typically contributes only small corrections to the sum of the rates of the other two branches and will be ignored in the theoretical rates used here for comparison. For low values of Z, the E_1 branch dominates, while, for $Z \geqslant 19$, the M_2 rate is larger. Although the lifetime measurements for the 2^3P_2 state are not as precise as those for the 2^3S_1 state, measurements at the 1.5% level of accuracy were

TABLE XIV
Comparison of Experimental E_1 Decay Rates A^{exp} for 2^3P_1 States with the Corresponding Calculated Decay Rates A^{th}, Which Are Sums of the Rates to the 2^3S_1 and 1^1S_0 States

Z	A^{exp}	A^{th}	Unit (s^{-1})	Reference
6	8.85 ± 0.39	8.478	10^7	78
7	2.04 ± 0.12	2.071	10^8	78
8	6.58 ± 0.35	6.292	10^8	66
9	1.88 ± 0.07	1.924	10^9	66
12	3.45 ± 0.18	3.388	10^{10}	77
13	7.81 ± 0.43	7.533	10^{10}	77
14	1.57 ± 0.08	1.572	10^{11}	76
16	6.37 ± 0.73	5.820	10^{11}	76

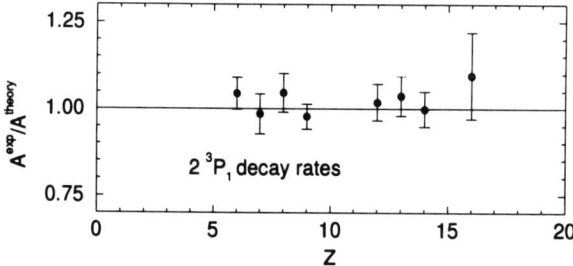

Fig. 5. Ratio of experiment E_1 decay rates of the 2^3P_1 state A^{exp} to those from the present calculation A^{theory}.

carried out for O^{+6} and F^{+7} [66]. The resulting decay rates for both of these ions agree with the present calculation to within the experimental errors. Comparisons of experimental decay rates with the present calculations for M_2 transitions are given in Table XIII and shown graphically in Fig. 4.

Measurements of the rate of the intercombination transition $2^3P_1 \to 1^1S_0$ (E_1) at low values of Z have also been made. These experimental measurements are compared with the present calculations in Table XIV and shown in Fig. 5.

C. Compilation of Transition Rates

Finally, in Table XV, we present a summary of our results for E_1 transitions in the range $Z = 2-100$, and, in Table XVI, we present a summary of our M_1 and M_2 calculations over the same range. In these tables, we list transition wavelengths λ (Å), transition multiplets, angular momenta of the initial and final states, Einstein A-coefficients, and line strengths, S. Additionally, for E_1 transitions, we tabulate (absorption) oscillator strengths, f. The line strength S (a.u.) is the square of the reduced matrix element of the multipole-moment operator. Thus,

$$S = \begin{cases} |\langle F\|Q_1^{(1)}\|I\rangle|^2, & \text{for } E_1 \text{ transitions} \\ |\langle F\|M_J\|I\rangle|^2, & \text{for } M_1 \text{ and } M_2 \text{ transitions.} \end{cases} \quad (91)$$

The Einstein A-coefficients (s^{-1}) listed in the tables were obtained in terms of line strengths using the relations

$$A_{ki} = \frac{2.02613 \times 10^{18}}{\lambda^3 [J_k]} S_{E_1}, \quad \text{for } E_1 \text{ transitions} \quad (92)$$

$$= \frac{2.69735 \times 10^{13}}{\lambda^3 [J_k]} S_{M_1}, \quad \text{for } M_1 \text{ transitions} \quad (93)$$

$$= \frac{1.49097 \times 10^{13}}{\lambda^5 [J_k]} S_{M_2}, \quad \text{for } M_2 \text{ transitions,} \quad (94)$$

where $[J_k] = 2J_k + 1$. Finally, the (dimensionless) oscillator strengths for E_1 transitions are given in terms of E_1 line strengths by

$$f_{ik} = \frac{303.756}{\lambda [J_i]} S_{E_1}. \quad (95)$$

In our numerical work, we have attempted to control the numerical accuracy of calculations of amplitudes to five significant figures. We believe that the transitions rates presented in Tables XV and XVI are accurate to 0.1% or better from the numerical point of view.

D. FUTURE PROSPECTS

As discussed above, all of the transition rates listed in Tables XV and XVI are expected to be accurate to better than 0.1%. At this level of accuracy it becomes important to consider contributions to the rates from other sources. The starting point of our calculations, the no-pair Hamiltonian, does not by itself include the effects of the finite mass of the nucleus or QED, although the effect of finite nuclear size has been included. However, all of these effects have been included as completely as possible in our previous calculations of $n = 1$ and $n = 2$ energy levels in the helium isoelectronic sequence [3,4]. By using these energies in the present calculation, we are partially accounting for recoil and QED. We now turn to the question of the importance of those recoil and QED corrections not accounted for by modifying the energies.

TABLE XV
E_1 Values

Z	Transition array	Multiplet	$\lambda(\text{Å})$	J_i-J_k	A_{ki} (s^{-1})	f_{ik}	S (a.u.)
2	$1s^2-1s2p$	$^1S-^3P$	591.33	0–1	1.787(2)	2.810(−8)	5.4698(−8)
		$^1S-^1P$	584.25	0–1	1.799(9)	2.761(−1)	5.3110(−1)
	$1s2s-1s2p$	$^3S-^3P$	10,832.	1–2	1.022(7)	2.995(−1)	3.2038(1)
			10,832	1–1	1.022(7)	1.797(−1)	1.9223(1)
			10,831.	1–0	1.022(7)	5.990(−2)	6.4077
		$^3S-^1P$	8,864.8	1–1	1.552	1.828(−8)	1.6006(−6)
		$^1S-^1P$	20,584.	0–1	1.976(6)	1.255(−1)	2.5516(1)
3	$1s^2-1s2p$	$^1S-^3P$	202.30	0–1	1.790(4)	3.295(−7)	2.1945(−7)
		$^1S-^1P$	199.26	0–1	2.556(10)	4.565(−1)	2.9947(−1)
	$1s2s-1s2p$	$^3S-^3P$	5485.6	1–2	2.276(7)	1.712(−1)	9.2730
			5486.2	1–1	2.276(7)	1.027(−1)	5.5637
			5484.7	1–0	2.278(7)	3.424(−2)	1.8546
		$^3S-^1P$	3879.7	1–1	4.190(1)	9.456(−8)	3.6232(−6)
		$^1S-^1P$	9583.4	0–1	5.148(6)	7.088(−2)	6.7089
4	$1s^2-1s2p$	$^1S-^3P$	101.68	0–1	4.002(5)	1.861(−6)	6.2305(−7)
		$^1S-^1P$	100.25	0–1	1.220(11)	5.513(−1)	1.8194(−1)
	$1s2s-1s2p$	$^3S-^3P$	3721.7	1–2	3.427(7)	1.186(−1)	4.3597
			3723.7	1–1	3.421(7)	7.112(−2)	2.6156
			3722.1	1–0	3.426(7)	2.372(−2)	8.7187(−1)
		$^3S-^1P$	2441.7	1–1	3.853(2)	3.444(−7)	8.3054(−6)
		$^1S-^1P$	6143.3	0–1	8.757(6)	4.955(−2)	3.0063
5	$1s^2-1s2p$	$^1S-^3P$	61.086	0–1	4.226(6)	7.093(−6)	1.4264(−6)
		$^1S-^1P$	60.311	0–1	3.719(11)	6.085(−1)	1.2081(−1)
	$1s2s-1s2p$	$^3S-^3P$	2822.3	1–2	4.557(7)	9.070(−2)	2.5281
			2826.5	1–1	4.536(7)	5.433(−2)	1.5165
			2825.2	1–0	4.542(7)	1.812(−2)	5.0551(−1)
		$^1S-^1P$	1771.9	1–1	2.061(3)	9.701(−7)	1.6977(−5)
		$^1S-^1P$	4491.8	0–1	1.262(7)	3.818(−2)	1.6936

Z							
6	$1s^2$–$1s2p$	^1S–^3P	40.728	0–1	2.830(7)	2.111(−5)	2.8309(−6)
		^1S–^1P	40.266	0–1	8.864(11)	6.463(−1)	8.5679(−2)
	$1s2s$–$1s2p$	^3S–^3P	2271.5	1–2	5.702(7)	7.352(−2)	1.6492
		^1S–^1P	2278.5	1–1	5.648(7)	4.396(−2)	9.8919(−1)
			2277.9	1–0	5.652(7)	1.466(−2)	3.2971(−1)
		^3S–^1P	1386.7	1–1	7.952(3)	2.293(−6)	3.1400(−5)
		^1S–^3P	3527.4	0–1	1.668(7)	3.111(−2)	1.0839
7	$1s^2$–$1s2p$	^1S–^3P	29.083	0–1	1.394(8)	5.303(−5)	5.0772(−6)
		^1S–^1P	28.786	0–1	1.806(12)	6.731(−1)	6.3789(−2)
	$1s2s$–$1s2p$	^3S–^3P	1896.7	1–2	6.891(7)	6.195(−2)	1.1604
		^1S–^1P	1907.3	1–1	6.774(7)	3.694(−2)	6.9586(−1)
			1907.6	1–0	6.770(7)	1.231(−2)	2.3194(−1)
8	$1s^2$–$1s2p$	^3S–^1P	1137.1	1–1	2.465(4)	4.778(−6)	5.3656(−5)
		^1S–^3P	2896.0	0–1	2.092(7)	2.630(−2)	7.5225(−1)
		^1S–^1P	21.803	0–1	5.499(8)	1.176(−4)	8.4392(−6)
	$1s2s$–$1s2p$	^3S–^3P	21.601	0–1	3.302(12)	6.929(−1)	4.9275(−2)
		^1S–^1P	1623.6	1–2	8.149(7)	5.368(−2)	8.6073(−1)
			1638.3	1–1	7.925(7)	3.189(−2)	5.1599(−1)
			1639.9	1–0	7.902(7)	1.062(−2)	1.7198(−1)
9	$1s^2$–$1s2p$	^3S–^1P	961.95	1–1	6.533(4)	9.063(−6)	8.6099(−5)
		^1S–^3P	2449.6	0–1	2.537(7)	2.283(−2)	5.5223(−1)
		^1S–^1P	16.950	0–1	1.833(9)	2.369(−4)	1.3219(−5)
	$1s2s$–$1s2p$	^1S–^1P	16.806	0–1	5.574(12)	7.080(−1)	3.9173(−2)
		^3S–^3P	1414.4	1–2	9.506(7)	4.752(−2)	6.6375(−1)
			1433.8	1–1	9.113(7)	2.809(−2)	3.9774(−1)
			1436.9	1–0	9.054(7)	9.342(−3)	1.3258(−1)
10	$1s^2$–$1s2p$	^3S–^1P	832.19	1–1	1.539(5)	1.598(−5)	1.3130(−4)
		^1S–^3P	2116.1	1–1	3.011(7)	2.021(−2)	4.2245(−1)
		^1S–^1P	13.553	0–1	5.356(9)	4.425(−4)	1.9742(−5)
	$1s2s$–$1s2p$	^3S–^3P	13.447	0–1	8.851(12)	7.198(−1)	3.1865(−2)
		^1S–^1P	1248.1	1–2	1.099(8)	4.278(−2)	5.2738(−1)
			1272.8	1–1	1.034(8)	2.513(−2)	3.1585(−1)
			1277.7	1–0	1.023(8)	8.344(−3)	1.0530(−1)
		^3S–^1P	731.96	1–1	3.307(5)	2.656(−5)	1.9200(−4)
		^1S–^1P	1856.3	0–1	3.521(7)	1.819(−2)	3.3345(−1)

TABLE XV Continued

Z	Transition array	Multiplet	λ(Å)	J_i–J_k	A_{ki} (s^{-1})	f_{ik}	S (a.u.)
11	$1s^2$–$1s2p$	^1S–^3P	11.083	0–1	1.405(10)	7.764(−4)	2.8328(−5)
		^1S–^1P	11.003	0–1	1.339(13)	7.291(−1)	2.6408(−2)
	$1s2s$–$1s2p$	^3S–^3P	1111.8	1–2	1.265(8)	3.908(−2)	4.2907(−1)
			1142.3	1–1	1.163(8)	2.276(−2)	2.5678(−1)
			1149.2	1–0	1.143(8)	7.544(−3)	8.5629(−2)
		^3S–^1P	652.03	1–1	6.600(5)	4.207(−5)	2.7089(−4)
		^1S–^1P	1647.1	0–1	4.077(7)	1.658(−2)	2.6976(−1)
12	$1s^2$–$1s2p$	^1S–^3P	9.2310	0–1	3.375(10)	1.293(−3)	3.9307(−5)
		^1S–^1P	9.1685	0–1	1.948(13)	7.364(−1)	2.2227(−2)
	$1s2s$–$1s2p$	^3S–^3P	997.46	1–2	1.453(8)	3.612(−2)	3.5587(−1)
			1034.3	1–1	1.299(8)	2.083(−2)	2.1275(−1)
			1043.3	1–0	1.266(8)	6.889(−3)	7.0983(−2)
		^3S–^1P	586.59	1–1	1.240(6)	6.399(−5)	3.7069(−4)
		^1S–^1P	1473.9	0–1	4.696(7)	1.529(−2)	2.2262(−1)
13	$1s^2$–$1s2p$	^1S–^3P	7.8068	0–1	7.519(10)	2.061(−3)	5.2970(−5)
		^1S–^1P	7.7571	0–1	2.742(13)	7.420(−1)	1.8950(−2)
	$1s2s$–$1s2p$	^3S–^3P	899.65	1–2	1.669(8)	3.375(−2)	2.9990(−1)
			943.16	1–1	1.441(8)	1.922(−2)	1.7903(−1)
			954.38	1–0	1.393(8)	6.343(−3)	5.9784(−2)
		^3S–^1P	531.85	1–1	2.217(6)	9.400(−5)	4.9377(−4)
		^1S–^1P	1327.3	0–1	5.393(7)	1.424(−2)	1.8673(−1)
14	$1s^2$–$1s2p$	^1S–^3P	6.6881	0–1	1.570(11)	3.159(−3)	6.9556(−5)
		^1S–^1P	6.6478	0–1	3.755(13)	7.463(−1)	1.6333(−2)
	$1s2s$–$1s2p$	^3S–^3P	814.70	1–2	1.919(8)	3.183(−2)	2.5614(−1)
			865.15	1–1	1.592(8)	1.786(−2)	1.5262(−1)
			878.65	1–0	1.524(8)	5.880(−3)	5.1029(−2)
		^3S–^1P	485.22	1–1	3.796(6)	1.340(−4)	6.4205(−4)
		^1S–^1P	1200.9	0–1	6.191(7)	1.339(−2)	1.5876(−1)

15	$1s^2-1s2p$	$^1S-^3P$	5.7933	0–1	3.099(11)	4.678(−3)	8.9226(−5)
		$^1S-^1P$	5.7601	0–1	5.021(13)	7.492(−1)	1.4208(−2)
	$1s2s-1s2p$	$^3S-^3P$	739.90	1–2	2.214(8)	3.028(−2)	2.2128(−1)
			797.52	1–1	1.751(8)	1.670(−2)	1.3153(−1)
			813.29	1–0	1.659(8)	5.485(−3)	4.4055(−2)
			444.88	1–1	6.265(6)	1.859(−4)	8.1675(−4)
			1090.1	0–1	7.118(7)	1.268(−2)	1.3651(−1)
16	$1s^2-1s2p$	$^1S-^3P$	5.0664	0–1	5.818(11)	6.717(−3)	1.1203(−4)
		$^1S-^1P$	5.0386	0–1	6.576(13)	7.509(−1)	1.2456(−2)
	$1s2s-1s2p$	$^3S-^3P$	673.41	1–2	2.562(8)	2.903(−2)	1.9307(−1)
			738.33	1–1	1.920(8)	1.569(−2)	1.1440(−1)
			756.31	1–0	1.799(8)	5.142(−3)	3.8410(−2)
			409.48	1–1	1.001(7)	2.517(−4)	1.0181(−3)
			991.61	0–1	8.208(7)	1.210(−2)	1.1850(−1)
17	$1s^2-1s2p$	$^1S-^3P$	4.4679	0–1	1.044(12)	9.374(−3)	1.3788(−4)
		$^1S-^1P$	4.4444	0–1	8.458(13)	7.514(−1)	1.0994(−2)
	$1s2s-1s2p$	$^3S-^3P$	613.78	1–2	2.978(8)	2.803(−2)	1.6991(−1)
			686.05	1–1	2.097(8)	1.480(−2)	1.0027(−1)
			706.13	1–0	1.944(8)	4.843(−3)	3.3776(−2)
			378.07	1–1	1.556(7)	3.335(−4)	1.2452(−3)
			903.23	0–1	9.505(7)	1.163(−2)	1.0371(−1)
18	$1s^2-1s2p$	$^1S-^3P$	3.9693	0–1	1.799(12)	1.274(−2)	1.6654(−4)
		$^1S-^1P$	3.9490	0–1	1.070(14)	7.506(−1)	9.7583(−3)
	$1s2s-1s2p$	$^3S-^3P$	559.98	1–2	3.477(8)	2.724(−2)	1.5066(−1)
			639.56	1–1	2.284(8)	1.401(−2)	8.8473(−2)
			661.57	1–0	2.094(8)	4.580(−3)	2.9925(−2)
19	$1s^2-1s2p$	$^3S-^1P$	349.90	1–1	2.358(7)	4.329(−4)	1.4958(−3)
		$^1S-^1P$	823.20	0–1	1.106(8)	1.124(−2)	9.1391(−2)
	$1s^2-1s2p$	$^1S-^3P$	3.5495	0–1	2.984(12)	1.691(−2)	1.9759(−4)
		$^1S-^1P$	3.5318	0–1	1.334(14)	7.486(−1)	8.7039(−3)
	$1s2s-1s2p$	$^3S-^3P$	511.25	1–2	4.079(8)	2.664(−2)	1.3450(−1)
			598.05	1–1	2.479(8)	1.329(−2)	7.8502(−2)
			621.77	1–0	2.250(8)	4.346(−3)	2.6691(−2)
		$^3S-^1P$	324.39	1–1	3.494(7)	5.512(−4)	1.7661(−3)
		$^1S-^1P$	750.19	0–1	1.296(8)	1.093(−2)	8.1010(−2)

TABLE XV Continued

Z	Transition array	Multiplet	λ(Å)	J_i–J_k	A_{ki} (s^{-1})	f_{ik}	S (a.u.)
20	$1s^2$–$1s2p$	^1S–^3P	3.1927	0–1	4.782(12)	2.192(−2)	2.3044(−4)
		^1S–^1P	3.1771	0–1	1.642(14)	7.453(−1)	7.7958(−3)
	$1s2s$–$1s2p$	^3S–^3P	466.91	1–2	4.809(8)	2.619(−2)	1.2079(−1)
			560.75	1–1	2.681(8)	1.264(−2)	6.9990(−2)
			585.94	1–0	2.412(8)	4.138(−3)	2.3947(−2)
		^3S–^1P	301.13	1–1	5.073(7)	6.896(−4)	2.0510(−3)
		^1S–^1P	683.34	0–1	1.528(8)	1.069(−2)	7.2171(−2)
21	$1s^2$–$1s2p$	^1S–^3P	2.8869	0–1	7.421(12)	2.782(−2)	2.6437(−4)
		^1S–^1P	2.8730	0–1	1.996(14)	7.409(−1)	7.0074(−3)
	$1s2s$–$1s2p$	^3S–^3P	426.48	1–2	5.697(8)	2.589(−2)	1.0906(−1)
			527.11	1–1	2.890(8)	1.204(−2)	6.2660(−2)
			553.48	1–0	2.581(8)	3.952(−3)	2.1601(−2)
		^3S–^1P	279.79	1–1	7.228(7)	8.483(−4)	2.3441(−3)
		^1S–^1P	622.00	0–1	1.812(8)	1.051(−2)	6.4577(−2)
22	$1s^2$–$1s2p$	^1S–^3P	2.6229	0–1	1.117(13)	3.458(−2)	2.9855(−4)
		^1S–^1P	2.6104	0–1	2.399(14)	7.352(−1)	6.3184(−3)
	$1s2s$–$1s2p$	^3S–^3P	389.58	1–2	6.782(8)	2.572(−2)	9.8954(−2)
			496.70	1–1	3.103(8)	1.148(−2)	5.6299(−2)
			523.97	1–0	2.758(8)	3.783(−3)	1.9578(−2)
		^3S–^1P	260.09	1–1	1.013(8)	1.027(−3)	2.6384(−3)
		^1S–^1P	565.58	0–1	2.165(8)	1.038(−2)	5.8002(−2)
23	$1s^2$–$1s2p$	^1S–^3P	2.3933	0–1	1.636(13)	4.215(−2)	3.3211(−4)
		^1S–^1P	2.3819	0–1	2.855(14)	7.285(−1)	5.7127(−3)
	$1s2s$–$1s2p$	^3S–^3P	355.86	1–2	8.109(8)	2.566(−2)	9.0179(−2)
			468.96	1–1	3.323(8)	1.096(−2)	5.0745(−2)
			497.02	1–0	2.941(8)	3.631(−3)	1.7822(−2)
		^3S–^1P	241.85	1–1	1.397(8)	1.225(−3)	2.9262(−3)
		^1S–^1P	513.84	0–1	2.602(8)	1.030(−2)	5.2271(−2)

Z							
24	$1s^2$–$1s2p$	^1S–^3P	2.1925	0–1	2.334(13)	5.046(−2)	3.6419(−4)
		^1S–^1P	2.1820	0–1	3.366(14)	7.208(−1)	5.1780(−3)
	$1s2s$–$1s2p$	^3S–^3P	325.06	1–2	9.735(8)	2.570(−2)	8.2513(−2)
		^1S–^1P	444.07	1–1	3.538(8)	1.046(−2)	4.5870(−2)
		^1S–^3P	472.24	1–0	3.134(8)	3.492(−3)	1.6288(−2)
25		^3S–^1P	224.85	1–1	1.901(8)	1.441(−3)	3.2003(−3)
	$1s^2$–$1s2p$	^1S–^3P	466.10	0–1	3.151(8)	1.026(−2)	4.7251(−2)
		^1S–^1P	2.0158	0–1	3.249(13)	5.937(−2)	3.9400(−4)
	$1s2s$–$1s2p$	^3S–^3P	2.0061	0–1	3.935(14)	7.123(−1)	4.7041(−3)
		^1S–^1P	296.91	1–2	1.173(9)	2.584(−2)	7.5776(−2)
		^1S–^3P	421.24	1–1	3.756(8)	9.993(−3)	4.1573(−2)
26		^3S–^1P	449.42	1–0	3.335(8)	3.366(−3)	1.4940(−2)
	$1s^2$–$1s2p$	^1S–^1P	209.02	1–1	2.554(8)	1.673(−3)	3.4541(−3)
		^1S–^3P	422.45	0–1	3.837(8)	1.027(−2)	4.2833(−2)
	$1s2s$–$1s2p$	^1S–^1P	1.8595	0–1	4.421(13)	6.876(−2)	4.2092(−4)
		^3S–^3P	1.8504	0–1	4.566(14)	7.031(−1)	4.2830(−3)
		^3S–^1P	271.19	1–2	1.419(9)	2.607(−2)	6.9824(−2)
27		^1S–^3P	400.42	1–1	3.973(8)	9.551(−3)	3.7772(−2)
	$1s^2$–$1s2p$	^1S–^1P	428.32	1–0	3.545(8)	3.250(−3)	1.3749(−2)
		^3S–^3P	194.26	1–1	3.392(8)	1.919(−3)	3.6822(−3)
	$1s2s$–$1s2p$	^3S–^1P	382.55	0–1	4.697(8)	1.030(−2)	3.8931(−2)
		^1S–^3P	1.7206	0–1	5.893(13)	7.846(−2)	4.4445(−4)
		^1S–^1P	1.7120	0–1	5.260(14)	6.934(−1)	3.9080(−3)
28		^1S–^3P	247.71	1–2	1.721(9)	2.638(−2)	6.4540(−2)
	$1s^2$–$1s2p$	^3S–^1P	381.36	1–1	4.189(8)	9.134(−3)	3.4402(−2)
		^1S–^1P	408.74	1–0	3.765(8)	3.144(−3)	1.2691(−2)
	$1s2s$–$1s2p$	^1S–^3P	180.47	1–1	4.459(8)	2.177(−3)	3.8807(−3)
		^3S–^1P	346.15	0–1	5.777(8)	1.038(−2)	3.5476(−2)
		^1S–^3P	1.5966	0–1	7.705(13)	8.834(−2)	4.6429(−4)
		^1S–^1P	1.5884	0–1	6.022(14)	6.833(−1)	3.5735(−3)
		^3S–^3P	226.29	1–2	2.092(9)	2.677(−2)	5.9827(−2)
		^3S–^1P	363.90	1–1	4.402(8)	8.739(−3)	3.1407(−2)
		^1S–^3P	390.53	1–0	3.996(8)	3.046(−3)	1.1748(−2)
		^3S–^1P	167.60	1–1	5.806(8)	2.445(−3)	4.0471(−3)
		^1S–^1P	313.06	0–1	7.134(8)	1.048(−2)	3.2408(−2)

TABLE XV Continued

Z	Transition array	Multiplet	λ(Å)	J_i–J_k	A_{ki} (s^{-1})	f_{ik}	S (a.u.)
29	$1s^2$–$1s2p$	^1S–^3P	1.4854	0–1	9.899(13)	9.823(−2)	4.8033(−4)
		^1S–^1P	1.4776	0–1	6.855(14)	6.731(−1)	3.2743(−3)
	$1s2s$–$1s2p$	^3S–^3P	206.76	1–2	2.549(9)	2.723(−2)	5.5606(−2)
			347.86	1–1	4.611(8)	8.365(−3)	2.8740(−2)
			373.54	1–0	4.239(8)	2.956(−3)	1.0904(−2)
		^3S–^1P	155.58	1–1	7.498(8)	2.721(−3)	4.1806(−3)
		^1S–^1P	283.05	0–1	8.839(8)	1.062(−2)	2.9680(−2)
30	$1s^2$–$1s2p$	^1S–^3P	1.3853	0–1	1.251(14)	1.080(−1)	4.9257(−4)
		^1S–^1P	1.3778	0–1	7.763(14)	6.629(−1)	3.0067(−3)
	$1s2s$–$1s2p$	^3S–^3P	188.96	1–2	3.112(9)	2.776(−2)	5.1810(−2)
			333.09	1–1	4.818(8)	8.013(−3)	2.6361(−2)
			357.65	1–0	4.493(8)	2.872(−3)	1.0144(−2)
		^3S–^1P	144.37	1–1	9.610(8)	3.003(−3)	4.2815(−3)
		^1S–^1P	255.90	0–1	1.098(9)	1.078(−2)	2.7248(−2)
31	$1s^2$–$1s2p$	^1S–^3P	1.2949	0–1	1.559(14)	1.176(−1)	5.0116(−4)
		^1S–^1P	1.2877	0–1	8.752(14)	6.527(−1)	2.7669(−3)
	$1s2s$–$1s2p$	^3S–^3P	172.74	1–2	3.804(9)	2.836(−2)	4.8385(−2)
			319.46	1–1	5.021(8)	7.682(−3)	2.4239(−2)
			342.74	1–0	4.760(8)	2.794(−3)	9.4586(−3)
		^3S–^1P	133.92	1–1	1.223(9)	3.290(−3)	4.3510(−3)
		^1S–^1P	231.37	0–1	1.367(9)	1.097(−2)	2.5077(−2)
32	$1s^2$–$1s2p$	^1S–^3P	1.2131	0–1	1.916(14)	1.268(−1)	5.0640(−4)
		^1S–^1P	1.2060	0–1	9.824(14)	6.427(−1)	2.5517(−3)
	$1s2s$–$1s2p$	^3S–^3P	157.98	1–2	4.654(9)	2.902(−2)	4.5283(−2)
			306.84	1–1	5.221(8)	7.370(−3)	2.2336(−2)
			328.74	1–0	5.040(8)	2.722(−3)	8.8379(−3)
		^3S–^1P	124.19	1–1	1.549(9)	3.581(−3)	4.3921(−3)
		^1S–^1P	209.27	0–1	1.705(9)	1.119(−2)	2.3136(−2)

Z	Transition	Term	λ				
33	$1s^2-1s2p$	$^1S-{}^3P$	1.1386	0–1	2.327(14)	1.357(−1)	5.0858(−4)
		$^1S-{}^1P$	1.1318	0–1	1.099(15)	6.329(−1)	2.3582(−3)
	$1s2s-1s2p$	$^3S-{}^3P$	144.55	1–2	5.697(9)	2.974(−2)	4.2465(−2)
			295.15	1–1	5.419(8)	7.077(−3)	2.0630(−2)
			315.55	1–0	5.335(8)	2.655(−3)	8.2741(−3)
		$^1S-{}^1P$	115.14	1–1	1.950(9)	3.875(−3)	4.4070(−3)
34	$1s^2-1s2p$	$^1S-{}^3P$	189.38	0–1	2.128(9)	1.144(−2)	2.1396(−2)
		$^1S-{}^1P$	1.0707	0–1	2.795(14)	1.441(−1)	5.0802(−4)
			1.0641	0–1	1.224(15)	6.235(−1)	2.1840(−3)
	$1s2s-1s2p$	$^3S-{}^3P$	132.33	1–2	6.977(9)	3.053(−2)	3.9898(−2)
			284.28	1–1	5.614(8)	6.802(−3)	1.9097(−2)
			303.11	1–0	5.646(8)	2.592(−3)	7.7604(−3)
35	$1s^2-1s2p$	$^3S-{}^1P$	106.73	1–1	2.443(9)	4.173(−3)	4.3988(−3)
		$^1S-{}^1P$	171.48	0–1	2.656(9)	1.171(−2)	1.9833(−2)
		$^1S-{}^3P$	1.0087	0–1	3.324(14)	1.521(−1)	5.0510(−4)
		$^1S-{}^1P$	1.0022	0–1	1.360(15)	6.143(−1)	2.0268(−3)
	$1s2s-1s2p$	$^3S-{}^3P$	121.22	1–2	8.543(9)	3.137(−2)	3.7552(−2)
			274.15	1–1	5.807(8)	6.543(−3)	1.7717(−2)
			291.37	1–0	5.972(8)	2.534(−3)	7.2910(−3)
36	$1s^2-1s2p$	$^3S-{}^1P$	98.933	1–1	3.048(9)	4.473(−3)	4.3707(−3)
		$^1S-{}^1P$	155.41	0–1	3.316(9)	1.201(−2)	1.8427(−2)
		$^1S-{}^3P$	0.95181	0–1	3.917(14)	1.596(−1)	5.0016(−4)
		$^1S-{}^1P$	0.94540	0–1	1.506(15)	6.056(−1)	1.8848(−3)
	$1s2s-1s2p$	$^3S-{}^3P$	111.11	1–2	1.046(10)	3.226(−2)	3.5403(−2)
			264.69	1–1	5.999(8)	6.301(−3)	1.6472(−2)
			280.25	1–0	6.316(8)	2.479(−3)	6.8611(−3)
37	$1s^2-1s2p$	$^3S-{}^1P$	91.699	1–1	3.789(9)	4.776(−3)	4.3255(−3)
		$^1S-{}^1P$	140.97	0–1	4.137(9)	1.233(−2)	1.7160(−2)
		$^1S-{}^3P$	0.89955	0–1	4.579(14)	1.666(−1)	4.9352(−4)
		$^1S-{}^1P$	0.89323	0–1	1.664(15)	5.972(−1)	1.7561(−3)
	$1s2s-1s2p$	$^3S-{}^3P$	101.91	1–2	1.280(10)	3.321(−2)	3.3429(−2)
			255.84	1–1	6.190(8)	6.074(−3)	1.5347(−2)
			269.71	1–0	6.678(8)	2.427(−3)	6.4662(−3)
		$^3S-{}^1P$	84.998	1–1	4.692(9)	5.082(−3)	4.2663(−3)
		$^1S-{}^1P$	128.00	0–1	5.158(9)	1.267(−2)	1.6015(−2)

303

TABLE XV Continued

Z	Transition array	Multiplet	λ(Å)	J_i–J_k	A_{ki} (s^{-1})	f_{ik}	S (a.u.)
38	$1s^2$–$1s2p$	^1S–^3P	0.85141	0–1	5.313(14)	1.732(−1)	4.8549(−4)
		^1S–^1P	0.84518	0–1	1.834(15)	5.892(−1)	1.6394(−3)
	$1s2s$–$1s2p$	^3S–^3P	93.547	1–2	1.565(10)	3.422(−2)	3.1612(−2)
			247.53	1–1	6.380(8)	5.861(−3)	1.4327(−2)
			259.71	1–0	7.059(8)	2.379(−3)	6.1027(−3)
		^3S–^1P	78.798	1–1	5.791(9)	5.391(−3)	4.1954(−3)
		^1S–^1P	116.35	0–1	6.422(9)	1.303(−2)	1.4977(−2)
39	$1s^2$–$1s2p$	^1S–^3P	0.80697	0–1	6.122(14)	1.793(−1)	4.7634(−4)
		^1S–^1P	0.80082	0–1	2.016(15)	5.815(−1)	1.5332(−3)
	$1s2s$–$1s2p$	^3S–^3P	85.934	1–2	1.912(10)	3.527(−2)	2.9936(−2)
			239.73	1–1	6.570(8)	5.660(−3)	1.3401(−2)
			250.20	1–0	7.460(8)	2.334(−3)	5.7674(−3)
		^3S–^1P	73.063	1–1	7.126(9)	5.703(−3)	4.1154(−3)
		^1S–^1P	105.88	0–1	7.985(9)	1.342(−2)	1.4035(−2)
40	$1s^2$–$1s2p$	^1S–^3P	0.76586	0–1	7.011(14)	1.850(−1)	4.6632(−4)
		^1S–^1P	0.75978	0–1	2.212(15)	5.742(−1)	1.4363(−3)
	$1s2s$–$1s2p$	^3S–^3P	79.004	1–2	2.333(10)	3.638(−2)	2.8386(−2)
			232.37	1–1	6.760(8)	5.472(−3)	1.2558(−2)
			241.16	1–0	7.884(8)	2.291(−3)	5.4573(−3)
		^3S–^1P	67.763	1–1	8.743(9)	6.019(−3)	4.0281(−3)
		^1S–^1P	96.473	0–1	9.912(9)	1.383(−2)	1.3178(−2)
41	$1s^2$–$1s2p$	^1S–^3P	0.72777	0–1	7.983(14)	1.902(−1)	4.5562(−4)
		^1S–^1P	0.72174	0–1	2.421(15)	5.673(−1)	1.3479(−3)
	$1s2s$–$1s2p$	^3S–^3P	72.694	1–2	2.843(10)	3.754(−2)	2.6950(−2)
			225.45	1–1	6.948(8)	5.295(−3)	1.1789(−2)
			232.57	1–0	8.328(8)	2.251(−3)	5.1701(−3)
		^3S–^1P	62.867	1–1	1.070(10)	6.338(−3)	3.9354(−3)
		^1S–^1P	88.011	0–1	1.228(10)	1.426(−2)	1.2396(−2)

Z	Transition	Term	λ				
42	$1s^2$–$1s2p$	$1S$–$3P$	0.69239	0–1	9.043(14)	1.950(−1)	4.4446(−4)
		$1S$–$1P$	0.68642	0–1	2.645(15)	5.606(−1)	1.2668(−3)
	$1s2s$–$1s2p$	$3S$–$3P$	66.943	1–2	3.460(10)	3.875(−2)	2.5617(−2)
			218.88	1–1	7.140(8)	5.128(−3)	1.1086(−2)
			224.34	1–0	8.799(8)	2.213(−3)	4.9035(−3)
43	$1s^2$–$1s2p$	$3S$–$1P$	58.343	1–1	1.306(10)	6.662(−3)	3.8389(−3)
		$1S$–$3P$	80.387	0–1	1.519(10)	1.471(−2)	1.1681(−2)
		$1S$–$1P$	0.65948	0–1	1.020(15)	1.994(−1)	4.3297(−4)
	$1s2s$–$1s2p$	$1S$–$3P$	0.65355	0–1	2.885(15)	5.542(−1)	1.1925(−3)
		$3S$–$3P$	61.701	1–2	4.205(10)	4.000(−2)	2.4377(−2)
			212.67	1–1	7.331(8)	4.971(−3)	1.0441(−2)
			216.51	1–0	9.295(8)	2.177(−3)	4.6557(−3)
44	$1s^2$–$1s2p$	$3S$–$1P$	54.167	1–1	1.589(10)	6.991(−3)	3.7400(−3)
		$1S$–$1P$	73.517	1–1	1.874(10)	1.519(−2)	1.1026(−2)
		$1S$–$3P$	0.62881	0–1	1.144(15)	2.035(−1)	4.2128(−4)
		$1S$–$1P$	0.62292	0–1	3.141(15)	5.481(−1)	1.1241(−3)
	$1s2s$–$1s2p$	$3S$–$3P$	56.918	1–2	5.104(10)	4.131(−2)	2.3223(−2)
			206.76	1–1	7.525(8)	4.823(−3)	9.8489(−3)
			209.01	1–0	9.819(8)	2.144(−3)	4.4248(−3)
45	$1s^2$–$1s2p$	$3S$–$1P$	50.311	1–1	1.930(10)	7.325(−3)	3.6395(−3)
		$1S$–$1P$	67.318	0–1	2.308(10)	1.568(−2)	1.0424(−2)
		$1S$–$3P$	0.60019	0–1	1.279(15)	2.072(−1)	4.0950(−4)
		$1S$–$1P$	0.59433	0–1	3.413(15)	5.423(−1)	1.0610(−3)
	$1s2s$–$1s2p$	$3S$–$3P$	52.552	1–2	6.183(10)	4.267(−2)	2.2145(−2)
			201.16	1–1	7.719(8)	4.683(−3)	9.3036(−3)
			201.84	1–0	1.037(9)	2.112(−3)	4.2094(−3)
46	$1s^2$–$1s2p$	$3S$–$1P$	46.751	1–1	2.339(10)	7.664(−3)	3.5385(−3)
		$1S$–$1P$	61.720	0–1	2.835(10)	1.619(−2)	9.8700(−3)
		$1S$–$3P$	0.57343	0–1	1.425(15)	2.107(−1)	3.9771(−4)
		$1S$–$1P$	0.56760	0–1	3.704(15)	5.367(−1)	1.0028(−3)
	$1s2s$–$1s2p$	$3S$–$3P$	48.564	1–4	7.479(10)	4.407(−2)	2.1138(−2)
			195.83	1–1	7.915(8)	4.550(−3)	8.8005(−3)
			194.99	1–0	1.095(9)	2.081(−3)	4.0081(−3)
		$3S$–$1P$	43.465	1–1	2.827(10)	8.008(−3)	3.4377(−3)
		$1S$–$1P$	56.658	0–1	3.475(10)	1.673(−2)	9.3588(−3)

TABLE XV *Continued*

Z	Transition array	Multiplet	λ(Å)	J_i–J_k	A_{ki} (s^{-1})	f_{ik}	S (a.u.)
47	$1s^2$–$1s2p$	^1S–^3P	0.54838	0–1	1.581(15)	2.138(−1)	3.8600(−4)
		^1S–^1P	0.54258	0–1	4.012(15)	5.312(−1)	9.4893(−4)
	$1s2s$–$1s2p$	^3S–^3P	44.919	1–2	9.030(10)	4.552(−2)	2.0196(−2)
			190.74	1–1	8.112(8)	4.425(−3)	8.3354(−3)
			188.42	1–0	1.157(9)	2.053(−3)	3.8198(−3)
		^3S–^1P	40.430	1–1	3.411(10)	8.359(−3)	3.3376(−3)
		^1S–^1P	52.075	0–1	4.250(10)	1.728(−2)	8.8863(−3)
48	$1s^2$–$1s2p$	^1S–^3P	0.52489	0–1	1.749(15)	2.167(−1)	3.7442(−4)
		^1S–^1P	0.51911	0–1	4.340(15)	5.260(−1)	8.9896(−4)
	$1s2s$–$1s2p$	^3S–^3P	41.584	1–2	1.088(11)	4.702(−2)	1.9313(−2)
			185.90	1–1	8.310(8)	4.305(−3)	7.9045(−3)
			182.13	1–0	1.222(9)	2.025(−3)	3.6433(−3)
		^3S–^1P	37.629	1–1	4.106(10)	8.715(−3)	3.2388(−3)
		^1S–^1P	47.921	0–1	5.185(10)	1.785(−2)	8.4486(−3)
49	$1s^2$–$1s2p$	^1S–^3P	0.50284	0–1	1.928(15)	2.193(−1)	3.6302(−4)
		^1S–^1P	0.49708	0–1	4.688(15)	5.210(−1)	8.5254(−4)
	$1s2s$–$1s2p$	^3S–^3P	38.530	1–2	1.309(11)	4.857(−2)	1.8483(−2)
			181.27	1–1	8.509(8)	4.192(−3)	7.5046(−3)
			176.10	1–0	1.290(9)	2.000(−3)	3.4776(−3)
		^3S–^1P	35.040	1–1	4.932(10)	9.078(−3)	3.1417(−3)
		^1S–^1P	44.152	0–1	6.311(10)	1.844(−2)	8.0423(−3)
50	$1s^2$–$1s2p$	^1S–^3P	0.48210	0–1	2.120(15)	2.216(−1)	3.5177(−4)
		^1S–^1P	0.47637	0–1	5.057(15)	5.161(−1)	8.0937(−4)
	$1s2s$–$1s2p$	^3S–^3P	35.731	1–2	1.573(11)	5.017(−2)	1.7704(−2)
			176.82	1–1	8.713(8)	4.084(−3)	7.1327(−3)
			170.29	1–0	1.363(9)	1.975(−3)	3.3219(−3)
		^3S–^1P	32.648	1–1	5.911(10)	9.446(−3)	3.0460(−3)
		^1S–^1P	40.726	0–1	7.664(10)	1.906(−2)	7.6650(−3)

51	$1s^2$–$1s2p$	^1S–^3P	0.46259	0–1	2.325(15)	2.238(−1)	3.4082(−4)
		^1S–^1P	0.45687	0–1	5.447(15)	5.113(−1)	7.6908(−4)
	$1s2s$–$1s2p$	^3S–^3P	33.166	1–2	1.885(11)	5.181(−2)	1.6971(−2)
			172.59	1–1	8.914(8)	3.981(−3)	6.7862(−3)
			164.75	1–0	1.439(9)	1.952(−3)	3.1755(−3)
52	$1s^2$–$1s2p$	^3S–^1P	30.438	1–1	7.072(10)	9.823(−3)	2.9529(−3)
		^1S–^1P	37.610	0–1	9.284(10)	1.969(−2)	7.3130(−3)
	$1s2s$–$1s2p$	^1S–^3P	0.44420	0–1	2.544(15)	2.258(−1)	3.3014(−4)
		^1S–^1P	0.43850	0–1	5.859(15)	5.067(−1)	7.3146(−4)
		^3S–^3P	30.810	1–2	2.256(11)	5.350(−2)	1.6279(−2)
			168.52	1–1	9.120(8)	3.883(−3)	6.4628(−3)
			159.40	1–0	1.519(9)	1.929(−3)	3.0375(−3)
53	$1s^2$–$1s2p$	^3S–^1P	28.394	1–1	8.444(10)	1.021(−2)	2.8622(−3)
		^1S–^1P	34.771	0–1	1.122(11)	2.034(−2)	6.9844(−3)
	$1s2s$–$1s2p$	^1S–^3P	0.42685	0–1	2.776(15)	2.275(−1)	3.1973(−4)
		^1S–^1P	0.42116	0–1	6.295(15)	5.022(−1)	6.9626(−4)
	$1s2s$–$1s2p$	^3S–^3P	28.645	1–2	2.694(11)	5.524(−2)	1.5628(−2)
			164.61	1–1	9.329(8)	3.789(−3)	6.1605(−3)
			154.26	1–0	1.605(9)	1.908(−3)	2.9074(−3)
54	$1s^2$–$1s2p$	^3S–^1P	26.503	1–1	1.006(11)	1.060(−2)	2.7739(−3)
		^1S–^1P	32.182	0–1	1.353(11)	2.101(−2)	6.6773(−3)
		^1S–^3P	0.41046	0–1	3.024(15)	2.291(−1)	3.0961(−4)
		^1S–^1P	0.40478	0–1	6.754(15)	4.977(−1)	6.6329(−4)
	$1s2s$–$1s2p$	^3S–^3P	26.655	1–2	3.212(11)	5.702(−2)	1.5012(−2)
			160.84	1–1	9.540(8)	3.700(−3)	5.8775(−3)
			149.31	1–0	1.695(9)	1.888(−3)	2.7845(−3)
55	$1s^2$–$1s2p$	^3S–^1P	24.752	1–1	1.197(11)	1.100(−2)	2.6882(−3)
		^1S–^1P	29.817	0–1	1.628(11)	2.170(−2)	6.3897(−3)
	$1s2s$–$1s2p$	^1S–^3P	0.39497	0–1	3.286(15)	2.305(−1)	2.9978(−4)
		^1S–^1P	0.38930	0–1	7.239(15)	4.934(−1)	6.3234(−4)
	$1s2s$–$1s2p$	^3S–^3P	24.823	1–2	3.823(11)	5.886(−2)	1.4430(−2)
			157.22	1–1	9.753(8)	3.614(−3)	5.6121(−3)
			144.54	1–0	1.790(9)	1.869(−3)	2.6684(−3)
		^3S–^1P	23.131	1–1	1.421(11)	1.140(−2)	2.6049(−3)
		^1S–^1P	27.655	0–1	1.954(11)	2.241(−2)	6.1200(−3)

TABLE XV Continued

Z	Transition array	Multiplet	λ(Å)	J_i–J_k	A_{ki} (s^{-1})	f_{ik}	S (a.u.)
56	$1s^2$–$1s2p$	^1S–^3P	0.38030	0–1	3.564(15)	2.318(−1)	2.9024(−4)
		^1S–^1P	0.37464	0–1	7.749(15)	4.891(−1)	6.0327(−4)
	$1s2s$–$1s2p$	^3S–^3P	23.135	1–2	4.542(11)	6.074(−2)	1.3879(−2)
			153.73	1–1	9.969(8)	3.532(−3)	5.3628(−3)
			139.96	1–0	1.891(9)	1.851(−3)	2.5585(−3)
		^3S–^1P	21.630	1–1	1.685(11)	1.182(−2)	2.5243(−3)
		^1S–^1P	25.676	0–1	2.341(11)	2.314(−2)	5.8667(−3)
57	$1s^2$–$1s2p$	^1S–^3P	0.36640	0–1	3.858(15)	2.330(−1)	2.8101(−4)
		^1S–^1P	0.36074	0–1	8.285(15)	4.849(−1)	5.7592(−4)
	$1s2s$–$1s2p$	^3S–^3P	21.580	1–2	5.386(11)	6.267(−2)	1.3357(−2)
			150.37	1–1	1.019(9)	3.453(−3)	5.1285(−3)
			135.53	1–0	1.997(9)	1.834(−3)	2.4544(−3)
		^3S–^1P	20.238	1–1	1.993(11)	1.224(−2)	2.4462(−3)
		^1S–^1P	23.862	0–1	2.798(11)	2.388(−2)	5.6284(−3)
58	$1s^2$–$1s2p$	^1S–^3P	0.35322	0–1	4.170(15)	2.340(−1)	2.7206(−4)
		^1S–^1P	0.34757	0–1	8.849(15)	4.808(−1)	5.5015(−4)
	$1s2s$–$1s2p$	^3S–^3P	20.144	1–2	6.376(11)	6.465(−2)	1.2862(−2)
			147.14	1–1	1.041(9)	3.377(−3)	4.9077(−3)
			131.28	1–0	2.109(9)	1.817(−3)	2.3557(−3)
		^3S–^1P	18.947	1–1	2.354(11)	1.267(−2)	2.3706(−3)
		^1S–^1P	22.197	0–1	3.337(11)	2.465(−2)	5.4040(−3)
59	$1s^2$–$1s2p$	^1S–^3P	0.34070	0–1	4.498(15)	2.348(−1)	2.6341(−4)
		^1S–^1P	0.33506	0–1	9.441(15)	4.767(−1)	5.2584(−4)
	$1s2s$–$1s2p$	^3S–^3P	18.819	1–2	7.535(11)	6.667(−2)	1.2392(−2)
			144.00	1–1	1.063(9)	3.304(−3)	4.6996(−3)
			127.18	1–0	2.228(9)	1.801(−3)	2.2621(−3)
		^3S–^1P	17.748	1–1	2.775(11)	1.311(−2)	2.2975(−3)
		^1S–^1P	20.668	0–1	3.972(11)	2.544(−2)	5.1924(−3)

Z	Transition	Term	Wavelength	i-j	Value 1	Value 2
60	$1s^2$–$1s2p$	^1S–^3P	0.32881	0–1		2.5505(−4)
	$1s^2$–$1s2p$	^1S–^1P	0.32318	0–1		5.0289(−4)
	$1s2s$–$1s2p$	^3S–^3P	17.593	1–2		1.1945(−2)
			141.00	1–1		4.5031(−3)
			123.23	1–0		2.1732(−3)
	$1s^2$–$1s2p$	^3S–^1P	16.636	1–1		2.2269(−3)
61	$1s^2$–$1s2p$	^1S–^1P	19.262	0–1		4.9925(−3)
	$1s2s$–$1s2p$	^1S–^3P	0.31751	0–1	2.356(−1)	2.4696(−4)
	$1s2s$–$1s2p$	^1S–^1P	0.31188	0–1	4.727(−1)	4.8118(−4)
	$1s2s$–$1s2p$	^3S–^3P	16.461	1–2	6.875(−2)	1.1520(−2)
			138.17	1–1	3.234(−3)	4.3173(−3)
			119.48	1–0	1.786(−3)	2.0886(−3)
	$1s^2$–$1s2p$	^3S–^1P	15.603	1–1	1.355(−2)	2.1587(−3)
	$1s^2$–$1s2p$	^1S–^1P	17.969	0–1	2.624(−1)	4.8035(−3)
62	$1s^2$–$1s2p$	^1S–^3P	0.30674	0–1	2.363(−1)	2.3915(−4)
	$1s^2$–$1s2p$	^1S–^1P	0.30112	0–1	4.687(−1)	4.6064(−4)
	$1s2s$–$1s2p$	^3S–^3P	15.412	1–2	7.086(−2)	1.1116(−2)
			135.31	1–1	3.164(−3)	4.1415(−3)
			115.76	1–0	1.770(−3)	2.0082(−3)
	$1s^2$–$1s2p$	^3S–^3P	14.642	1–1	1.401(−2)	2.0929(−3)
	$1s^2$–$1s2p$	^1S–^1P	16.776	0–1	2.707(−1)	4.6246(−3)
63	$1s^2$–$1s2p$	^1S–^3P	0.29649	0–1	2.368(−1)	2.3160(−4)
	$1s^2$–$1s2p$	^1S–^1P	0.29087	0–1	4.647(−1)	4.4117(−4)
	$1s2s$–$1s2p$	^3S–^3P	14.438	1–2	7.303(−2)	1.0731(−2)
			132.63	1–1	3.099(−3)	3.9750(−3)
			112.24	1–0	1.756(−3)	1.9316(−3)
	$1s^2$–$1s2p$	^3S–^1P	13.748	1–1	1.447(−2)	2.0293(−3)
	$1s^2$–$1s2p$	^1S–^1P	15.675	0–1	2.791(−1)	4.4551(−3)
64	$1s^2$–$1s2p$	^1S–^3P	0.28671	0–1	2.373(−1)	2.2431(−4)
	$1s^2$–$1s2p$	^1S–^1P	0.28109	0–1	4.607(−1)	4.2270(−4)
	$1s2s$–$1s2p$	^3S–^3P	13.536	1–2	7.525(−2)	1.0364(−2)
			129.97	1–1	2.974(−3)	3.8171(−3)
			108.79	1–0	1.730(−3)	1.8586(−3)
	$1s^2$–$1s2p$	^3S–^1P	12.916	1–1	1.543(−2)	1.9679(−3)
	$1s^2$–$1s2p$	^1S–^1P	14.658	0–1	2.966(−2)	4.2942(−3)

TABLE XV Continued

Z	Transition array	Multiplet	λ(Å)	J_i–J_k	A_{ki} (s^{-1})	f_{ik}	S (a.u.)
65	$1s^2$–$1s2p$	^1S–^3P	0.27739	0–1	6.875(15)	2.379(−1)	2.1727(−4)
		^1S–^1P	0.27177	0–1	1.363(16)	4.529(−1)	4.0517(−4)
	$1s2s$–$1s2p$	^3S–^3P	12.698	1–2	1.982(12)	7.984(−2)	1.0013(−2)
			127.40	1–1	1.198(9)	2.914(−3)	3.6671(−3)
			105.47	1–0	3.090(9)	1.718(−3)	1.7890(−3)
		^3S–^1P	12.140	1–1	7.204(11)	1.592(−2)	1.9086(−3)
		^1S–^1P	13.718	0–1	1.083(12)	3.057(−2)	4.1415(−3)
66	$1s^2$–$1s2p$	^1S–^3P	0.26848	0–1	7.345(15)	2.381(−1)	2.1047(−4)
		^1S–^1P	0.26287	0–1	1.445(16)	4.489(−1)	3.8851(−4)
	$1s2s$–$1s2p$	^3S–^3P	11.920	1–2	2.316(12)	8.221(−2)	9.6788(−3)
			124.93	1–1	1.221(9)	2.857(−3)	3.5246(−3)
			102.27	1–0	3.263(9)	1.706(−3)	1.7226(−3)
		^3S–^1P	11.418	1–1	8.401(11)	1.642(−2)	1.8514(−3)
		^1S–^1P	12.848	0–1	1.273(12)	3.149(−2)	3.9962(−3)
67	$1s^2$–$1s2p$	^1S–^3P	0.25998	0–1	7.838(15)	2.383(−1)	2.0392(−4)
		^1S–^1P	0.25436	0–1	1.529(16)	4.450(−1)	3.7267(−4)
	$1s2s$–$1s2p$	^3S–^3P	11.197	1–2	2.702(12)	8.463(−2)	9.3591(−3)
			122.54	1–1	1.244(9)	2.800(−3)	3.3890(−3)
			99.174	1–0	3.446(9)	1.694(−3)	1.6592(−3)
		^3S–^1P	10.744	1–1	9.782(11)	1.693(−2)	1.7962(−3)
		^1S–^1P	12.042	0–1	1.492(12)	3.244(−2)	3.8580(−3)
68	$1s^2$–$1s2p$	^1S–^3P	0.25184	0–1	8.354(15)	2.383(−1)	1.9757(−4)
		^1S–^1P	0.24623	0–1	1.618(16)	4.411(−1)	3.5757(−4)
	$1s2s$–$1s2p$	^3S–^3P	10.524	1–2	3.147(12)	8.710(−2)	9.0533(−3)
			120.24	1–1	1.267(9)	2.745(−3)	3.2600(−3)
			96.200	1–0	3.638(9)	1.683(−3)	1.5986(−3)
		^3S–^1P	10.115	1–1	1.137(12)	1.745(−2)	1.7429(−3)
		^1S–^1P	11.295	0–1	1.747(12)	3.340(−2)	3.7263(−3)

69	$1s^2$–$1s2p$	^1S–^3P	0.24406	0–1	8.895(15)	2.383(−1)	1.9146(−4)
		^1S–^1P	0.23845	0–1	1.710(16)	4.372(−1)	3.4320(−4)
	$1s2s$–$1s2p$	^3S–^3P	9.8976	1–2	3.661(12)	8.962(−2)	8.7607(−3)
			118.02	1–1	1.289(9)	2.691(−3)	3.1370(−3)
			93.333	1–0	3.839(9)	1.671(−3)	1.5406(−3)
		^3S–^1P	9.5283	1–1	1.321(12)	1.797(−2)	1.6914(−3)
70	$1s^2$–$1s2p$	^1S–^1P	10.602	0–1	2.041(12)	3.439(−2)	3.6008(−3)
		^1S–^3P	0.23661	0–1	9.460(15)	2.382(−1)	1.8555(−4)
		^1S–^1P	0.23099	0–1	1.805(16)	4.333(−1)	3.2950(−4)
	$1s2s$–$1s2p$	^3S–^3P	9.3141	1–2	4.253(12)	9.219(−2)	8.4804(−3)
			115.90	1–1	1.310(9)	2.638(−3)	3.0197(−3)
			90.582	1–0	4.049(9)	1.660(−3)	1.4851(−3)
71		^3S–^1P	8.9801	1–1	1.531(12)	1.851(−2)	1.6417(−3)
		^1S–^1P	9.9578	0–1	2.381(12)	3.539(−2)	3.4810(−3)
	$1s^2$–$1s2p$	^1S–^3P	0.22947	0–1	1.005(16)	2.381(−1)	1.7984(−4)
		^1S–^1P	0.22385	0–1	1.905(16)	4.294(−1)	3.1641(−4)
	$1s2s$–$1s2p$	^3S–^3P	8.7696	1–2	4.934(12)	9.481(−2)	8.2119(−3)
			113.81	1–1	1.332(9)	2.587(−3)	2.9077(−3)
			87.892	1–0	4.273(9)	1.650(−3)	1.4320(−3)
72		^3S–^1P	8.4673	1–1	1.773(12)	1.906(−2)	1.5937(−3)
		^1S–^1P	9.3588	0–1	2.774(12)	3.642(−2)	3.3665(−3)
	$1s^2$–$1s2p$	^1S–^3P	0.22263	0–1	1.067(16)	2.379(−1)	1.7433(−4)
		^1S–^1P	0.21701	0–1	2.009(16)	4.254(−1)	3.0393(−4)
	$1s2s$–$1s2p$	^3S–^3P	8.2617	1–2	5.716(12)	9.749(−2)	7.9543(−3)
			111.80	1–1	1.354(9)	2.536(−3)	2.8008(−3)
			85.308	1–0	4.507(9)	1.639(−3)	1.3810(−3)
73		^3S–^1P	7.9878	1–1	2.051(12)	1.961(−2)	1.5474(−3)
		^1S–^1P	8.8015	0–1	3.226(12)	3.747(−2)	3.2572(−3)
	$1s^2$–$1s2p$	^1S–^3P	0.21606	0–1	1.132(16)	2.376(−1)	1.6900(−4)
		^1S–^1P	0.21044	0–1	2.116(16)	4.215(−1)	2.9199(−4)
	$1s2s$–$1s2p$	^3S–^3P	7.7872	1–2	6.614(12)	1.002(−1)	7.7071(−3)
			109.86	1–1	1.375(9)	2.487(−3)	2.6985(−3)
			82.808	1–0	4.754(9)	1.629(−3)	1.3322(−3)
		^3S–^1P	7.5389	1–1	2.369(12)	2.018(−2)	1.5026(−3)
		^1S–^1P	8.2825	0–1	3.747(12)	3.854(−2)	3.1525(−3)

TABLE XV Continued

Z	Transition array	Multiplet	λ(Å)	J_i–J_k	A_{ki} (s^{-1})	f_{ik}	S (a.u.)
74	$1s^2$–$1s2p$	^1S–^3P	0.20976	0–1	1.199(16)	2.373(−1)	1.6386(−4)
		^1S–^1P	0.20414	0–1	2.227(16)	4.175(−1)	2.8058(−4)
	$1s2s$–$1s2p$	^3S–^3P	7.3439	1–2	7.642(12)	1.030(−1)	7.4698(−3)
			107.99	1–1	1.395(9)	2.438(−3)	2.6007(−3)
			80.399	1–0	5.011(9)	1.619(−3)	1.2854(−3)
		^3S–^1P	7.1185	1–1	2.732(12)	2.076(−2)	1.4593(−3)
		^1S–^1P	7.7988	0–1	4.346(12)	3.963(−2)	3.0523(−3)
75	$1s^2$–$1s2p$	^1S–^3P	0.20371	0–1	1.269(16)	2.369(−1)	1.5888(−4)
		^1S–^1P	0.19809	0–1	2.343(16)	4.135(−1)	2.6966(−4)
	$1s2s$–$1s2p$	^3S–^3P	6.9291	1–2	8.821(12)	1.058(−1)	7.2418(−3)
			106.15	1–1	1.416(9)	2.391(−3)	2.5070(−3)
			78.047	1–0	5.287(9)	1.609(−3)	1.2404(−3)
		^3S–^1P	6.7243	1–1	3.149(12)	2.134(−2)	1.4175(−3)
		^1S–^1P	7.3472	0–1	5.034(12)	4.074(−2)	2.9563(−3)
76	$1s^2$–$1s2p$	^1S–^3P	0.19789	0–1	1.343(16)	2.365(−1)	1.5407(−4)
		^1S–^1P	0.19227	0–1	2.463(16)	4.095(−1)	2.5920(−4)
	$1s2s$–$1s2p$	^3S–^3P	6.5410	1–2	1.017(13)	1.087(−1)	7.0226(−3)
			104.38	1–1	1.435(9)	2.345(−3)	2.4173(−3)
			75.783	1–0	5.574(9)	1.600(−3)	1.1973(−3)
		^3S–^1P	6.3548	1–1	3.624(12)	2.194(−2)	1.3771(−3)
		^1S–^1P	6.9258	0–1	5.823(12)	4.187(−2)	2.8642(−3)
77	$1s^2$–$1s2p$	^1S–^3P	0.19230	0–1	1.419(16)	2.360(−1)	1.4942(−4)
		^1S–^1P	0.18667	0–1	2.587(16)	4.055(−1)	2.4918(−4)
	$1s2s$–$1s2p$	^3S–^3P	6.1775	1–2	1.171(13)	1.116(−1)	6.8118(−3)
			102.68	1–1	1.454(9)	2.299(−3)	2.3313(−3)
			73.596	1–0	5.875(9)	1.590(−3)	1.1558(−3)
		^3S–^1P	6.0082	1–1	4.167(12)	2.255(−2)	1.3380(−3)
		^1S–^1P	6.5320	0–1	6.727(12)	4.303(−2)	2.7759(−3)

Z	Transition	Term					
78	$1s^2$–$1s2p$	^1S–^3P	0.18692	0–1	1.499(16)	2.355(−1)	1.4492(−4)
		^1S–^1P	0.18129	0–1	2.715(16)	4.014(−1)	2.3958(−4)
	$1s2s$–$1s2p$	^3S–^3P	5.8370	1–2	1.347(13)	1.146(−1)	6.6088(−3)
			101.06	1–1	1.471(9)	2.253(−3)	2.2488(−3)
			71.494	1–0	6.187(9)	1.580(−3)	1.1159(−3)
79	$1s^2$–$1s2p$	^3S–^1P	5.6828	1–1	4.785(12)	2.317(−1)	1.3002(−3)
		^1S–^1P	6.1639	1–1	7.761(12)	4.420(−1)	2.6911(−3)
		^1S–^3P	0.18174	0–1	1.581(16)	2.349(−1)	1.4057(−4)
	$1s2s$–$1s2p$	^1S–^1P	0.17611	0–1	2.848(16)	3.973(−1)	2.3037(−4)
		^3S–^3P	5.5177	1–2	1.547(13)	1.177(−1)	6.4133(−3)
			99.493	1–1	1.488(9)	2.208(−3)	2.1696(−3)
			69.458	1–0	6.515(9)	1.571(−3)	1.0776(−3)
80	$1s^2$–$1s2p$	^3S–^1P	5.3772	1–1	5.489(12)	2.379(−2)	1.2636(−3)
		^1S–^1P	5.8194	1–1	8.943(12)	4.540(−2)	2.6095(−3)
		^1S–^3P	0.17676	0–1	1.668(16)	2.343(−1)	1.3636(−4)
		^1S–^1P	0.17112	0–1	2.986(16)	3.932(−1)	2.2153(−4)
	$1s2s$–$1s2p$	^3S–^3P	5.2182	1–2	1.775(13)	1.208(−1)	6.2250(−3)
			98.029	1–1	1.501(9)	2.162(−1)	2.0935(−3)
			67.511	1–0	6.852(9)	1.561(−3)	1.0406(−3)
81	$1s^2$–$1s2p$	^3S–^1P	5.0901	1–1	6.290(12)	2.443(−2)	1.2282(−3)
		^1S–^1P	5.4970	1–1	1.029(13)	4.662(−2)	2.5311(−3)
		^1S–^3P	0.17195	0–1	1.757(16)	2.337(−1)	1.3229(−4)
		^1S–^1P	0.16632	0–1	3.128(16)	3.891(−1)	2.1304(−4)
	$1s2s$–$1s2p$	^3S–^3P	4.9368	1–2	2.035(13)	1.239(−1)	6.0434(−3)
			96.554	1–1	1.516(9)	2.119(−3)	2.0204(−3)
			65.597	1–0	7.215(9)	1.551(−3)	1.0051(−3)
82	$1s^2$–$1s2p$	^3S–^1P	4.8199	1–1	7.202(12)	2.508(−2)	1.1940(−3)
		^1S–^1P	5.1946	1–1	1.183(13)	4.787(−2)	2.4557(−3)
		^1S–^3P	0.16732	0–1	1.850(16)	2.330(−1)	1.2834(−4)
		^1S–^1P	0.16168	0–1	3.274(16)	3.849(−1)	2.0489(−4)
	$1s2s$–$1s2p$	^3S–^3P	4.6715	1–2	2.333(13)	1.272(−1)	5.8682(−3)
			94.713	1–1	1.550(9)	2.085(−3)	1.9501(−3)
			63.553	1–0	7.663(9)	1.547(−3)	9.7084(−4)
		^3S–^1P	4.5647	1–1	8.242(12)	2.575(−2)	1.1608(−3)
		^1S–^1P	4.9113	0–1	1.359(13)	4.913(−2)	2.3831(−3)

TABLE XV Continued

Z	Transition array	Multiplet	λ(Å)	J_i-J_k	A_{ki} (s^{-1})	f_{ik}	S (a.u.)
83	$1s^2$–$1s2p$	^1S–^3P	0.16286	0–1	1.947(16)	2.323(−1)	1.2452(−4)
		^1S–^1P	0.15721	0–1	3.425(16)	3.807(−1)	1.9706(−4)
	$1s2s$–$1s2p$	^3S–^3P	4.4243	1–2	2.667(13)	1.304(−1)	5.6992(−3)
			93.893	1–1	1.536(9)	2.030(−3)	1.8824(−3)
			62.005	1–0	7.971(9)	1.531(−3)	9.3783(−4)
		^3S–^1P	4.3267	1–1	9.411(12)	2.641(−3)	1.1286(−3)
		^1S–^1P	4.6453	0–1	1.559(13)	5.042(−2)	2.3131(−3)
84	$1s^2$–$1s2p$	^1S–^3P	0.15855	0–1	2.047(16)	2.315(−1)	1.2082(−4)
		^1S–^1P	0.15290	0–1	3.581(16)	3.765(−1)	1.8952(−4)
	$1s2s$–$1s2p$	^3S–^3P	4.1908	1–2	3.048(13)	1.338(−1)	5.5359(−3)
			92.701	1–1	1.541(9)	1.985(−3)	1.8173(−3)
			60.318	1–0	8.365(9)	1.521(−3)	9.0601(−4)
		^3S–^1P	4.1016	1–1	1.074(13)	2.709(−2)	1.0975(−3)
		^1S–^1P	4.3956	0–1	1.786(13)	5.173(−2)	2.2457(−3)
85	$1s^2$–$1s2p$	^1S–^3P	0.15439	0–1	2.151(16)	2.307(−1)	1.1723(−4)
		^1S–^1P	0.14874	0–1	3.741(16)	3.723(−1)	1.8228(−4)
	$1s2s$–$1s2p$	^3S–^3P	3.9711	1–2	3.480(13)	1.371(−1)	5.3782(−3)
			91.597	1–1	1.542(9)	1.939(−3)	1.7546(−3)
			58.698	1–0	8.769(9)	1.510(−3)	8.7531(−4)
		^3S–^1P	3.8894	1–1	1.225(13)	2.778(−2)	1.0673(−3)
		^1S–^1P	4.1612	0–1	2.044(13)	5.306(−2)	2.1806(−3)
86	$1s^2$–$1s2p$	^1S–^3P	0.15038	0–1	2.259(16)	2.298(−1)	1.1376(−4)
		^1S–^1P	0.14472	0–1	3.907(16)	3.680(−1)	1.7531(−4)
	$1s2s$–$1s2p$	^3S–^3P	3.7644	1–2	3.970(13)	1.406(−1)	5.2257(−3)
			90.664	1–1	1.535(9)	1.892(−3)	1.6941(−3)
			57.175	1–0	9.168(9)	1.498(−3)	8.4569(−4)
		^3S–^1P	3.6896	1–1	1.396(13)	2.849(−2)	1.0380(−3)
		^1S–^1P	3.9409	0–1	2.337(13)	5.441(−2)	2.1178(−3)

Z			λ	J–J'			
87	$1s^2$–$1s2p$	^1S–^3P	0.14649	0–1	2.372(16)	2.289(−1)	1.1039(−4)
		^1S–^1P	0.14083	0–1	4.077(16)	3.637(−1)	1.6860(−4)
	$1s2s$–$1s2p$	^3S–^3P	3.5693	1–2	4.525(13)	1.441(−1)	5.0782(−3)
			89.657	1–1	1.533(9)	1.847(−3)	1.6358(−3)
			55.649	1–0	9.607(9)	1.487(−3)	8.1709(−4)
		^3S–^1P	3.5008	1–1	1.589(13)	2.920(−2)	1.0096(−3)
		^1S–^1P	3.7334	0–1	2.670(13)	5.579(−3)	2.0572(−3)
88	$1s^2$–$1s2p$	^1S–^3P	0.14274	0–1	2.488(16)	2.280(−1)	1.0712(−4)
		^1S–^1P	0.13707	0–1	4.252(16)	3.593(−1)	1.6214(−4)
	$1s2s$–$1s2p$	^3S–^3P	3.3856	1–2	5.154(13)	1.476(−1)	4.9355(−3)
			88.824	1–1	1.522(9)	1.801(−3)	1.5796(−3)
			54.214	1–0	1.004(10)	1.474(−3)	7.8946(−4)
		^3S–^1P	3.3229	1–1	1.808(13)	2.992(−2)	9.8205(−4)
		^1S–^1P	3.5382	0–1	3.047(13)	5.719(−2)	1.9986(−3)
89	$1s^2$–$1s2p$	^1S–^3P	0.13911	0–1	2.608(16)	2.270(−1)	1.0395(−4)
		^1S–^1P	0.13344	0–1	4.432(16)	3.549(−1)	1.5591(−4)
	$1s2s$–$1s2p$	^3S–^3P	3.2124	1–2	5.864(13)	1.512(−1)	4.7973(−3)
			88.089	1–1	1.507(9)	1.753(−3)	1.5254(−3)
			52.837	1–0	1.048(10)	1.462(−3)	7.6277(−4)
		^3S–^1P	3.1548	1–1	2.055(13)	3.066(−2)	9.5530(−4)
		^1S–^1P	3.3544	0–1	3.475(13)	5.862(−2)	1.9419(−3)
90	$1s^2$–$1s2p$	^1S–^3P	0.13560	0–1	2.732(16)	2.260(−1)	1.0088(−4)
		^1S–^1P	0.12993	0–1	4.616(16)	3.505(−1)	1.4991(−4)
	$1s2s$–$1s2p$	^3S–^3P	3.0489	1–2	6.668(13)	1.549(−1)	4.6635(−3)
			87.437	1–1	1.488(9)	1.706(−3)	1.4730(−3)
			51.510	1–0	1.093(10)	1.449(−3)	7.3697(−4)
		^3S–^1P	2.9961	1–1	2.334(13)	3.141(−2)	9.2933(−4)
		^1S–^1P	3.1812	0–1	3.959(13)	6.006(−2)	1.8871(−3)
91	$1s^2$–$1s2p$	^1S–^3P	0.13221	0–1	2.861(16)	2.249(−1)	9.7889(−5)
		^1S–^1P	0.12652	0–1	4.806(16)	3.460(−1)	1.4412(−4)
	$1s2s$–$1s2p$	^3S–^3P	2.8946	1–2	7.575(13)	1.586(−1)	4.5339(−3)
			86.947	1–1	1.462(9)	1.656(−3)	1.4224(−3)
			50.253	1–0	1.137(10)	1.435(−3)	7.1202(−4)
		^3S–^1P	2.8461	1–1	2.649(13)	3.216(−2)	9.0411(−4)
		^1S–^1P	3.0179	0–1	4.507(13)	6.153(−2)	1.8341(−3)

TABLE XV Continued

Z	Transition array	Multiplet	λ(Å)	J_i–J_k	A_{ki} (s^{-1})	f_{ik}	S (a.u.)
92	$1s^2$–$1s2p$	^1S–^3P	0.12892	0–1	2.994(16)	2.238(–1)	9.4994(–5)
		^1S–^1P	0.12323	0–1	5.001(16)	3.415(–1)	1.3855(–4)
	$1s2s$–$1s2p$	^3S–^3P	2.7488	1–2	8.600(13)	1.624(–1)	4.4082(–3)
			86.565	1–1	1.430(9)	1.607(–3)	1.3735(–3)
			49.047	1–0	1.181(10)	1.420(–3)	6.8789(–4)
		^3S–^1P	2.7043	1–1	3.004(13)	3.293(–2)	8.7960(–4)
		^1S–^1P	2.8638	0–1	5.126(13)	6.303(–2)	1.7827(–3)
93	$1s^2$–$1s2p$	^1S–^3P	0.12573	0–1	3.132(16)	2.227(–1)	9.2176(–5)
		^1S–^1P	0.12004	0–1	5.200(16)	3.370(–1)	1.3316(–4)
	$1s2s$–$1s2p$	^3S–^3P	2.6111	1–2	9.757(13)	1.662(–1)	4.2862(–3)
			86.331	1–1	1.392(9)	1.556(–3)	1.3263(–3)
			47.899	1–0	1.225(10)	1.405(–3)	6.6453(–4)
		^3S–^1P	2.5702	1–1	3.404(13)	3.371(–2)	8.5577(–4)
		^1S–^1P	2.7184	0–1	5.826(13)	6.455(–2)	1.7330(–3)
94	$1s^2$–$1s2p$	^1S–^3P	0.12265	0–1	3.274(16)	2.215(–1)	8.9444(–5)
		^1S–^1P	0.11694	0–1	5.405(16)	3.324(–1)	1.2798(–4)
	$1s2s$–$1s2p$	^3S–^3P	2.4809	1–2	1.106(14)	1.701(–1)	4.1679(–3)
			86.327	1–1	1.344(9)	1.502(–3)	1.2806(–3)
			46.827	1–0	1.267(10)	1.388(–3)	6.4192(–4)
		^3S–^1P	2.4433	1–1	3.855(13)	3.450(–2)	8.3261(–4)
		^1S–^1P	2.5812	0–1	6.616(13)	6.609(–2)	1.6847(–3)
95	$1s^2$–$1s2p$	^1S–^3P	0.11965	0–1	3.422(16)	2.203(–1)	8.6783(–5)
		^1S–^1P	0.11394	0–1	5.614(16)	3.278(–1)	1.2296(–4)
	$1s2s$–$1s2p$	^3S–^3P	2.3577	1–2	1.253(14)	1.741(–1)	4.0530(–3)
			86.407	1–1	1.294(9)	1.449(–3)	1.2364(–3)
			45.783	1–0	1.309(10)	1.371(–3)	6.2002(–4)
		^3S–^1P	2.3231	1–1	4.364(13)	3.531(–2)	8.1007(–4)
		^1S–^1P	2.4514	0–1	7.509(13)	6.765(–2)	1.6379(–3)

96	$1s^2$–$1s2p$	^1S–^3P	0.11675	0–1	3.573(16)	2.191(−1)	8.4201(−5)
		^1S–^1P	0.11103	0–1	5.828(16)	3.232(−1)	1.1812(−4)
	$1s2s$–$1s2p$	^3S–^3P	2.2411	1–2	1.419(14)	1.781(−1)	3.9414(−3)
			86.782	1–1	1.233(9)	1.393(−3)	1.1936(−3)
			44.819	1–0	1.348(10)	1.353(−3)	5.9881(−4)
		^3S–^1P	2.2093	1–1	4.936(13)	3.612(−2)	7.8815(−4)
		^1S–^1P	2.3288	0–1	8.516(13)	6.924(−2)	1.5925(−3)
97	$1s^2$–$1s2p$	^1S–^3P	0.11393	0–1	3.730(16)	2.178(−1)	8.1686(−5)
		^1S–^1P	0.10821	0–1	6.047(16)	3.185(−1)	1.1345(−4)
	$1s2s$–$1s2p$	^3S–^3P	2.1308	1–2	1.605(14)	1.821(−1)	3.8329(−3)
			87.441	1–1	1.164(9)	1.334(−3)	1.1521(−3)
			43.923	1–0	1.383(10)	1.333(−3)	5.7825(−4)
		^3S–^1P	2.1016	1–1	5.580(13)	3.694(−2)	7.6680(−4)
		^1S–^1P	2.2129	0–1	9.651(13)	7.085(−2)	1.5484(−3)
98	$1s^2$–$1s2p$	^1S–^3P	0.11121	0–1	3.892(16)	2.165(−1)	7.9244(−5)
		^1S–^1P	0.10547	0–1	6.271(16)	3.137(−1)	1.0894(−4)
	$1s2s$–$1s2p$	^3S–^3P	2.0263	1–2	1.815(14)	1.863(−1)	3.7274(−3)
			88.394	1–1	1.087(9)	1.274(−3)	1.1120(−3)
			43.082	1–0	1.415(10)	1.312(−3)	5.5832(−4)
		^3S–^1P	1.9994	1–1	6.304(13)	3.778(−2)	7.4602(−4)
		^1S–^1P	2.1032	0–1	1.093(14)	7.248(−2)	1.5056(−3)
99	$1s^2$–$1s2p$	^1S–^3P	0.10855	0–1	4.059(16)	2.151(−1)	7.6867(−5)
		^1S–^1P	0.10281	0–1	6.500(16)	3.090(−1)	1.0458(−4)
	$1s2s$–$1s2p$	^3S–^3P	1.9274	1–2	2.052(14)	1.904(−1)	3.6249(−3)
			89.866	1–1	9.986(8)	1.209(−3)	1.0731(−3)
			42.345	1–0	1.438(10)	1.289(−3)	5.3900(−4)
		^3S–^1P	1.9026	1–1	7.117(13)	3.862(−2)	7.2577(−4)
		^1S–^1P	1.9994	1–1	1.237(14)	7.414(−2)	1.4640(−3)
100	$1s^2$–$1s2p$	^1S–^3P	0.10597	0–1	4.231(16)	2.137(−1)	7.4554(−5)
		^1S–^1P	0.10022	0–1	6.733(16)	3.042(−1)	1.0037(−4)
	$1s2s$–$1s2p$	^3S–^3P	1.8336	1–2	2.317(14)	1.947(−1)	3.5251(−3)
			91.813	1–1	9.035(8)	1.142(−3)	1.0354(−3)
			41.673	1–0	1.457(10)	1.264(−3)	5.2026(−4)
		^3S–^1P	1.8108	1–1	8.031(13)	3.948(−2)	7.0604(−4)
		^1S–^1P	1.9011	0–1	1.399(14)	7.582(−2)	1.4235(−3)

TABLE XVI
M_1 AND M_2 VALUES

Z	Transition array	Multiplet	λ(Å)	J_i–J_k	A_{ki} (s^{-1})	Type	S (a.u.)
2	$1s^2$–$1s2s$	^1S–^3S	625.48	0–1	1.266(−4)	M_1	3.4468(−9)
	$1s^2$–$1s2p$	^1S–^3P	591.33	0–2	3.271(−1)	M_2	7.9312
3	$1s^2$–$1s2s$	^1S–^3S	210.05	0–1	2.035(−2)	M_1	2.0980(−8)
	$1s^2$–$1s2p$	^1S–^3P	202.30	0–2	3.498(1)	M_2	3.9755
4	$1s^2$–$1s2s$	^1S–^3S	104.54	0–1	5.615(−1)	M_1	7.1343(−8)
	$1s^2$–$1s2p$	^1S–^3P	101.68	0–2	6.171(2)	M_2	2.2497
5	$1s^2$–$1s2s$	^1S–^3S	62.436	0–1	6.696	M_1	1.8127(−7)
	$1s^2$–$1s2p$	^1S–^3P	61.085	0–2	5.014(3)	M_2	1.4299
6	$1s^2$–$1s2s$	^1S–^3S	41.470	0–1	4.860(1)	M_1	3.8550(−7)
	$1s^2$–$1s2p$	^1S–^3P	40.726	0–2	2.622(4)	M_2	9.8495(−1)
7	$1s^2$–$1s2s$	^1S–^3S	29.533	0–1	2.537(2)	M_1	7.2683(−7)
	$1s^2$–$1s2p$	^1S–^3P	29.081	0–2	1.030(5)	M_2	7.1824(−1)
8	$1s^2$–$1s2s$	^1S–^3S	22.097	0–1	1.047(3)	M_1	1.2561(−6)
	$1s^2$–$1s2p$	^1S–^3P	21.800	0–2	3.308(5)	M_2	5.4628(−1)
9	$1s^2$–$1s2s$	^1S–^3S	17.152	0–1	3.621(3)	M_1	2.0323(−6)
	$1s^2$–$1s2p$	^1S–^3P	16.947	0–2	9.155(5)	M_2	4.2912(−1)
10	$1s^2$–$1s2s$	^1S–^3S	13.699	0–1	1.092(4)	M_1	3.1226(−6)
	$1s^2$–$1s2p$	^1S–^3P	13.550	0–2	2.257(6)	M_2	3.4577(−1)
11	$1s^2$–$1s2s$	^1S–^3S	11.191	0–1	2.952(4)	M_1	4.6022(−6)
	$1s^2$–$1s2p$	^1S–^3P	11.080	0–2	5.079(6)	M_2	2.8440(−1)
12	$1s^2$–$1s2s$	^1S–^3S	9.3141	0–1	7.293(4)	M_1	6.5546(−6)
	$1s^2$–$1s2p$	^1S–^3P	9.2280	0–2	1.060(7)	M_2	2.3792(−1)
13	$1s^2$–$1s2s$	^1S–^3S	7.8720	0–1	1.672(5)	M_1	9.0715(−6)
	$1s^2$–$1s2p$	^1S–^3P	7.8037	0–2	2.080(7)	M_2	2.0188(−1)
14	$1s^2$–$1s2s$	^1S–^3S	6.7402	0–1	3.598(5)	M_1	1.2253(−5)
	$1s^2$–$1s2p$	^1S–^3P	6.6849	0–2	3.873(7)	M_2	1.7338(−1)

Z	Transition	Term					
15	$1s^2-1s2s$	$^1S-^3S$	5.8357	0-1	7.333(5)	M_1	1.6208(−5)
	$1s^2-1s2p$	$^1S-^3P$	5.7900	0-2	6.895(7)	M_2	1.5045(−1)
16	$1s^2-1s2s$	$^1S-^3S$	5.1014	0-1	1.426(6)	M_1	2.1054(−5)
	$1s^2-1s2p$	$^1S-^3P$	5.0631	0-2	1.181(8)	M_2	1.3174(−1)
17	$1s^2-1s2s$	$^1S-^3S$	4.4972	0-1	2.661(6)	M_1	2.6915(−5)
	$1s^2-1s2p$	$^1S-^3P$	4.4645	0-2	1.955(8)	M_2	1.1626(−1)
18	$1s^2-1s2s$	$^1S-^3S$	3.9941	0-1	4.787(6)	M_1	3.3926(−5)
	$1s^2-1s2p$	$^1S-^3P$	3.9658	0-2	3.141(8)	M_2	1.0332(−1)
19	$1s^2-1s2s$	$^1S-^3S$	3.5707	0-1	8.341(6)	M_1	4.2231(−5)
	$1s^2-1s2p$	$^1S-^3P$	3.5459	0-2	4.914(8)	M_2	9.2385(−2)
20	$1s^2-1s2s$	$^1S-^3S$	3.2110	0-1	1.412(7)	M_1	5.1981(−5)
	$1s^2-1s2p$	$^1S-^3P$	3.1891	0-2	7.510(8)	M_2	8.3067(−2)
21	$1s^2-1s2s$	$^1S-^3S$	2.9028	0-1	2.328(7)	M_1	6.3336(−5)
	$1s^2-1s2p$	$^1S-^3P$	2.8832	0-2	1.123(9)	M_2	7.5060(−2)
22	$1s^2-1s2s$	$^1S-^3S$	2.6368	0-1	3.750(7)	M_1	7.6467(−5)
	$1s^2-1s2p$	$^1S-^3P$	2.6191	0-2	1.648(9)	M_2	6.8130(−2)
23	$1s^2-1s2s$	$^1S-^3S$	2.4056	0-1	5.913(7)	M_1	9.1552(−5)
	$1s^2-1s2p$	$^1S-^3P$	2.3895	0-2	2.377(9)	M_2	6.2092(−2)
24	$1s^2-1s2s$	$^1S-^3S$	2.2034	0-1	9.143(7)	M_1	1.0878(−4)
	$1s^2-1s2p$	$^1S-^3P$	2.1886	0-2	3.373(9)	M_2	5.6800(−2)
25	$1s^2-1s2s$	$^1S-^3S$	2.0255	0-1	1.389(8)	M_1	1.2835(−4)
	$1s^2-1s2p$	$^1S-^3P$	2.0118	0-2	4.718(9)	M_2	5.2135(−2)
26	$1s^2-1s2s$	$^1S-^3S$	1.8682	0-1	2.075(8)	M_1	1.5046(−4)
	$1s^2-1s2p$	$^1S-^3P$	1.8554	0-2	6.510(9)	M_2	4.8002(−2)
27	$1s^2-1s2s$	$^1S-^3S$	1.7284	0-1	3.053(8)	M_1	1.7534(−4)
	$1s^2-1s2p$	$^1S-^3P$	1.7164	0-2	8.873(9)	M_2	4.4324(−2)
28	$1s^2-1s2s$	$^1S-^3S$	1.6036	0-1	4.431(8)	M_1	2.0321(−4)
	$1s^2-1s2p$	$^1S-^3P$	1.5923	0-2	1.195(10)	M_2	4.1035(−2)
29	$1s^2-1s2s$	$^1S-^3S$	1.4917	0-1	6.346(8)	M_1	2.3431(−4)
	$1s^2-1s2p$	$^1S-^3P$	1.4811	0-2	1.594(10)	M_2	3.8084(−2)
30	$1s^2-1s2s$	$^1S-^3S$	1.3911	0-1	8.981(8)	M_1	2.6889(−4)
	$1s^2-1s2p$	$^1S-^3P$	1.3809	0-2	2.104(10)	M_2	3.5425(−2)

TABLE XVI Continued

Z	Transition array	Multiplet	λ(Å)	J_i–J_k	A_{ki} (s^{-1})	Type	S (a.u.)
31	$1s^2$–$1s2s$	^1S–^3S	1.3002	0–1	1.257(9)	M_1	3.0719(−4)
	$1s^2$–$1s2p$	^1S–^3P	1.2905	0–2	2.751(10)	M_2	3.3020(−2)
32	$1s^2$–$1s2s$	^1S–^3S	1.2179	0–1	1.740(9)	M_1	3.4949(−4)
	$1s^2$–$1s2p$	^1S–^3P	1.2086	0–2	3.567(10)	M_2	3.0840(−2)
33	$1s^2$–$1s2s$	^1S–^3S	1.1430	0–1	2.385(9)	M_1	3.9608(−4)
	$1s^2$–$1s2p$	^1S–^3P	1.1341	0–2	4.587(10)	M_2	2.8856(−2)
34	$1s^2$–$1s2s$	^1S–^3S	1.0748	0–1	3.239(9)	M_1	4.4721(−4)
	$1s^2$–$1s2p$	^1S–^3P	1.0661	0–2	5.855(10)	M_2	2.7046(−2)
35	$1s^2$–$1s2s$	^1S–^3S	1.0124	0–1	4.360(9)	M_1	5.0323(−4)
	$1s^2$–$1s2p$	^1S–^3P	1.0040	0–2	7.420(10)	M_2	2.5390(−2)
36	$1s^2$–$1s2s$	^1S–^3S	0.95525	0–1	5.822(9)	M_1	5.6442(−4)
	$1s^2$–$1s2p$	^1S–^3P	0.94710	0–2	9.341(10)	M_2	2.3871(−2)
37	$1s^2$–$1s2s$	^1S–^3S	0.90272	0–1	7.714(9)	M_1	6.3110(−4)
	$1s^2$–$1s2p$	^1S–^3P	0.89479	0–2	1.168(11)	M_2	2.2475(−2)
38	$1s^2$–$1s2s$	^1S–^3S	0.85434	0–1	1.015(10)	M_1	7.0363(−4)
	$1s^2$–$1s2p$	^1S–^3P	0.84661	0–2	1.453(11)	M_2	2.1188(−2)
39	$1s^2$–$1s2s$	^1S–^3S	0.80969	0–1	1.325(10)	M_1	7.8231(−4)
	$1s^2$–$1s2p$	^1S–^3P	0.80214	0–2	1.796(11)	M_2	1.9999(−2)
40	$1s^2$–$1s2s$	^1S–^3S	0.76840	0–1	1.719(10)	M_1	8.6756(−4)
	$1s^2$–$1s2p$	^1S–^3P	0.76099	0–2	2.208(11)	M_2	1.8900(−2)
41	$1s^2$–$1s2s$	^1S–^3S	0.73012	0–1	2.217(10)	M_1	9.5972(−4)
	$1s^2$–$1s2p$	^1S–^3P	0.72286	0–2	2.701(11)	M_2	1.7880(−2)
42	$1s^2$–$1s2s$	^1S–^3S	0.69459	0–1	2.842(10)	M_1	1.0592(−3)
	$1s^2$–$1s2p$	^1S–^3P	0.68745	0–2	3.289(11)	M_2	1.6933(−2)
43	$1s^2$–$1s2s$	^1S–^3S	0.66153	0–1	3.622(10)	M_1	1.1663(−3)
	$1s^2$–$1s2p$	^1S–^3P	0.65451	0–2	3.985(11)	M_2	1.6052(−2)
44	$1s^2$–$1s2s$	^1S–^3S	0.63073	0–1	4.592(10)	M_1	1.2816(−3)
	$1s^2$–$1s2p$	^1S–^3P	0.62382	0–2	4.808(11)	M_2	1.5230(−2)

Z							
45	$1s^2-1s2s$	$^1S-^3S$	0.60198	0–1	5.792(10)	M_1	1.4054(−3)
	$1s^2-1s2p$	$^1S-^3P$	0.59517	0–2	5.775(11)	M_2	1.4464(−2)
46	$1s^2-1s2s$	$^1S-^3S$	0.57511	0–1	7.270(10)	M_1	1.5381(−3)
	$1s^2-1s2p$	$^1S-^3P$	0.56838	0–2	6.910(11)	M_2	1.3747(−2)
47	$1s^2-1s2s$	$^1S-^3S$	0.54996	0–1	9.083(10)	M_1	1.6803(−3)
	$1s^2-1s2p$	$^1S-^3P$	0.54331	0–2	8.237(11)	M_2	1.3076(−2)
48	$1s^2-1s2s$	$^1S-^3S$	0.52637	0–1	1.130(11)	M_1	1.8324(−3)
	$1s^2-1s2p$	$^1S-^3P$	0.51979	0–2	9.781(11)	M_2	1.2447(−2)
49	$1s^2-1s2s$	$^1S-^3S$	0.50423	0–1	1.399(11)	M_1	1.9948(−3)
	$1s^2-1s2p$	$^1S-^3P$	0.49772	0–2	1.157(12)	M_2	1.1856(−2)
50	$1s^2-1s2s$	$^1S-^3S$	0.48342	0–1	1.726(11)	M_1	2.1681(−3)
	$1s^2-1s2p$	$^1S-^3P$	0.47697	0–2	1.365(12)	M_2	1.1301(−2)
51	$1s^2-1s2s$	$^1S-^3S$	0.46384	0–1	2.120(11)	M_1	2.3529(−3)
	$1s^2-1s2p$	$^1S-^3P$	0.45744	0–2	1.605(12)	M_2	1.0778(−2)
52	$1s^2-1s2s$	$^1S-^3S$	0.44538	0–1	2.595(11)	M_1	2.5495(−3)
	$1s^2-1s2p$	$^1S-^3P$	0.43903	0–2	1.881(12)	M_2	1.0286(−2)
53	$1s^2-1s2s$	$^1S-^3S$	0.42796	0–1	3.164(11)	M_1	2.7586(−3)
	$1s^2-1s2p$	$^1S-^3P$	0.42166	0–2	2.197(12)	M_2	9.8219(−3)
54	$1s^2-1s2s$	$^1S-^3S$	0.41151	0–1	3.846(11)	M_1	2.9807(−3)
	$1s^2-1s2p$	$^1S-^3P$	0.40526	0–2	2.560(12)	M_2	9.3835(−3)
55	$1s^2-1s2s$	$^1S-^3S$	0.39596	0–1	4.658(11)	M_1	3.2165(−3)
	$1s^2-1s2p$	$^1S-^3P$	0.38974	0–2	2.974(12)	M_2	8.9691(−3)
56	$1s^2-1s2s$	$^1S-^3S$	0.38124	0–1	5.625(11)	M_1	3.4665(−3)
	$1s^2-1s2p$	$^1S-^3P$	0.37506	0–2	3.446(12)	M_2	8.5770(−3)
57	$1s^2-1s2s$	$^1S-^3S$	0.36729	0–1	6.771(11)	M_1	3.7313(−3)
	$1s^2-1s2p$	$^1S-^3P$	0.36115	0–2	3.983(12)	M_2	8.2057(−3)
58	$1s^2-1s2s$	$^1S-^3S$	0.35406	0–1	8.126(11)	M_1	4.0117(−3)
	$1s^2-1s2p$	$^1S-^3P$	0.34795	0–2	4.592(12)	M_2	7.8537(−3)
59	$1s^2-1s2s$	$^1S-^3S$	0.34151	0–1	9.725(11)	M_1	4.3082(−3)
	$1s^2-1s2p$	$^1S-^3P$	0.33542	0–2	5.281(12)	M_2	7.5198(−3)
60	$1s^2-1s2s$	$^1S-^3S$	0.32958	0–1	1.161(12)	M_1	4.6215(−3)
	$1s^2-1s2p$	$^1S-^3P$	0.32352	0–2	6.060(12)	M_2	7.2027(−3)

TABLE XVI Continued

Z	Transition array	Multiplet	λ(Å)	J_i–J_k	A_{ki} (s^{-1})	Type	S (a.u.)
61	$1s^2$–$1s2s$	^1S–^3S	0.31824	0–1	1.382(12)	M_1	4.9523(−3)
	$1s^2$–$1s2p$	^1S–^3P	0.31220	0–2	6.939(12)	M_2	6.9015(−3)
62	$1s^2$–$1s2s$	^1S–^3S	0.30744	0–1	1.640(12)	M_1	5.3014(−3)
	$1s^2$–$1s2p$	^1S–^3P	0.30143	0–2	7.927(12)	M_2	6.6150(−3)
63	$1s^2$–$1s2s$	^1S–^3S	0.29715	0–1	1.943(12)	M_1	5.6695(−3)
	$1s^2$–$1s2p$	^1S–^3P	0.29116	0–2	9.038(12)	M_2	6.3422(−3)
64	$1s^2$–$1s2s$	^1S–^3S	0.28735	0–1	2.295(12)	M_1	6.0574(−3)
	$1s^2$–$1s2p$	^1S–^3P	0.28138	0–2	1.028(13)	M_2	6.0825(−3)
65	$1s^2$–$1s2s$	^1S–^3S	0.27799	0–1	2.706(12)	M_1	6.4659(−3)
	$1s^2$–$1s2p$	^1S–^3P	0.27204	0–2	1.168(13)	M_2	5.8349(−3)
66	$1s^2$–$1s2s$	^1S–^3S	0.26906	0–1	3.183(12)	M_1	6.8959(−3)
	$1s^2$–$1s2p$	^1S–^3P	0.26312	0–2	1.324(13)	M_2	5.5988(−3)
67	$1s^2$–$1s2s$	^1S–^3S	0.26053	0–1	3.736(12)	M_1	7.3478(−3)
	$1s^2$–$1s2p$	^1S–^3P	0.25461	0–2	1.498(13)	M_2	5.3735(−3)
68	$1s^2$–$1s2s$	^1S–^3S	0.25237	0–1	4.376(12)	M_1	7.8233(−3)
	$1s^2$–$1s2p$	^1S–^3P	0.24646	0–2	1.691(13)	M_2	5.1582(−3)
69	$1s^2$–$1s2s$	^1S–^3S	0.24457	0–1	5.115(12)	M_1	8.3225(−3)
	$1s^2$–$1s2p$	^1S–^3P	0.23867	0–2	1.907(13)	M_2	4.9526(−3)
70	$1s^2$–$1s2s$	^1S–^3S	0.23709	0–1	5.968(12)	M_1	8.8466(−3)
	$1s^2$–$1s2p$	^1S–^3P	0.23121	0–2	2.146(13)	M_2	4.7560(−3)
71	$1s^2$–$1s2s$	^1S–^3S	0.22993	0–1	6.950(12)	M_1	9.3969(−3)
	$1s^2$–$1s2p$	^1S–^3P	0.22406	0–2	2.412(13)	M_2	4.5679(−3)
72	$1s^2$–$1s2s$	^1S–^3S	0.22307	0–1	8.079(12)	M_1	9.9738(−3)
	$1s^2$–$1s2p$	^1S–^3P	0.21721	0–2	2.706(13)	M_2	4.3879(−3)
73	$1s^2$–$1s2s$	^1S–^3S	0.21649	0–1	9.375(12)	M_1	1.0579(−2)
	$1s^2$–$1s2p$	^1S–^3P	0.21063	0–2	3.032(13)	M_2	4.2154(−3)
74	$1s^2$–$1s2s$	^1S–^3S	0.21017	0–1	1.086(13)	M_1	1.1213(−2)
	$1s^2$–$1s2p$	^1S–^3P	0.20432	0–2	3.392(13)	M_2	4.0503(−3)

Z	Transition	Term	λ				
75	$1s^2$–$1s2s$	^1S–^3S	0.20410	0–1	1.256(13)	M_1	1.1877(−2)
	$1s^2$–$1s2p$	^1S–^3P	0.19826	0–2	3.789(13)	M_2	3.8919(−3)
76	$1s^2$–$1s2s$	^1S–^3S	0.19827	0–1	1.450(13)	M_1	1.2572(−2)
	$1s^2$–$1s2p$	^1S–^3P	0.19243	0–2	4.226(13)	M_2	3.7400(−3)
77	$1s^2$–$1s2s$	^1S–^3S	0.19266	0–1	1.672(13)	M_1	1.3300(−2)
	$1s^2$–$1s2p$	^1S–^3P	0.18683	0–2	4.708(13)	M_2	3.5942(−3)
78	$1s^2$–$1s2s$	^1S–^3S	0.18727	0–1	1.925(13)	M_1	1.4061(−2)
	$1s^2$–$1s2p$	^1S–^3P	0.18145	0–2	5.237(13)	M_2	3.4543(−3)
79	$1s^2$–$1s2s$	^1S–^3S	0.18208	0–1	2.213(13)	M_1	1.4857(−2)
	$1s^2$–$1s2p$	^1S–^3P	0.17626	0–2	5.819(13)	M_2	3.3200(−3)
80	$1s^2$–$1s2s$	^1S–^3S	0.17708	0–1	2.540(13)	M_1	1.5689(−2)
	$1s^2$–$1s2p$	^1S–^3P	0.17127	0–2	6.457(13)	M_2	3.1909(−3)
81	$1s^2$–$1s2s$	^1S–^3S	0.17226	0–1	2.913(13)	M_1	1.6558(−2)
	$1s^2$–$1s2p$	^1S–^3P	0.16645	0–2	7.157(13)	M_2	3.0668(−3)
82	$1s^2$–$1s2s$	^1S–^3S	0.16762	0–1	3.335(13)	M_1	1.7466(−2)
	$1s^2$–$1s2p$	^1S–^3P	0.16181	0–2	7.923(13)	M_2	2.9476(−3)
83	$1s^2$–$1s2s$	^1S–^3S	0.16314	0–1	3.813(13)	M_1	1.8414(−2)
	$1s^2$–$1s2p$	^1S–^3P	0.15734	0–2	8.761(13)	M_2	2.8329(−3)
84	$1s^2$–$1s2s$	^1S–^3S	0.15882	0–1	4.355(13)	M_1	1.9404(−2)
	$1s^2$–$1s2p$	^1S–^3P	0.15302	0–2	9.676(13)	M_2	2.7224(−3)
85	$1s^2$–$1s2s$	^1S–^3S	0.15465	0–1	4.968(13)	M_1	2.0437(−2)
	$1s^2$–$1s2p$	^1S–^3P	0.14885	0–2	1.068(14)	M_2	2.6163(−3)
86	$1s^2$–$1s2s$	^1S–^3S	0.15063	0–1	5.660(13)	M_1	2.1513(−2)
	$1s^2$–$1s2p$	^1S–^3P	0.14483	0–2	1.176(14)	M_2	2.5141(−3)
87	$1s^2$–$1s2s$	^1S–^3S	0.14673	0–1	6.442(13)	M_1	2.2637(−2)
	$1s^2$–$1s2p$	^1S–^3P	0.14094	0–2	1.295(14)	M_2	2.4155(−3)
88	$1s^2$–$1s2s$	^1S–^3S	0.14297	0–1	7.325(13)	M_1	2.3808(−2)
	$1s^2$–$1s2p$	^1S–^3P	0.13718	0–2	1.425(14)	M_2	2.3207(−3)
89	$1s^2$–$1s2s$	^1S–^3S	0.13933	0–1	8.319(13)	M_1	2.5029(−2)
	$1s^2$–$1s2p$	^1S–^3P	0.13354	0–2	1.565(14)	M_2	2.2292(−3)
90	$1s^2$–$1s2s$	^1S–^3S	0.13581	0–1	9.439(13)	M_1	2.6300(−2)
	$1s^2$–$1s2p$	^1S–^3P	0.13002	0–2	1.718(14)	M_2	2.1411(−3)

TABLE XVI Continued

Z	Transition array	Multiplet	λ(Å)	J_i–J_k	A_{ki} (s^{-1})	Type	S (a.u.)
91	$1s^2$–$1s2s$	^1S–^3S	0.13241	0–1	1.070(14)	M_1	2.7626(−2)
	$1s^2$–$1s2p$	^1S–^3P	0.12662	0–2	1.884(14)	M_2	2.0561(−3)
92	$1s^2$–$1s2s$	^1S–^3S	0.12911	0–1	1.212(14)	M_1	2.9005(−2)
	$1s^2$–$1s2p$	^1S–^3P	0.12332	0–2	2.064(14)	M_2	1.9743(−3)
93	$1s^2$–$1s2s$	^1S–^3S	0.12592	0–1	1.371(14)	M_1	3.0442(−2)
	$1s^2$–$1s2p$	^1S–^3P	0.12012	0–2	2.260(14)	M_2	1.8952(−3)
94	$1s^2$–$1s2s$	^1S–^3S	0.12282	0–1	1.550(14)	M_1	3.1935(−2)
	$1s^2$–$1s2p$	^1S–^3P	0.11703	0–2	2.471(14)	M_2	1.8190(−3)
95	$1s^2$–$1s2s$	^1S–^3S	0.11982	0–1	1.751(14)	M_1	3.3493(−2)
	$1s^2$–$1s2p$	^1S–^3P	0.11402	0–2	2.700(14)	M_2	1.7454(−3)
96	$1s^2$–$1s2s$	^1S–^3S	0.11691	0–1	1.976(14)	M_1	3.5110(−2)
	$1s^2$–$1s2p$	^1S–^3P	0.11111	0–2	2.948(14)	M_2	1.6745(−3)
97	$1s^2$–$1s2s$	^1S–^3S	0.11408	0–1	2.228(14)	M_1	3.6794(−2)
	$1s^2$–$1s2p$	^1S–^3P	0.10829	0–2	3.216(14)	M_2	1.6059(−3)
98	$1s^2$–$1s2s$	^1S–^3S	0.11135	0–1	2.510(14)	M_1	3.8543(−2)
	$1s^2$–$1s2p$	^1S–^3P	0.10555	0–2	3.506(14)	M_2	1.5398(−3)
99	$1s^2$–$1s2s$	^1S–^3S	0.10868	0–1	2.827(14)	M_1	4.0362(−2)
	$1s^2$–$1s2p$	^1S–^3P	0.10288	0–2	3.819(14)	M_2	1.4760(−3)
100	$1s^2$–$1s2s$	^1S–^3S	0.10610	0–1	3.181(14)	M_1	4.2252(−2)
	$1s^2$–$1s2p$	^1S–^3P	0.10029	0–2	4.156(14)	M_2	1.4143(−3)

The highest-order calculations of radiative corrections to decay rates have been carried out for the exotic atom positronium [36]. The first-order corrections to the decay of orthopositronium are quite large [37],

$$\Gamma^1 = -10.2866(6)\frac{\alpha}{\pi}\Gamma^0, \tag{96}$$

a 2.4% correction. This correction has been confirmed experimentally [38], although a notable feature of the decay is that higher-order corrections must account for a 0.1% discrepancy between theory and experimentation. Remarkably little is known about radiative corrections to decay rates of nonexotic atoms. Part of the reason for this is that, until recently, there has not been as much experimental impetus to carry out these difficult calculations. It is unusual to measure a decay rate at under the 1% level, and even corrections as large as those in orthopositronium would be difficult to detect for most decays. There are, however, important exceptions to this statement, most prominently in the case of helium-like ions. As discussed above, the M_1 decay of helium-like carbon has been measured to 0.2% [55]. The agreement with our theory then definitely rules out a radiative correction as large as that in orthopositronium, yet a correction of the order of the electron anomalous magnetic moment, $\alpha/2\pi$, which enters at the 0.1% level, cannot be ruled out. For the case of hydrogen, QED corrections have been treated in a general manner by Barbieri and Sucher [6], and, for the specific case of M_1 decays, by Lin and Feinberg [39]. It was shown in the latter work that possible corrections of order $\alpha \ln \alpha$ and α vanish for the decay $2S_{1/2} \to 1S_{1/2}$ in hydrogenic ions. This result appears to be specific to hydrogen, and the more complicated helium-like calculation has not been carried out. Inasmuch as at high Z, helium-like ions can be treated as perturbations to the hydrogenic case, a possible α/Z correction may be expected, which will be beyond our claimed accuracy. In addition to the leading order α corrections, there are also corrections of order $\alpha(Z\alpha)^2$, which could enter at the 0.1% level at high Z depending on the coefficient. We note that while the E_1 transition rates are in general poorly known, a high-precision (0.5%) measurement of lifetimes of 3^3S_1 and 3^3D_J levels in neutral helium [40] agrees with the calculations of Kono and Hattori [22]. Again, while this rules out large radiative corrections, smaller corrections could easily be present.

Another high-accuracy experiment that sheds light on the size of radiative corrections is the measurement of the He$^+$ lifetime by Drake et al. [41]. This experiment quotes an error of 0.075%, and agrees with a theoretical decay rate that does not include QED. This again rules out a large coefficient of α/π, and is close to ruling out even an $\alpha/2\pi$ correction. Finally, we note that a number of experiments accurate at the few tenths of a percent level have been carried out on alkali atoms [42–44], some of which disagree with theory by almost a percent. Because of the different electronic configuration, we do not see that the small radiative correction inferred from He$^+$ need carry over to these atoms and believe it possible that larger radiative corrections may be present in the case of the alkalis. In any case, it seems clear that the theory of radiative corrections to atomic decay rates is, with the exception of orthopositronium, a relatively unexplored field of QED. It cannot be overemphasized that experiments of the highest possible accuracy are essential for stimulating progress on these demanding calculations. What we hope we have provided in this chapter is a set of calculations of such high accuracy that any deviation from such experiments can be unambiguously interpreted as due to radiative corrections. We consider the complete calculation of radiative corrections the main theoretical challenge remaining for our theoretical understanding of decay rates of helium and helium-like ions.

Appendix: Useful Identities

In this appendix, we prove that the commutator $[V, \Xi]$ vanishes if the sum over intermediate states in Eq. (49) is extended to include negative-energy states. As before, we may write the potential operator as

$$V = \frac{1}{2} \sum_{ijkl} v_{ijkl} a_i^\dagger a_j^\dagger a_l a_k,$$

$$\Xi = \sum_{pq} \xi_{pq} a_p^\dagger a_q,$$

where

$$v_{ijkl} = \langle ij|V(\mathbf{x}_1, \mathbf{x}_2)|kl\rangle.$$

Here V is the sum of the Coulomb and instantaneous Breit interactions, and

$$\xi_{ij} = \langle i|\xi(\mathbf{x})|j\rangle.$$

The sums are restricted to positive-energy states only. With some algebra, the commutator can be put in the form

$$[V, \Xi] = \frac{1}{2} \sum_{ijkl} O_{ijkl} a_i^\dagger a_j^\dagger a_l a_k, \tag{97}$$

where

$$O_{ijkl} = \sum_r \{v_{ijrl}\xi_{rk} + v_{ijkr}\xi_{rl} - \xi_{ir}v_{rjkl} - \xi_{jr}v_{irkl}\}. \tag{98}$$

Let us introduce the positive-energy projection operator

$$\Lambda_+(\mathbf{x}, \mathbf{x}') = \sum_r \psi_r(\mathbf{x})\psi_r^\dagger(\mathbf{x}'). \tag{99}$$

We can then write Eq. (98) as

$$O_{ijkl} = \int d^3x_1 d^3x_2 d^3x_3 d^3x_4 \psi_i^\dagger(\mathbf{x}_1)\psi_j^\dagger(\mathbf{x}_2)O(\mathbf{x}_1,\mathbf{x}_2;\mathbf{x}_3,\mathbf{x}_4)\psi_k(\mathbf{x}_3)\psi_l(\mathbf{x}_4), \quad (100)$$

with

$$O(\mathbf{x}_1,\mathbf{x}_2;\mathbf{x}_3,\mathbf{x}_4)$$
$$= \Lambda_+(\mathbf{x}_1,\mathbf{x}_3)\delta(\mathbf{x}_2,\mathbf{x}_4)[V(\mathbf{x}_1,\mathbf{x}_2)\xi(\mathbf{x}_3) - \xi(\mathbf{x}_1)V(\mathbf{x}_3,\mathbf{x}_4)]$$
$$+ \Lambda_+(\mathbf{x}_2,\mathbf{x}_4)\delta(\mathbf{x}_1,\mathbf{x}_3)[V(\mathbf{x}_1,\mathbf{x}_2)\xi(\mathbf{x}_4) - \xi(\mathbf{x}_2)V(\mathbf{x}_3,\mathbf{x}_4)]. \quad (101)$$

If the sum in Eq. (99) is extended over both positive- and negative-energy states, then $\Lambda_+(\mathbf{x},\mathbf{x}') \to \delta(\mathbf{x},\mathbf{x}')$, and it follows from Eq. (101) that $O(\mathbf{x}_1,\mathbf{x}_2;\mathbf{x}_3,\mathbf{x}_4) = 0$. This, in turn, implies $[V,\Xi] = 0$. In a similar way, the terms on the first two lines of Eq. (84) in Section IV.A can be shown to vanish if the sum over intermediate states i extends over both positive- and negative-energy states.

Acknowledgments

The work of DP and JS was supported by NSF Grant PHY-92-04089. The work of WRJ was supported in part by NSF Grant PHY-92-04089 and in part by the Alexander von Humboldt foundation at the Technical University of Dresden. The authors owe thanks to Y.-K. Kim, P. Mohr, and G. Soff for useful suggestions and to A. E. Livingston and H. G. Berry for a careful reading of the manuscript. Thanks are also due to J. Fuhr for helpful discussions on the NIST database.

References

[1] J. Sucher, *Phys. Rev. A* **22**, 348 (1980).
[2] M. H. Mittleman, *Phys. Rev. A* **4**, 893 (1971); *Phys. Rev. A* **5**, 2395 (1972); *Phys. Rev. A* **24**, 1167 (1981).
[3] M. H. Chen, K. T. Cheng, and W. R. Johnson, *Phys. Rev. A* **47**, 3692 (1993).
[4] D. R. Plante, W. R. Johnson, and J. Sapirstein, *Phys. Rev. A* **49**, 3519 (1994).
[5] J. Hiller, J. Sucher, G. Feinberg, and B. Lynn, *Ann. Phys. (N.Y.)* **127**, 149 (1980).
[6] R. Barbieri and J. Sucher, *Nucl. Phys. B* **134**, 155 (1978); see also P. J. Mohr, in *Relativistic, Quantum Electrodynamic, and Weak Interaction Effects in Atoms* (Johnson, W., Mohr, P., and Sucher, J., eds.), AIP Conference Proceeding, No. 189, AIP, New York, 1989.
[7] M. Gell-Mann and F. Low, *Phys. Rev.* **84**, 350 (1951).
[8] E. Lindroth and S. Salomonson, *Phys. Rev. A* **41**, 4659 (1990).

[9] W. R. Johnson and C.-p. Lin, *Phys. Rev. A* **9**, 1486 (1974).
[10] W. R. Johnson and C. D. Lin, *Phys. Rev. A* **14**, 565 (1976); C. D. Lin, W. R. Johnson, and A. Dalgarno, *Phys. Rev. A* **15**, 154 (1977).
[11] W. R. Johnson, S. A. Blundell, and J. Sapirstein, *Phys. Rev. A* **37**, 2764 (1988).
[12] A. I. Akhiezer and V. B. Berestetskii, *Quantum Electrodynamics*, p. 28, Wiley Interscience, New York, 1965.
[13] B. W. Shore and D. H. Menzel, *Principles of Atomic Spectra*, p. 435, Wiley, New York, 1968; M. Mizushima, *Quantum Mechanics of Atomic Spectra and Atomic Structure*, p. 88, Benjamin, New York, 1970; I. I. Sobelman, *Atomic Spectra and Radiative Transitions*, Sect. 9.3, Springer-Verlag, Berlin, 1992.
[14] See e.g., J. R. Fuhr, G. A. Martin, and W. L. Wiese, *J. Phys. Chem. Ref. Data* **17**, Suppl. 4 (1988).
[15] P. Indelicato and P. J. Mohr, *Theor. Chim. Acta* **80**, 207 (1991).
[16] J. Sucher, *Phys. Rev.* **107**, 1448 (1957).
[17] A. Desiderio and W. R. Johnson, *Phys. Rev. A* **3**, 1267 (1971).
[18] Deleted in proof.
[19] Deleted in proof.
[20] Deleted in proof.
[21] B. Schiff, C. L. Pekeris, and Y. Accad, *Phys. Rev. A* **4**, 885 (1971).
[22] A. Kono and S. Hattori, *Phys. Rev. A* **29** 2981 (1984).
[23] F. C. Sanders and R. F. Knight, *Phys. Rev. A* **39**, 4387 (1989).
[24] N. M. Cann and A. J. Thakkar, *Phys. Rev. A* **46**, 5397 (1992).
[25] C. Froese Fischer, *Nucl. Instrum. Methods Phy. Res., B* **31**, 265 (1988).
[26] G. W. F. Drake, *Phys. Rev. A* **19**, 1387 (1979).
[27] J. Hata and I. P. Grant, *J. Phys. B* **14**, 2111 (1981).
[28] J. Krause, *Phys. Rev. A* **34**, 3692 (1986).
[29] G. Feinberg and J. Sucher, *Phys. Rev. Lett.* **26**, 681 (1971).
[30] R. Marrus and P. J. Mohr, *Adv. At. Mol. Phys.* **14** 181 (1978).
[31] A. H. Gabriel and Carol Jordan, *Nature (London)* **221**, 947 (1969); *Mon. Not. R. Astron. Soc.* **145**, 241 (1969); *Phys. Lett. A* **32**, 166 (1970).
[32] G. W. F. Drake, *Phys. Rev. A* **3**, 908 (1971).
[33] I. L. Beĭgman and U. I. Safronova, *Zh. Eksp. Teor. Fiz.* **60**, 2045 (1971) [*Sov. Phys.—JETP (Engl. Transl.)* **33**, 1102 (1971)].
[34] G. W. F. Drake, *Ap. J.* **158**, 1199 (1969); G. W. F. Drake, *Ap. J.* **163**, 439 (1971).
[35] B. Kundu, P. K. Mukherjee, and H. P. Roy, *Phys. Scr.* **39**, 722 (1989).
[36] P. Labelle, G. P. Lepage, and U. Magnea, *Phys. Rev. Lett.* **72**, 2006 (1994).
[37] G. S. Adkins, A. A. Salahuddin, and K. E. Schalm, *Phys. Rev. A* **45**, 7774 (1992).
[38] J. S. Nico, D. W. Gidley, A. Rich, and P. W. Zitzewitz, *Phys. Rev. Lett.* **65**, 1344 (1990).
[39] D. L. Lin and G. Feinberg, *Phys. Rev. A* **10**, 1425 (1974).
[40] U. Volz, D. Marger, H. Roth, and H. Schmoranzer, *J. Phys. B* **28**, 579 (1995).
[41] G. W. F. Drake, J. Kwela, and A. van Wijngaarden, *Phys. Rev. A* **46**, 113 (1992).
[42] A. Gaupp, R. Kuske, and H. J. Andraä, *Phys. Rev. A* **26**, 3351 (1982).
[43] C. Tanner, *Atomic Physics* (Wineland, D. J., Wieman, C. E., and Smith, S. J., eds.), Vol. 14, p. 130, AIP Press, New York, 1995.
[44] J. Jin and D. A. Church, *Phys. Rev. A* **49**, 3463 (1994).
[45] H. Gould, R. Marrus, and R. W. Schmieder, *Phys. Rev. Lett.* **31**, 504 (1973).
[46] H. Gould, R. Marrus, and P. J. Mohr, *Phys. Rev. Lett.* **33** 676 (1974).
[47] H. W. Moos and J. R. Woodworth, *Phys. Rev. A* **12** 2455 (1975).
[48] J. A. Bednar, C. L. Cocke, B. Curnutte, and R. Randall, *Phys. Rev. A* **11**, 460 (1975).
[49] R. D. Knight and M. H. Prior, *Phys. Rev. A* **21**, 179 (1980).
[50] G. Hurbricht amd E. Träbert, *Z. Phys. D* **7**, 243 (1987).
[51] R. Marrus, P. Charles, P. Indelicato, L. de Billy, C. Tazi, J. P. Briand, A. Simionovici, D. D. Dietrich, F. Bosch, and D. Liesen, *Phys. Rev. A* **39**, 3725 (1989).

[52] R. W. Dunford, D. A. Church, C. J. Liu, H. G. Berry, M. L. Raphaelian, M. Hass, and L. J. Curtis, *Phys. Rev. A* **41**, 4109 (1990).
[53] B. J. Wargelin, P. Beiersdorfer, and S. M. Kahn, *Phys. Rev. Lett.* **71**, 2196 (1993).
[54] B. B. Birkett, J. P. Briand, P. Charles, D. D. Dietrich, K. Finlayson, P. Indelicato, D. Liesen, R. Marrus, and A. Simionovici, *Phys. Rev. A* **47**, R2454 (1993).
[55] H. T. Schmidt, P. Forck, M. Grieser, D. Habs, J. Kenntner, G. Miersch, R. Repnow, U. Schramm, T. Schüssler, D. Schwalm, and A. Wolf, *Phys. Rev. Lett.* **72**, 1616 (1994).
[56] S. Cheng, R. W. Dunford, C. J. Liu, B. J. Zabransky, A. E. Livingston, and L. J. Curtis, *Phys. Rev. A* **49**, 2347 (1994).
[57] A. Simionovici, B. B. Birkett, R. Marrus, P. Charles, P. Indelicato, D. D. Dietrich, and K. Finlayson, *Phys. Rev. A* **49**, 3553 (1994).
[58] R. Marrus and R. W. Schmieder, *Phys. Rev. A* **5**, 1160 (1972).
[59] C. L. Cocke, B. Curnutte, and R. Randall, *Phys. Rev. A* **9**, 1823 (1974).
[60] C. L. Cocke, B. Curnutte, J. R. MacDonald, and R. Randall, *Phys. Rev. A* **9**, 57 (1974).
[61] H. Gould, R. Marrus, and P. J. Mohr, *Phys. Rev. Lett.* **33** 676 (1974).
[62] J. R. Mowat, P. M. Griffin, H. H. Haselton, R. Laubert, D. J. Pegg, R. S. Peterson, I. A. Sellin, and R. S. Thoe, *Phys. Rev. A* **11**, 2198 (1975).
[63] W. A. Davis and R. Marrus, *Phys. Rev. A* **15**, 1963 (1977).
[64] R. Marrus and P. J. Mohr, *Adv. At. Mol. Phys.* **14**, 181 (1978).
[65] B. Denne, S. Huldt, J. Pihl, and R. Hallin, *Phys. Scr.* **22**, 45 (1980).
[66] L. Engström, C. Jupén, B. Denne, S. Huldt, W. T. Meng, P. Kaijser, J. O. Ekberg, U. Litzén, and I. Martinson, *Phys. Scr.* **22**, 570 (1981).
[67] P. Deschepper, P. Lebrun, L. Palffy, and P. Pellegrin, *Phys. Rev. A* **26**, 1271 (1982).
[68] A. E. Livingston and S. J. Hinterlong, *Nucl. Instrum. Methods* **202**, 103 (1982).
[69] H. D. Dohmann, R. Mann, and E. Pfeng, *Z. Phys. A.* **309**, 101 (1982).
[70] J. P. Buchet, M. C. Buchet-Poulizac, A. Denis, J. Désequelles, M. Dreutta, J. P. Grandin, M. Huet, X. Husson, and D. Lecler, *Phys. Rev. A* **30**, 309 (1984).
[71] R. Hutton, N. Reistad, L. Engström, and S. Huldt, *Phys Scr.* **31**, 506 (1985).
[72] J. P. Buchet, M. C. Buchet-Poulizac, A. Denis, J. Désesquelles, M. Dreutta, J. P. Grandin, X. Husson, D. Lecler, and H. F. Beyer, *Nucl. Instrum. Methods B* **9**, 645 (1985).
[73] R. W. Dunford, C. J. Liu, J. Last, N. Berrah-Mansour, R. Vondrasek, D. A. Church, and L. J. Curtis, *Phys. Rev. A* **44**, 764 (1991).
[74] A. Simionovici, B. B. Birkett, J. P. Briand, P. Charles, D. D. Dietrich, K. Finlayson, P. Indelicato, D. Liesen, and R. Marrus, *Phys. Rev. A* **48**, 1695 (1993).
[75] Deleted in proof.
[76] S. L. Varghese, C. L. Cocke, and B. Curnutte, *Phys. Rev. A* **14**, 1729 (1976).
[77] I. A. Armour, J. D. Silver, and E. Träbert, *J. Phys. B* **14**, 3563 (1981).
[78] R. Hutton, N. Reistad, L. Engström, and S. Huldt, *Phys. Scr.* **31**, 506 (1985).

ROTATIONAL ENERGY TRANSFER IN SMALL POLYATOMIC MOLECULES[1]

HENRY O. EVERITT

U.S. Army Research Office, Research Triangle Park, North Carolina

and

FRANK C. DE LUCIA

Department of Physics, Ohio State University, Columbus, Ohio

I. Introduction to Rotational Energy Transfer	332
A. Applications	332
B. Theoretical Considerations	334
C. Rotational Energy Transfer Experiments	344
D. Model Building and Parameterization	346
E. The Relation between Experiment and Theory — What Can You Know?	355
II. State-Specific Rotational Energy Transfer — Principal Pathways	356
A. Dipole–Dipole Processes	357
B. $\Delta J = n$ Processes — Fitting Law Descriptions	357
C. Summary	365
III. Transfer to Nonprincipal Pathways: The Grouping of States	365
A. Symmetric Tops	367
B. Asymmetric Tops	370
C. Spherical Tops	370
D. Summary	371
IV. Near-Resonant Ro-Vibrational Energy Transfer	372
A. Symmetric and Spherical Tops	372
B. Asymmetric Tops	375
C. Summary	376
V. The Physical Basis of Rotational Energy Transfer	376
A. Selection and Propensity Rules	377
B. The Rotational Energy Transfer Processes — Molecular Constants	379
VI. The Future?	394
A. Advances in Experiments	394
B. Advances in Computing	396
C. What Will We Find?	396
References	397

---

[1] For Rodney I. McCormick (1946–1994), a leader, a scholar, and a friend.

I. Introduction to Rotational Energy Transfer

Significant progress has been made in the study of state-specific collision-induced rotational energy transfer (RET) in polyatomic molecules since the early studies of Wilson and co-workers (Cox *et al.*, 1965; Ronn and Wilson, 1967), Unland and Flygare (1966), and Oka (1966, 1967a, b, 1969). In an earlier contribution to this series, Oka (1973) reviewed the general issue of RET and summarized the field to that date. Additionally, there has been extensive work on simpler collisions, especially those between homonuclear diatomic molecules and atoms (Brunner and Pritchard, 1982; McCaffery *et al.*, 1986). This work, which explored the relationships among formal scattering theories, semiempirical scaling and fitting rules, and modern state-resolved experimental results, provides a useful basis for the development of an understanding of the more complex RET problem of polyatomic molecules.

Because the subject of RET has become very broad, we have chosen to focus on incoherent state-to-state rates which result from collisions between like molecules and on the relationship of these rates to fundamental molecular parameters. In general, we have excluded discussions of experimental methods, collisions with foreign gases, velocity subset effects (Shin *et al.*, 1991; Collins *et al.*, 1993), nonlinear processes (Seligson *et al.*, 1977; Temkin, 1977; Reiser and Steinfeld, 1981), polarization (Lees, 1975; Feuillade and Bottcher, 1981), and coherent effects (Brewer and Shoemaker, 1971; Levy *et al.*, 1972; Adam *et al.*, 1985).

A. APPLICATIONS

It is appropriate to begin with a brief discussion of the applications of RET studies to other areas of science and technology. Indeed, these applications have become so numerous they can be only incompletely summarized and referenced here. Applications range from very fundamental inquiries into the nature of intermolecular potentials; through the development of models of planetary atmospheres, the interstellar medium, and star formation; to remote sensing studies of ozone chemistry in the upper atmosphere and the diagnostics of molecular lasers.

The study of the relationship between the intermolecular potential (IMP) and a wide range of important molecular phenomena is a holy grail of physical chemistry. Because collisions directly sample the IMP, a considerable amount of the RET work to date has been either directly or indirectly motivated by this quest. In the context of RET, the goal of this quest would be a fully invertible theory which connects experimental observations of RET (as well as other observations) to the molecular parameters which describe the IMP. As is well known, even in the simplest systems this is a difficult problem. Ho and Rabitz (1993) have explored some of the basis for such an effort as well as the challenges to be met.

The importance of spectroscopic-based remote sensing has grown enormously in recent years. Ordinarily, accurate deconvolution of data from a remote physical system requires *a priori* knowledge of a number of molecular parameters. For systems in thermal equilibrium these include the energy level structure, line positions, and broadening parameters. However, there are a number of important circumstances in which rotational equilibrium is not achieved and RET parameters are required to model the state of the system. Because low pressures and energetic environments favor nonequilibrium, many of these examples are found in astronomy. Perhaps the most well known and important example is found in the study of interstellar clouds of gas and dust (Herbst and Millar, 1991). Other examples are to be found in the study of planetary atmospheres (Varanasi, 1988).

Another area in which RET plays a role is that of laser isotope separation (Harradine *et al.*, 1984). One scheme which has received considerable attention uses selective multiple infrared photon dissociation (MIRPD) for deuterium separation. It has been observed (Herman and Marling, 1979) that the addition of buffer gases increases the dissociation yield by large amounts because they significantly increase the coupling of the laser to the spectroscopic gas. A detailed knowledge of the RET process is important to optimize these processes. Similarly, MIRPD can play a role in chemical vapor deposition for amorphous silicon thin films (Millot *et al.*, 1988). Here a high-power CO_2 laser is used to decompose silane (SiH_4). Again, RET plays a critical role in the coupling of the laser power into the molecular system.

Molecular gas lasers are complex devices, and as a result the large majority have been discovered rather than invented. Indeed, even today the critical internal RET mechanisms of many, especially those in the far infrared (FIR), are poorly known. The most complex of these are the FIR discharge lasers, which have the additional complexity of the discharge process itself, as well as the chemistry induced by this discharge (Skatrud and De Lucia, 1985). Additionally, optically pumped lasers have played an important role in the development of the FIR region of the electromagnetic spectrum. Although these lasers have been extensively studied both experimentally and theoretically (Chang and Bridges, 1970; Tobin *et al.*, 1982), the models used to describe their molecular collision dynamics ordinarily have been oversimplified. For example, it has been shown (Everitt *et al.*, 1986) that a more careful examination of molecular collision dynamics indicates that the well-known "vibrational bottleneck" is an artifact of the model typically used to describe these laser systems and that the resultant "pressure cutoff" can be circumvented by choice of appropriate design and operating conditions. In fact, the relationship between RET studies and FIR laser development is symbiotic, with many of the CO_2 laser pump coincidences used in RET studies having previously been discovered in searches for new FIR lasers.

B. THEORETICAL CONSIDERATIONS

The size and dimensionality of the general RET problem in a polyatomic molecule are very large. In a molecule containing N levels, the levels are connected by $N(N-1)/2$ rates, numbers which routinely exceed 10^3 and 10^6, respectively. Moreover, the large number of levels within the thermally populated rotational manifold leads to a multitude of *open* collision channels. Thus, the dimensionality of the problem is increased further, with each of the calculations of each of the $N(N-1)/2$ rates requiring the inclusion of the N^2 combinations of initial and final states of the collision partners. This results in an overall dimensionality for the computational problem of $\sim N^4$.

1. The Exact Quantal Problem

Currently, for collisions among polyatomic molecules near room temperature the complexities and dimensionality are too great for an "exact" quantal treatment. However, it is instructive to explore the nature of such an exact solution and regimes in which such solutions have been successful. The formalism for this problem was developed for molecules by Arthurs and Dalgarno (1960) and has been developed to its current form in large numerical codes such as MOLSCAT (Green, 1980) by many workers.

In principle, it is possible to do such an exact calculation of any RET rate given knowledge of the intermolecular potential and the internal energy level structure of the collision partners. In these calculations a set of coupled differential equations is solved numerically. For all of the molecules to be discussed in this section, this presents a formidable computational challenge because of the dimensionality of the problem at or near room temperature. Additionally, the *ab initio* calculation and subsequent numerical description of the IMP for the complex combinations of geometries between two polyatomic molecules present a further challenge.

As a consequence, such calculations have been restricted to relatively simple systems at low temperature. For example, both the RET and the pressure broadening problems for the CO–He system at low temperature have been studied (Green and Thaddeus, 1976; Green, 1985; Palma and Green, 1986). Perhaps the most striking results of these calculations are the large resonances, due to the formation of quasibound states, found in the collisional cross-sections. These occur for systems whose collision energies are smaller than or comparable to the well depths in the IMP. Such quantal calculations have found wide applicability in the calculation of energy transfer in the cold interstellar medium, but have not been widely tested experimentally.

2. Semiclassical Elements—Energy, Angular Momentum, and Symmetry

Although it is possible to speculate about the rate at which computational advances will overtake the complexity of the exact quantal solution to the polyatomic room temperature problem, the conclusion of Rabitz and Gordon (1970a, b) many years ago that the computationally simpler semiclassical approach is appropriate still holds true for all of the systems to be discussed here. Furthermore, these semiclassical approaches can lift the veil from the underlying physics which is often buried in the numerical analysis of the fully quantal calculations. While the desire for physically intuitive models may simply be a prejudice of the authors, it seems to us that physical intuition is desirable because it can lead both experimenters and theorists to the most interesting problems and suggest lines of attack which lead to their solution.

While the semiclassical problem is simpler, in its most general form it is still formidable and contains all of the dimensionality of the fully quantal problem, albeit in simpler mathematical form. *If all of the $N(N-1)/2$ rates (and the resulting N-state populations) are truly independent, the RET of polyatomic molecules is a very complex problem and the experimental and theoretical solutions are insurmountably time consuming, independent of the method of attack. If not, then the challenge is to discover either experimentally or theoretically the simplifying principles.*

a. Symmetry, Angular Momentum, and the Intermolecular Potential. The most powerful simplifying principles involve symmetry and the rules of angular momentum theory (Edmonds, 1957). Their application to the derivation of selection rules has been nicely developed by Oka (1973) in his earlier contribution. Central to all considerations in RET, including those of symmetry, is the characterization of the IMP. The development of an appropriate form in the context of a multipole expansion has been considered by Gray (1968) and in the notation of Oka is given by

$$\bar{V} = \sum_{ik} \frac{e_i e_k}{r_{ik}}$$
$$= 4\pi \sum_{l_1 l_2} \sum_{m_1 m_2} C_{l_1 l_2} \begin{pmatrix} l_1 & l_2 & l \\ m_1 & m_2 & m \end{pmatrix} Y^*_{lm}(\mathbf{R}) \sum_{ik} e_i e_k r_i^{l_1} r_k^{l_2} Y_{l_1 m_1}(\mathbf{r}_i) Y_{l_2 m_2}(\mathbf{r}_k) / R^{l+1}, \tag{1}$$

where e_i and e_k are the electronic and nuclear charges of the first and second molecules, respectively, r_{ik} is the distance between these charges, $\mathbf{R}$ is the vector between the centers of mass of the molecules; $\mathbf{r}_i$ and $\mathbf{r}_k$ are the vectors from the center of mass of each molecule to its respective electrons, Y_{lm} are the spherical harmonics, the total angular momentum $l = l_1 + l_2$, and

$$C_{l_1 l_2} = (-1)^{l_1} [4\pi(2l+1)!/(2l_1+1)!(2l_2+1)!]^{1/2}/(2l+1)^{1/2}. \tag{2}$$

From this it is straightforward to derive angular momentum selection rules for the transition $JK \rightarrow J'K'$ as a function of the order of the multiple moments of the potential expansion responsible for the RET:

$$\Delta K = K - K' \leqslant l$$
$$|\Delta J| = |J - J'| \leqslant l$$
$$J + J' \geqslant l. \tag{3}$$

The effect of these relations is to impose limits on the collision-induced change of angular momenta as a function of the responsible multipole term in the potential expansion. Unfortunately, while there are many cases in which the dominant multipole contribution can be attributed to the dipole–dipole term of Eq. (1), there are virtually none which can be attributed uniquely to a particular higher order term. In fact, attempts at relating specific RET processes to specific higher order terms have been largely unsuccessful (Abel et al., 1992). *This lack of convergence of the multipole expansion, combined with the difficulty of ab initio calculations of the intermolecular potential for the complex geometries of polyatomic molecules, is one of the major obstacles to quantitative RET calculations in polyatomic molecules.*

As a result more empirical characterizations of the IMP have been developed. Many authors (see Oka, 1973) have divided collisions into two classes, usually termed *soft* and *hard*. The characteristics of soft collisions typically include dominance by a single, well-characterized interaction term of Eq. (1) (usually the dipole–dipole), collision-induced changes of kinetic energy small in comparison with kT, and well-defined selection rules. In the other limit, hard collisions are characterized not by the terms of Eq. (1) (too many would be required), but rather by radii (inside which the colliding molecules lose all or most of the knowledge of their initial states), have broad final state distributions, and result in substantial changes in the kinetic energy of the collision partners. This hard shell limit was considered 50 years ago by Van Vleck and Weisskopf (1945), who concluded that the final population of states was simply proportional to their Boltzmann factor, independent of the initial states.

While this division into hard and soft collisions may seem so arbitrary as to be of little quantitative utility, it will be shown that in many cases it can lead to at least a semiquantitative understanding of the RET process. The physical basis for this lies in the nature of the IMP as expressed by the expansion of Eq. (1). Briefly stated, for species with moderate to large electric dipole moments, the cross-section due to the dipole–dipole term of the expansion is large in comparison with the sum contributed by all other terms. It is also large in comparison with the physical size of the molecule as characterized by either the gas kinetic or the Lennard–Jones cross-section. The resulting soft collisions are particularly amenable to perturbation-theory-based calculations because of their relatively small energy interaction and the well-known and mathematically tractable form of the IMP in this case.

Additionally, the rest of the IMP can be approximated by interactions between hard spheres or other relatively simple geometric shapes. For these hard collisions, more general rules associated with symmetry, angular momentum, and energy defect dominate the description of the RET process. The most striking effects of symmetry and angular momentum are those associated with identical particles and nuclear spin statistics. Because the nuclear moments are coupled very weakly to the external collisional environment, true interconversion between these symmetries is very rare (Curl et al., 1967). Symmetry selection rules have been observed in many systems including spherical tops such as SiH_4 (Millot et al., 1988; Hetzler and Steinfeld, 1990) and $^{13}CD_4$ (Laux et al., 1984); symmetric tops such as CH_3F (Matteson and De Lucia, 1983; Shin and Schwendeman, 1991), CH_3Cl (Pape et al., 1994), CDF_3 (Harradine et al., 1984), and NH_3 (Oka, 1968; Abel et al., 1992); as well as asymmetric rotors such as CH_3OH (Lees and Oka, 1969) and D_2CO (Bewick et al., 1988). So strong are these selection rules that much of the interest in this subject is now focused on the study of parallel, near-resonant exchange channels which, although based on different physics, affect the same population transfers.

b. Energy Considerations. Many of the early RET experiments were based on microwave-microwave double-resonance measurements in which the energy defect was ordinarily small in comparison with kT (Oka, 1973). As a result these studies focused on propensity and selection rules associated with symmetry and angular momentum rather than energy defect. However, as the energy defects studied have become larger, their role has become more obvious and additional attention has been given to their impact on the RET process.

Many of the foundations of the calculation of quantitative rates via semiclassical theory can be traced to Anderson's pressure broadening theory (1949), which was expanded by Tsao and Curnutte (1961) and applied specifically to the RET problem by Rabitz and Gordon (1970a, b). Townes and Schalow (1955) have given a particularly intuitive form which can be used to illustrate a number of important factors in RET. Consider Fig. 1 which shows the collision between an idealized molecule, M', with two states and transition frequency ω_{ab} and a collision partner, M. Here M passes M' with a relative velocity, v, along z and with impact parameter b. A general interaction potential can be written in the form

$$V(t) = \frac{K}{R(t)^n}, \qquad (4)$$

where $R(t) = (z^2 + b^2)^{1/2}$ is the time varying distance between M and M', and K and n are constants characteristic of the multipole moment(s)

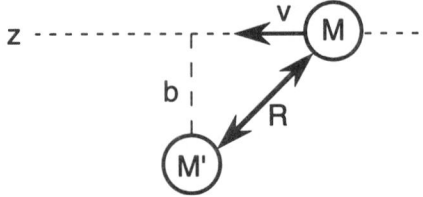

FIG. 1. The basic geometry of a collision in Anderson theory, showing the impact parameter b.

involved. With the substitutions

$$x \equiv \frac{vt}{b} \quad k \equiv \frac{b}{v}\omega_{ab} \tag{5}$$

into Eq. (4), time-dependent perturbation theory gives the matrix element for the transition probability as

$$\langle a|P|b\rangle = \frac{2\pi K}{hb^{n-1}v} \int_{-\infty}^{+\infty} \frac{e^{ikx}}{(1+x^2)^{n/2}}\,dx. \tag{6}$$

Equation (6) is in the form of a Fourier integral and shows the interaction of the multipole moments of the molecule M' with the spectrum of frequencies produced by the collision. For $k \gg 1$, the Fourier components of the radiation produced in the collision do not reach the transition frequency ω_{ab}, and the contribution to the transition probability is small. Implicit in Eq. (6) is an important correlation between the range of the multipole interaction and the extent of the Fourier spectrum. For large n the range of the interaction is small, making the interaction time short and the Fourier spectrum extend to high energy. Thus, there is an *inverse* relationship between the size of the cross-section for a particular interaction and the number of states energetically accessible for an *average* collision within the cross-section. More exact treatments include the internal energy levels and multipole moments of the collision partner as well as orientation effects.

In the simplest interpretation of this model, the velocity is simply the relative translational velocity. However, as pointed out many years ago by Moore (1965), for molecules with small moments of inertia due to reduced masses dominated by hydrogen, the velocities of these hydrogens can be several times higher than the relative translational velocity of the molecules. This concept has been incorporated in a more recent theory by Miklavc (1983), who reduces the complex problem of the interaction of two polyatomic molecules to one involving only one generalized coordinate; but one in which the effective collision velocity depends on many factors, including those related to this rotational effect. This modification is significant because it provides a mechanism to populate more highly excited levels.

c. *Near-Resonant Ro-Vibrational Energy Transfer.* Vibrational state-changing processes occur with varying degrees of rapidity, depending primarily upon two factors, the energy defect and the strength of the transition moment. Although most vibrational processes take place on the order of tens to thousands of gas kinetic collisions, nearly resonant ($\Delta E \ll kT$) ro-vibrational energy transfer collisions like

$$M(V=0) + M'(V=1) \leftrightarrow M(V=1) + M'(V=0) + \Delta E \quad (7)$$

have been observed with cross-sections approaching those of gas kinetic collisions. Because their rate is comparable in speed to that of rotational transitions, such near-resonant vibrational transitions must be considered in any discussion of RET. Moreover, such processes effectively transcend rotational and symmetry selection rules by placing a mirror image of the rotationally (and translationally) thermal ground vibrational state into an excited vibrational state as

$$M(V=0, A) + M'(V=1, E) \leftrightarrow M(V=1, A) + M'(V=0, E) + \Delta E$$
$$M(V=0, E) + M'(V=1, A) \leftrightarrow M(V=1, E) + M'(V=0, A) + \Delta E$$
$$M(V=0, A) + M'(V=1, A) \leftrightarrow M(V=1, A) + M'(V=0, A) + \Delta E$$
$$M(V=0, E) + M'(V=1, E) \leftrightarrow M(V=1, E) + M'(V=0, E) + \Delta E, \quad (8)$$

where A and E are symmetry-state labels. Therefore, the near-resonant vibrational exchange may serve as an effective rotational (and translational) thermalization mechanism and is best observed in time-resolved studies of rotational states with different symmetry than that of the directly pumped state.

When the near-resonant vibrational processes are mediated by long-range transition moments, a version of Anderson's semiclassical theory can be used to predict their cross-sections (Sharma and Brau, 1969). The calculation begins with a Taylor series expansion of the dipole moment matrix elements

$$\langle \psi_{V'} | \mu | \psi_V \rangle = \mu_0 \delta_{V',V} + \sum_{k=1}^{3N-6} \left(\frac{d\mu}{dQ_k} \right)_0 \langle \psi_{V+1} | Q_k | \psi_V \rangle + \cdots, \quad (9)$$

which reveals that to first order the vibrational state changing transition depends upon the first derivative of the dipole moment with respect to the generalized vibrational coordinate Q_k. Rotatonal transitions are also allowed and follow the traditional dipolar selection rules $\Delta J = 0, \pm 1, \Delta K = 0$. The calculation then proceeds much along the lines of Eqs. (4)–(6) and yields a state-specific transition probability for molecules with mass m of

$$P_{i \to f}(b, \omega_{ab}, T) = \frac{4m}{3\hbar^2 b^4 kT} (M_2)^2 \Gamma(b, \omega_{ab}, T) \qquad (10)$$

after averaging over a Maxwell–Boltzmann distribution of velocities. The cumulative rotational and vibrational energy defect of the transition $\Delta E = \hbar \omega_{ab}$ is contained in the Γ resonance term which, before velocity averaging, effectively limits the interaction to values of $\omega_{ab} b/v < 5$. Because of this, rotationally and vibrationally resonant ($\Delta E = 0$) collisions constitute a large portion of the total cross-section.

The transition probability also depends strongly upon the strength of the dipole derivative transition matrix element

$$(M_2)^2 = (\langle J'_1, K_1, V'_1 | \mu | J_1, K_1, V_1 \rangle \langle J'_2, K_2, V'_2 | \mu | J_2, K_2, V_2 \rangle)^2, \qquad (11)$$

which, for homomolecular collisions, contains $(d\mu/dQ_k)^4$.

To calculate the total probability for near-resonant vibrational collisions, all possible combinations of initial and final states for both colliding molecules must be summed, weighted explicitly by their relative populations and implicitly by their energy defect through the Γ term. Thus, near-resonant vibrational energy transfer is preferred between molecular vibrational levels with a large $d\mu/dQ_k$, small ro-vibrational energy defect, and large rotational partition function.

d. Scaling and Fitting Laws. Numerous modifications of the exact close coupling theory commonly used for simple molecules at low temperature have been developed in order to provide numerically tractable models of collisions between more complex molecular systems at higher temperatures. Among these are the "coupled states" (CS) (McGuire and Kouri, 1974), "infinite order sudden" (IOS) (Pack, 1974), and "energy-corrected sudden" (ECS) (DePristo et al., 1979) approximations. Goldflam et al. (1977) have studied the fundamental relations among these approximations and exact calculations in the context of a number of specific examples.

Consider as an example the application of IOS theory to calculate the rate constant for the transition $\Delta J = n$, $\Delta K = 0$ which results from a collision between an atom and a symmetric top molecule. For $J_i > J_f$

$$k_{J_i \to J_f} = (2J_f + 1) \sum_{L=|J_i - J_f|}^{J_i + J_f} (2L + 1) \begin{pmatrix} J_i & J_f & L \\ -K & K & 0 \end{pmatrix}^2 k_{L \to 0}, \qquad (12)$$

where $(2L+1)k_{L \to 0}$, the dynamical term, is the actual rate constant for the transition $J = L \to J = 0$ within $K = 0$, and (:::) is a 3-j symbol. The rate of the reverse process is given by the same relation with the additional term $\exp(-\Delta E/kT)$, included to preserve detailed balance. The ECS version of Eq. (12) improves upon the IOS approximation in that it allows for molecular rotation during the collision, with a characteristic time, τ_c, which in principle is determined by the collision geometry and in practice is an adjustable parameter.

For collisions between two polyatomic molecules, the equivalent relation is much more complex in form because it must account for the RET in both collision partners (Everitt, 1990). It is also much more complex in application because its use requires averaging the more complex form over the distribution of collision partners. As a result Eq. (12) is ordinarily used, with the average over the collision partners understood to be contained in the base rates $k_{L \to 0}$. Clearly, the impact of rotational resonances and the overall energy level structure of the collision partner on the energy defect make this effective average a significant approximation. The success of this simplified approach is due in no small part to the smoothing effect that results from the average over the large number of thermally populated rotational states. This smoothing and success might be expected to be lessened both for light species with widely spaced rotational levels and for studies at lower temperature.

Approximations such as IOS and ECS do not directly resolve the difficulties associated with the characterization of the intermolecular potential; rather, they transfer these problems to the calculation of the dynamical terms. However, they have the significant attribute that they separate those terms which depend only on the intermolecular potential (dynamical terms) from those which depend upon the states involved (spectroscopic terms). This factorization suggests that relationships exist among the $N(N-1)/2$ "independent" RET channels, and since it is possible to measure the base rates $k_{L \to 0}$ directly, this relationship may be obtained without a prerequisite knowledge of the intermolecular potential.

However, the number of required base rates can be quite large, and generally they have not been measured. As a result of this and of the difficulties in characterizing the intermolecular potential, phenomenological scaling laws have been developed to characterize the observed rate constants in terms of fewer parameters. The simplest of these laws, the exponential gap (EG) and power gap (PG) laws, relate the observed rate constants only to the rotational energy change of the observed molecule. Implicit in these scaling laws is the omission of explicit consideration of the collision-induced change in internal energy of the collision-inducing molecule. Although more detailed theories show that such considerations can make significant contributions to the rates, the density of thermally populated states in polyatomic molecules makes it possible to absorb these effects in the free parameters of the scaling law along with effects of the intermolecular potential, angular momentum considerations, etc.

The EG scaling law postulates that the cross-section decreases exponentially with increasing rotational energy change (Polyanyi and Woodall, 1972) according to

$$k_{J_i \to J_f} = k_0 (2J_f + 1) e^{-C(\Delta E/kT)} \qquad (13)$$

for collisions with $J_i > J_f$, where $\Delta E = (E_i - E_f)$ and where k_0 and C are adjustable parameters. Although this relation has been shown to have a limited range of applicability (Alexander et al., 1980; Garg and Agrawal, 1986), its appropriateness in some situations is suggested by surprisal theory (Procaccia and Levine, 1975; Rubinson and Steinfeld, 1975).

Alternatively, a scaling law which has a power law dependence upon the change in rotational energy has been proposed (Brunner et al., 1978) and used with wide success. In its simplest form the PG law predicts the rate constant

$$k_{J_i \to J_f} = k_0(2J_f + 1)\left(\frac{\Delta E}{kT}\right)^{-\alpha} \qquad (14)$$

for collisions with $J_i > J_f$, where $\Delta E = (E_i - E_f)$ and where k_0 and α are adjustable parameters. Although this law, too, has limited applicability, Smith and Pritchard (1981) have shown that the PG scaling law may be derived from IOS/ECS theories. In order for this connection to be established, an assumption about the behavior of the base rates must be made. Smith and Pritchard used an energy-like scaling law for the base rates

$$k_{L \to 0} = k_0 [L(L+1)]^{-\gamma} \qquad (15)$$

in Eq. (12) to derive Eq. (14) in the regime where $|\Delta J|/(J_i + J_f)$ and $|\Delta E|/kT$ are small compared with 1, and for values of γ near 1. The scaling coefficients are related by $(2\gamma - 1) = \alpha$.

Both scaling laws have been applied extensively to collisions involving diatomic molecules (Smith and Pritchard, 1981; Brunner et al., 1981; Wilkins and Kwok, 1983; Copeland and Crim, 1984). These applications have been successful in characterizing a large body of experimental data in terms of only a few parameters; however, because the parameters are empirical, it has proven difficult to extrapolate them to new systems (Alexander et al., 1980).

e. Scaling of Fitting Parameters — The Relation to Molecular Parameters. Equations (4)–(6) and (12)–(15) provide a means for examining the relationships which might be expected between the parameters of molecules or isotopes of similar geometries and IMPs. These relationships are based on factors associated with the degeneracy of states, rotational partition functions, geometric size, and molecular velocity.

The degeneracy of the final state in the RET process is important. Most theories, of which the IOS of Eq. (12) is an example, explicitly include a $(2J_f + 1)$ factor to account for this degeneracy. The physical basis of this factor is perhaps most easily understood in the context of a thermal process which populates some large manifold of states according to a Boltzmann distribution, filling each level according to its degeneracy and energy.

Because of the generality of this effect, a normalized rate, defined by

$$k_{J_i \to J_f} = (2J_f + 1)k_{\Delta J} \tag{16}$$

where $k_{\Delta J}$ represents the normalized rate, can be used to remove the contribution of the degeneracy. It should be noted, however, that the dependence of rates on the degeneracy of the target states is more complicated than might be inferred from this useful definition. For example, in the context of the IOS theory of Eq. (12), if the first term in the sum is much larger than the other terms, the target state dependence drops out. This is because the expansion of the 3-j symbol with the element $L = |J_i - J_f| = \Delta J$ has the term $1/(2J_f + 1)$ in it. The $\Delta J = \pm 1$ transition satisfies this criterion, and the spectroscopic term for the $L = \Delta J = 1$ is the well-known dipole matrix element. In practice, this criterion is satisfied best for the dipole–dipole interactions because the dynamical coefficients for $L > 2$ are much smaller than the one for $L = 1$. For ΔJ transitions with larger values of n, the target scaling implied by Eqs. (12) and (16) becomes stronger.

If comparisons are to be made among different molecules, it is also necessary to consider the rotational partition functions of the available states. Returning again to a thermal process, the transfer rate to a particular rotational state depends not only on its degeneracy and energy, but also on the number of states which must share the transferred population. For more complex processes, it is necessary to consider more carefully what portion of the rotational manifold is available. If a rotational manifold is clearly divided into symmetry species by nuclear spin statistics, only those states of the same symmetry should be included in the calculation of the partition function available for RET. However, for processes with less well-defined selection rules, the factors involved are more difficult.

Clearly, RET rates must have some proportionality to geometric size, but one which must be examined in the context of the particular physical process. If, for example, long-range dipole–dipole interactions are involved, the rate of the interaction is determined primarily by the strengths of the dipole moments involved, not by the geometrical sizes of the molecules. However, many process are clearly the result of fairly close, hard encounters, and the size of the gas kinetic or Lennard–Jones cross-section should be quantitatively related to the observed RET. Thus, it is reasonable to expect to be able to associate a molecular size with the total RET rate and to compare it quantitatively with an appropriate physical measure of the molecule.

The RET processes can be characterized by its transition probability P (per "collision"), rate k (s^{-1}), rate constant k_0 (ms^{-1} mTorr^{-1}), or cross-section σ(Å^2). We will primarily use the cross-section σ because it is most easily compared with gross molecular dimensions such as gas kinetic cross-section σ_{GK} and Lennard–Jones cross-section σ_{LJ}. As defined by Hirschfelder *et al.* (1964), the gas kinetic collision cross-section $\sigma_{GK} = \pi d^2$ is defined by assuming the molecule is an impenetrable sphere of diameter d. Unlike the temperature-insensitive gas kinetic cross-section, the Lennard–Jones cross-section includes the effects of a spherically averaged potential well through the famous 6–12 potential. The Lennard–Jones cross-section is related to the gas kinetic cross-section through

$$\sigma_{LJ} = \sigma_{GK}\Omega^{(2,2)*}(kT/\varepsilon), \qquad (17)$$

where $\Omega^{(2,2)*}(kT/\varepsilon)$, tabulated in Hirschfelder *et al.* (1964), shows that the greater the well depth ε is in comparison with kT, the larger the Lennard–Jones cross-section will be.

Finally, for geometrically similar molecules the collision velocity can vary either because of a change in temperature or because of a difference in mass. Additionally, the rotational energy level structure with which the spectrum of the collision must interact may vary (e.g., a hydrogen/deuterium isotopic substitution can change rotational constants by a factor of 2). Here the density of states relative to kT and the Fourier components of Eq. (6) is important. In the limit of very small rotational constants, even a rather modest change in energy corresponds to a large change in angular momentum. The selection rules of Eq. (3) show that limits on the changes in angular momentum exist, determined by the order of the multipole moments which participate significantly in the collision. Consequently, for this limit it should be expected that restrictions on angular momentum transfer will preclude transfers to energetically available states. In the opposite limit of large rotational constants and few thermally open channels for the products of the collision to enter, the total cross-section for RET will be reduced. This is because the "optical" strength of the transition allowed by angular momentum will be wasted on energetically unavailable channels. Additionally, it is less likely that "classical" averages over many channels and states will occur, and more quantal RET should be observable.

C. ROTATIONAL ENERGY TRANSFER EXPERIMENTS

Although is is not the purpose of this chapter to review experimental techniques in any detail, it is important to understand some of their attributes, specifically those attributes that determine sensitivity to the different kinds of RET which might be observed. This understanding is especially important because the complexity of RET has led to a focus on the search for propensity rules. Thus, it becomes important to separate propensity rules that are general properties of the RET process from those based on a correlation with a type of experimental observation.

Most of the work discussed here is based on double resonance (DR) techniques involving various combinations of optical, infrared, and microwave pumps and probes. Most of the early experiments were CW or quasi-CW (Oka, 1969; Ronn and Lide, 1967; Frenkel *et al.*, 1971), whereas in recent years most have been time resolved. Although kinetic information can be obtained from either, time-resolved experiments, especially those with resolution on the time scale of the phenomena, in general give results which are less model dependent. Similar conclusions have been arrived at by Gordon (1967) in his analysis of modulated microwave double-resonance techniques and later by Rohlfing et al. (1987) in their study of RET in diatomic molecules.

As an example, consider an initially empty system of N states which contains a state, p, into which a δ function of population is placed at $t = 0$. The population in each state, n_i, subsequently evolves away from p according to the coupled set of equations

$$\frac{dn_i}{dt} = \sum_{j \neq i} n_j k_{ji} - n_i \sum_{j \neq i} k_{ij}. \tag{18}$$

The evolution in time of this system of N states and $N(N-1)/2$ distinct k_{ji} rates is complex. Similarly, the steady-state populations in a CW experiment depend upon the k_{ji} rates in an equally complex fashion. However, if in a time-resolved experiment it is possible to make observations at very early times, all the n_j except n_p are zero, and Eqs. (18) decouple to become

$$\left.\frac{dn_i}{dt}\right|_{t=0} = n_p k_{pi}. \tag{19}$$

In this limit, the k_{ji} can be simply recovered from the slope of the observed data at $t = 0$ without any model assumptions. As will be discussed below, many different time scales exist for RET processes, even within a given molecule. As a result the meaning of the limit $t \to 0$ must be defined in the context of the process being studied.

Next consider the significance of the use of the several spectral regions for pumps and probes. The most significant effects involve the relative energy of the pump/probe photon, the thermal quanta kT, and the energy level spacings of vibrational and rotational states. Although there are many possible combinations, in general microwave (MW), millimeter-wave (MM), and submillimeter-wave (SUBMM) pumps/probes are comparable in energy to rotational energy level splittings and are smaller than vibrational energy level splittings or kT; infrared (IR) pump/probes are comparable to vibrational splittings, but are larger than either rotational splittings or kT; and optical/ultraviolet (O/UV) pump/probes are large in comparison with rotational splittings, vibrational splittings, and kT.

The frequencies of pumps which have been used in MWMWDR are typically near $1\,\text{cm}^{-1}$ ($kT \sim 200\,\text{cm}^{-1}$). Thus, even at saturation they modify the populations of the pumped levels by less than 1%. Infrared pumps have photon energies far greater than kT. For the typical example of a CO_2 laser pump (whose photons have energies of $\sim 1000\,\text{cm}^{-1}$), the thermal populations of the rotational levels of the pumped ro-vibrational state are less than 1% that of the ground state, and a saturated pump transition changes the population by ~ 100 times (10,000%). Although the differences between MW and IR pumps have obvious sensitivity implications, another important effect also exists. As we will see below, many RET experiments based on IR pumps have shown the existence of significant, but rather unspecific, collisional transfers of population to relatively large manifolds of states. Because MW pumps change the population of a single state by such a small amount, if a fraction of this population is transferred to a large manifold of states *unselectively*, it is very difficult to detect this effect. As a consequence systems based on MW pumps have tended to observe only strongly state-specific energy transfer processes, especially those associated with the electric dipole moment.

Probe effects are also important. Because IR and O/UV probes are between rotational levels in different vibrational states, they essentially measure the change in the absolute population of the rotational level in the pumped vibrational state. MW/MM/SUBMM probes directly measure the *difference* in populations of the rotational states. As shown in Eq. (19), the initial slope of the MW/MM/SUBMM probe directly gives the difference between upper and lower RET states, whereas (at least in the case of a pump which does not disturb the lower probed state) an infrared probe directly provides the population transfer rate. However, the infrared probe is not nearly as sensitive to the nature of the population distribution of the rotational states within vibrational manifolds which are near thermal equilibrium. Returning to the MW example at $1\,\text{cm}^{-1}$, a 1% addition to the population of the lower probed state will increase the absorption by 300%, whereas a 1% addition to the upper state will change the observed absorption into an equally large emission. To distinguish between these two cases qualitatively with an infrared probe requires a subtraction accurate to better than 1%; to measure this quantitatively to 1%, a subtraction accuracy of better than 0.01% is required. Thus, in general IR probes more directly give the absolute population in the probed state, whereas MW/MM/SUBMM probes more accurately show the population distribution.

D. MODEL BUILDING AND PARAMETERIZATION

At the most fundamental level RET in even the simplest polyatomic molecules is almost hopelessly complicated. However, as will be discussed below, dominant processes can be identified and isolated, and quantitative rates can be experimentally measured. Moreover, it has been possible to do this with success comparable to that of much simpler molecular systems.

An important issue is how to bring together these individual processes, each of which may be no more complex than the dominant process for a simpler system, into a single description of RET for a polyatomic molecule. Additionally, it is important to seek not only an empirical description of these cross-sections, but also a parametrization in terms of fundamental molecular parameters.

1. Time Scales, Selective Processes, and Pools

a. Pools of States — General Considerations. The first thing that must be done is to investigate if the number of independent state populations, N, and independent rates that couple these states, $N(N-1)/2$, can be reduced to a number that is more consistent with the number of observations that can be made and the size of the numerical model that is computationally practical. Although this is clearly desirable, there is no *a priori* reason that this has to be possible. However, extensive evidence that has been obtained by several groups shows it is possible and suggests a method of attack. If so, this reduction might be accomplished by (1) collecting many individual states together into *pools*, (2) including in the model only the subset of the states which respond most strongly to the pump, or (3) some combination of the two.

First, consider the well-known example in which an excess of molecules is placed into a single rotational level of an excited vibrational state and allowed to relax toward equilibrium. Since vibrational relaxation is typically orders of magnitude slower than rotational relaxation, the rotational nonequilibrium relaxes long before the excess population in the vibrational state. After this rotational relaxation, the entire rotational manifold can be treated as a single *thermal* pool. This is, of course, the standard assumption which allows the study of vibrational relaxation without the explicit inclusion of the rotational substructure of the problem.

The question at hand, then, is whether or not a larger hierarchy of rates, and the pool concept which results, can be extended back in time through the rotational substructure of the problem, thereby subdividing the rotational problem according to the time scales of the RET phenomena and making it tractable. If the rates which connected the rotational states are random and arbitrary, this is clearly not possible. However, we know on both experimental and theoretical grounds that this is not the case. For example, strong symmetry selection rules based on nuclear spin statistics effectively isolate rotational manifolds into separate pools. These rules result in interconversion rates which are much slower than the vibrational rates discussed above, and the experimental methodology ordinarily applied to the study of vibrational rates allows direct symmetry interconversion to be neglected.

The distribution of population among the states within a pool is not restricted to the thermal relationship of the vibrational manifold in the example cited above. Reduction in the dimensionality of the problem can result from any appropriate grouping of related states. For example, all states of a pool can be initially populated equally or via a scaling law. In some cases these pool states serve as unobserved reservoirs of molecules, and widely varying assumptions about the distribution of population among the states all result in similar effects on the experimentally observed states. In other cases, these pool states are observed, and they can be grouped together according to some more well-founded strategy. Thus, in some cases clear experimental information shows the appropriate relation; in others any of a variety of grouping schemes are experimentally indistinguishable.

The complexity of such *hierarchical models* depends upon the time scale and required detail. At long times large pools, perhaps comprising entire vibrational states, exist and the models are simple. Likewise, at short times, electric dipole processes may dominate, and models may be truncated to only the few states connected by this process. Between these limits, the models which will be required are more complex, but still of a tractable scale.

b. Pools of Rotational States — $^{13}CH_3F$ *as an Example.* Because the symmetries, selection rules, and relative rates of collision-induced energy transfer can vary widely among molecules, it is possible to be specific only in the case of a particular molecule. Because the energy level structure of symmetric tops is well known, time-resolved studies of $^{13}CH_3F$ will be used as an example. Although the details are molecule specific, the hierarchical structure of the evolution of the several degrees of freedom is remarkably general. Figure 2 shows a hierarchical model which will be used as a basis for considering the state of the system as a function of time, starting with

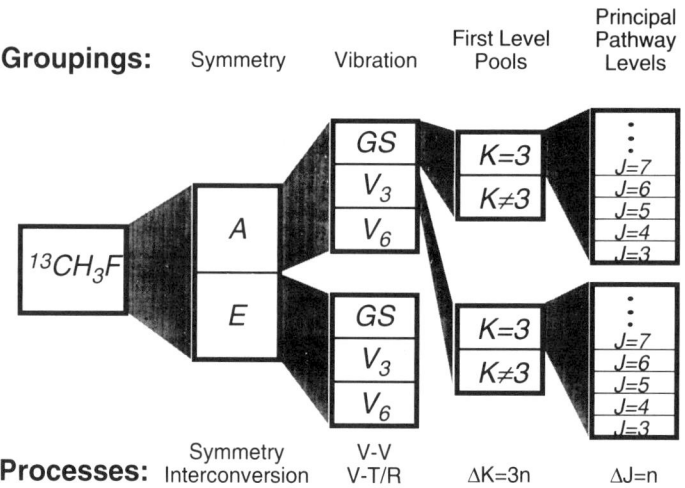

FIG. 2. The hierarchical model for RET in $^{13}CH_3F$.

the longest time scale and progressing to ever shorter times until the initial excitation of a single state by the pump laser.

The development of a model must be based on a proper combination of theoretical expectation and experimental observation. In this case, the strong theoretical expectation of a factoring of the rotational manifold into those states belonging to A ($K = 3n$) and E ($K \neq 3n$) symmetry provided the first factorization in the hierarchy. For the time scales of the experiments here, the separation is absolute. As will be shown in the sections below, this symmetry factorization has been observed in a wide variety of molecular systems and its generality is firmly established.

Next, the expectation that vibrational transitions whose energy defects are large or comparable to kT are slow in comparison with RET suggests the factorization according to vibrational state as shown in Fig. 2. These vibrational transfer rates have been measured by Flynn and co-workers (Weitz et al., 1972; Sheorey and Flynn, 1980), with the fundamental decay of the v_3 vibrational state ($\Delta E \sim 1000 \text{ cm}^{-1}$) and the v_3 to v_6 transfer ($\Delta E \sim 150 \text{ cm}^{-1}$) requiring $\sim 10{,}000$ and ~ 15 gas kinetic collisions, respectively. All other vibrational states are energetically inaccessible and are not included in the kinetic model.

If a vibrational process is resonant or nearly resonant, then the transfer rate can be comparable to that of pure rotational processes. For example, Flynn and co-workers (Sheorey et al., 1978) have shown that the process

$$\text{CH}_3\text{F}(V_3=1) + \text{CH}_3\text{F}'(V_3=1) \rightarrow \text{CH}_3\text{F}(V_3=2) + \text{CH}_3\text{F}'(V_3=0) + \Delta E, \quad (20)$$

which converts two molecules in the v_3 state into one in $2v_3$ and one in the ground state, occurs about every three gas kinetic collisions, largely due to the small energy defect (which arises from the anharmonicity of the vibrational mode) of 18 cm^{-1}. However, because the population in $1v_3$ is ordinarily small in time-resolved experiments, the probability for this process is very low and it may be omitted from the model.

In rotationally resolved studies a similar process which transcends the symmetry selection rule and effectively moves population from one symmetry species to the other via a swapping process is revealed. In $^{13}\text{CH}_3\text{F}$ a molecule in one symmetry of the excited state collides with a molecule of the other symmetry in the ground state (8), and they exchange vibrational quanta as

$$\text{CH}_3\text{F}(V_3=1) + \text{CH}_3\text{F}'(V_3=0) \rightarrow \text{CH}_3\text{F}(V_3=0) + \text{CH}_3\text{F}'(V_3=1) + \Delta E. \quad (21)$$

While the existence of this process does not require a modification of the hierarchy of pools developed to this point, it does introduce a new process which effectively moves molecules between the A and E pools of a given vibrational state at a rate faster than the pure vibrational rates. The inclusion of this process in the hierarchal model for CH_3F is a result of its experimental observation. In many other species either the matrix elements (the electric dipole derivatives) which govern this rate are much smaller or the experiments are less sensitive to its observation. In these cases, the model can be factored according to vibrational state (or effectively by symmetry species), thereby reducing its dimensionality.

Next, the distribution of the population of rotational states within each pool of the hierarchy defined by the vibrational states and symmetry species must be examined. With one exception, discussed in the next paragraph, it is observed experimentally that within each pool all rotational transitions have an identical time-resolved absorption signature. This observation leads to the modeling of the relative population of the rotational states within the symmetry/vibrational hierarchical pools according to a thermal, Boltzmann distribution. Because the process which distributes population among these pools does so according to the symmetry selection rule, it can be referred to as a $\Delta K = 3n$ process.

In addition to these rather nonspecific processes and the pools of states associated with them, faster relaxation has been observed along principal pathways from the pumped state to specific states. For example, in $^{13}CH_3F$ within the same $K = 3$ stack as the level pumped by the CO_2 laser, it is possible to observe experimentally at a very early time a specific $\Delta J = n$, $\Delta K = 0$ transfer process. Even for relatively large values of $\Delta J(\sim 10)$ these rates exceed the less specific rates by a large margin. These process have clear analogs in linear or diatomic molecules and have been the subject of extensive study (Brunner et al., 1981; Copeland and Crim, 1984; Rohlfing et al., 1987). As a result, each of the J states within the pumped K stack are treated as separate repositories of population within the hierarchical scheme and are provided with energy transfer pathways as shown in the figure.

Because of the speed of the electric-dipole-allowed $\Delta J = \pm 1, 0$ process in comparison with that of all others, it is possible to build a simple model from the fast end of the hierarchal model. Here all states and pools which are not connected by the electric dipole matrix elements are eliminated. Jetter et al. (1973) exploited the mathematical simplicity of this limit to build an elegant model of relaxation based on the mapping between this simple model and magnetic resonance theory.

Thus, for the processes of concern to us here (symmetry species interconversion, the fundamental vibrational decay, the $v_3 \rightarrow v_6$ vibrational transfer, the vibrational swap, the $\Delta K = 3n$ process, the $\Delta J = n$ processes, and the electric dipole $\Delta J = \pm 1, 0$ process) the relative rates (in terms of gas kinetic collisions) are approximately ∞, 10,000, 15, 2, 1, 1 $\rightarrow \infty$, and 0.1. Because the last two rates are not summed over many target states, they are in some sense proportionately faster than the other rates which are summed over many target states.

In this context it is useful to specify the state of this system as a function of the number of gas kinetic collisions. After more than 10,000 collisions all excitation provided by the pump will have decayed and the system will be in thermal equilibrium. In the time regime between 15 and 10,000 collisions, all of the ro-vibrational states above the ground state will be in equilibrium, and a study of the time-dependent population of any of them will be a study of the fundamental ($v_3 \rightarrow 0$) vibrational rate. Between 15 and 2 collisions the v_6 state will have not yet reached equilibrium with v_3, and a study of any rotational state within v_6 will reveal the v_3 to v_6 transfer rate. Between 2 and 1 collisions the vibrational swapping mechanism will have not yet equilibrated the A and E symmetries, and in this time interval a study of any state in either the A or the E symmetry will reveal this rate. At times which correspond to about one gas kinetic collision, the $\Delta K = 3n$ process which distributes population among states of the same symmetry as the pump becomes important. Finally, at very short times the specific $\Delta J = n$ rates will dominate, although for large enough n the rates are slower than those of some of the other processes.

Because of this hierarchical distribution of rates, which is especially favorable but hardly unique in $^{13}CH_3F$, the modeling of the RET in this species is vastly simplified. At least as importantly, by selecting the pump/probe time interval and probed state carefully, it is possible to observe rather directly the individual rates without dependence on model assumptions as part of a deconvolution process.

2. Numerical Simulations

Modeling approaches to quantitative descriptions of rotational energy transfer vary according to the complexity of the problem. The simplest begin by assuming a two-level system and introducing concepts such as T_1 and T_2. These are most appropriate for circumstances in which there are two relatively isolated levels (i.e., the inversion spectrum of NH_3; Oka, 1973) or in which the relaxation is dominated by a single strong relaxation (i.e., the dominant dipole–dipole relaxation in CH_3F (Jetter et al., 1973)). Additionally, Green (1978) has considered in the context of IOS theory the relation of observed T_1 and T_2 rates to the more general RET problem.

Although analytical techniques can be applied to somewhat more complex cases, when a global representation of rotational relaxation is desired, the generality associated with numerical models is required. This has led to development of numerical models for the characterization of RET in a number of polyatomic molecules (Matteson and De Lucia, 1983; McCormick et al., 1987a, b; Foy et al., 1988; Bewick et al., 1988; Everitt and De Lucia, 1989; Abel et al., 1992).

In their most general form, these numerical models are conceptually straightforward. The population of each state i is given by N_i, and the RET rates among them are governed by a rate matrix whose elements are k_{ji}.

Because it is desirable that systems evolve toward thermal equilibrium, models ordinarily include detailed balance which, after accounting for degeneracy, reduces the upward rates relative to the downward rates by a factor $k_u/k_d = e^{-\Delta E/kT}$. Additionally, detailed balance reduces the number of independent k_{ji} from $N(N-1)$ to $N(N-1)/2$. The evolution of the system is then governed by the coupled set of Eqs. (18), which can be solved by standard numerical techniques.

The first practical concern is the criterion for the truncation of the number of states (which is infinite in the number both rotational and vibrational states). Simple energetics limit the number of thermally available rotational states to typically $\sim 10^3$ and the number of interacting vibrational states to a countable few. Because this still leaves the scale of the problem unmanageably large, numerical models must be based on the simplifying principles discussed above. An approach is to incorporate formally a hierarchical model into a general RET code which can be applied to any species whose observed RET behavior is hierarchical (Everitt, 1990; Pape, 1993).

The specific hierarchical model shown in Fig. 2 for CH_3F can be extended and incorporated in a general numerical code applicable to any system which can be organized hierarchically. The generalized dimensions are: (1) the number of separate symmetry species, (2) the number of vibrational states per symmetry species, (3) the number of pools of rotational states within each vibrational state, (3a) the number of subpools, (3b) the number of subpools of subpools,..., and (4) the number of principal pathway rotational levels in each pool. This nesting of hierarchies is shown in Fig. 3.

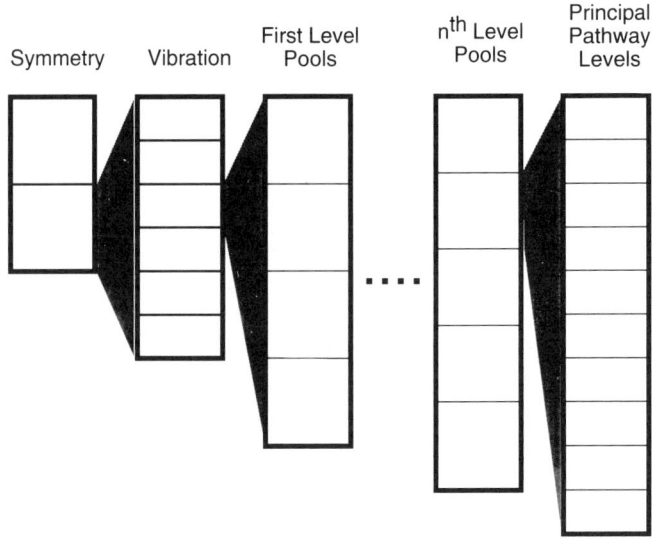

FIG. 3. A generalized hierarchical model for RET in polyatomic molecules.

Although this nesting of hierarchies has been presented here in terms of specific processes, numerically nothing distinguishes one level of the hierarchy from another.

As in angular momentum coupling schemes, the basis represented by this hierarchy represents a simplification, both numerically and conceptually, only if the organization of the levels has a corresponding hierarchical set of rates. If such a correspondence exists (as it does for the CH_3F example), then not only does this hierarchical grouping of states represent a reduction in the dimensionality of the problem, but also it results in a factorable overall rate matrix, with many of the off-diagonal elements being small.

It is often advantageous to include explicitly the effect of the pump laser in the model. By doing so it is possible to include effects associated with pump saturation (which can be very large) and velocity subclass effects. Since these subclasses persist, especially along the principal pathways during RET, it is sometimes necessary to subdivide similarly the levels of the principal pathways.

Finally, whether or not the state represented by the cells of the lowest level of the hierarchy are also included in the next level and so forth changes the physical meaning of the calculated constants. For example, in $^{13}CH_3F$ if the A pool *excludes* the $K = 3$, $J = n$ levels (which include the states along the principal pathway connected by the $\Delta J = n$ process), all of the molecules transferred along the J states of $K = 3$ will be ascribed to the $\Delta J = n$ process and none to the $\Delta K = 3n$ process. Numerically, the models are identical; however, the calculated rates differ, as do their meaning.

Although a number of approaches can be taken for the extraction of the energy transfer parameters from the data, the most direct is a nonlinear least-squares fit to the experimentally obtained time-resolved spectra (Everitt and De Lucia, 1989). In this approach initial estimates of the parameters which govern the energy transfer are made, and the numerical model is used to calculate its estimate of the observed time-resolved signals. Then an iterative nonlinear least-squares adjustment is used to improve the initial estimates. Although in this approach all of the data are fit and all of the parameters are adjusted simultaneously, its power is not always necessary. Because of the selection rules and hierarchical nature of the problem, it is often easy to identify the data that in first order determine the different parameters. In such cases trial adjustments of parameters and visual comparison between calculated and observed time responses can be used successfully (Hetzler and Steinfeld, 1990). Using $^{12}CH_3F$ as an example: The $\Delta J = n$ rates are primarily determined in the fit to the model by the initial slope of the probe signal for transitions within the same K state as the pump. For example, in Fig. 4, the initial slope (which at $t = 0$ is determined solely by single-collision processes) of the $J = 4 - 5$, $K = 2$ transition in $^{12}CH_3F$ can be calculated from the difference between the rates which populate the upper ($\Delta J = 7$) and lower ($\Delta J = 8$) levels from the pumped

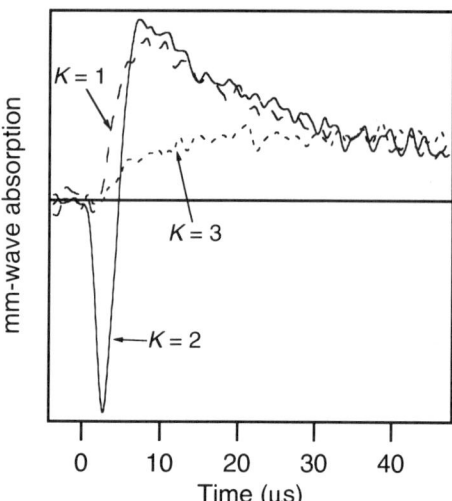

FIG. 4. Time-resolved data for the $J = 4$–5 v_3 transition in $^{12}CH_3F$ at 20 mTorr and 300 K. Near $t = 0$ the different K states provide independent information about the $\Delta K = 3n(K = 1)$, $\Delta J = n(K = 2)$, and vibrational swap ($K = 3$) processes. Beyond 30 μs all of these states come into equilibrium.

$J = 12$, $K = 2$ state. The rate of the $\Delta K = 3n$ process is calculated by fitting any number of equivalent data from J and K states of the same symmetry as the pump but of different K (e.g., $K = 1$). In each of these, the $\Delta K = 3n$ rate is calculated from the initial rise of the population in these states, which is a first-order measure of the transition rate. Likewise, the vibrational swap rate is most readily calculated from the initial slope of the population in any of the equivalent states of opposite symmetry (e.g., $K = 3$). Finally, energy transfer between v_3 and v_6 is observed by directly monitoring the arrival of that population in v_6 (Fig. 5).

In each of these cases, the parameters which govern the observed energy transfer can be obtained by measurement of the initial slope at $t = 0$; complex deconvolution of later time data is not required. Nevertheless, the agreement at later times between the experiment and the simulation is strong evidence of the validity of the model which underlies the simulation. Additionally, there are subtle sum rules and second-order effects which serve as checks on the validity of the model. For example, although the vibrational swap rate is most simply observed in the rate of increase of population in states of the opposite symmetry ($K = 3$ in Fig. 4), it is also measurable from the decrease of population in the states of the same symmetry ($K = 1$ in Fig. 4) and even more subtly from the differences in time response between the A and the E symmetry species in v_6. Likewise, the sum of all rates should be the total depopulation rate.

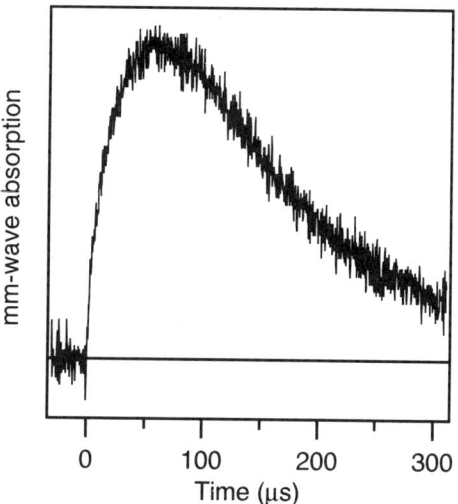

FIG. 5. Time-resolved data for the $J = 4-5, K = 2, l = -1$ transition in the v_6 vibrational state of $^{13}CH_3F$ at 40 mTorr and 300 K. These data provide direct information about the $v_3 \to v_6$ vibrational transfer.

E. THE RELATION BETWEEN EXPERIMENT AND THEORY —WHAT CAN YOU KNOW?

In the sections which follow, we shall see that the factors discussed above have led the field to a description of the RET problem which depends on symmetry, angular momentum, and energy for the necessary simplification, with minimal reference to the details of the IMP. Wherever available, symmetry and other "absolute" selection rules provide an initial factorization, and other angular momentum considerations often separate some states which are closely coupled to the pumped state from other much larger groupings of states. In some systems a single long-range interaction (typically the dipole–dipole) is dominant, and the problems associated with the nonconvergence of the IMP as characterized by Eq. (1) are avoided. In others, energy gap laws, whose physics has its origins in the Fourier integral of Eq. (6) and the averaging over many open collision channels, are also important contributors to the formulation of tractable solutions for the characterization for RET. Indeed, it might be argued that the many averages (over the state of the collision partner, thermal velocity distributions, multiple relaxation paths, etc.) implicit in the parameters which have been used to describe the RET processes also average away the detailed relationships between the data and more fundamental molecular parameters.

A complex relation between theory and experimentation has developed in RET. As in most fields, the early experiments were difficult and only a very limited set of the questions which could be asked were answerable. Additionally, the dimensionality of the theory is such that the unfactored problem is enormous. Although very significant advances have been made on both fronts, the generality of both the experimental questions which can be answered and the calculations which can be done is still limited. As a result the relationships among the results of many studies are not well developed. More specifically, RET in polyatomic molecules currently is far from being an *invertible* problem, lacking either the prospect of starting with the IMP and calculating the RET rates *or* the inverse. Ho and Rabitz (1993) have considered this issue in some detail with particular emphasis on the difference between "considering the experimental data as a *functional* of the potential, instead of fitting the data to a constrained potential form having a few parameters."

We will later return to a narrower question: whether it is possible (or desirable) to build a "spectroscopic" theory of RET for polyatomic molecules. In some sense such models are the basis of most current spectroscopic studies of the energy levels in polyatomic molecules because the link to many of the underlying molecular parameters is tenuous at best. While such an approach might be considered a retreat from a fully invertible relation between the IMP and the manifold of RET rates, the collision problem is far more complicated than the energy level problem. Such a spectroscopic approach eliminates the need for *ab initio* knowledge of the intermolecular potential or, alternatively, a numerically invertible theory and a data set complete enough to support such an inversion. This puts the collisional problem on the same philosophical footing as the energy level problem, which also typically lacks a sufficiently accurate ($\sim 1/10^7$) *ab initio* knowledge of the energy level structure of polyatomic molecules or a numerically invertible theory and data set.

II. State-Specific Rotational Energy Transfer — Principal Pathways

In the large majority of RET experiments, energy transfer to a relatively small fraction of the energetically accessible states has been observed to be significantly faster than that to the remaining states. The states which are so connected can be referred to as members of the *principal pathways* for RET. For molecules with dipole moments, the physical basis for the large $\Delta J = 0$, ± 1 rates is easily understood. However, many other principal pathways have been identified experimentally and will be discussed in this section. The states associated with these principal pathways belong to the first hierarchical level of Fig. 3 and in general have a dimensionality comparable to the number of thermally accessible J levels.

A. Dipole–Dipole Processes

In collisions between polar molecules, by far the fastest RET process is the $\Delta J = 0, \pm 1$ associated with the dipole–dipole interaction, which has been discussed by Oka (1973). As a result the number of studies of dipole–dipole processes in collisions between polyatomic molecules is very large, and only a small portion can be listed here. Among the species which have been studied are NH_3 (Oka, 1973; Klaassen et al., 1982), H_2CO (Oka, 1973; Andrews, 1980), CH_3OH (Lees and Haque, 1974), CH_3F (Jetter et al., 1973; Brewer et al., 1974; Shoemaker et al., 1974; Brechignac, 1982; McCormick et al., 1987a,b), CH_3Cl (Frenkel et al., 1971; Pape et al., 1994), CH_3Br (Herlemont et al., 1976), and CH_3I (Arimondo et al., 1978; Schrepp and Dreizler, 1981; Glorieux et al., 1983).

Because of the strength and specificity of this process, much of the study of dipole–dipole processes has been focused on an array of interesting and important topics which are beyond the scope of this chapter. These include coherent effects, velocity subsets, and molecular reorientation. Furthermore, because the long-range dipole–dipole interaction is often an accurate characterization of the intermolecular potential, it has been possible to make good comparisons between experimentation and theory.

B. $\Delta J = n$ Processes—Fitting Law Descriptions

Experiments on a number of polyatomic molecules have shown that, in addition to the dipole allowed $\Delta J = 0, \pm 1$ process, a number of other state-specific processes are also important. Because these principal pathways ordinarily involve significant changes in J, with various symmetry and angular momentum selection rules governing the changes in the other quantum numbers, these processes are perhaps best referred to as $\Delta J = n$ processes. Furthermore, it has been shown that in many cases fitting laws can adequately represent the observed $\Delta J = n$ rates. Closely related $\Delta J = n$ processes have been observed for collisions between pairs of diatomic molecules and between diatomic molecules and atoms (Smith and Pritchard, 1981; Brunner et al., 1981; Wilkins and Kwok, 1983; Copeland and Crim, 1984). Much of our understanding of the $\Delta J = n$ process in the more complex polyatomic molecules considered here rests upon the foundations provided by this work.

The polyatomic species in which $\Delta J = n$ processes have been studied include spherical tops, symmetric tops, and asymmetric rotors. Table I shows the fitting law parameters which describe this process for several species. For each of the molecular types there are many sublevels of the same J described by additional quantum numbers. However, the dominance of $\Delta J = n$ processes in many species has led to the development of kinetic

TABLE I
ROTATIONAL ENERGY TRANSFER ALONG PRINCIPAL PATHWAYS

	PG		IOS		EG		
	$\sigma_0(\text{Å}^2)$	γ^a	$\sigma_0(\text{Å}^2)$	γ^a	$\sigma_0(\text{Å}^2)$	C	Reference
Spherical							
$^{13}\text{CD}_4$					0.2	0.8	Foy et al. (1988).
SiH_4 (A)					0.49	0.075	Hetzler and Steinfeld (1990).
SiH_4 (E)					0.69	0.05	Hetzler and Steinfeld (1990)
SiH_4 (F)					0.42	0.05	Hetzler and Steinfeld (1990)
Prolate							
CH_3F	0.221	1.22	287	1.3	6.06	4.60	Everitt and De Lucia (1990).
$^{13}\text{CH}_3\text{F}$	0.221	1.22	287	1.3	6.06	4.60	Everitt and De Lucia (1990).
CH_3Cl	0.074	1.12	134	1.23			Pape et al. (1994).
Oblate							
NH_3						1.8	Abel et al. (1992).
Asymmetric							
D_2CO	4.3	1.425			215	5.0	Bewick et al. (1988).

$^a\gamma$ is related to α [Eq. (14)] by $\alpha = 2\gamma - 1$.

models which place them first in the hierarchy of processes. Subsequent or parallel transfers to the J sublevels described by the other quantum numbers of the several molecular types will be discussed in Section III.

1. Symmetric Tops

The methyl halides are an illustrative example. They have well-known energy level structures, and it has been possible to show simple relationships among the results of a number of different RET experiments. The description of the energy level structure of symmetric tops in nondegenerate vibrational states is well known. In addition to the total angular momentum quantum number J, its projection onto the molecular symmetry axis K is also an important quantum number. For prolate rigid rotors, $E = BJ(J + 1) + (A - B)K^2$, where A and B are the rotational constants associated with the moments of inertia about the a and b axes. For many of the species studied, A is much greater than B, leading to large changes in energy for changes in K. Additionally, Oka (1973) has shown the selection rule $\Delta K = 3n$ for molecules of C_{3v} symmetry. Thus, especially for prolate

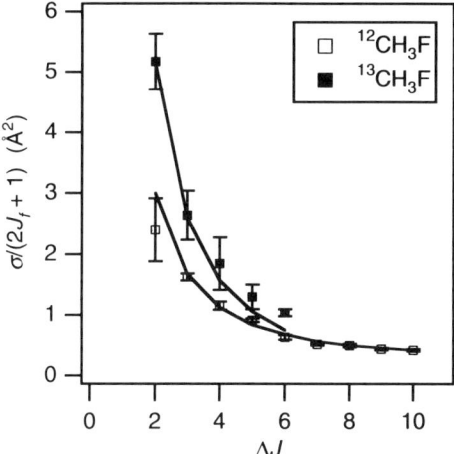

FIG. 6. Normalized cross-sections recovered from the fit of the numerical simulation to the time-resolved data plotted as a function of ΔJ. The solid lines represent the results of the theoretical calculation.

species, both energy and symmetry considerations favor changes in J over changes in K.

Changes of J within a particular K manifold are by far the fastest RET processes in CH_3F. Not only are these induced by the long-range electric dipole moment, but also the long–narrow geometry of the molecule suggests the presence of other interactions which would lead to $\Delta K = 0$, $\Delta J = n$ transitions. It has been found that the scaling and fitting laws discussed in Section I provide a good means to characterize the J changing processes in these molecules. For $^{13}CH_3F$ and $^{12}CH_3F$ (Everitt and De Lucia, 1990), the normalized rates of Eq. (16) are plotted in Fig. 6 as a function of $\Delta J = n$ and in Fig. 7 as a function of the energy gaps. Although Fig. 6 suggests a unique family of points for each isotopic species, Fig. 7 reveals that, when the normalized rates are plotted as a function of energy gap, only one family of points is found. In other words, the variation in normalized rates is not due to an isotopic variation of the dynamical coefficients. Instead, the variation in the normalized rates arises primarily from the differing energy gaps which resulted from the different pumped states ($J = 5$, $K = 3$ for $^{13}CH_3F$ and $J = 12$, $K = 2$ for $^{12}CH_3F$).

The application of the PG [Eq. (14)] fitting law to $^{13}CH_3F$ and $^{12}CH_3F$ is shown by the solid line in Fig. 7. Alternatively, it is possible to account for the time-resolved data by substituting Eq. (15) directly into Eq. (12) to provide the IOS–P relationship between the rates and fitting parameters k_0 and γ. Fits based on the exponential gap model were also tried but were found to be less successful.

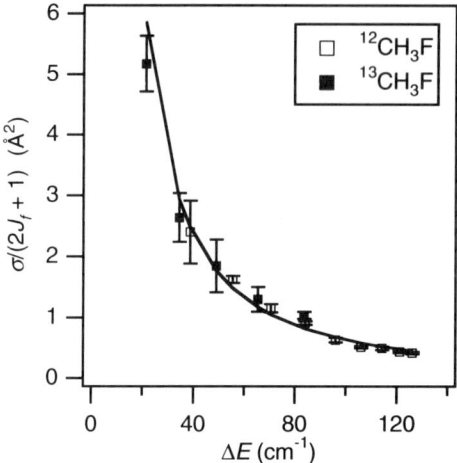

FIG. 7. Normalized cross-sections recovered from the fit of the numerical simulation to the time-resolved data plotted as a function of ΔE. The solid line represents the results of the theoretical calculation.

In characterizing any system which requires a complex model, it is important to consider alternatives. For processes which change the J quantum number, the alternative which has been considered most often uses successive $\Delta J = \pm 1$ steps (Oka, 1967a; Fabis and Oka, 1983; Matsuo and Schwendeman, 1989) as a substitute for large $\Delta J = n$ processes. Although it can be difficult to separate these alternatives in CW experiments, the time-resolved experiments on CH_3F provide unambiguous evidence. Consider Fig. 8. In this figure the time-resolved experimental data for J changing (up to $\Delta J = 10$) are shown for $^{12}CH_3F$. It can be seen that for all states the population begins to arrive immediately, without the long delay required for a series of $\Delta J = \pm 1$ steps random walking population from the pumped state. Figure 8 also shows a comparison between the experimental data and a model based only on successive $\Delta J = \pm 1$ steps. In contrast to the experimental results for transitions well removed from the pumped state, this model predicts no significant transfer other than a more general thermal absorption.

Methyl chloride (CH_3Cl) is geometrically similar to methyl fluoride and has also been studied by time-resolved double-resonance techniques (Pape et al., 1994). As in the case of CH_3F, it was possible to fit the observed $\Delta J = n$ processes using either the PG or the IOS–P procedure. Additionally, the effect of fitting for a separate dipole–dipole rate or requiring the scaling law to account for all $\Delta J = n$ rates, including the $n = 1$ dipole rate, was carefully evaluated. Both procedures produced satisfactory fits to the experimental data and similar RET parameters. Because CH_3Cl and CH_3F have similar geometries, the pair should represent a good opportunity to test

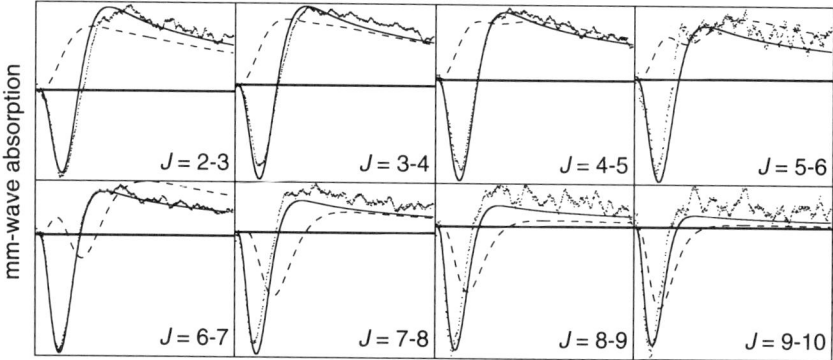

FIG. 8. Observed (points) and calculated (solid and dashed lines represent the $\Delta J = n$ and successive $\Delta J = 0, \pm 1$ models, respectively) time-resolved data for the principal pathway ($K = 2$) states $^{12}CH_3F$. Each figure covers $20\,\mu s$ and the pressure is $20\,mTorr$.

the validity of the separation between dynamical variables and spectroscopic variables as well as the parameter scaling discussed in Section I. This will be discussed in more detail in Section V.

Similar, but less quantitative results have been obtained in a CW IRMW double-resonance study of CH_3Br by Herlemont et al. (1976). In this study they observed that within the same K stack as the pump level, transferred populations decreased monotonically from the pumped $J = 7$ level. Although it was not possible to make quantitative measurements, again the larger rates associated with the $\Delta K = 0$, $\Delta J = n$ were clearly observed. Although these CW measurements were not inconsistent with successive $\Delta J = \pm 1$ steps, it would be surprising if the basic behavior of this species would differ greatly from that of CH_3F or CH_3Cl.

Rotational energy transfer in methyl iodide (CH_3I) has been studied by the technique of radio-frequency spectroscopy inside the cavity of an infrared laser (Glorieux et al., 1983). In these CW experiments, collision-induced resonances with ΔJ up to 9 were observed and interpreted as being the result of successive $\Delta J = \pm 1$ steps. In a MW–MW double-resonance experiment Schrepp and Dreizler (1981) have also studied this molecule. Their results also show a $\Delta K = 0$ selection rule and population transfer for which $\Delta J > 1$, but they conclude that the data do not allow a determination between successive $\Delta J = \pm 1$ processes and higher order processes.

Although some of the earlier CW observations of the methyl halides have been interpreted as being the result of successive $\Delta J = \pm 1$ steps rather than $\Delta J = n$ processes, it seems in light of more recent observations that it is unlikely that the $\Delta J = \pm 1, 0$ dipole processes are the dominant mechanism for population transfer to states removed from the pumped state by more than a few J. A good test would be to study in a time-resolved experiment one of the heavier species (perhaps CH_3I or CH_3Br, in which J states to ~ 100 are thermally populated) for which the thermalization time due to successive randomized $\Delta J = \pm 1$ steps would be of the order of 10,000 collisions. Because many vibrational transition rates are faster than this, an expecially interesting mix of rotational and vibrational energy transfer would result if the $\Delta J = 0, \pm 1$ dipole processes are the dominant RET mechanism for large changes in J.

Ammonia (NH_3) is an oblate symmetric rotor in which nonthermal RET has been carefully studied (Abel et al., 1992). Again it was found that the $\Delta J = n$, $\Delta K = 0$ processes could be characterized by a fitting law, but this time by the EG relation. It is significant that the rotational constants of NH_3 are much larger than those of the methyl halides. As a result the number of parallel RET paths is significantly reduced, and individual $\Delta J = n$ rates, which could not be described by a fitting law, for $|\Delta K| = 3$ were observed. Although these rates are about an order of magnitude slower than the $\Delta J = n$, $\Delta K = 0$ rates, because they are fitted separately they too could be considered to be states of the principal pathway. Abel et al. (1992) also address the issue of model dependence, especially the question of the separation of the $\Delta J = n$, $\Delta K = 0$ rates from successive, fast $\Delta J = \pm 1$ steps. Again, the time-resolved results are similar to those shown in Fig. 8 for CH_3F: the observation of population in states far removed in J from the pump at very early times precludes a mechanism based on a random walk $\Delta J = \pm 1$ mechanism. Additionally, several $\Delta J = -1$ and $\Delta J = -2$ rates have been measured by Chardon et al. (1983) via molecular beam techniques.

Fluoroform (CDF_3) is fundamentally different from most of the species for which RET has been observed because it is an oblate rotor in which the spacing between K levels is much closer than the spacing between J levels. As such, it represents a case in which the largest multipole moment, the dipole, seeks to induce transitions which are energetically less advantageous than competing processes which depend on higher order moments. In this case, large $\Delta K = 3n$ (n as large as perhaps 8 or 9) were observed, with slower $\Delta J = \pm 1$ steps. It was concluded that RET in this species followed dipole-like propensity rules (Harradine et al., 1984; Foy et al., 1988). It would be very interesting to study this and similar oblate species in more detail in order to see if a thermal distribution of population, or a population described by a PG, EG, or some other fitting law, is appropriate for the description of the $\Delta K = 3n$ process in these molecules. Similarly, it would be useful to observe this species in a fully time-resolved experiment in order to see if the population transfer to J states far removed from the pumped states proceeds via the $\Delta J = \pm 1$ process or via higher order processes. Again, if the $\Delta J = \pm 1$ is dominant, it would be especially interesting because this would provide an example of a system in which fundamental vibrational relaxations are faster than the time required for rotational

ROTATIONAL ENERGY TRANSFER IN SMALL POLYATOMIC MOLECULES 363

equilibration. In any event, a proper generalization (and perhaps a more appropriate designation) of the $\Delta J = n$ process awaits the results of such experiments.

2. Asymmetric Tops

D_2CO is a slightly asymmetric prolate rotor whose density and structure of states are similar to those of the methyl halides discussed above. However, its asymmetry leads to families of relatively small energy splittings which correspond to transitions in the conventional microwave region below about 30 GHz. As a consequence of their strength and convenient spectral location, the transitions between these states have been the subject of a number of CW MWMWDR experiments in the ground vibrational and electronic state (Oka, 1967a) as well as observations in excited electronic states and highly excited ($\sim 10^4 \text{cm}^{-1}$) vibrational states of the ground electronic state (Vaccaro et al., 1985; Temps et al., 1987).

More recently, Bewick et al. (1988) have used IRUV time-resolved double-resonance methods to study RET in this species. Because the infrared pump produced a large nonequilibrium population in the $(J, K_a) = (18, 11)$ state and the UV probe can observe many states, it was possible to observe changes in J up to 7 and to compare these results with the EG and PG laws as well as to a model based on successive $\Delta J = \pm 1$ steps. They found that the PG law was marginally better than the EG law in fitting their data, but that both were substantially superior to fits based on a $\Delta J = \pm 1$ propensity rule. They concluded that the propensity rule observed in the earlier MWMWDR experiments was primarily a product of the insensitivity of those experiments to higher order $\Delta J = n$ processes, combined with the expectation of the predominance of the $\Delta J = \pm 1$ process.

3. Spherical Tops

Rotational energy transfer in spherical tops, including SF_6 (Dubs et al., 1982), $^{13}CD_4$ (Laux et al., 1984; Foy et al., 1985, 1988), and SiH_4 (Hetzler and Steinfeld, 1990) has also been studied. In the more recent of these studies, it has been possible to establish that $\Delta J = n$ principal pathways play a central role in the RET of these species. It has been found that these principal pathways are associated with a $\Delta(J - R) = 0$ propensity rule which is closely analogous to the $\Delta K = 0$ selection rule of symmetric tops.

The energy level structure of spherical tops in excited vibrational states is more complex than that of the symmetric and near-symmetric tops discussed above. Coriolis coupling between the vibrational angular momentum l and the total angular momentum J causes the vibrational mode to break up into Coriolis subbands. In these, the angular momentum of the molecular frame is equal to the total angular momentum minus the vibrational angular momentum ($R = J - l$). In analogy with the symmetric top quantum number K, a quantum number, K_R, may be associated with R. However, at this level of approximation the z axis to which the K_R is referenced is arbitrary because the Hamiltonian still has spherical symmetry.

The spherical symmetry is broken by interactions which cause the Coriolis subbands to break up into fine structure levels, much as the difference between the rotational constants in a symmetic top split the rotational levels according to the K projection quantum number. Unfortunately, the quantitative descriptions of these levels are much more complicated because of the mixing of all of the states involved, especially at low J. However, it has been shown that at high J (Harter and Patterson, 1977, 1979) the fine structure levels form clusters, with the individual states described by K_R. Physically, this clustering can be associated with distortion along its axis of rotation, with this distortion slowly tunneling among the equivalent symmetry axes. In this picture, because this tunneling is slow on the time scale of a collision, collisions between spherical tops can be considered on the same basis as collisions between symmetric tops. Interested readers are referred to Appendix A of Hetzler and Steinfeld (1990) for a particularly lucid discussion of the spectroscopy and level notation for spherical tops, which has been developed by Champion and his colleagues (Champion, 1977; Champion and Pierre, 1980).

Methane is a spherical top and as such does not have a permanent dipole moment. In the earliest observation by Laux et al. (1984) of RET in methane ($^{13}CD_4$), population transfer with values of ΔJ up to 15 was observed. However, it was not possible to identify the mechanisms responsible or to quantify the rates associated with them. In a later time-resolved experiment, Foy et al. (1988) were able to show directly the existence of a large $\Delta J = n$ process which they quantitatively modeled with the EG law. Because fits for which $\Delta J = n$, $n > 5$ were indistinguishable from those for which the maximum n was limited to 5, they chose the latter limit in their model. Parson (1990) has shown that the principal pathways in methane can be predicted with remarkable accuracy by extension of the well-known symmetric top selection rules. It is important to note that these predictions are much more specific than the $\Delta(J - R) = 0$ propensity rule because the latter predicts many more states to be part of principal pathways than are experimentally observed.

Another spherical top which has received considerable attention is silane (SiH_4). In order to represent their data, as in methane Hetzler and Steinfeld (1990) have used an EG dependence with ΔJ limited to 5. Also as in methane and the symmetric and asymmetric tops, the early arrival of population in states far removed from the pump directly showed the importance of large $\Delta J = n$ processes. For SiH_4 the CO_2 laser pump places population in the $J = 13$ states of the A, E, and F symmetry species. Because of the strong symmetry selection rule, the population in each symmetry relaxes independently. As can be seen from the parameters in Table I, the $\Delta J = n$ rates are not strong functions of symmetry state. Parson (1990) has also considered the application of his theory which maps symmetric top RET selection rules onto the spherical tops to SiH_4. Although SiH_4 represents a somewhat less favorable case than $^{13}CD_4$ because of its greater rotational and vibrational state mixing, the success of this theory while less complete is encouraging. It would appear that the underlying physical principles are correct and that a more extensive consideration of mixing and energy gaps would provide an excellent agreement between experimentation and theory.

C. SUMMARY

For all of the molecules which have been studied in fully time-resolved experiments, we see that distinct, principal pathways for RET exist. Although the details vary with molecular type, all may be described as $\Delta J = n$ processes. Furthermore, the rates for these processes can be described by fitting laws using only a small number of adjustable parameters, with the rates associated with the dipole-allowed $\Delta J = \pm 1$ playing no special role. Additionally, it would appear that the selection rules for these principal pathways can be understood on the basis of the symmetries of the molecules and their wave functions. In all of the quantitative numerical models which have been developed for these molecules, the rotational states of these principal pathways form the first level of the hierarchy.

III. Transfer to Nonprincipal Pathways: The Grouping of States

With the advent of infrared laser pumps, photon energies capable of inducing both vibrational state nonequilibria and observable transfers to large manifolds of rotational states became available. In this section we will consider the processes which transfer populations to these groups of states or *pools*. Within these groupings the relative populations of individual states may be characterized in any of a number of ways: it is in some cases equally divided according to degeneracy, in others partitioned according to an energy based scaling law, or assigned a Boltzmann distribution. Examples of these distributions are shown in Table II for a number of molecules. Sometimes experimental evidence which directly leads to one of these choices is available. In other cases, no experimental information which might prompt a choice is available, making the choice either arbitrary or based on some theoretical expectation.

TABLE II
ROTATIONAL ENERGY TRANSFER ALONG NONPRINCIPAL PATHWAYS

	Pathway	J scaling	$\sigma(\text{Å}^2)$	Exponent	Reference
Spherical					
$^{13}CD_4$	$F_{PP}^+ \to F^+$	None	8.0	—	Foy et al. (1988).
$^{13}CD_4$	$F^+ \to F^0, F^-$	EGL	0.04	0.8	Foy et al. (1988).
SiH_4	$A_{PP}^- \to A^-$	None	866	—	Hetzler et al. (1990).
SiH_4	$E_{PP}^- \to E^-$	None	69	—	Hetzler et al. (1990).
SiH_4	$F_{PP}^- \to F^-$	None	445	—	Hetzler et al. (1990).
SiH_4	$A^- \to A^0, A^+$	EGL	7.9	0.075	Hetzler et al. (1990).
SiH_4	$E^- \to E^0, E^+$	EGL	0.18	0.05	Hetzler et al. (1990).
SiH_4	$F^- \to F^0, F^+$	EGL	0.12	0.05	Hetzler et al. (1990).
Prolate					
CH_3F	$\Delta K = 3n$	Thermal	137	—	Everitt and De Lucia (1989), Everitt (1990).
$^{13}CH_3F$	$\Delta K = 3n$	Thermal	137	—	Everitt and De Lucia (1989), Everitt (1990).
CH_3Cl	$\Delta K = 3n$	Thermal	69	—	Pape et al. (1994).
Oblate					
NH_3	$\Sigma_{\Delta K \neq 0}$	State specific	62	—	Abel et al. (1992).
Asymmetric					
D_2CO	$\Sigma_{\Delta K \neq 0}$	None	150	—	Bewick et al. (1988).

If the population is distributed according to a Boltzmann distribution, a *thermal pool* results. Such a pool represents a collection of states in thermal equilibrium with each other, but whose total population, as well as the temperature which describes the distribution, is free to rise and fall as a function of time. These pools exist either because of the nonselectivity of the RET processes into them or because of internal thermalization on a time scale that is fast compared with that of the experimental observation. The simplest, but not exclusive, signature for a pool of this type is a large number of probed transitions with identical time evolutions.

Alternatively, fitting law distribution based on energy gaps initially place the largest populations in states nearest the pumped state, thereby initially producing population inversions or reduced absorptions between all pairs of states below the pump. Although for most values of the exponents C [Eq. (13)] or α [*Eq.* (14)] the difference in population between adjacent states is relatively small, the impact on the pure rotational spectrum can be very large because the difference in population which leads to the equilibrium absorption spectrum is also typically small (~ 1–5%). After a period of time determined by the RET rates, detailed balance will convert the initial scaling law distribution into a thermal distribution.

A. SYMMETRIC TOPS

In Section II, the fastest group of processes in CH_3F, the $\Delta J = n$, connected states of the same K as the pumped state together in the first hierarchy for RET. Experimentally, it was observed that the next fastest time responses in the data belonged to the states of different K, which had the same symmetry and vibrational states as the pumped state. For these states it was observed that the time-resolved data are all identical (Everitt and De Lucia, 1989), confirming the earlier CW results of Matteson and De Lucia (1983). Thus, the measurement of any of these probe data can be used to determine the rate of population transfer from the pumped state and the other rotational levels of the first level of the hierarchy to the pool of states of the second level of the hierarchy. Because of the selection rule for this process, it has been called the $\Delta K = 3n$ process. In this case, the dimensionality of the second hierarchical level is 1 because only one pool is required to describe the observed data. In contrast, the dimensionality of the first level hierarchy was ~ 20, the number of distinct thermally accessible J states.

Figure 9 shows the results of measurements of the cross-sections for the $\Delta K = 3n$ process between 100 and 450 K (Goyette et al., 1992). The most striking feature of this is the rapid rise from essentially the gas kinetic cross-section which begins at about 200 K. Although upon first consideration these appear to be large cross-sections for a nonelectric dipole transition, the rate is defined as the sum of all rates from a particular state to all other thermally populated states of the same symmetry. Since the partition function of the pool is of order 500, the specific state-to-state rates are smaller by the same factor.

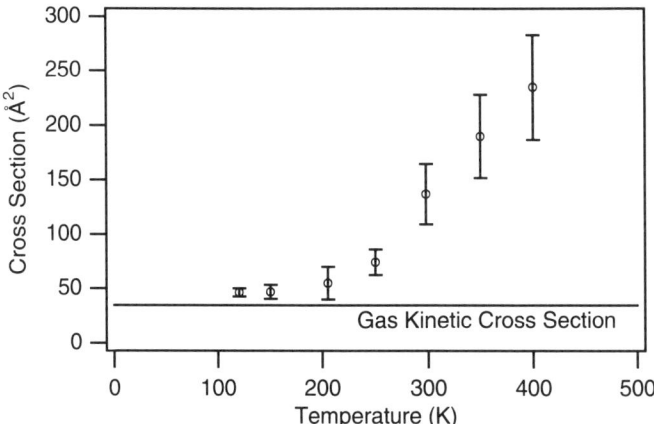

FIG. 9. Cross-section as a function of the temperature of the $\Delta K = 3n$ process in v_3 of $^{13}CH_3F$.

CH_3Cl represents an opportunity to consider this process in the context of a molecule of very similar geometry but more closely spaced rotational states. Additionally, there are two pumped states of substantially different rotational energy (85 and 285 cm^{-1}). In this case raw experimental results that were significantly different from those of CH_3F were obtained. The time-resolved data of the $J = 8-9$ transitions are shown in Fig. 10 for four different combinations of pump and K-stack probe (Pape et al., 1994). The K stack which includes the pumped state is not included in this figure. Each of these four traces represents a family of identical time responses for which $J \to J'$ varies. Even casual inspection reveals significantly different early time behavior as a function of either pumped state or observed transition.

Although these experimental observations would seem to show in CH_3Cl that the second hierarchical level should be broken up into separate pools for each K, such an increase in complication is not necessary. Identical time responses represent only the simplest case of a thermal pool, one in which the RET process populates the pool at a rotational temperature identical to the translational temperature. If, instead, the rotational temperature of the pool varies according to

$$T(t) = T_e + (T_0 - T_e)e^{-ct}, \qquad (22)$$

where T_e is the equilibrium translational temperature of the gas; T_0, the initial rotational temperature of the pool; and c, the rate constant of the relaxation of the rotational temperature, then the time–response curves will be a function of the energy of the probed state. As a reference, Fig. 10 shows as a dashed curve an attempt to fit the CH_3Cl data to a model with a fixed rotational temperature, whereas the solid line allows this temperature to vary according to Eq. (22). As can be seen, the single pool model based on Eq. (22) fits the data extremely well because it accounts for the initial excess population in the low energy states (such as $K = 1$) relative to higher energy states (such as $K = 5$). It is interesting that the initial temperature calculated from the fit to the data ($T_0 = 138 K$ (6, 4 pumped state) and $T_0 = 374 K$ (11, 6 pumped state)) is very close to the temperature equivalent of the energy of the pumped state.

Thus we see that the observed $\Delta K \neq 0$ time-resolved data of states of the same symmetry and vibrational state can be described in terms of thermal pools for both CH_3F and CH_3Cl. However, for the latter case the initial temperature which describes the population is more closely related to the pump energy than to the translational energy. We will return to this issue as well as the relation between these results in Section V.

Similar, but less quantitative, observations have been made in CH_3Br (Herlemont et al., 1976) by use of CW IRMW double-resonance techniques. In this work, in addition to the $\Delta J = n$ transitions, a general population transfer, without selection rules, was observed to the other states of the same symmetry and vibrational state.

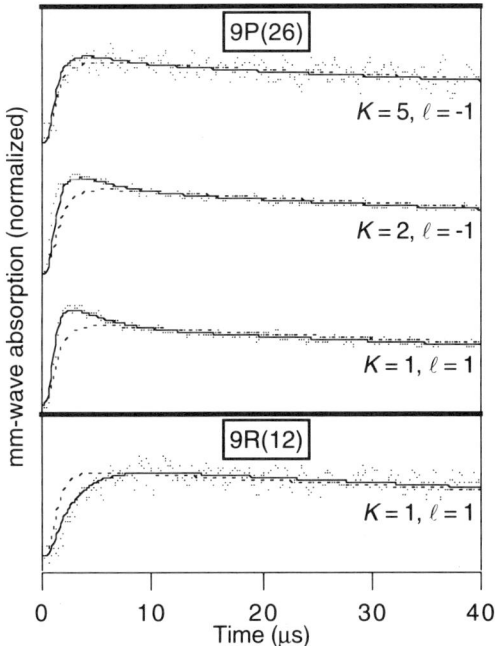

FIG. 10. Comparison of the results of the generalized model for K-changing collisions in CH_3Cl at 50 mTorr with those of the fixed-temperature model for four combinations of pumps and probes. The dashed lines represent the results of a model with a fixed rotational temperature, and the solid lines, a model in which the temperature can vary according to Eq. 22.

NH_3 has substantially larger rotational constants than the species discussed above and as a result the observed RET in NH_3 is considerably more state specific. As discussed in Section II, Abel et al. (1992) have found significant $\Delta J = 1 \rightarrow 4$, $|\Delta K| = 3$ rates. However, unlike the $\Delta K = 0$ rates they observed, it was not possible to account for them in the context of any fitting law. Instead, each rate was determined individually from the experimental data. However, in their model all states further removed from the pump were collected together in baths according to K. Because molecules once placed in these baths were not allowed to feed back into the observed states, it was not necessary to define the distribution of the molecules among the individual rotational levels of the baths. Thus NH_3 represents a sparse state case in which most of the rates were modeled either according to a fitting law (the $|\Delta K| = 0$ transitions discussed in Section II), fitted individually (the $\Delta J \leq 4$, $\Delta K = 3$), or grouped together to provide pathways to baths. Because all of these states and baths are directly connected and have rates that connect them and that are faster than other processes in the hierarchy, it is possible to truncate the model at what is in effect the first hierarchical level.

B. ASYMMETRIC TOPS

Bewick et al. (1988) have studied RET in the slightly asymmetric rotor D_2CO. Although the principal focus of their study was the $\Delta K_a = 0$, $\Delta J = n$ process discussed in Section II, they also observed and considered transitions to other K_a states by means of a grouping process. In their model (in which the pumped state was J, $K_a = 18, 11$) a parameter, k_K, was used to transfer population to $K_a = 9$ ($J = 16-26$) and $K_a = 13$ ($J = 13-22$), and a second parameter, k_D, was used to remove population from the states considered individually to an unspecified collection of states belonging to other families of K_a or vibrational states. Although in the master equation calculations these parameters were treated as fixed, a change of a factor of 2 in k_D resulted in a significant deterioration in the quality of the fit of the time-resolved data. Furthermore, it was found that their sum was relatively well determined at about three times the gas kinetic collision rate, although the values of the two parameters were not expecially well determined individually.

As in the case of NH_3, this study represents a good example of a case in which only a part of the thermally accessible manifold is modeled in any detail. However, the $\Delta J = n$ processes of interest are dominant enough that detailed modeling of the molecules which might cycle back into the states of interest can be ignored on *the time scale of the experiment*.

C. SPHERICAL TOPS

In their work on $^{13}CD_4$, Laux et al. (1984) observed up to 15 J-changing collisions from the pumped $J = 11$ state of the F^+ Coriolis sublevel. The population appeared with comparable rapidity over this wide range of states and was interpreted as indicating no restriction on the change of the quantum number for RET in this species. Subsequently, Foy et al. (1988) studied the same RET system in $^{13}CD_4$ in more detail and developed a kinetic model to account for the observed energy transfer. As discussed in Section II, the first level of their hierarchal model was defined by a $\Delta J = n$ process which preferentially deposited population along principal pathways within the pumped F^+ Coriolis sublevel into states of particular J_R, according to the selection rule $\Delta(J - R) = 0$ or the more specific selection rule of Parson (1990).

Based on the experimental observation of similar RET probe data for groups of states which did not include the principal pathways, a grouping procedure was adopted to describe the subsequent RET. It was found for each of the three Coriolis sublevels that it was possible to first group together all of the fine-structure states with the same J_R value (excluding the states which are included in the principal pathway). Transfer from states in the principal pathway to the groups in the same Coriolis sublevel is governed by a single parameter, k_1, with subsequent transfer to each of the other two Coriolis substates reduced by a factor, p, from the principal pathway scaling law rate. In this case the dimensionality of the first hierarchical level is equal to the number of J_R states (~ 25), as is the second level, with one grouping of fine-structure levels for each J_R.

Rotational energy transfer in SiH_4 (Millot et al., 1988; Hetzler and Steinfeld, 1990) has much in common with that of $^{13}CD_4$. However, the CO_2 laser pump places population into all three symmetry species (A, E, F), and it is possible to observe the independent evolution of each. In each symmetry, a $J = 13$ level of a Coriolis subgroup is excited by the pump, and the subsequent RET follows the same basic structure as that in $^{13}CD_4$. However, the RET is somewhat surprisingly a strong function of the symmetry species, as can be seen in Table II. Because the observed RET in SiH_4 was found to be experimentally more complex than that of $^{13}CD_4$, it was necessary to provide separately adjustable constants for the scaling laws which governed energy transfer along the principal pathways (k_0') and for transfer among the groups of states (k_0). Additionally, because the transfers between the different Coriolis levels are slower, their rates are reduced by an additional factor, p, as was the case for $^{13}CD_4$. The large variations by symmetry species for the constants which describe transfer among the groupings of states have no obvious physical cause and are model dependent. Consequently, they should not be considered rate coefficients but rather model parameters (Hetzler and Steinfeld, 1990).

D. Summary

For the species for which detailed kinetic models have been developed, some form of a hierarchical model has been employed. This has been possible because the experimental results have shown both fast principal pathways to individual states and large groupings of states with either similar or identical time-response signatures, for which the RET is slower. In the limit that the relaxation along the principal pathways is fast in comparison with the processes which transfer populations to the next higher level of the hierarchy, the relative populations of the groups of the higher levels of the hierarchy would be a thermal distribution because the detailed balance would thermalize the population of the principal pathway before significant population transfer to the grouped states. In the alternative limit that the transfer of population within a given J is fast compared with the principal pathways, the populations of the groups would simply mirror that of the principal pathways, with early time population inversions prevalent throughout the molecule. In Section VI we will return to the question of how fundamental the pools of states are and if they might be broken down into additional hierarchical levels.

IV. Near-Resonant Ro-Vibrational Energy Transfer

Two types of near-resonant ro-vibrational energy transfer have been observed: rotationally state-specific transfer between two nearby vibrational levels that are coupled through a Fermi resonance or Coriolis coupling and state-nonspecific transfer through near-resonant V–V exchange, typically between an excited and the ground vibrational levels. The former class of processes has been reviewed by Orr and Smith (1987) and will not be discussed here. The latter, sometimes referred to as V-swap [Eq. (7)], has been recognized as an important process in several molecules. Table III summarizes much of the work that has been done in this area. The calculated cross-sections were obtained using Sharma and Brau theory, outlined in Section I, or are from the cited reference.

A. Symmetric and Spherical Tops

The most extensively studied of the V-swap processes (Everitt and De Lucia, 1989, 1993) connect the $V_3 = 1$ and ground vibrational states of the ^{12}C and ^{13}C species of CH_3F as

$$CH_3F(V_3=1) + CH_3F'(V_3=0) \rightarrow CH_3F(V_3=0) + CH_3F'(V_3=1) + \Delta E, \quad (23)$$

where ΔE reflects the changes in rotational state of the collision partners. The large value of $d\mu/dQ_3$ (0.28D) and relatively large rotational partition function (~ 1000) combine to give a calculated 300-K V-swap cross-section of 27 Å^2. This compares favorably with the experimentally measured value of 21 Å^2, especially given that $d\mu/dQ_3$ has a 5% uncertainty and is raised to the fourth power in the calculation. This process has been measured in both species by observations of: (1) the decrease in population of states of the same symmetry as the pumped state and (2) the increase in population of states of the opposite symmetry. Moreover, the cross-section is comparable in size to fast rotational energy processes, making this process observable and important in RET analyses.

Studies of many rotational states in $V_3 = 1$ of CH_3F reveal the rotationally thermal nature of the V-swap process, and other work has demonstrated its translationally thermal nature as well (Shin et al., 1991). Both are expected because both are derived from the thermal nature of the ground vibrational state. Within the limits of experimental error, the cross-section was observed to be independent of symmetry type, pumped state, and isotope. The temperature dependence of the $V_3 = 1$ CH_3F cross-section has been measured (Everitt and De Lucia, 1993). The experimental results, shown in Fig. 11, reveal a cross-section which decreases with increasing temperature as $1/T$.

TABLE III
Near-Resonant Vibrational Exchange

	Vibrational states	Vibrational defect (cm^{-1})	$d\mu/dQ_k$ (D)	Observed σ_{vs}[a] (Å^2)	Calculated σ_{vs}[a] (Å^2)	Reference
Spherical						
^{13}CD$_4$	GS $\leftrightarrow v_4$	0	0.05		0.0082	Foy et al. (1988).
SiH$_4$	GS $\leftrightarrow v_4$	0	0.232		6.0	Fox and Person (1976).
SF$_6$	GS $\leftrightarrow v_3$	0	0.437		400	Kim et al. (1980).
Prolate						
CH$_3$F	GS $\leftrightarrow v_3$	0	0.2756	18.9	27	Kondo et al. (1986), Everitt and De Lucia (1993).
	$v_3 + v_3 \leftrightarrow$ GS $+ 2v_3$	18		12	50	Sheorey and Flynn (1980); Weitz and Flynn (1973).
^{13}CH$_3$F	GS $\leftrightarrow v_3$	0		21.0	27	Everitt and De Lucia (1993).
	Isotopic swap	12		7.8[b]	26	Preses and Flynn (1977).
CH$_3$Cl	GS $\leftrightarrow v_6$	0	0.0384	1.18	0.014	Pape (1993).
	Isotopic swap	≈ 0		0.92	0.014	Pape (1993).
CD$_3$Cl	GS $\leftrightarrow v_2$	0	0.0972	3.88	0.68	Kondo et al. (1985), Doyennette and Menard-Bourcin (1988).
CD$_3$Cl	Isotopic swap	≈ 0		3.88	0.68	Doyennette and Menard-Bourcin (1988).
Oblate						
NH$_3$	GS $\leftrightarrow v_2$	0	0.24	2.5	5.3	Danagher and Reid (1987).
CD$_3$H	GS $\leftrightarrow v_3$	0		0.88		Menard-Bourcin and Doyennette (1988).
Asymmetric						
O$_3$	(001) + (010) $\leftrightarrow$ (000) + (011)	17	[c]	5.15	3.0	Boursier et al. (1993), Menard-Bourcin et al. (1990).

[a] 300-K cross-sections, unless otherwise noted. Calculations are based on Sharma–Brau theory.
[b] 260-K cross-section.
[c] Contained in Menard-Bourcin et al. (1990).

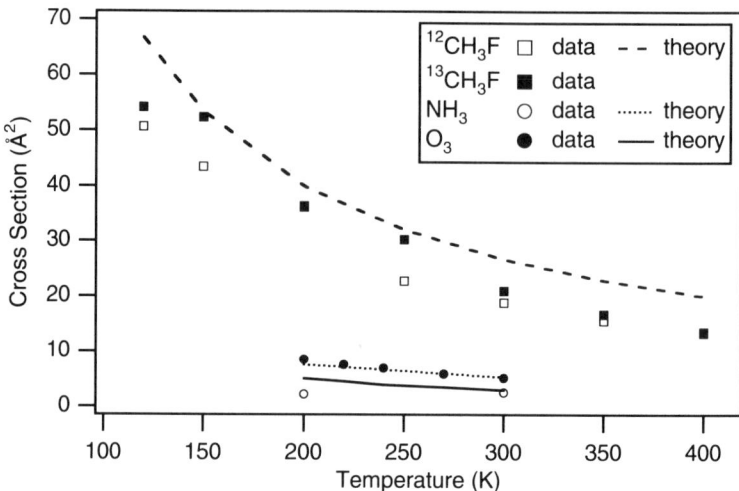

FIG. 11. Vibrational swap cross-sections as a function of temperature.

In contrast, Chesnokov and Panfilov (1977) reported a direct $A \leftrightarrow E$ symmetry conversion process in CH_3F with a cross-section of 4.7 Å². Since this measurement was based on the observation of an unassigned line, is much smaller than the cross-section measured by Everitt and De Lucia (1989), and would require unknown physical processes to account for its speed, it seems more likely that it is the rate for some other process, perhaps one of the faster vibrational transfer processes.

The V-swap process has also been observed in $V_6 = 1$ of CH_3Cl (Pape et al., 1994). Because $d\mu/dQ_6 = 0.038$ D in CH_3Cl is smaller than $d\mu/dQ_3$ in CH_3F, it was expected that V-swap would be much less important in CH_3Cl. Experimental evidence bears this prediction out: the measured V-swap cross-section in $V_6 = 1\, CH_3Cl$ (1.2 Å²) is smaller than in $V_3 = 1\, CH_3F$ (21 Å²).

Observations of a related process, the isotopic V-swap, can be compared with the pure V-swap. Because the vibrational energy shifts that occur between isotopic variants of a species are typically small ($<20\,\text{cm}^{-1}$), it is expected that pure V-swap and isotopic V-swap have nearly identical cross-sections. Early studies of isotopic V-swap (Preses and Flynn, 1977; Sheorey and Flynn, 1980) used IR-induced fluorescence to measure the cross-section of

$$^{12}CH_3F(V_3 = 1) + {}^{13}CH_3F(V_3 = 0) \rightarrow {}^{12}CH_3F(V_3 = 0)$$
$$+ {}^{13}CH_3F(V_3 = 1) + \Delta E \qquad (24)$$

to be 7.1 Å² at 260 K. Although this is smaller than the comparable V-swap cross-section at 250 K (26 Å^2), some of this difference can be accounted for by the 12-cm^{-1} vibrational energy defect in Eq. (24). More recent studies of the isotopic V-swap in CH$_3$Cl have provided even more compelling agreement between the two processes. Pape et al. (1994) have observed the $V_6 = 1$ isotopic V-swap between the ^{35}Cl and the ^{37}Cl variants of CH$_3$Cl in all four permutations of pumped and probed isotopic species. The pure V-swap cross-section was observed to be 1.2 Å^2, in good agreement with the isotopic V-swap cross-section of 0.92 Å^2. Menard-Bourcin and Doyennette (1988) observed both the pure and the isotopic V-swap cross-sections in $V_2 = 1$ CD$_3$ ^{35}Cl and CD$_3$ ^{37}Cl to be 3.9 Å^2. In both cases, the small isotopic vibrational energy defect accounts for the equivalence of the observed cross-sections. The smaller value of $d\mu/dQ_6 = 0.038$ D in CH$_3$Cl versus $d\mu/dQ_2 = 0.097$ D for CD$_3$Cl accounts for the smaller cross-section in the former.

The pure V-swap process has also been observed in the oblate rotors NH$_3$ (Danagher and Reid, 1987) and isotopically substituted methane (CD$_3$H) (Menard-Bourcin and Doyennette, 1988). In NH$_3$ $d\mu/dQ_2$ (0.24 D) is only slightly smaller than $d\mu/dQ_3$ in CH$_3$F (0.2756 D), yet the 300-K NH$_3$ cross-section is smaller by almost a factor of 10 (2.5 Å^2). According to theory, the difference can be traced to the much larger rotational constant B for NH$_3$ (298 vs 25 GHz) which reduces the rotational partition function (~ 100 vs ~ 1000) and reduces the opportunity for rotationally resonant collisions. Experimental results and predictions of theory for NH$_3$ at 200 and 300 K are shown in Fig. 11. An isotopic V-swap was observed between ^{14}N and ^{15}N versions of ammonia (Kano et al., 1976). Although the rate constant was too fast to be measured in that experiment, the estimated lower limit of about 2 Å^2 is consistent with the values presented above. This relatively slow V-swap made possible the truncation of the model for NH$_3$, discussed in Sections II and III.

The cross-section in $V_3 = 1$ of CD$_3$H was observed to be much slower (~ 40 gas kinetic collisions), likely due to the small value of $d\mu/dQ$ (Menard-Bourcin and Doyennette, 1988). In fact, the V-swap process is not observed in $V_4 = 1$ of ^{13}CD$_4$, whose $d\mu/dQ_4 \approx 0.05$ D (Foy et al., 1988). It would be interesting to see if SiH$_4$ and SF$_6$, which have smaller B values (85 and 2.7 GHz vs 150 GHz) and much larger $d\mu/dQ$ (0.23 and 0.44 D vs 0.05 D), show a V-swap process in $V_4 = 1$ and $V_3 = 1$, respectively.

B. ASYMMETRIC TOPS

A fairly complete vibrational energy transfer map has been obtained for the asymmetric rotor O$_3$ (Menard-Bourcin et al., 1990). Among the many vibrational processes measured, a V-swap process involving collisions between two molecules that are not in the ground state has also been observed

(Boursier et al., 1993). The mechanism

$$O_3(001) + O_3(010) \rightarrow O_3(000) + O_3(011) + 17\,\text{cm}^{-1} \qquad (25)$$

is similar to a class of near-resonant "ladder climbing" processes,

$$M(V=n) + M'(V=m) \rightarrow M(V=0) + M'(V=n+m) + \Delta E, \qquad (26)$$

the most common of which involves $n = m = 1$. The energy defect ΔE in these processes arise from the anharmonicity of the vibrational motion due to the nonparabolic nature of the potential. The cross-section of the ladder climbing processes, which is related to the V-swap cross-section through ΔE and the vibrational matrix element (fundamental vibrational frequency ω_k and $V = n, m$)

$$\left(\frac{d\mu}{dQ_k}\right)^2 \frac{\hbar}{2\omega_k}(V+1), \qquad (27)$$

has been measured in CH_3F (Sheorey and Flynn, 1980) and O_3 (Menard-Bourcin et al., 1990), among others. The temperature dependence of the cross-section of Eq. (25) was measured to be $T^{-1.24}$. A modification of Sharma and Brau theory reproduced the temperature dependence of this cross-section (see Fig. 11) and was about as accurate in absolute terms as the unmodified version was for CH_3F.

C. Summary

Vibrational swap processes, and their derivatives, the isotopic swap and the ladder climbing processes, are significant in many molecules. The size of the cross-sections were observed to be largest for molecules with large values of the vibrational matrix element (especially $d\mu/dQ_k$), small vibrational energy defects, and small rotational constants. When the temperature dependence of the cross-section was measured, it was seen to decrease with increasing temperature as $T^{-\alpha}$, where $\alpha \sim 1$.

V. The Physical Basis of Rotational Energy Transfer

Ultimately, it is the goal of science to understand observed phenomena in terms of physical cause. In RET, as well as in most fields with complex relationships between experimental observations and the basic physical processes, intermediate empirical or semiempirical parameterizations have been developed and have come to play an important role. In this context, it has been established that the RET observed in polyatomic molecules can be described by models which include: (1) energy transfer along principal pathways, which often can be described by one of several fitting laws; (2) subsequent transfer to pools of states; and (3) fast vibrational interactions with nearby states. In this section we will consider these models and the parameters which describe RET in an attempt to relate them to more basic physical mechanisms.

A. SELECTION AND PROPENSITY RULES

First it is useful to investigate a number of selection and propensity rules which have been used to help define and describe the RET process.

1. Nuclear Spin

As we have seen above, the symmetry associated with identical nuclei represents the only "absolute" selection rule of RET. Although in principle interconversion is possible, especially in the context of magnetic interactions, in all of the RET experiments of which we are aware it is negligibly slow. Thus, based on both the strong evidence of experimental observations and theoretical expectations, a nuclear-spin-based selection rule which precludes collision-induced transitions between different spin species exists. As a result, the RET of polyatomic molecules is ordinarily factored according to symmetry species, thereby reducing the dimensionality of the problem.

2. The $\Delta J = 0, \pm 1$ Propensity

One of the early themes in the study of RET in polyatomic molecules was the experimental establishment of the theoretically expected $\Delta J = 0, \pm 1$ dipole-allowed propensity rule. In retrospect, it would appear that the difficulty of characterizing the higher order terms of the intermolecular potential along with the emphasis on the dipole–dipole interaction in the calculation of pressure broadening coefficients led to overdependence on this mechanism for the description of RET. This was especially true in the development of models for CW and quasi-CW experiments which do not provide direct information on which a choice between models, based on many successive $\Delta J = \pm 1$ or a few $\Delta J = n$ processes, can be made. This early work stands in contrast to the energy gap scaling and fitting laws which have found widespread acceptance both in studies of diatomic species and in time-resolved studies of RET in polyatomic molecules. Thus, it is reasonable to conclude for collisions between species with large electric dipole moments that the $\Delta J = 0 \pm 1$ rate is substantially the largest single rate. However, for energy transfer to states removed by even a few J, the $\Delta J = n$ processes dominate.

3. Parity

Many studies have considered the existence of a parity propensity rule. In early MWMWDR studies, especially those of NH_3, H_2CO and other molecules in which closely spaced, dipole-selection-rule-connected doublets exist in an otherwise sparse spectra, such a propensity rule was observed. However, it should be noted that, although these species were selected for the experimental convenience afforded by these doublets, this selection criterion also introduced a correlation for the study of transitions which follow the parity propensity rule. In a number of experiments for which a broader range of states were probed, a general propensity rule was *not* observed. These include NH_3 (Abel et al., 1992), SiH_4 (Hetzler et al., 1990), and $^{13}CD_4$ (Foy et al., 1988). Thus, it would appear that the physical basis for the parity propensity rule lies primarily in its correlation with the strong electric-dipole-allowed $\Delta J = 0, \pm 1$ process.

4. Vibrational

Although this chapter is primarily concerned with rotational energy transfer, we have seen in a number of circumstances that fast, near-resonant vibrational processes interact intimately with the RET process. Indeed, when vibrational states are in close coincidence, the mixing of ro-vibrational states can render separate discussions of rotational and vibrational paths meaningless.

Observations of vibrational energy transfer can be divided into three classes, those in which (1) the vibrational defect is large compared with kT, (2) the vibrational defect is small or comparable to kT, but no rotational state-specific transfer is observed; and (3) the vibrational defect is small compared with kT and there is strong mixing of rotational states which couple the manifolds at a detailed state-to-state level. The separation of membership between classes 2 and 3 can depend either on the physics of the system or on experimental selection effects. In some sense all molecules belong to group 1 in that they have many vibrational states for which the energy defects are large in comparison with kT and the RET problem can be factored according to the selection rule $\Delta V = 0$. Examples of case 2 are the vibrational swapping in CH_3F, CH_3Cl, NH_3, CD_3H and O_3 or transitions between closely spaced vibrational levels such as v_2 and v_4 in $^{13}CD_4$ or v_3 and v_6 in CH_3F and CH_3Cl. In these cases two separate RET manifolds connected only by a general, nonrotational state-specific vibrational energy transfer term are appropriate and have been successfully used in RET kinetic models. Case 3 is represented by the state-specific mixing in HDCO and HCN.

In HCN this mixing provides the rotational state-specific inversion responsible for FIR discharge lasers. For such lasing transitions the photon energy $h\nu$ is much less than kT, and it is necessary to invoke an inversion mechanism more selective than that provided by resonance and energy defect considerations. Furthermore, because this lasing requires a strong state-specific mixing, more examples are to be found among the discharge lasers (Skatrud and De Lucia, 1985) than in experiments specifically designed to study RET. This is another example of an experimental selection effect at work which potentially biases conclusions about propensity rules in RET.

Because of the relatively close spacing ($\sim 60\,\text{cm}^{-1}$) between v_2 and v_4 in SiH_4 and the resulting overlap and mixing of rotational states, it might be expected that rotationally state-specific transfer between these vibrational states might be observed, as in H_2CO. However, it has not been possible to observe definitively such preferred pathways (Hetzler *et al.*, 1989), although it was concluded that the rotational character provided by the mixing made a major contribution to the observed fast V–V rate. In this regard it is significant that the SiH_4 $v_2 \to v_4$ rate is about an order of magnitude faster than that observed for the same transition in $^{13}CD_4$ (Foy *et al.*, 1988). This difference can be interpreted in terms of the much larger Coriolis mixing in SiH_4 (Hetzler *et al.*, 1989). Similarly, Menard-Bourcin and Doyennette (1988) have observed "vibrational" transfer between the strongly Coriolis-coupled v_3 and v_6 in CD_3H that approaches the gas kinetic cross-section.

B. The Rotational Energy Transfer Processes — Molecular Constants

In the sections above the variety of observed RET processes has been reviewed, the numerical modeling of these observations has been described, and kinetic rate constants for many of the RET processes have been obtained. Here the relationships among these rate constants and fundamental molecular properties will be explored. Although the development of many of these relationships is still in its infancy and many of the calculated rate constants are model dependent, it is still possible to develop a number of quantitative or semiquantitative relationships.

An important result of time-resolved double-resonance studies of RET in polyatomic molecules is that the total cross-section for large changes of angular momentum in a single collision is substantial. Because this requires participation of correspondingly high-order moments of the IMP, success associating RET rates with particular terms in the multipole expansion of the IMP above the dipole should not be expected. Indeed, at some level of approximation a good model is that the slowly converging sum of the multipole moments effectively represents a hard collision in which detailed information about the initial state is lost. Similar conclusions have been

TABLE IV
Spectroscopic Parameters

	A (GHz)	B (GHz)	C (GHz)	Mass (amu)
Spherical				
CH_4		157		16
$^{13}CD_4$		79.0		21
SiH_4		85.8		32
SF_6		2.7		146
Prolate				
CH_3F	155.4	25.5		34
$^{13}CH_3F$	155.5	24.9		35
CH_3Cl	156.2	13.3		50
$CH_3{}^{37}Cl$	156.5	13.1		52
CD_3Cl	78.4	10.8		53
CH_3Br	155	9.57		95
CH_3I	150	7.50		142
CD_3I	150	6.04		145
Oblate				
NH_3		298	189	17
CD_3H		100	80	19
CHF_3		10.35	5.67	70
CDF_3		9.92	5.67	71
Asymmetric				
H_2CO	282.0	38.8	34.0	30
D_2CO	141.7	32.3	26.2	32
O_3	106.5	13.3	11.8	48

reached by Abel et al. (1992) in their study of NH_3 and by Foy et al. (1988) in their study of $^{13}CD_4$. In order to provide a basis for a discussion of these results, important physical and spectroscopic parameters for a number of them have been collected in Table IV.

1. The $\Delta J = 0, \pm 1$ Process

A continuous thread throughout the development of RET studies is the dominance of the dipole–dipole-induced $\Delta J = 0, \pm 1$ transitions which result from collisions between polar molecules and the qualitative difference between these collisions and those which result from other combinations of collision partners. Globally, this difference is manifest in the ratio, which typically approaches an order of magnitude, between the pressure broadening coefficients (which are closely related to total RET depopulation rates) of these two classes of molecular collisions.

Early experimental studies were dominated by considerations and manifestations of the dipole–dipole process and confirmed its expected dominance. These have been discussed in Section II and in the review of Oka (1973). Because the interaction is long range and involves energy transfers which are small in comparison with the kinetic energy, models based on perturbation theory can be quite good. As a result a well-defined molecular parameter, the electric dipole moment, quantitatively describes this RET process. Moreover, because this parameter can be accurately measured in Stark effect experiments, this part of the characterization of RET approaches being a *predictive* theory.

2. The $\Delta J = n$ Processes

Significant $\Delta J = n$ processes were observed in all of the experiments discussed above which were sensitive to their existence. Moreover, those which resulted in the calculation of quantitative rates had values in good agreement with simple fitting laws. However, it might be expected that deviations from simple energy gap characterization will be found as more experimental results are obtained for RET in light polyatomic species such as NH_3 and H_2O. For example, in both HCl and HF these fitting laws have been found to represent the data poorly (Copeland and Crim, 1983, 1984; Rohlfing et al., 1987). In these cases the smaller number of energetically open collision channels reduces the "classical" smoothing that results from averaging over the detailed dynamics (rotational resonances, quantum number effects, etc.) of the individual channels.

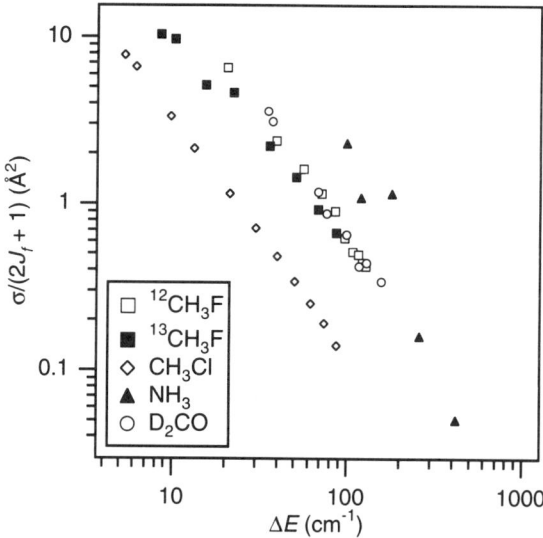

FIG. 12. Normalized cross-sections of symmetric and asymmetric tops for the $\Delta J = n$ process as a function of the energy gap.

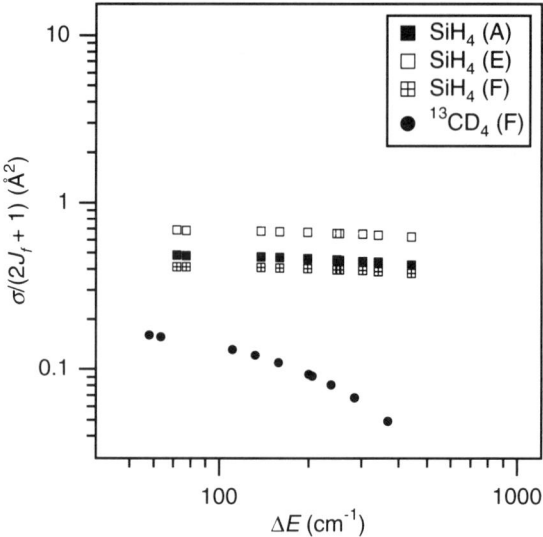

FIG. 13. Normalized cross-sections of spherical tops for the $\Delta J = n$ process as a function of the energy gap.

Cross-sections as a function of energy gap for a number of species in which it has been possible to make quantitative measurements are shown in Figs. 12 and 13 for polar and nonpolar species, respectively. Following the discussion of Section I, these have been normalized according to the factor $(2J_f + 1)$. In both figures the *downward* rates have been plotted. For RET to states at higher energy, detailed balance reduces these rates by the factor $\exp(-\Delta E/kT)$.

These figures show a number of interesting and perhaps unanticipated results. First, all of the systems containing large electric dipole moments [CH$_3$F (Everitt and De Lucia, 1990), CH$_3$Cl (Pape et al., 1994), NH$_3$ (Abel et al., 1992), and D$_2$CO (Bewick et al., 1988)] show no extraordinary propensity for the $\Delta J = \pm 1$ dipole-allowed transition. Indeed, as shown in Fig. 12 the $\Delta J = \pm 1$ rates fit smoothly with the higher order $\Delta J = n$ rates in the context of the several fitting laws. Similar results have been obtained for the nonpolar species [^{13}CD$_4$ (Foy et al., 1988) and SiH$_4$ (Hetzler and Steinfeld, 1990)], with the major qualitative differences between the polar and the nonpolar species being (1) the size of the maximum rate and (2) the slope of the decrease of this rate with increasing energy.

Next, it is perhaps surprising that the normalized rates for ^{13}CH$_3$F and ^{12}CH$_3$F belong to the same family of points in Fig. 12 because states differing substantially in both energy and angular momentum are pumped ($J = 5$, $K = 3$ and $J = 12$, $K = 2$ for ^{13}CH$_3$F and ^{12}CH$_3$F, respectively). Although the scaling laws explicitly include the degeneracy factor $2J_f + 1$

(a fitting law which did not contain this factor would fail dramatically here), it is significant that other angular momentum factors do not play a larger role. A similar observation can be made about CH_3Cl because as in CH_3F the points in Fig. 12 come from measurements of energy transfer from two different pumped states. More remarkably, the D_2CO points also belong to the same family as the CH_3F data. Similar molecular mass (32 and 35 amu for D_2CO and CH_3F, respectively), density of rotational states (the rotational constants differ only by about 10%), and overall size and shape undoubtedly underlie this similarity in RET cross sections.

The difference between D_2CO and CH_3F on one hand and CH_3Cl on the other can at least qualitatively be understood in the same context. As discussed in Section I, the larger rotational partition function of CH_3Cl increases the number of available states by about a factor of 2, thereby reducing the $\Delta J = n$ rates by the same factor. While some of the difference which remains can properly be attributed to differences in the IMP, it is probable that much of the difference can be attributed to its lower molecular velocities and larger size and impact parameter b. These factors reduce the extent of the Fourier components of Eq. (6) and diminish the transition probabilities with increasing energy defect. Although it would be reasonably straightforward to model this conjecture, the relatively small correction that is required, its simple functional form, and the multitude of physically reasonable IMPs which could be chosen almost certainly would lead to a multiplicity of satisfactory solutions.

Also shown in Fig. 12 are a few data for NH_3. Although these points lie fairly close to the family of points for D_2CO and CH_3F, this is most likely due to an accidental compensation among several large factors. The translational velocity of NH_3 (17 amu) is higher, thereby significantly altering the exponential in Eq. (6). Additionally, the partition function is much smaller, a factor which will significantly increase the individual rates. However, NH_3 is smaller, has a different shape, and has more widely spaced rotational levels, factors which will all reduce the individual cross-sections.

For the spherical tops $^{13}CD_4$ and SiH_4, the results, while similar, vary substantially more than might be expected. As in the case of NH_3 there are compensating factors. The larger mass of SiH_4 results in a lower translational velocity and less efficient RET excitation. However, its larger size should result in an increase in cross-section. The differences in the cross-sections for the several symmetry species of SiH_4 are probably another indicator of the substantial mixing in these species and a warning against a too detailed physical interpretation of the model parameters.

Finally, because the values of rates obtained from complex numerical models can be model dependent, it is important to note that the models of time-resolved experiments *clearly* eliminate the possible equivalence between successive $\Delta J = 1$ steps and $\Delta J = n$ processes (Bewick *et al.*, 1988; Everitt and De Lucia, 1990; Abel *et al.*, 1992). An independent experimental

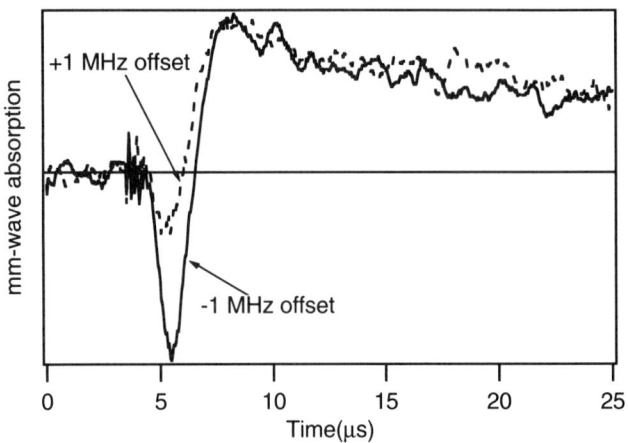

FIG. 14. Time-resolved data for the $J = 4-5, K = 2$ transition in $^{12}CH_3F$ at 20 mTorr with the probe offset from the rotational line center by ± 1 MHz.

demonstration of the importance of the $\Delta J = n$ process exists for CH_3F. Everitt (1990) has observed in a time-resolved experiment some preservation of the pumped velocity subclass following a $\Delta J = n$ collision. Figure 14 shows the $J = 4-5$, $K = 2$ time-resolved data in $^{12}CH_3F$ ($J = 12$ pumped) with ± 1 MHz SUBMM probe offsets from rotational line center. The pumped velocity subclass in $J = 12$ was offset from line center by about -600 kHz. The remnants of the pumped velocity subclass are clearly visible in the asymmetry at early time in the $J = 4-5$ data. Because $(12 - 5)$ 7 successive $\Delta J = -1$ collisions could not produce an effect this soon after the pump pulse, the asymmetry conclusively shows that higher order $\Delta J = n$ processes (namely, $\Delta J = 7, 8$) are responsible for the data. Moreover, the data reveal that some velocity information survives the collision which produces this large change in J.

Serendipitously, Shin et al. (1991) have measured with an infrared probe the Δv_{rms} for these collisions in a low-pressure CW experiment. Because of the large doppler broadening in the infrared, this experiment was an especially good probe of velocity subset effects. Several velocity components were observed, but the one most strongly correlated to the $\Delta J = n$ process was the narrowest. Plotting the Δv_{rms} of the narrow component as a function of ΔE in Fig. 15 reveals a strong correlation, suggesting that the more energetic the collision, the "harder" the impact and the greater the change in molecular speed, direction, or both. Figures 7 and 15 can be combined to show in Fig. 16 the relation between the normalized cross-section $\sigma_{\Delta J = n}/(2J_f + 1)$ and Δv_{rms}. Not surprisingly, the large cross-sections, which are associated with softer collisions, result in smaller velocity changes.

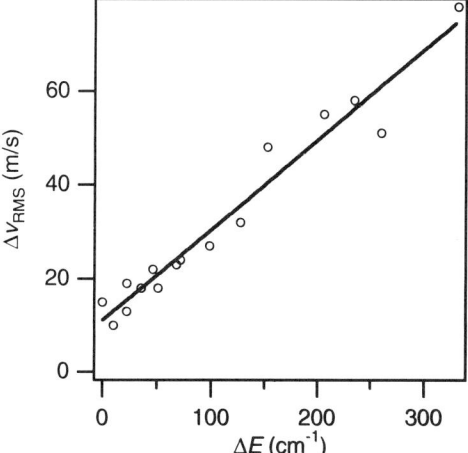

FIG. 15. Collision-induced velocity change as a function of energy change in CH_3F.

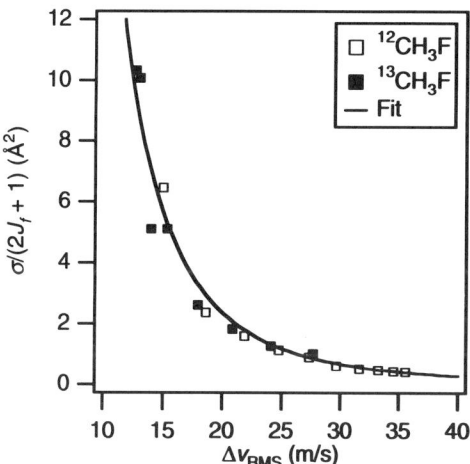

FIG. 16. Normalized cross-section for $\Delta J = n$ processes as a function of velocity change for CH_3F.

3. Transfer to Nonprincipal Pathways: The Grouping of States

Although many experiments lend themselves to the study of energy transfer along principal pathways, processes which induce transfers from these states to groupings of other states are also important. Since many of the species in which these processes have been studied are symmetric or slightly asymmetric tops, the relations between the experimentally obtained cross-sections and molecular parameters for these examples will be considered first.

In Table II the cross-sections which characterize the nonprincipal pathways for the symmetric and near-symmetric tops $^{12}CH_3F$, $^{13}CH_3F$, $^{12}CH_3\,^{35}Cl$, NH_3, and D_2CO were shown. In the cases of CH_3Cl and CH_3F populations were observed both leaving the pumped K stack *and* arriving in the other K states. Because the arrival was observed by monitoring the rotational spectral absorptions which are very sensitive to thermal nonequilibria, it was possible to establish experimentally that the target states were populated according to a Boltzmann distribution at very early time. For D_2CO, the distribution in the target states was not measured quantitatively.

Figure 9 shows that the cross-sections for K-changing collisions can be a relatively strong function of the temperature or, more precisely in the context of the Fourier integral of Eq (6), the ratio v/b. Since for all of these species the effective b (i.e., the distance perpendicular to the z axis that the hydrogen extends) for collisions which change K is about the same, only the differences in molecular velocity caused by the molecular mass need be considered for the comparison.

Figure 17 replots the temperature-dependent K-changing cross-sections for CH_3F as a function of the relative molecular velocity and adds the 300-K CH_3Cl and D_2CO results. The CH_3F results are straightforward to understand in the context of the ideas presented in Section I: a rough hard sphere, whose size is comparable to the gas kinetic cross-section, provides the observed thermal distribution at low temperature. At higher temperature (and higher molecular velocity), the longer range forces begin to have Fourier components which can effectively induce RET. The CH_3Cl K-changing cross-section also falls along the locus of points defined by the CH_3F data. This consistency results from (1) defining the cross-section as the *leaving* cross-section (rather than as the directly observable, but unphysical and widely varying, arrival cross-section which could be obtained from

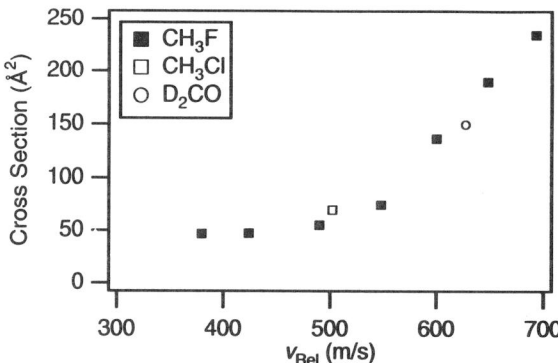

FIG. 17. Total K-changing cross-sections as function of relative velocity for CH_3F, CH_3Cl, and D_2CO.

the several K of Fig. 10) and (2) plotting the cross-sections according to their average molecular velocity, the parameter which determines the extent of the Fourier components. Similarly, the K-changing cross-section for D_2CO also falls along the locus of points defined by the CH_3F and CH_3Cl.

NH_3 is much lighter than CH_3F, CH_3Cl, or D_2CO and changes of only a few J or K can correspond to energy which is comparable to kT. Consequently, significantly more quantal effects, with less averaging over a multitude of open channels to produce classical results, should be expected. The results discussed in Section III are fully consistent with this expectation. There it was found that some of the cross-sections identified with the K-changing cross-section could be assigned to individual K-changing rates and the rest associated with "baths" of states. Although Abel et al. (1992) did not focus on obtaining a total K-changing rate, it is possible to infer from the measured rates that this total cross-section is similar to the gas kinetic cross-section. Although it is tempting to simply declare a general agreement with the results obtained for the other symmetric and near-symmetric tops, it is more likely that the much higher frequency Fourier components which result from light weight and high velocity are approximately compensated for by the much wider rotational energy level spacing in NH_3.

Because CDF_3 is an oblate species, it presents an opportunity to study a situation in which the energetically more favorable pathways (ΔK) do not correspond to those defined by the dipole moment (ΔJ). Unfortunately, only semiquantitative results have been obtained for this species. Energy transfer through at least $\Delta K = 18$ was observed. However, because of the small moments of inertia of this species, this corresponds to an energy transfer of significantly less than kT. It is reasonable to conclude that limitations on angular momentum transfer play an important role here as they do for J-changing collisions in CH_3Cl. It would be especially interesting in this oblate rotor to see if this ΔK transfer can be described by a process analogous to the fitting laws which have worked so well for ΔJ processes in prolate tops.

Finally, consider the spherical tops $^{13}CD_4$ and SiH_4. In the most detailed work on these species, nonspecific transfer from states associated with the principal pathways defined by $\Delta(J - R) = 0$ has been modeled in the context of a multistep process which first groups together the fine-structure states according to Coriolis substate, with subsequent transfer to other Coriolis substates. However, as shown in Table II the constants which describe these processes show much wider variations than have been observed in either symmetric or asymmetric tops. This is true even for different symmetry species of the same molecules. Although a detailed analysis would be required for a definitive explanation, it would appear that, because the vibrational states studied are in many cases strongly mixed with other nearby states, the relationships between the model parameter and more fundamental molecular properties are more complicated than those in the symmetric and asymmetric tops. In this context, it has been noted that some of the constants which were necessary to bring the kinetic model into agreement with the data are unphysically large and should not be considered actual rate constants (Hetzler and Steinfeld, 1990).

Thus, for molecular species with rotational states which are not significantly mixed, the cross-sections which govern transfer from the principal pathways to the more general pools can be related to the physical sizes of the molecules. Additionally, the dependence of these cross-sections on the Fourier components of the collision are remarkably uniform and provide a consistent picture of the variation of the cross-section with collision energy.

4. Total Depopulation Cross-Sections

The sum of all cross-sections for RET from a given state is the total depopulation cross-section. Table V shows these cross-sections along with a number of other measures of total molecular cross-section for several of the species considered here. In principle, the total depopulation cross-sections can be obtained either by direct observation of a pumped state or by summing the appropriate rates in a kinetic model. In complete models, which conserve molecules, these are not independent measures.

Although in principle both the total depopulation cross-section and the pressure broadening cross-section are functions of rotational state, they are in fact slowly varying, and repesentative values can be shown in the table. Even in very light molecules such as HCl and NH_3, for which the largest variations might be expected, the total depopulation cross-sections vary only by $\pm 20\%$ over the thermally populated rotational states. These slow variations are closely related to a well-known effect in "optical" spectroscopy, the independence of the total integrated cross-section from the internal details of the molecule. For example, it is easy to show in a symmetric-top basis that the sum of the strengths of the electric dipole transitions out of a particular level sum to a constant, independent of initial J and K (Gordy and Cook, 1984).

However, even in optical spectroscopy complex mixings can lead to intensities for individual transitions which differ enormously from the intensities of the basis states. Nevertheless, the sum of the transition strength is preserved. The same holds for "collisional" spectroscopy. SiH_4 is an especially good illustration of this. Inspection of Table II, which shows the details of the RET rates in each of the symmetry species, reveals the enormous differences among the detailed rate constants. However, the total depopulation cross-sections are almost independent of the symmetry species and about one and one-half to two times the gas kinetic σ_{GK} or Lennard–Jones σ_{LJ} cross-sections. A similar ratio is observed for $^{13}CD_4$.

TABLE V
TOTAL DEPOPULATION CROSS-SECTIONS

	μ^a (D)	σ_{GK}^b (Å²)	σ_{LJ}^b (Å²)	σ_{PB} (Å²)	PB method	σ_{depop} (Å²)	Reference
Spherical							
CH₄	0.	46	53	76	IR, VIS		Kvijin et al. (1993).
¹³CD₄	0.	43	50	84.2	IR	41.9	Kistemaker et al. (1974), and references therein, Foy et al. (1988), Foy et al. (1985).
¹³CD₄(ν₄)	0.	43	50			66.5	Foy et al. (1985).
SiH₄(ν₄)	0.	52	70			104	Hetzler et al. (1989), Hetzler and Steinfeld (1990), Millot et al. (1988).
SF₆	0.	79	117	350	IR	225	Dubs et al. (1982), Reiser and Steinfeld (1981).
Prolate							
CH₃F(ν₃)	1.85	35	47	424	MW		T. M. Goyette, private communication (1994).
¹³CH₃F	1.85	35	47			493	Jetter et al. (1973).
¹³CH₃F(ν₃)	1.85	35	47	478	MW	737	T. M. Goyette, private communication (1994), This work.
CH₃Cl(ν₆)	1.87	54	94	554	MW	502	Pape et al. (1993).
Oblate							
NH₃	1.47	31	54	472	MW	291	Markov et al. (1993), and references therein.
NH₃(ν₂)				391	MW, IR	123	Markov et al. (1993), and references therein.
CDF₃	1.65	61	71			298	Harradine et al. (1984).
Asymmetric							
D₂CO	2.33	49	65	720	MW	670	Bewick et al. (1988), and references therein.
O₃	0.532	53	67	148	IR	121	Hartmann et al. (1988), and references therein. Flannery and Steinfeld (1992), and references therein.

[a]Nelson et al. (1967).
[b]Hirschfelder et al. (1964).

It is possible to make comparisons between total depopulation rates and pressure broadening parameters by use of the relation (DePristo and Rabitz, 1978)

$$\gamma/P\,(\mathrm{MHz/Torr}) = (1/2\pi)(k_u + k_l)/2, \qquad (28)$$

where the total depopulation rates of the upper (k_u) and lower (k_l) states are in units of ms^{-1} mTorr^{-1}. This relation does not include contributions to the pressure broadening which might arise from non-state-changing effects and as a result should provide a lower limit on the pressure broadening parameter.

The results for CH_3F, while in general agreement, differ by more than the expected experimental error. Additionally, the total depopulation cross-section is a very stable model parameter because these experiments observed both the molecules leaving the state and their arrival in the target states. Furthermore, σ_{depop} is greater than σ_{PB} so the difference cannot be attributed to additional contributions to the pressure broadening cross-section from non-state-changing effects. In contrast, σ_{depop} and σ_{PB} for CH_3Cl are almost identical. The difference between these two results is subtle, which is illustrative of the problems associated with relating kinetic model parameters with molecular constants. In the CH_3F experiments, which grew out of an experiment designed to model the optically pumped CH_3F laser (Matteson and DeLucia, 1983; McCormick et al., 1987a,b), the model parameters include contributions from molecules leaving both the state *and* the velocity subclass. In contrast the CH_3Cl experiment (Pape et al., 1994) was much less sensitive to the velocity subclass effects. A similar good agreement between the total depopulation rates and pressure broadening parameters has been observed in D_2CO and O_3.

In NH_3, σ_{depop} in both the ground and the v_2 vibrational states underestimate σ_{PB}. Abel et al. (1992) have shown that this effect is due to non-state-changing effects, a result which is to be expected because the larger energy defects associated with the more widely spaced rotational levels reduce the efficiency of RET processes. Likewise in the spherical tops $^{13}CD_4$ and SiH_4, σ_{depop} is somewhere smaller than σ_{PB}. All of these species have more widely spaced rotational levels than the species discussed in the previous paragraph.

Thus, we come to a relatively straightforward understanding of the relationship between the total depopulation rate and other measures of molecular size. For those species without electric dipole moments, the total depopulation cross-section is somewhat larger than the physical cross-sections. Interestingly, this is about the same result as that obtained for the K-changing cross-sections in the symmetric tops. Presumably, this is not a coincidence and the physical mechanisms are similar. For species with electric dipole moments, the total depopulation cross-sections are much larger, typically an order of magnitude greater than the physical size, and related to the electric dipole moment. The collisions associated with nonpolar species can be characterized as hard or short range, whereas for polar species most of the cross-section is soft and of much longer range. As a result, for both polar and nonpolar molecules with large rotational constants, the energetic inaccessibility of states reduces the total depopulation rates, but more so for the soft collisions between polar species.

5. The Vibrational Swapping Process

The relationship between the V-swap process and molecular parameters is similar to the relationship between the pure rotational $\Delta J = 0, \pm 1$ process and the electric dipole moment. As has been shown above in cases with relatively large V-swap rates, it is possible to calculate *predictively* the rate and its temperature dependence from a perturbative theory. However, as for the pure rotational transitions, when the dipole transition element becomes small, higher order effects which are not included in the theory become important.

Studies of many rotational states in $V_3 = 1$ of CH_3F reveal the rotationally thermal nature of the V-swap process, and other work has demonstrated its translationally thermal nature as well (Shin et al., 1991). Both are expected because both are derived from the thermal nature of the ground vibrational state through Eq. (8). However, it is possible for a rotational nonequilibrium to be created by the V-swap process. If a significant amount of population resides in a specific rotational state, only molecules in a limited number of rotational states may have the near-resonant collisions required for vibrational exchange to take place. For example, immediately after the pump pulse, states energetically near the pumped rotational state receive an excess of population, relative to their thermal share, at the expense of states energetically removed from the pumped state. Because of fast $R-R$ equilibrium processes, this effect has not been seen experimentally and thermal distributions have been observed.

The temperature dependence of the $V_3 = 1\,CH_3F$ cross-section has been measured (Everitt and DeLucia, 1993). Sharma-Brau theory contains temperature dependence in Eq. (10) in both the leading term and the resonance term Γ. As the temperature is increased, the resonance condition is relaxed ($\omega_{ab}b/v$ is reduced) but the rotational partition function increases (fewer resonant collisions occur). Detailed calculations for CH_3F reveal that these counterpoised mechanisms balance over the temperature range of the study, an understandable result for a process whose cross-section is derived in large part from nearly resonant collisions. In such cases, the resonance function becomes a temperature-independent constant, and the V-swap

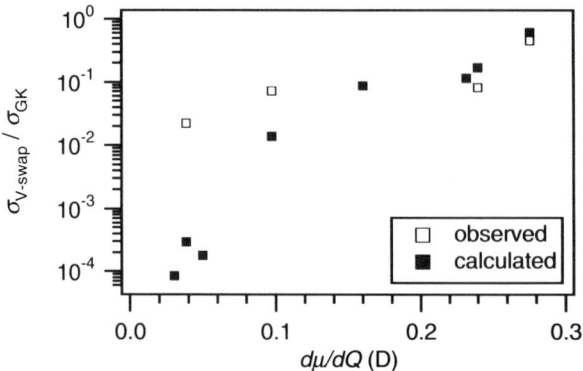

FIG. 18. Normalized vibrational swap cross-sections as a function of the dipole derivative.

cross-section varies as $1/T$. The experimental results for CH_3F shown in Fig. 11 reveal a decreasing cross-section with increasing temperature, which indeed follows the $1/T$ dependence.

Thus, Sharma–Brau theory has proven predictive in regimes where the value of $d\mu/dQ_k$ dominates the values of other terms in the multipolar expansion. In other regimes, the dipole-based derivation of Section I is incomplete, and only the propensity for a V-swap collision can be predicted. For example, $d\mu/dQ_6$ is much smaller in CH_3Cl, suggesting that V-swap is much less important in the pumped vibrational level of CH_3Cl than in CH_3F. Experimental evidence bears this prediction out: the measured V-swap cross-section in $V_6 = 1\,CH_3Cl$ (1.2 Å^2) is smaller than that in $V_3 = 1\,^{13}CH_3F$ (21 Å^2). However, theory substantially underpredicts the $V_6 = 1\,CH_3Cl$ V-swap cross-section (0.014 Å^2), suggesting that higher order multipolar terms and their derivatives contribute.

Figure 18 compares for a number of molecules the observed V-swap cross-sections with the cross-sections calculated via the theory of Sharma–Brau. Individual molecules may be identified in this figure by referring to the data in Table III. Although the trend of decreasing cross-section with decreasing $d\mu/dQ_k$ is clear in both cases, the decrease of the calculated values is more rapid. This trend is due to the omission of higher order moments in the calculation, moments which substantially contribute when $d\mu/dQ_k$ is small. Additionally, the somewhat smaller values for both the experimental and the calculated cross-sections of NH_3 can be attributed directly to the widely spaced energy level structure of this species and the resultant increases in the energy defect. Thus, the correlation between observed V-swap cross section and $d\mu/dQ_k$ is established, and for large $d\mu/dQ_k$ the *predictive* calculations based on the molecular constant are accurate.

6. A Spectroscopic Theory?

From the point of view of those who would like to be able to characterize the RET process, in an ideal circumstance a "spectroscopic-like" theory would emerge in which a well-defined relationship could be developed between a limited number of parameters and a much larger set of observable cross-sections. However, from the perspective of those who seek to use observations of these cross-sections to understand the details of the IMP, the realization of such an objective would show that there is little independent information to be obtained from the experiments.

In the analogous problem of molecular spectroscopy, as ever more complex spectra have been studied, an interesting mix of physically well-founded and empirical parameters have evolved as a means of characterizing the spectral energy levels. Some of the parameters, such as inertial rotational constants and the vibrational fundamentals, are readily associated with physical properties of the molecule. Others, such as higher order distortion parameters and mixing terms, are at best "in principle" related to molecular properties. Nevertheless, the success of these spectroscopic models in characterizing these enormously complex spectra with precision is remarkable. However, even here what might appear to be some of the most straightforward relationships of an invertible theory are difficult to establish. For example, in light asymmetric rotors such as H_2O and H_2S, the relationships among data, distortion parameters, and vibrational frequencies are unclear (Cook *et al.*, 1974).

A similar, but more primitive, spectroscopy for the characterization of collision-induced RET in polyatomic molecules has, in fact, emerged. The kinetic models and their parameters play a role similar to that of the more familiar spectroscopic models. As in the spectroscopic models, the parameters can be determined from the experimental data without recourse to their underlying physical basis, but an understanding of angular momentum, selection rules, and energy is required for a proper structuring of the problem. As in spectroscopy, it is often not necessary to understand the relationships between the basic physics and the observed data, but it is useful to be able to do so. In this context it has been shown that (1) the total depopulation rate is closely related to the "size" of the molecule, as determined by either its gas kinetic size or its electric dipole moment; (2) the rates associated with the principal pathways are determined by only a few constants, and at least in the cases of the symmetric and slightly asymmetric tops simple relations exist among them; (3) the cross-sections for transfers to nonprincipal pathways in symmetric and near-symmetric tops are also closely related to the gas kinetic cross-section; (4) vibrational swapping is well described by a predictive theory based on independently measured dipole derivative matrix elements; and (5) for molecules with widely spaced rotational levels, the inability of the Fourier components of Eq. (6) to excite fully all of the allowed transitions causes the RET cross-sections to be reduced below those expected from simple classical arguments.

As RET experiments become more sophisticated and as more data are accumulated, it is almost certain that the amount of independent RET information which they contain will also grow, as will the required parametrization of the problem. In our view, there is a good prospect that a theory of intermediate complexity and likewise of intermediate power and rigor can be developed. Such an approach might use as its basis the parameters of theories like the IOS. In such an approach, the base rates would play the role of the adjustable spectroscopic parameters, which in principle are traceable to the IMP. One of the most interesting questions for RET in the years ahead is whether or not the rather modest absolute accuracy ($\sim 1\%$) required for most applications of interest will make such a theory attainable and practical.

VI. The Future?

What does the future hold for RET? Currently, the quantitative description of RET in polyatomic molecules is based on numerical kinetic models with a remarkably small number of adjustable parameters. Furthermore, it is possible to relate many of the parameters to basic molecular properties such as dipole moments or molecular size. In some spectroscopic sense this represents a great triumph, the development of a methodology for the description of a seemingly complex and diverse set of experimental data in terms of only a few parameters. Conversely, the amount of fundamental detailed information that has been obtained is relatively small. What are the prospects for obtaining additional detailed information about the RET process? If this is possible, will it lead down a path to undesirable complexity or will it be possible to characterize in detail the RET process in an intuitive and physically satisfying manner?

A. ADVANCES IN EXPERIMENTS

Although there have been enormous advances in the power of RET experiments, the experimental database is still very small and severely hampered by experimental selection effects as well as the model dependence of many of the parameters. Although a number of tunable probe techniques have been used to sample the target states in the rotational manifold, the use of fixed-frequency pumps (most often line-tunable gas lasers) has severely restricted the number of pumped states, usually to only one or two per molecule. In some sense this has limited the number of directly measured rates to of the order N of the $N(N-1)/2$ independent rates of a rotational manifold. We do not mean to suggest that it is either necessary or desirable to observe directly all of these rates; however, a wider range (both in energy and in quantum number) of pumped states is highly desirable, especially to eliminate the model dependence currently encountered because of the correlation of the data. Furthermore, most pump coincidences have been found in searches for either FIR lasers or isotope separation systems. This has led to a selection effect which favors the pumping of rotational levels well within the thermally populated manifold.

An increased capability for the study of very early time system responses is also desirable. Not only will this provide a direct means of establishing elements of the rate matrix *without* any model assumptions (see Section I), but it also makes it possible to probe deeper into the hierarchical structure of the problem. Because there are so many parallel pathways in the RET problem for polyatomic molecules, detailed balance quickly brings states into thermal equilibrium. This is especially true because hard collisions can produce large changes in energy and angular momentum, thereby providing significant population at very early time to large portions of the rotational manifold. Likewise, it will be important that experiments be designed to characterize accurately the relative populations of these states at early time before they become thermalized.

Additionally, measurements made as a function of temperature have been extremely valuable. At some fundamental level, varying the temperature is a means of tuning the probe energy as represented by the Fourier components of the collision in Eq. (6) to investigate the collision. In this view collisions provide a means of conducting spectroscopy on and characterizing the higher order moments of the IMP, whereas energy level spectroscopy ordinarily probes only dipole-allowed transitions. Also, increases in sensitivity will provide for the possibility of designating the effects of the collision partner as a function of its state, thereby removing the average of the thermal distribution of collision partners.

There are at least two molecular types for which it would be useful to have detailed, time-resolved RET data: a relatively heavy oblate rotor such as CDF_3 and a strongly asymmetric rotor with electric dipole moments along at least two axes such as HNO_3. In CDF_3 ΔK transitions are energetically more favorable than ΔJ. From this it can be inferred that principal pathways, which have been successfully modeled with fitting laws, exist for K-changing collisions. However, what is almost certainly the fastest rate, that associated with the electric dipole moment, produces change in J. Thus, it would appear that CDF_3 represents a case in which the fastest transitions effect small changes in J, whereas the next fastest rates effect large changes in K. If so, the kinetic modeling strategies for the prolate rotors will not simply map onto the oblate case. A strongly asymmetric rotor with significant dipole moments along at least two principal axes represents a similar situation, one in which a single principal pathway does not exist and the evolution of the system after the dominance by the electric dipole moment is unclear.

B. Advances in Computing

If, as expected, experimental advances significantly increase the amount of available independent RET information, then more detailed theoretical approaches will be required. While it is probable that RET can be modeled under these circumstances by the introduction of ever more empirical fittings laws, constants, etc., it is not particularly desirable

Above we have remarked on the almost insurmountable complexity of the "exact" quantal solution to the polyatomic RET problem near ambient temperatures. However, it does not require much prescience to predict that computational power will rapidly overtake the complexity of even this brute force approach to the RET problem. In our view it is unlikely that this approach will ever be generally attractive for much the same reasons that *ab initio* calculations of molecular properties are not today the starting points for analyses of molecular spectra.

It is much more likely that kinetic models parameterized in terms of constants which at least in principle can be calculated from fundamental molecular properties will become the norm for the analyses of RET data rich in independent information. However, there remains the problem of the characterization of the intermolecular potential and the calculation of the relationships of the IMP to the constants of the kinetic models. Clearly, as the data sets become richer and more accurate, more exact techniques will be required for the inversion problem, most of which will be numerically intensive.

C. What Will We Find?

It is always dangerous to speculate about the future, especially when there is a strong chance that the speculation could be experimentally tested relatively soon. Nevertheless, we will do so briefly:

(1) The results of a much broader set of pump/probe combinations will lead to significant increases in the amount of *independent* experimental information available.
(2) This independent information will be greatest for lighter species with widely spaced rotational levels, at low temperature, or both because under these circumstances classical averaging will be minimized and the quantal nature of the problem revealed.

(3) Advances in the characterization of the population distributions within pools, especially at early time, will require more detailed modeling of the systems. This may result in either the addition of levels of hierarchy of pools or in some cases a detailed state-by-state accounting. In either case the amount of independent RET information will increase significantly.

(4) For those molecules for which RET can be observed over a wider range of temperature, additional *independent* information will be obtained and this information will be especially valuable in testing the physical foundations of proposed models.

(5) Alternative methods for the description of the complex geometries of collisions between polyatomic molecules that can be parameterized in the context of physically meaningful quantities will be developed.

Acknowledgments

We thank all of these who have helped us in the preparation of this chapter and with the work on which it is based, especially Chris Ball, Travis Pape, David Skatrud, and Pam Watts. We also appreciate the support of the Army Research Office, the National Science Foundation, and NASA in this work over the years.

References

Abel, B., Coy, S. L., Klassen, J. J., and Steinfeld, J. I. (1992). *J. Chem. Phys.* **96**, 8236.
Adam, A. G., Gough, T. E., and Lewin, A. K. (1985). *Chem. Phys. Lett.* **113**, 520.
Alexander, M. H., Jendrek, E. F., and Dagdigian, P. J. (1980). *J. Chem. Phys.* **73**, 3739.
Anderson, P. W. (1949). *Phys. Rev.* **76**, 647.
Andrews, D. A. (1980). *Chem. Phys.* **56**, 231.
Arimondo, E., Glorieux, P., and Oka, T. (1978). *Phys. Rev. A* **17**, 1375.
Arthurs, A. M., and Dalgarno, A. (1960). *Proc. R. Soc. London, Ser. A* **256**, 540.
Bewick, C. P., Haub, J. G., Hynes, R. G., Martins, J. F., and Orr, B. J. (1988). *J. Chem. Phys.* **88**, 6350.
Boursier, C., Menard-Bourcin, F., Menard, J., and Doyennette, L. (1993). *J. Chem. Phys.* **99**, 5905.
Brechignac, P. (1982). *J. Chem. Phys.* **76**, 3389.
Brewer, R. G., and Shoemaker, R. L. (1971). *Phys. Rev. Lett.* **27**, 631.
Brewer, R. G., Shoemaker, R. L., and Stenhom, S. (1974). *Phys. Rev. Lett.* **33**, 63.
Brunner, T. A., and Pritchard, D. E. (1982). In *Dynamics of the Excited State* (Lawley, K. P., ed.), Wiley, Chichester, England.
Brunner, T. A., Driver, R. D., Smith, N., and Pritchard, D. E. (1978). *Phys. Rev. Lett.* **41**, 816.
Brunner, T. A., Smith, N., Karp, A. W., and Pritchard, D. E. (1981). *J. Chem. Phys.* **74**, 3324.
Champion, J. P. (1977). *Can. J. Phys.* **55**, 1802.

Champion, J. P., and Pierre, G. (1980). *J. Mol. Spectrosc.* **79**, 255.
Chang, T. Y., and Bridges, T. J. (1970). *Opt. Commun.* **1**, 432.
Chardon, J.-C., Genty, C., Hill, E. K., and Labrune, J.-C. (1983). *J. Phys. (Paris)* **44**, 1149.
Chesnokov, E. N., and Panfilov, V. N. (1977). *Sov. Phys.*—JETP *(Engl. Transl.)* **46**, 1112.
Collins, T. L. D., McCaffery, A. J., Richardson, J. P., and Wynn, M. J. (1993). *Phys. Rev. Lett.* **70**, 3392.
Cook, R. L., De Lucia, F. C., and Helminger, P. (1974). *J. Mol. Spectrosc.* **53**, 62.
Copeland, R. A., and Crim, F. F. (1983). *J. Chem. Phys.* **78**, 5551.
Copeland, R. A., and Crim, F. F. (1984). *J. Chem. Phys.* **81**, 5819.
Cox, A. P., Flynn, G. W., and Wilson, E. B., Jr. (1965). *J. Chem. Phys.* **42**, 3094.
Curl, R. F., Jr., Kasper, J. V. V., and Pitzer, K. S. (1967). *J. Chem. Phys.* **46**, 3220.
Danagher, D. J., and Reid, J. (1987). *J. Chem. Phys.* **86**, 5449.
DePristo, A. E., and Rabitz, H. (1978). *J. Chem. Phys.* **69**, 902.
DePristo, A. E., Augustin, S. D., Ramaswamy, R., and Rabitz, H. (1979). *J. Chem. Phys.* **71**, 850.
Doyennette, L., and Menard-Bourcin F. (1988). *J. Chem. Phys.* **89**, 5578.
Dubs, M., Harradine, D., Schweitzer, E., and Steinfeld, J. I. (1982). *J. Chem. Phys.* **77**, 3824.
Edmonds, A. R. (1957). *Angular Momentum in Quantum Mechanics.* Princeton Univ. Press, Princeton, New Jersey.
Everitt, H. O. (1990) Ph.D. Thesis, Duke Univ., Durham, North Carolina.
Everitt, H. O. and De Lucia, F. C. (1989). *J. Chem. Phys.* **90**, 3520.
Everitt, H. O., and De Lucia, F. C. (1990). *J. Chem. Phys.* **92**, 6480.
Everitt, H. O., and De Lucia, F. C. (1993). *Mol. Phys.* **79**, 1087.
Everitt, H. O., Skatrud, D. D., and De Lucia, F. C. (1986). *Appl. Phys. Lett.* **49**, 995.
Fabis, A. R., and Oka, T. (1983). *J. Chem. Phys.* **78**, 3462.
Feuillade, C., and Bottcher, C. (1981). *Chem. Phys.* **62**, 67.
Flannery, C.C. and Steinfeld, J. I. (1992). *J. Chem. Phys.* **96**, 8157.
Fox, K. and Person, W. B. (1976). *J. Chem. Phys.* **64**, 5218.
Foy, B., Laux, L., Kable, S., and Steinfeld, J. I. (1985). *Chem. Phys. Lett.* **118**, 464.
Foy, B., Hetzler, J., Millot, G., and Steinfeld, J. I. (1988). *J. Chem. Phys.* **88**, 6838.
Frenkel, L., Marantz, H., and Sullivan, T. (1971). *Phys. Rev. A* **3**, 1640.
Garg, V., and Agrawal, P. M. (1986). *Acta Phys. Polon. A* **69**, 103.
Glorieux, P., Arimondo, E., and Oka, T. (1983). *J. Chem. Phys.* **87**, 2133.
Goldflam, R., Green, S., and Kouri, D. J. (1977). *J. Chem. Phys.* **67**, 4149.
Gordon, R. G. (1967). *J. Chem. Phys.* **46**, 4399.
Gordy, W., and Cook, R. L. (1984). *Microwave Molecular Spectra.* Wiley (Interscience), New York.
Goyette, T. M., McCormick, R. I., DeLucia, F. C., and Everitt, H. O. (1992). *J. Mol. Spectrosc.* **153**, 324.
Gray, C. G. (1968). *Can. J. Phys.* **46**, 135.
Green, S. (1978). *J. Chem. Phys.* **69**, 4076.
Green, S. (1980). *National Resource for Computation in Chemistry Software Catalogue.* Lawrence Berkeley Lab., Berkeley, California.
Green, S. (1985). *J. Chem. Phys.* **82**, 4548.
Green, S., and Thaddeus, P. (1976). *Astophys. J.* **205**, 766.
Harradine, D., Foy, B., Laux, L., Dubs, M., and Steinfeld, J. I. (1984). *J. Chem. Phys.* **81**, 4267.
Harter, W. G., and Patterson, C. W. (1977). *J. Chem. Phys.* **66**, 4872.
Harter, W. G., and Patterson, C. W. (1979). *Phys. Rev. A* **10**, 2277.
Hartmann, J. M., Camy-Peyret, C., Flaud, J. M., Bonamy, J., and Robert, D. (1988). *J. Quantum Spectrosc. Radiat. Transfer* **40**, 489.
Herbst, E., and Millar, T. J. (1991). In *Molecular Clouds,* (James, R. A., and Millar, T. J., eds.), Cambridge Univ. Press, Cambridge, England.
Herlemont, F., Thibault, J., and Lemaire, J. (1976). *Chem. Phys. Lett.* **41**, 466.
Herman, I. P., and Marling, J. B. (1979). *Chem. Phys. Lett.* **64**, 75.

Hetzler, J. R., Millot, G., and Steinfeld, J. I. (1989). *J. Chem. Phys.* **90**, 5434.
Hetzler, J. R., and Steinfeld, J. I. (1990). *J. Chem. Phys.* **92**, 7135.
Hirschfelder, J. O., Curtis, C. F., and Bird, R. B. (1964). *Molecular Theory of Gases and Liquids* Wiley, New York.
Ho, T.-S., and Rabitz, H. (1993). *J. Phys. Chem.* **97**, 13447.
Jetter, H., Pearson, E. F., Norris, C. L., McGurk, J. C., and Flygare, W. H. (1973). *J. Chem. Phys.* **59**, 1796.
Kano, S., Amano, T., and Shimizu, T. (1976). *J. Chem. Phys.* **64**, 4711.
Kim, D., McDowell, R. S., and King, W. T. (1980). *J. Chem. Phys.* **73**, 36.
Kistenmaker, P. G., Hanna, M. M., and DeVries, A. E. (1974). *Physica (Amsterdam)* **78**, 457.
Klaassen, D. B. M., ter Meulen, J. J., and Dymanus, A. (1982). *J. Chem. Phys.* **77**, 4972.
Kondo, S., Koga, Y., and Nakanaga, T. (1985). *Bull. Chem. Soc. Jpn.* **58**, 65.
Kondo, S., Koga, Y., and Nakanaga, T. (1986). *J. Chem. Phys.* **90**, 1519.
Kvijin, P. V., Wells, W. K., Mendas, J. K., Delaney, J. K., Lunine, J. I., Hunten, D. M., and Atkinson, G. H. (1993). *J. Quant. Spectrosc. Radiat. Transfer* **49**, 639.
Laux, L., Foy, B., Harradine, D., and Steinfeld, J. I. (1984). *J. Chem. Phys.* **80**, 3499.
Lees, R. M. (1975). *Can. J. Phys.* **53**, 2593.
Lees, R. M. and Oka, T. (1969). *J. Chem. Phys.* **51**, 3027.
Levy, J. M., Wang, J. H.-S., Kukolich, S. G., and Steinfeld, J. I. (1972). *Phys. Rev. Lett.* **29**, 395.
Markov, V. N., Pine, A. S., Buffa, G., and Tarrini, O. (1993). *J. Quant. Spectrosc. Radiat. Transfer* **50**, 167.
Matsuo, Y., and Schwendeman, R. H. (1989). *J. Chem. Phys.* **91**, 3966.
Matteson, W. H., and De Lucia, F. C. (1983). *IEEE J. Quantum Electron.* **QE-19**, 1284.
McCaffery, A. J., Proctor, M. J., and Whitaker, B. J. (1986). In *Annual Review of Physical Chemistry* (Strauss, H. L., Babcock, G. T., and Moore, C. B., eds), Annual Reviews, Palo Alto, California.
McCormick, R. I., De Lucia, F. C., and Skatrud, D. D. (1987a). *IEEE J. Quantum Electron.* **QE-23**, 2060.
McCormick, R. I., Everitt, H. O., De Lucia, F. C., and Skatrud, D. D. (1987b). *IEEE J. Quantum Electron.* **QE-23**, 2069.
McGuire, P., and Kouri, D. J. (1974). *J. Chem. Phys.* **60**, 2488.
Menard-Bourcin, F., and Doyennette, L. (1988). *J. Chem. Phys.* **88**, 5506.
Menard-Bourcin, F., Doyennette, L., and Menard, J. (1990). *J. Chem. Phys.* **92**, 4212.
Miklavc, A. (1983). *J. Chem. Phys.* **78**, 4502.
Millot, G., Hetzler, J., Foy, B., and Steinfeld, J. I. (1988). *J. Chem. Phys.* **88**, 6742.
Moore, C. B. (1965). *J. Chem. Phys.* **43**, 2979.
Nelson, R. D., Lide, D. R., and Maryott, A. A. (1967). *Natl. Stand. Ref. Data Ser. (U.S. Natl. Bur. Stand.)* **NSRDS–NBS 10**.
Oka, T. (1966). *J. Chem. Phys.* **45**, 753.
Oka, T. (1967a). *J. Chem. Phys.* **47**, 13.
Oka, T. (1967b). *J. Chem. Phys.* **47**, 4852.
Oka, T. (1968). *J. Chem. Phys.* **49**, 3135.
Oka, T. (1969). *Can. J. Phys.* **47**, 2343.
Oka, T. (1973). *Adv. At. Mol. Phys.* **9**, 127.
Orr, B. J., and Smith, I. W. M. (1987). *J. Chem. Phys.* **91**, 6106.
Pack, R. T. (1974). *J. Chem. Phys.* **60**, 633.
Palma, A., and Green, S. (1986). *J. Chem. Phys.* **85**, 1333.
Pape, T. W. (1993). Ph.D. Thesis, Duke Univ., Durham, North Carolina.
Pape, T. W., De Lucia, F. C., and Skatrud, D. D. (1994). *J. Chem. Phys.* **100**, 5666.
Parson, R. (1990). *J. Chem. Phys.* **93**, 8731.
Polanyi, J. C., and Woodall, K. B. (1972). *J. Chem. Phys.* **56**, 1563.
Preses, J. M., and Flynn, G. W. (1977). *J. Chem. Phys.* **66**, 3112.
Procaccia, I., and Levine, R. D. (1975). *J. Chem. Phys.* **63**, 4261.

Rabitz, H. A., and Gordon, R. G. (1970a). *J. Chem. Phys.* **53**, 1815.
Rabitz, H. A., and Gordon, R. G. (1970b). *J. Chem. Phys.* **53**, 1831.
Reiser, C., and Steinfeld, J. I. (1981). *J. Chem. Phys.* **74**, 2189.
Rohlfing, E. A., Chandler, D. W., and Parker, D. H. (1987). *J. Chem. Phys.* **87**, 5229.
Ronn, A. M., and Lide, D. R., Jr., (1967). *J. Chem. Phys.* **47**, 3669.
Ronn, A. M., and Wilson, E. B., Jr. (1967). *J. Chem. Phys.* **46**, 3262.
Rubinson, M., and Steinfeld, J. I. (1975). *Chem. Phys.* **4**, 167.
Schrepp, W., and Dreizler, H. (1981). *Z. Naturforsch.* **86a**, 654.
Seligson, D., Ducloy, M., Leite, J. R. R., Sanchez, A., and Feld, M. S. (1977). *IEEE J. Quant. Electron.* **QE-13**, 468.
Sharma, R. D., and Brau, C. A. (1969). *J. Chem. Phys.* **50**, 924.
Sheorey, R. S., and Flynn, G. W. (1980). *J. Chem. Phys.* **72**, 1175.
Sheorey, R. S., Slater, R. C., and Flynn, G. W. (1978). *J. Chem. Phys.* **68**, 1058.
Shin, U., and Schwendeman, R. H. (1991). *J. Chem. Phys.* **94**, 7560.
Shin, U., Song, Q., and Schwendeman, R. H. (1991). *J. Chem. Phys.* **95**, 3965.
Shoemaker, R. L., Stenholm, S., and Brewer, R. G. (1974). *Phys. Rev. A* **6**, 2037.
Skatrud, D. D., and De Lucia, F. C. (1985). *Appl. Phys. Lett.* **46**, 631.
Smith, N., and Pritchard, D. E. (1981). *J. Chem. Phys.* **74**, 3939.
Temkin, R. J. (1977). *IEEE J. Quant. Electron.* **QE-13**, 450.
Temps, F., Halle, S., Vaccaro, P. H., Field, R. W., and Kinsey, J. L. (1987). *J. Chem. Phys.* **89**, 1895.
Tobin, M. S., Sattler, J. P., and Daley, T. W. (1982). *IEEE J. Quant. Electron.* **QE-18**, 79.
Townes, C. H., and Schawlow, A. L. (1955). *Microwave Spectroscopy*. McGraw-Hill, New York.
Tsao, C. J., and Curnutte, B. (1961). *J. Quant. Spectrosc. Radiat. Transfer* **2**, 41.
Unland, U. L., and Flygare, W. H. (1966). *J. Chem. Phys.* **45**, 2421.
Vaccaro, P. H., Redington, R. L., Schmidt, J., Kinsey, J. L., and Field, R. W. (1985). *J. Chem. Phys.* **82**, 5755.
Van Vleck, J. H., and Weisskopf, V. F. (1945). *Rev. Mod. Phys.* **17**, 227.
Varanasi, P. (1988). *J. Quant. Spectrosc. Radiat. Transfer*, **39**, 13.
Weitz, E., and Flynn, G. (1973). *J. Chem. Phys.* **58**, 2781.
Weitz, E., Flynn, G. W., and Ronn, A. M. (1972). *J. Chem. Phys.* **56**, 6060.
Wilkins, R. L., and Kwok, M. A. (1983). *J. Chem. Phys.* **78**, 7153.

Index

A

Above-threshold ionization
  definition, 79
  electron energy distribution, 92–94, 102–103, 108
  vs. OHG, 108
Acetone–ethylene, spectra and structure, 136–139
Acetonitrile, spectra and structure, 146–148
Amplitudes, *see* Transition amplitudes
ATI, *see* Above-threshold ionization
Atomic beams
  deceleration, 13–17
    chirp slowing, 14–15, 17
    Zeeman slowing, 14, 15, 17
  deflection, 18–20
  dipole force traps and, 31–34
  MOT characteristics and, 21–26
  slowing using deceleration, 13–17
  slowing using deflection, 17–20
  slowing using MOTs, 26–30
Atomic fountains, 26–27, 47
Atomic hydrogen, *see* Hydrogenic targets
Atom optics, 32–34
Atoms
  channeling of, 32
  cooling, 21–30
  degenerated ground state, 11–12
  ionization dynamics in strong fields, 79–117
  laser manipulation, *see* Laser manipulation, of atoms
  single, trapping, 37–39
  trapping, 20–30
  two-level, 6–11, 17
  ultracold collisions, 45–76
Atom traps
  applications, 35–39
  background, 20–21
  experiments, 35–39
  magnetooptical, *see* Magnetooptical traps
  optical Ramsey spectroscopy
    on Mg atoms, 35–37
  single, 37–39
  types, 21
  ultracold collisions and, 45–75
Auger decay, laser-assisted, 94–96

B

Backscattering, 105, 107
BEC, *see* Bose–Einstein condensation
Born approximations, 211–214
Bose–Einstein condensation, 46–47

C

CCC method, *see* Convergent close-coupling method
CCO method, *see* Coupled-channel optical method
Chirp slowing, 14–15, 17
Cluster physics, *see* Molecular clusters
Cold atoms, *see* Laser cooling; Ultracold collisions
Colliding pulse mode-locked ring dye lasers, 167–170
Collisions, ultracold, 45–76
  complex potentials, 71–73
  inelastic, optical control, 48–65
    introduction, 48–49
    optical suppression and shielding, 57–65
    photoassociation, 49–56
  introduction, 45–46
  and MOTs, 30
  optical Bloch equations, 73–76
  quantum Monte Carlo wave functions, 73–76
  scattering length, 46–47

Collisions, ultracold (*continued*)
  s-wave, 46–47
  theoretical developments, 71–76
Complex potentials, 71–73
Convergent close-coupling method
  angular correlation parameters, 235–236
  application to electron-sodium scattering, 242–250
  and Born-based approximations, 212
  conclusions, 250–251
  coupled integral equations, 228–234
  definition, 210
  electron–hydrogen scattering and, 219–234
  generation of target states, 221–222
  hydrogen-like target approximation, 220–221
  and ionization, 236–241
  Lippman–Schwinger equations, 232–234
  relationship to PSCC, 216–217
  Temkin–Poet model and, 234–235
  three-body scattering problem, 223–228
  V-matrix elements, 229–232
Coupled-channel optical method, 218–219
CPM lasers, *see* Colliding pulse mode-locked ring dye lasers

D

Damping, 4–5, 11
Deceleration, of atomic beams, 13–17
Deflection, of atomic beams, 18–20
Dipole force
  atom optics and, 32–34
  laser manipulation and, 7–8
  manipulation schemes based on, 31–34
  traps, 31–34
Dissociation, *see* Photodissociation
Doppler cooling, 2, 4–5; *see also* Sub-Doppler cooling
Double ionization, strong-field, 108–116
  multiphoton, 108–111
  tunneling, 111–113
Double resonances, 130–132

E

Earnshaw theorem, 21
Electromagnetic fields, ionization dynamics, 79–117
Electron energy distribution, 92–97
  Auger decay, 94–96
  classical ATI, 92–94
  large initial velocity, 94–96
  narrow phase distribution, 96–97
  small initial velocity, 96–97
  tunnel ionization, 96–97
  uniform phase distribution, 94–96
Electron–helium scattering, 241–242
Electron–hydrogen scattering, 234–241
  calculations on hydrogenic targets, 209–250
  convergent close-coupling method, 219–234
  Temkin–Poet model, 234–235
  theories for hydrogenic targets, 211–219
Electron scattering
  angular correlation parameters, 235–236
  Born-based approximations, 211–213
  calculations on hydrogenic targets, 209–250
  convergent close-coupling method, 219–234
  coupled-channel optical method, 218–219
  finite-element methods, 217
  on helium ion, 241–242
  hyperspherical-coordinate methods, 219
  intermediate-energy R-matrix method, 215–216
  J-matrix method, 217–218
  pseudostate-close-coupling methods, 216–217
  R-matrix method, 214–215
  second-order Born approximations, 213–214
  Temkin–Poet model, 234–235
Electron–sodium scattering, 242–250
  differential cross-sections, 242–243
  ionization, 247–250
  spin asymmetries, 244–246
  total cross-sections, 247–250
Ethylene, spectra and structure, 136
Ethylene–acetone, spectra and structure, 136–139
Excited state population trapping, 82–89

F

Femtosecond spectroscopy, 163–206
  CPM oscillator, 167–170
  dynamics of sodium resonance, 191–195
  experimental results in cluster physics, 188–206
  experimental results in molecular physics, 172–188
  experimental techniques, 165–172
  laser system description, 167–170

of molecules and clusters, 163–206
multiphoton ionization of sodium dimer,
    178–184
  high laser field effects, 186–188
  pump–probe experiments
    controlling sodium yield, 184–186
    experimental setup, 170–172
  Ti:sapphire oscillator, 170
  vibrational wave packet motion, 172–178
Finite-element analysis, 217
Fitting law descriptions, 357–365
Fountains, atomic, 26–27, 47
Free atoms, laser manipulation, 1–44
Free particles, trapping, 21–30
Fullerenes, 203–204, 206

H

Hartree–Fock approximation, 242–243
Helium ion electron scattering, 241–242
Helium isoelectronic sequence
  calculations of transition amplitudes,
    255–327
  compilation of transition rates, 294–295
  laboratory experiments, 291–294
  table of transition rates, 296–324
  theoretical methods of calculation,
    286–291
Helium-like ions, perturbation theory,
    276–286
High harmonic cutoff, 99–100, 108
High-resolution spectroscopy, 35–37
Hydrazine, spectra and structure, 143–146
Hydrogen, see Hydrogenic targets
Hydrogenic targets
  electron scattering, 234–241
    calculations, 209–250
    theories, 211–219
  ionization, 236–241
Hyperspherical-coordinate techniques, 219

I

IERM, see Intermediate-energy R-matrix
  method
IMP, see Intermolecular potential
Inelastic collisions, optical control, 48–68
  introduction, 48–49
  optical suppression and shielding, 57–65
    in photoassociative ionization,
      49, 58–61
    suppression of trap loss, 57–58, 63–65
    temperature controlled suppression,
      63–65

  in xenon and krypton collisional
    ionization, 62–63
  photoassociation, 49–56
    atomic lifetime determination, 55–56
    and ionization, 58–61
    ionization and, 49
    line shapes, 49–55
Infrared photodissociation, of molecular
  clusters, 127–130
Infrared spectroscopy, of molecular clusters,
    121–158
  conclusions, 155–158
  dissociation spectra, 128, 148, 149–150
  double-resonance experiments, 130–132
  excitation and decay mechanisms,
    148–151
  experimental methods, 124–132
  phase transitions, 151–155
  photodissociation, 127–130
  rare gas clusters, 158
  results, 136–155
  simulation of temperature effects, 135
  size selection methods, 124–127
  spectra and structure, 136–148
    acetone–ethylene, 136–139
    acetonitrile, 146–148
    ethylene, 136
    hydrazine, 143–146
    methanol, 139–143
  structure calculations, 132–134
  temperature calculations, 135
  theoretical methods, 132–135
  vibrational spectra, 134–135
Intense laser fields, ionization dynamics,
    79–117
Intermediate-energy R-matrix method,
    215–216
Intermolecular potential, 332–333,
    335–337
Ionization
  atomic-hydrogen, 236–241
  differential cross-section, 238–241
  and electron–sodium scattering, 247–250
  helium ion and, 242
  photoassociative, 49, 58–61
  strong-field, 79–117
    above-threshold ionization, definition,
      79
    double, 108–116
    multiphoton ionization, definition, 79
    multiphoton mechanism, 89–92,
      108–111
    multiphoton vs. tunneling, 89–92

Ionization (*continued*)
  optical harmonic generation, definition, 79
  transient resonances, 82–89
  tunneling mechanism, 89–92, 111–113
  total cross-section, 237–238
Ion traps, 20

J
J-matrix method, 217–218

K
Krypton collisions, optical suppression and shielding, 62–63

L
Laser-assisted Auger decay, 94–96
Laser cooling
  applications, 35–39
  atomic beam deceleration and, 13–17
  atomic beam deflection and, 18–20
  background, 1–3
  capture range limitations, 13
  counterpropagating beams
    deceleration and, 13–14
    MOTs and, 27
    optical molasses and, 8–10
  Doppler mechanisms, 2, 4–5
  general principles, 3–13
  limits, 12–13
  sub-Doppler mechanisms, 4–6, 12, 23–24
  sub-recoil temperatures, 12–13
Laser deflection, 18–20
Laser manipulation, of atoms, 1–44
  applications, 35–39
  atomic beam deceleration, 13–17
  atomic beam deflection, 18–20
  confinement of atoms by radiation fields in traps, 27
  dipole force and, 31–34
  general principles, 3–13
  high-resolution spectroscopy and, 35–37
  introduction, 1–3
Lasers
  colliding pulse mode-locked ring dye lasers, 167–170
  femtosecond system, 165–172
  manipulation of atoms, *see* Laser manipulation, of atoms
  strong-field, ionization dynamics, 79–117
  Ti:sapphire lasers, 170

Laser-trapped atoms
  high-resolution spectroscopy and, 35–37
  single atoms, 37–39
Line shift calculations, 128, 134–135, 136–148
Lippmann–Schwinger equations, 232–234
Lithium, trap loss in, 68–69

M
Magnesium atoms
  optical high-resolution spectroscopy and, 35–37
  single-trapped, 37–39
Magnetooptical traps
  basic principle, 21–23
  BEC conditions and, 46–47
  development, 21
  manipulation techniques, 26–30
  properties of trapped ensembles, 23
  single atoms and, 39
  sub-Doppler cooling and, 23–24, 29
  temperature measurement techniques for ensembles, 24–25
  trap dynamics, 23, 25–26
Manipulation, laser, *see* Laser manipulation, of atoms
Matter wave optics, 32–34
Mercury clusters, 201–203, 205
Metal cluster physics, 164, 165, 205
  decay dynamics, 195–200
  resonance dynamics, 191–195
  sodium cluster resonances, 195–200
Methanol clusters
  excitation and decay mechanisms, 149
  isomeric transitions, 151–155
  spectra and structure, 139–143
Mixed-cluster system, spectra and structure, 136–139
Molecular/cluster beams, 165–167
Molecular clusters
  femtosecond spectroscopy, 163–206
    data acquisition, 172
    dynamics of photofragmentation, 188–191
    experimental results, 188–206
    experimental techniques, 165–172
    introduction, 163–165
    laser system description, 167–170
    multiphoton ionization dynamics, 178–184
    pump–probe delay lines, 170–172
  fullerene experiments, 203–206

infrared spectroscopy, 121–158
  conclusions, 155–158
  dissociation spectra, 128, 148, 149–150
  double-resonance experiments, 130–132
  excitation and decay mechanisms, 148–151
  experimental methods, 124–132
  intermolecular potential models, 133–134
  phase transitions, 151–155
  photodissociation, 127–130
  of rare gases, 158
  results, 136–155
  simulation of temperature effects, 135
  size selection methods, 124–127
  spectra and structure, 136–148
    acetone–ethylene, 136–139
    acetonitrile, 146–148
    ethylene, 136
    hydrazine, 143–146
    methanol, 139–143
  structure calculations, 132–134
  temperature calculations, 135
  theoretical methods, 132–135
  vibrational spectra, 134–135
mercury experiments, 201–203
Molecules
  diatomic, femtosecond spectroscopy
    controlling chemical reactions, 184–186
    experimental results, 172–188
    experimental techniques, 165–172
    multiphoton ionization of sodium dimer, 178–184
      high laser field effects, 186–188
    vibrational wave packet motion, 172–178
  polyatomic, rotational energy transfer, 331–397
Momenta of light quanta, 1
MOTs, see Magnetooptical traps
MPI, see Multiphoton ionization
Multiphoton ionization
  controlling chemical reactions, 184–186
  definition, 79
  double, 108–111
  dynamics of small molecules, 178–184
  femtosecond pump–probe techniques, 184–186
  fullerenes, 203–204, 206
  of mercury clusters, 201–203
  of sodium dimer, 178–184
    high laser field effects, 186–188
  in strong fields, 79, 82–83, 84

O

OBE, see Optical Bloch equations
OHG, see Optical harmonic generation
One-electron ions
  S-matrix theory for decay rates, 270–276
  transition amplitudes, 270–276
Optical Bloch equations, 73–74, 75
Optical harmonic generation, 79, 108
Optical molasses, 8–10
  dipole force traps and, 31
  limitations, 13
  MOTs and, 24, 27, 30
  sub-Doppler cooling, 12
Optical pumping
  complications, 16–17
  degenerated ground state atoms and, 11
  Doppler cooling and, 4–5
  sub-recoil cooling and, 12
Optical suppression and shielding, 57–65
  in photoassociative ionization, 49, 58–61
  temperature controlled suppression of trap loss, 63–65
  of trap loss, 57–58
  in xenon and krypton collisional ionization, 62–63
Optical tweezers, 32

P

PAI, see Photoassociative ionization
Particles
  collisionally interacting, 45–76
  single, trapping, 37–39
  trapping, 21–30
  ultracold collisions and, 45–76
Perturbation theory
  angular reduction, 279–282
  application to helium-like ions, 276–286
  basic equations, 276–279
  second-order amplitudes, 282–286
PES, see Photoelectron energy spectroscopy
Photoassociation, 49–56
  atomic lifetime determination, 55–56
  line shapes, 49–55
Photoassociative ionization, 49, 58–61
Photodissociation, infrared, 127–130
Photoelectron energy spectroscopy
  and evolution from multiphoton to tunneling ionization, 89–90
  measurements, 100–103
  and transient resonances, 83–84
Photoelectrons

Photoelectrons (*continued*)
  effects of rescattering on energy and momentum, 97–108
  energy distribution, 92–97
  momentum characteristics, 103–108
  scattering rings and, 103–108
Ponderomotive shifts, 81, 83, 87, 88
Pools, thermal, 347–351, 365–371
Population trapping, 82–89
PSCC, *see* Pseudostate-close-coupling methods
Pseudostate-close-coupling methods, 217–218
Pump–probe experiments
  controlling sodium yield, 184–186
  experimental setup, 170–172

R

Ramsey fringes, 36
Ramsey spectroscopy, 35–37
Recoil cooling limit, 12–13
Rescattering
  ATI vs. OHG, 108
  effects on photoelectron energy and momentum, 97–108
  electron distribution experiment, 100–103
  experiments, 82, 99–108
  high harmonic cutoff experiment, 99–100
  scattering rings experiment, 103–108
  theory, 97–99
  two-step mechanism, 97–99, 111, 113, 114–116
RET, *see* Rotational energy transfer
Rings, scattering, 103–108
R-matrix method, 214–216
  intermediate-energy, 215–216
Rotational energy transfer
  angular momentum, 335–337
  applications, 332–334
  computational advances, 396
  description of problem, 355–356
  energy considerations, 337–338
  exact quantal problem, 334
  experimental advances, 394–396
  experiments, 344–346
  experiment vs. theory, 355–356
  future direction, 394–397
  grouping of states, 367–371, 385–388
  intermolecular potential, 332–333, 335–337
  introduction, 332–356

  model building and parameterization, 346–355
  numerical simulations, 351–355
  pools of rotational states, 348–351
  pools of states, 347–348
  near-resonant ro-vibrational energy transfer
    theoretical considerations, 339–340
    V-swap process, 372–376
  physical basis, 377–394
    $\Delta J = n$ processes, 381–385
    $\Delta J = 0, \pm 1$ *process*, 380–381
    $\Delta J = 0, \pm 1$ *propensity*, 377
    grouping of states, 385–388
    molecular constants, 379–394
    nuclear spin, 377
    parity, 378
    selection and propensity rules, 377–379
    spectroscopic issues, 393–394
    total depopulation cross-sections, 389–391
    vibrational processes, 378–379
    vibrational swapping process, 391–392
  scaling and fitting laws, 340–342
  scaling of fitting parameters, 342–344
  semiclassical elements, 335–344
  in small polyatomic molecules, 331–397
  state-specific, 356–365
    $\Delta J = n$ processes, 358–365
    dipole–dipole processes, 357
  symmetry, 335–337
  theoretical considerations, 334–344
Rubidium, trap loss in, 65–68
Running molasses, *see* Optical molasses

S

Scaling laws, 341–342
Scattering, *see also* Electron scattering
  molecular cluster size selection and, 124–127
Scattering length
  in alkali systems, 47
  magnitude, 46–47
  sign of, 46–47
  in two-body collisions, 46–47
  in ultracold collisions, 46–47
Scattering rings, 103–108
Schrodinger equation, 81, 85, 92, 99, 163, 217
Simpleman model, 92, 97
Single atoms, trapping, 37–39
S-matrix theory, 270–276
Sodium

atomic beam deceleration, 13–15, 16
cluster resonances, 195–200
decay dynamics, 195–200
dynamics of photofragmentation, 188–191
dynamics of resonance, 191–195
electron scattering, 242–250
femtosecond spectroscopy, 188–200
multiphoton ionization, 178–184
trap loss, 69–70
Spectral hole burning, 130
Spectrometers, time-of-flight, 165–172
Spectroscopy, see Femtosecond spectroscopy; High-resolution spectroscopy; Infrared spectroscopy; Photoelectron energy spectroscopy; Ramsey spectroscopy
Spin asymmetries, 244–246
Stark-induced resonances, 82–83, 84–86, 87, 89
Strong-field ionization
bound-free step, 80–81, 82–92
double, 82, 108–116
electron energy distributions and, 92–97
free-free step, 80, 81, 92–108
multiphoton, 108–111
multiphoton vs. tunneling, 89–92
steps in, 80
tunneling, 111–113
Sub-Doppler cooling, 2, 3, 4–6
magnetooptical traps and, 23–24, 29
in optical molasses, 12
optical pumping and, 4–5
Sub-recoil temperatures, 12–13
s-wave collisions, scattering length, 46–47

T

TDSE, see Time-dependent Schrödinger equation
Temkin–Poet model, 234–235
Thermal pools, 347–351, 365–371
Time-dependent Schrödinger equation, 81, 85, 92, 99
Time-independent Schrödinger equation, 163
Time-of-flight spectrometers, 165–167
Ti:sapphire lasers, 170
TOF spectrometers, see Time-of-flight spectrometers
Transient resonances
diabatic vs. adiabatic crossings, 87–88
experiments, 87–89
large ionization cross-section, 87

nonponderomotive shifts, 87
overview, 82–84
Stark-induced, 84–86
Transition amplitudes
future direction, 295, 325–326
for helium isoelectronic sequence, 255–327
no-pair, 258–270
commutator identity, 265–267
electromagnetic interaction Hamiltonian, 258–260, 261–262
multipole potentials, 262–265
numerical calculations, 267–270
tests of, 267–270
for one-electron ions, 270–276
rates of transition, 294–295
results and comparisons, 286–326
table of transition rates, 296–324
useful identities, 326–327
Trap loss
collisional processes, 65–70
in lithium, 68–69
in rubidium, 65–68
in sodium, 69–70
optical suppression, 57–58
temperature-controlled suppression, 63–65
Trapped atoms, see Atom traps; Magnetooptical traps
Tunnel ionization, 89–92, 96–97, 111–113
Tweezers, optical, 32
Two-body collisions, scattering length and, 46–47
Two-level atoms, 6–11, 17

U

Ultracold collisions, 45–76
complex potentials, 71–73
developments in theory, 71–76
future direction, 76
inelastic, optical control, 48–65
introduction, 48–49
optical suppression and shielding, 57–65
photoassociation, 49–56
introduction, 45–46
MOTs and, 30
optical Bloch equations, 73–76
quantum Monte Carlo wave functions, 73–76
scattering length, 46–47
s-wave, 46–47
theoretical developments, 71–76

V

van der Waals clusters, 136
Vibrational predissociation, 127–130, 148

W

Wave packets, 82, 107, 164, 172–178

X

Xenon collisions, optical suppression and shielding, 62–63

Z

Zacharias fountain, 26
Zeeman slowing, 14, 15, 17, 30

Contents of Volumes in This Serial

Volume 1

Molecular Orbital Theory of the Spin Properties of Conjugated Molecules, *G. G. Hall and A. T. Amos*

Electron Affinities of Atoms and Molecules, *B. L. Moiseiwitsch*

Atomic Rearrangement Collisions, *B. H. Bransden*

The Production of Rotational and Vibrational Transitions in Encounters between Molecules, *K. Takayanagi*

The Study of Intermolecular Potentials with Molecular Beams at Thermal Energies, *H. Pauly and J. P. Toennies*

High-Intensity and High-Energy Molecular Beams, *J. B. Anderson, R. P. Andres, and J. B. Fen*

Volume 2

The Calculation of van der Waals Interactions, *A. Dalgarno and W. D. Davison*

Thermal Diffusion in Gases, *E. A. Mason, R. J. Munn, and Francis J. Smith*

Spectroscopy in the Vacuum Ultraviolet, *W. R. S. Garton*

The Measurement of the Photoionization Cross Sections of the Atomic Gases, *James A. R. Samson*

The Theory of Electron–Atom Collisions, *R. Peterkop and V. Veldre*

Experimental Studies of Excitation in Collisions between Atomic and Ionic Systems, *F. J. de Heer*

Mass Spectrometry of Free Radicals, *S. N. Foner*

Volume 3

The Quantal Calculation of Photoionization Cross Sections, *A. L. Stewart*

Radiofrequency Spectroscopy of Stored Ions I: Storage, *H. G. Dehmelt*

Optical Pumping Methods in Atomic Spectroscopy, *B. Budick*

Energy Transfer in Organic Molecular Crystals: A Survey of Experiments, *H. C. Wolf*

Atomic and Molecular Scattering from Solid Surfaces, *Robert E. Stickney*

Quantum Mechanics in Gas Crystal-Surface van der Waals Scattering, *E. Chanoch Beder*

Reactive Collisions between Gas and Surface Atoms, *Henry Wise and Bernard J. Wood*

Contents of Volumes in This Serial

Volume 4

H. S. W. Massey—A Sixtieth Birthday Tribute, *E. H. S. Burhop*

Electronic Eigenenergies of the Hydrogen Molecular Ion, *D. R. Bates and R. H. G. Reid*

Applications of Quantum Theory to the Viscosity of Dilute Gases, *R. A. Buckingham and E. Gal*

Positrons and Positronium in Gases, *P. A. Fraser*

Classical Theory of Atomic Scattering, *A. Burgess and I. C. Percival*

Born Expansions, *A. R. Holt and B. L. Moiselwitsch*

Resonances in Electron Scattering by Atoms and Molecules, *P. G. Burke*

Relativistic Inner Shell Ionizations, *C. B. O. Mohr*

Recent Measurements on Charge Transfer, *J. B. Hasted*

Measurements of Electron Excitation Functions, *D. W. O. Heddle and R. G. W. Keesing*

Some New Experimental Methods in Collision Physics, *R. F. Stebbings*

Atomic Collision Processes in Gaseous Nebulae, *M. J. Seaton*

Collisions in the Ionosphere, *A. Dalgarno*

The Direct Study of Ionization in Space, *R. L. F. Boyd*

Volume 5

Flowing Afterglow Measurements of Ion-Neutral Reactions, *E. E. Ferguson, F. C. Fehsenfeld, and A. L. Schmeltekopf*

Experiments with Merging Beams, *Roy H. Neynaber*

Radiofrequency Spectroscopy of Stored Ions II: Spectroscopy, *H. G. Dehmelt*

The Spectra of Molecular Solids, *O. Schnepp*

The Meaning of Collision Broadening of Spectral Lines: The Classical Oscillator Analog, *A. Ben-Reuven*

The Calculation of Atomic Transition Probabilities, *R. J. S. Crossley*

Tables of One- and Two-Particle Coefficients of Fractional Parentage for Configurations $s^\lambda s'^\mu p^q$, *C. D. H. Chisholm, A. Dalgarno, and F. R. Innes*

Relativistic Z-Dependent Corrections to Atomic Energy Levels, *Holly Thomis Doyle*

Volume 6

Dissociative Recombination, *J. N. Bardsley and M. A. Biondi*

Analysis of the Velocity Field in Plasmas from the Doppler Broadening of Spectral Emission Lines, *A. S. Kaufman*

The Rotational Excitation of Molecules by Slow Electrons, *Kazuo Takayanagi, and Yukikazu Itikawa*

The Diffusion of Atoms and Molecules, *E. A. Mason and T. R. Marrero*

Theory and Application of Sturmian Functions, *Manuel Rotenberg*

Use of Classical Mechanics in the Treatment of Collisions between Massive Systems, *D. R. Bates and A. E. Kingston*

Volume 7

Physics of the Hydrogen Master, *C. Audoin, J. P. Schermann, and P. Grivet*

Molecular Wave Functions: Calculation and Use in Atomic and Molecular Processes, *J. C. Browne*

Localized Molecular Orbitals, *Harel Weinstein, Ruben Pauncz, and Maurice Cohen*

General Theory of Spin-Coupled Wave Functions for Atoms and Molecules, *J. Gerratt*

Diabatic States of Molecules—Quasi-Stationary Electronic States, *Thomas F. O'Malley*

Selection Rules within Atomic Shells, *B. R. Judd*

Green's Function Technique in Atomic and Molecular Physics, *Gy. Csanak, H. S. Taylor, and Robert Yaris*

Contents of Volumes in This Serial

A Review of Pseudo-Potentials with Emphasis on Their Application to Liquid Metals, *Nathan Wiser and A. J. Greenfield*

Volume 8

Interstellar Molecules: Their Formation and Destruction, *D. McNally*

Monte Carlo Trajectory Calculations of Atomic and Molecular Excitation in Thermal Systems, *James C. Keck*

Nonrelativistic Off-Shell Two-Body Coulomb Amplitudes, *Joseph C. Y. Chen and Augustine C. Chen*

Photoionization with Molecular Beams, *R. B. Cairns, Halstead Harrison, and R. I. Schoen*

The Auger Effect, *E. H. S. Burhop and W. N. Asaad*

Volume 9

Correlation in Excited States of Atoms, *A. W. Weiss*

The Calculation of Electron–Atom Excitation Cross Sections, *M. R. H. Rudge*

Collision-Induced Transitions between Rotational Levels, *Takeshi Oka*

The Differential Cross Section of Low-Energy Electron–Atom Collisions, *D. Andrick*

Molecular Beam Electric Resonance Spectroscopy, *Jens C. Zorn, and Thomas C. English*

Atomic and Molecular Processes in the Martian Atmosphere, *Michael B. McElroy*

Volume 10

Relativistic Effects in the Many-Electron Atom, *Lloyd Armstrong, Jr., and Serge Feneuille*

The First Born Approximation, *K. L. Bell and A. E. Kingston*

Photoelectron Spectroscopy, *W. C. Price*

Dye Lasers in Atomic Spectroscopy, *W. Lange, J. Luther, and A. Steudel*

Recent Progress in the Classification of the Spectra of Highly Ionized Atoms, *B. C. Fawcett*

A Review of Jovian Ionospheric Chemistry, *Wesley T. Huntress, Jr.*

Volume 11

The Theory of Collisions between Charged Particles and Highly Excited Atoms, *I. C. Percival and D. Richards*

Electron Impact Excitation of Positive Ions, *M. J. Seaton*

The *R*-Matrix Theory of Atomic Process, *P. G. Burke and W. D. Robb*

Role of Energy in Reactive Molecular Scattering: An Information-Theoretic Approach, *R. B. Bernstein, and R. D. Levine*

Inner Shell Ionization by Incident Nuclei, *Johannes M. Hansteen*

Stark Broadening, *Hans R. Griem*

Chemiluminescence in Gases, *M. F. Golde and B. A. Thrush*

Volume 12

Nonadiabatic Transitions between Ionic and Covalent States, *R. K. Janev*

Recent Progress in the Theory of Atomic Isotope Shift, *J. Bauche, and R.-J. Champeau*

Topics on Multiphoton Processes in Atoms, *P. Lambropoulos*

Optical Pumping of Molecules, *M. Broyer, G. Goudedard, J. C. Lehmann, and J. Vigué*

Highly Ionized Ions, *Ivan A. Sellin*

Time-of-Flight Scattering Spectroscopy, *Wilhelm Raith*

Ion Chemistry in the D Region, *George C. Reid*

Volume 13

Atomic and Molecular Polarizabilities—A Review of Recent Advances, *Thomas M. Miller and Benjamin Bederson*

Contents of Volumes in This Serial

Study of Collisions by Laser Spectroscopy, *Paul R. Berman*

Collision Experiments with Laser-Excited Atoms in Crossed Beams, *I. V. Hertel and W. Stoll*

Scattering Studies of Rotational and Vibrational Excitation of Molecules, *Manfred Faubel and J. Peter Toennies*

Low-Energy Electron Scattering by Complex Atoms: Theory and Calculations, *R. K. Nesbet*

Microwave Transitions of Interstellar Atoms and Molecules, *W. B. Somerville*

Volume 14

Resonances in Electron Atom and Molecule Scattering, *D. E. Golden*

The Accurate Calculation of Atomic Properties by Numerical Methods, *Brian C. Webster, Michael J. Jamieson, and Ronald F. Stewart*

(e, 2e) Collisions, *Erich Weigold and Ian E. McCarthy*

Forbidden Transitions in One- and Two-Electron Atoms, *Richard Marrus and Peter J. Mohr*

Semiclassical Effects in Heavy-Particle Collisions, *M. S. Child*

Atomic Physics Tests of the Basic Concepts in Quantum Mechanics, *Francis M. Pipkin*

Quasi-Molecular Interference Effects in Ion–Atom Collisions, *S. V. Bobashev*

Rydberg Atoms, *S. A. Edelstein and T. F. Gallagher*

UV and X-Ray Spectroscopy in Astrophysics, *A. K. Dupree*

Volume 15

Negative Ions, *H. S. W. Massey*

Atomic Physics from Atmospheric and Astrophysical Studies, *A. Dalgarno*

Collisions of Highly Excited Atoms, *R. F. Stebbings*

Theoretical Aspects of Positron Collisions in Gases, *J. W. Humberston*

Experimental Aspects of Positron Collisions in Gases, *T. C. Griffith*

Reactive Scattering: Recent Advances in Theory and Experiment, *Richard B. Bernstein*

Ion–Atom Charge Transfer Collisions at Low Energies, *J. B. Hasted*

Aspects of Recombination, *D. R. Bates*

The Theory of Fast Heavy Particle Collisions, *B. H. Bransden*

Atomic Collision Processes in Controlled Thermonuclear Fusion Research, *H. B. Gilbody*

Inner-Shell Ionization, *E. H. S. Burhop*

Excitation of Atoms by Electron Impact, *D. W. O. Heddle*

Coherence and Correlation in Atomic Collisions, *H. Kleinpoppen*

Theory of Low Energy Electron–Molecule Collisions, *P. G. Burke*

Volume 16

Atomic Hartree–Fock Theory, *M. Cohen and R. P. McEachran*

Experiments and Model Calculations to Determine Interatomic Potentials, *R. Düren*

Sources of Polarized Electrons, *R. J. Celotta and D. T. Pierce*

Theory of Atomic Processes in Strong Resonant Electromagnetic Fields, *S. Swain*

Spectroscopy of Laser-Produced Plasmas, *M. H. Key and R. J. Hutcheon*

Relativistic Effects in Atomic Collisions Theory, *B. L. Moiseiwitsch*

Parity Nonconservation in Atoms: Status of Theory and Experiment, *E. N. Fortson and L. Wilets*

Volume 17

Collective Effects in Photoionization of Atoms, *M. Ya. Amusia*

Contents of Volumes in This Serial

Nonadiabatic Charge Transfer, *D. S. F. Crothers*

Atomic Rydberg States, *Serge Feneuille and Pierre Jacquinot*

Superfluorescence, *M. F. H. Schuurmans, Q. H. F. Vrehen, D. Polder, and H. M. Gibbs*

Applications of Resonance Ionization Spectroscopy in Atomic and Molecular Physics, *M. G. Payne, C. H. Chen, G. S. Hurst, and G. W. Foltz*

Inner-Shell Vacancy Production in Ion–Atom Collisions, *C. D. Lin and Patrick Richard*

Atomic Processes in the Sun, *P. L. Dufton and A. E. Kingston*

Volume 18

Theory of Electron–Atom Scattering in a Radiation Field, *Leonard Rosenberg*

Positron–Gas Scattering Experiments, *Talbert S. Stein and Walter E. Kauppila*

Nonresonant Multiphoton Ionization of Atoms, *J. Morellec, D. Normand, and G. Petite*

Classical and Semiclassical Methods in Inelastic Heavy-Particle Collisions, *A. S. Dickinson and D. Richards*

Recent Computational Developments in the Use of Complex Scaling in Resonance Phenomena, *B. R. Junker*

Direct Excitation in Atomic Collisions: Studies of Quasi-One-Electron Systems, *N. Anderson and S. E. Nielsen*

Model Potentials in Atomic Structure, *A. Hibbert*

Recent Developments in the Theory of Electron Scattering by Highly Polar Molecules, *D. W. Norcross and L. A. Collins*

Quantum Electrodynamic Effects in Few-Electron Atomic Systems, *G. W. F. Drake*

Volume 19

Electron Capture in Collisions of Hydrogen Atoms with Fully Stripped Ions, *B. H. Bransden and R. K. Janev*

Interactions of Simple Ion–Atom Systems, *J. T. Park*

High-Resolution Spectroscopy of Stored Ions, *D. J. Wineland, Wayne M. Itano, and R. S. Van Dyck, Jr.*

Spin-Dependent Phenomena in Inelastic Electron–Atom Collisions, *K. Blum and H. Kleinpoppen*

The Reduced Potential Curve Method for Diatomic Molecules and Its Applications, *F. Jenč*

The Vibrational Excitation of Molecules by Electron Impact, *D. G. Thompson*

Vibrational and Rotational Excitation in Molecular Collisions, *Manfred Faubel*

Spin Polarization of Atomic and Molecular Photoelectrons, *N. A. Cherepkov*

Volume 20

Ion–Ion Recombination in an Ambient Gas, *D. R. Bates*

Atomic Charges within Molecules, *G. G. Hall*

Experimental Studies on Cluster Ions, *T. D. Mark and A. W. Castleman, Jr.*

Nuclear Reaction Effects on Atomic Inner-Shell Ionization, *W. E. Meyerhof and J.-F. Chemin*

Numerical Calculations on Electron-Impact Ionization, *Christopher Bottcher*

Electron and Ion Mobilities, *Gordon R. Freeman and David A. Armstrong*

On the Problem of Extreme UV and X-Ray Lasers, *I. I. Sobel'man and A. V. Vinogradov*

Radiative Properties of Rydberg States in Resonant Cavities, *S. Haroche and J. M. Raimond*

Rydberg Atoms: High-Resolution Spectroscopy and Radiation Interaction—Rydberg Molecules, *J. A. C. Gallas, G. Leuchs, H. Walther, and H. Figger*

Contents of Volumes in This Serial

Volume 21

Subnatural Linewidths in Atomic Spectroscopy, *Dennis P. O'Brien, Pierre Meystre, and Herbert Walther*

Molecular Applications of Quantum Defect Theory, *Chris H. Greene and Ch. Jungen*

Theory of Dielectronic Recombination, *Yukap Hahn*

Recent Developments in Semiclassical Floquet Theories for Intense-Field Multi-photon Processes, *Shih-I Chu*

Scattering in Strong Magnetic Fields, *M. R. C. McDowell and M. Zarcone*

Pressure Ionization, Resonances, and the Continuity of Bound and Free States, *R. M. More*

Volume 22

Positronium—Its Formation and Interaction with Simple Systems, *J. W. Humberston*

Experimental Aspects of Positron and Positronium Physics, *T. C. Griffith*

Doubly Excited States, Including New Classification Schemes, *C. D. Lin*

Measurements of Charge Transfer and Ionization in Collisions Involving Hydrogen Atoms, *H. B. Gilbody*

Electron–Ion and Ion–Ion Collisions with Intersecting Beams, *K. Dolder and B. Pearl*

Electron Capture by Simple Ions, *Edward Pollack and Yukap Hahn*

Relativistic Heavy-Ion–Atom Collisions, *R. Anholt and Harvey Gould*

Continued-Fraction Methods in Atomic Physics, *S. Swain*

Volume 23

Vacuum Ultraviolet Laser Spectroscopy of Small Molecules, *C. R. Vidal*

Foundations of the Relativistic Theory of Atomic and Molecular Structure, *Ian P. Grant and Harry M. Quiney*

Point-Charge Models for Molecules Derived from Least-Squares Fitting of the Electric Potential, *D. E. Williams and Ji-Min Yan*

Transition Arrays in the Spectra of Ionized Atoms, *J. Bauche, C. Bauche-Arnoult, and M. Klapisch*

Photoionization and Collisional Ionization of Excited Atoms Using Synchroton and Laser Radiation, *F. J. Wuilleumier, D. L. Ederer, and J. L. Picqué*

Volume 24

The Selected Ion Flow Tube (SIFT): Studies of Ion–Neutral Reactions, *D. Smith and N. G. Adams*

Near-Threshold Electron–Molecule Scattering, *Michael A. Morrison*

Angular Correlation in Multiphoton Ionization of Atoms, *S. J. Smith and G. Leuchs*

Optical Pumping and Spin Exchange in Gas Cells, *R. J. Knize, Z. Wu, and W. Happer*

Correlations in Electron–Atom Scattering, *A. Crowe*

Volume 25

Alexander Dalgarno: Life and Personality, *David R. Bates and George A. Victor*

Alexander Dalgarno: Contributions to Atomic and Molecular Physics, *Neal Lane*

Alexander Dalgarno: Contributions to Aeronomy, *Michael B. McElroy*

Alexander Dalgarno: Contributions to Astrophysics, *David A. Williams*

Dipole Polarizability Measurements, *Thomas M. Miller and Benjamin Bederson*

Flow Tube Studies of Ion-Molecule Reactions, *Eldon Ferguson*

Differential Scattering in He–He and He$^+$–He Collisions at KeV Energies, *R. F. Stebbings*

Atomic Excitation in Dense Plasmas, *Jon C. Weisheit*

Contents of Volumes in This Serial

Pressure Broadening and Laser-Induced Spectral Line Shapes, *Kenneth M. Sando and Shih-I Chu*

Model-Potential Methods, *G. Laughlin and G. A. Victor*

Z-Expansion Methods, *M. Cohen*

Schwinger Variational Methods, *Deborah Kay Watson*

Fine-Structure Transitions in Proton-Ion Collisions, *R. H. G. Reid*

Electron Impact Excitation, *R. J. W. Henry and A. E. Kingston*

Recent Advances in the Numerical Calculation of Ionization Amplitudes, *Christopher Bottcher*

The Numerical Solution of the Equations of Molecular Scattering, *A. C. Allison*

High Energy Charge Transfer, *B. H. Bransden and D. P. Dewangan*

Relativistic Random-Phase Approximation, *W. R. Johnson*

Relativistic Sturmian and Finite Basis Set Methods in Atomic Physics, *G. W. F. Drake and S. P. Goldman*

Dissociation Dynamics of Polyatomic Molecules, *T. Uzer*

Photodissociation Processes in Diatomic Molecules of Astrophysical Interest, *Kate P. Kirby and Ewine F. van Dishoeck*

The Abundances and Excitation of Interstellar Molecules, *John H. Black*

Volume 26

Comparisons of Positrons and Electron Scattering by Gases, *Walter E. Kauppila and Talbert S. Stein*

Electron Capture at Relativistic Energies, *B. L. Moiseiwitsch*

The Low-Energy, Heavy Particle Collisions—A Close-Coupling Treatment, *Mineo Kimura and Neal F. Lane*

Vibronic Phenomena in Collisions of Atomic and Molecular Species, *V. Sidis*

Associative Ionization: Experiments, Potentials, and Dynamics, *John Weiner, Françoise Masnou-Sweeuws, and Annick Giusti-Suzor*

On the β Decay of ^{187}Re: An Interface of Atomic and Nuclear Physics and Cosmochronology, *Zonghau Chen, Leonard Rosenberg, and Larry Spruch*

Progress in Low Pressure Mercury-Rare Gas Discharge Research, *J. Maya and R. Lagushenko*

Volume 27

Negative Ions: Structure and Spectra, *David R. Bates*

Electron Polarization Phenomena in Electron–Atom Collisions, *Joachim Kessler*

Electron–Atom Scattering, *I. E. McCarthy and E. Weigold*

Electron–Atom Ionization, *I. E. McCarthy and E. Weigold*

Role of Autoionizing States in Multiphoton Ionization of Complex Atoms, *V. I. Lengyel and M. I. Haysak*

Multiphoton Ionization of Atomic Hydrogen Using Perturbation Theory, *E. Karule*

Volume 28

The Theory of Fast Ion–Atom Collisions, *J. S. Briggs and J. H. Macek*

Some Recent Developments in the Fundamental Theory of Light, *Peter W. Milonni and Surendra Singh*

Squeezed States of the Radiation Field, *Khalid Zaheer and M. Suhail Zubairy*

Cavity Quantum Electrodynamics, *E. A. Hinds*

Volume 29

Studies of Electron Excitation of Rare-Gas Atoms into and out of Metastable Levels Using Optical and Laser Techniques, *Chun C. Lin and L. W. Anderson*

Cross Sections for Direct Multiphoton Ionization of Atoms, *M. V. Ammosov, N. B. Delone, M. Yu. Ivanov, I. I. Bondar, and A. V. Masalov*

Collision-Induced Coherences in Optical Physics, *G. S. Agarwal*

Muon-Catalyzed Fusion, *Johann Rafelski and Helga E. Rafelski*

Cooperative Effects in Atomic Physics, *J. P. Connerade*

Multiple Electron Excitation, Ionization, and Transfer in High-Velocity Atomic and Molecular Collisions, *J. H. McGuire*

Volume 30

Differential Cross Sections for Excitation of Helium Atoms and Helium-like Ions by Electron Impact, *Shinobu Nakazaki*

Cross-Section Measurements for Electron Impact on Excited Atomic Species, *S. Trajmar and J. C. Nickel*

The Dissociative Ionization of Simple Molecules by Fast Ions, *Colin J. Latimer*

Theory of Collisions between Laser Cooled Atoms, *P. S. Julienne, A. M. Smith, and K. Burnett*

Light-Induced Drift, *E. R. Eliel*

Continuum Distorted Wave Methods in Ion–Atom Collisions, *Derrick S. F. Crothers and Louis J. Dubé*

Volume 31

Energies and Asymptotic Analysis for Helium Rydberg States, *G. W. F. Drake*

Spectroscopy of Trapped Ions, *R. C. Thompson*

Phase Transitions of Stored Laser-Cooled Ions, *H. Walther*

Selection of Electronic States in Atomic Beams with Lasers, *Jacques Baudon, Rudolf Düren, and Jacques Robert*

Atomic Physics and Non-Maxwellian Plasmas, *Michèle Lamoureux*

Volume 32

Photoionization of Atomic Oxygen and Atomic Nitrogen, *K. L. Bell and A. E. Kingston*

Positronium Formation by Positron Impact on Atoms at Intermediate Energies, *B. H. Bransden and C. J. Noble*

Electron–Atom Scattering Theory and Calculations, *P. G. Burke*

Terrestrial and Extraterrestrial H_3^+, *Alexander Dalgarno*

Indirect Ionization of Positive Atomic Ions, *K. Dolder*

Quantum Defect Theory and Analysis of High-Precision Helium Term Energies, *G. W. F. Drake*

Electron–Ion and Ion–Ion Recombination Processes, *M. R. Flannery*

Studies of State-Selective Electron Capture in Atomic Hydrogen by Translational Energy Spectroscopy, *H. B. Gilbody*

Relativistic Electronic Structure of Atoms and Molecules, *I. P. Grant*

The Chemistry of Stellar Environments, *D. A. Howe, J. M. C. Rawlings, and D. A. Williams*

Positron and Positronium Scattering at Low Energies, *J. W. Humberston*

How Perfect are Complete Atomic Collision Experiments?, *H. Kleinpoppen and H. Hamdy*

Adiabatic Expansions and Nonadiabatic Effects, *R. McCarroll and D. S. F. Crothers*

Electron Capture to the Continuum, *B. L. Moiseiwitsch*

How Opaque Is a Star?, *M. J. Seaton*

Studies of Electron Attachment at Thermal Energies Using the Flowing Afterglow–Langmuir Technique, *David Smith and Patrik Španěl*

Exact and Approximate Rate Equations in Atom-Field Interactions, *S. Swain*

Atoms in Cavities and Traps, *H. Walther*

Some Recent Advances in Electron-Impact Excitation of $n = 3$ States of Atomic Hydrogen and Helium, *J. F. Williams and J. B. Wang*

Volume 33

Principles and Methods for Measurement of Electron Impact Excitation Cross Sections for Atoms and Molecules by Optical Techniques, *A. R. Filippelli, Chun C. Lin, L. W. Andersen, and J. W. McConkey*

Benchmark Measurements of Cross Sections for Electron Collisions: Analysis of Scattered Electrons, *S. Trajmar and J. W. McConkey*

Benchmark Measurements of Cross Sections for Electron Collisions: Electron Swarm Methods, *R. W. Crompton*

Some Benchmark Measurements of Cross Sections for Collisions of Simple Heavy Particles, *H. B. Gilbody*

The Role of Theory in the Evaluation and Interpretation of Cross-Section Data, *Barry I. Schneider*

Analytic Representation of Cross-Section Data, *Mitio Inokuti, Mineo Kimura, M. A. Dillon, Isao Shimamura*

Electron Collisions with N_2, O_2 and O: What We Do and Do Not Know, *Yukikazu Itikawa*

Need for Cross Sections in Fusion Plasma Research, *Hugh P. Summers*

Need for Cross Sections in Plasma Chemistry, *M. Capitelli, R. Celiberto, and M. Cacciatore*

Guide for Users of Data Resources, *Jean W. Gallagher*

Guide to Bibliographies, Books, Reviews, and Compendia of Data on Atomic Collisions, *E. W. McDaniel and E. J. Mansky*

Volume 34

Atom Interferometry, *C. S. Adams, O. Carnal, and J. Mlynek*

Optical Tests of Quantum Mechanics, *R. Y. Chiao, P. G. Kwiat, and A. M. Steinberg*

Classical and Quantum Chaos in Atomic Systems, *Dominique Delande and Andreas Buchleitner*

Measurements of Collisions between Laser-Cooled Atoms, *Thad Walker and Paul Feng*

The Measurement and Analysis of Electric Fields in Glow Discharge Plasmas, *J. E. Lawler and D. A. Doughty*

Polarization and Orientation Phenomena in Photoionization of Molecules, *N. A. Cherepkov*

Role of Two-Center Electron-Electron Interaction in Projectile Electron Excitation and Loss, *E. C. Montenegro, W. E. Meyerhof, and J. H. McGuire*

Indirect Processes in Electron Impact Ionization of Positive Ions, *D. L. Moores and K. J. Reed*

Dissociative Recombination: Crossing and Tunneling Modes, *David R. Bates*

Volume 35

Laser Manipulation of Atoms, *K. Sengstock and W. Ertmer*

Advances in Ultracold Collisions: Experiment and Theory, *J. Weiner*

Ionization Dynamics in Strong Laser Fields, *L. F. DiMauro and P. Agostini*

Infrared Spectroscopy of Size Selected Molecular Clusters, *U. Buck*

Femtosecond Spectroscopy of Molecules and Clusters, *T. Baumer and G. Gerber*

Calculation of Electron Scattering on Hydrogenic Targets, *I. Bray and A. T. Stelbovics*

Relativistic Calculations of Transition Amplitudes in the Helium Isoelectronic Sequence, *W. R. Johnson, D. R. Plante, and J. Sapirstein*

Rotational Energy Transfer in Small Polyatomic Molecules, *H. O. Everitt and F. C. De Lucia*